A Concise Introduction to Quantum Mechanics (Second Edition)

Online at: https://doi.org/10.1088/978-0-7503-5663-3

A Concise Introduction to Quantum Mechanics (Second Edition)

Mark S Swanson
Emeritus Professor of Physics, University of Connecticut, Storrs, CT, USA

IOP Publishing, Bristol, UK

ISBN 978-0-7503-5663-3 (ebook)
ISBN 978-0-7503-5661-9 (print)
ISBN 978-0-7503-5664-0 (myPrint)
ISBN 978-0-7503-5662-6 (mobi)

DOI 10.1088/978-0-7503-5663-3

Version: 20231001

IOP ebooks

British Library Cataloguing-in-Publication Data: A catalogue record for this book is available from the British Library.

Published by IOP Publishing, wholly owned by The Institute of Physics, London

IOP Publishing, No.2 The Distillery, Glassfields, Avon Street, Bristol, BS2 0GR, UK

US Office: IOP Publishing, Inc., 190 North Independence Mall West, Suite 601, Philadelphia, PA 19106, USA

In loving memory of my sister Joann Elaine Vest.

Contents

Preface

As the undeniable successes of the atomic model of matter led to its acceptance in the latter half of the 19th century, atomic level behavior became the focus of accurate physical experiments. By the early part of the 20th century the outcome of these experiments was both a surprise and an enigma. Despite the best efforts of physicists, the version of mechanics first formulated by Newton and his contemporaries was unable to explain the results of experiments where atomic level behavior clearly manifested itself. Physics faced a profound dilemma, which was how to explain and predict atomic level behavior without discarding the unquestionable successes of Newtonian physics at the everyday level. The solution to this dilemma culminated in the formal structure known as *quantum mechanics* to distinguish it from the *classical mechanics* of Newton, and launched a profound and ongoing revolution in the understanding of matter. It is safe to say that both the theoretical and experimental investigation of atomic and subatomic behavior has dominated the last one hundred years of physics.

The purpose of this monograph is to present a fairly concise yet reasonably thorough and logical introduction to the basic concepts, applications, and physical meaning of quantum mechanics. Its intended readership is anyone who wishes to transit from an understanding of basic Newtonian or classical physics to an understanding of how and why quantum mechanics supersedes classical physics in the microscopic realm. It assumes only a basic undergraduate level knowledge of classical mechanics, electrodynamics, and wave phenomena, although all three of these topics are reviewed in the first chapter. The mathematical level of the monograph assumes the reader is familiar with basic multivariable calculus and vector manipulation. It is also useful to have some knowledge of partial differential equations and their boundary conditions, but the content in this regard is self-contained. All other mathematics, including complex variables, Fourier series and function spaces, inner products, matrix and operator theory, and eigenvalue equations is presented concisely and without rigor, assuming a level of understanding appropriate to the intended audience. A more advanced reader may wish to skip those sections where familiar mathematical concepts are presented, while the introductory reader can obtain an overview of these essential areas of mathematics, identifying those topics that require further study.

Since there are many excellent introductory texts and monographs on quantum mechanics, it is natural to ask how this monograph differs from them. First, this monograph has purposely been written to remain concise, and so its content has been limited to the essential ideas of quantum mechanics with a handful of illustrative applications in order to hold the size down. Hopefully, this allows the reader to understand the fundamental structure of quantum mechanics without dealing with the myriad of refinements and applications that have taken place over the last century. In that regard, the monograph uses the Copenhagen interpretation of quantum mechanics and the wave function prior to discussing the Schrödinger equation, introducing the concepts and postulates regarding the wave function and observables that are required to formulate quantum mechanics as a logical outgrowth of experiment. Second, it develops formal quantum mechanics to be explicitly consistent with Galilean relativity and the principle of superposition as well as the formulation of random variables used in statistics. Combining these three requirements with the probabilistic interpretation of the wave function dictates the form of the Schrödinger equation. The general Hilbert space formulation of quantum mechanics then follows from the properties of the Schrödinger equation and its solutions. The nature of observation and its relation to the Heisenberg uncertainty principle are stressed. The text presents what the author believes are useful and critically important applications of quantum mechanics to a variety of problems, including wave packet spreading, barrier penetration, the simple harmonic oscillator, the hydrogen atom, a charge in a uniform magnetic field, two state oscillation, electron diffraction, angular momentum and its addition, and its extension to spin angular momentum. Finally, the second edition of the text has expanded the topics covered to include multiparticle states, quantum entanglement, Bell's theorem, quantum decoherence, perturbation theory, path integrals, and scattering theory.

These choices for content and level were made with the hope that this monograph will be useful *and accessible* to an audience of scientists, both in training and in practice, larger than advanced level physics students and specialists. This ranges from advanced undergraduates to practitioners in allied disciplines such as physical chemistry, electrical engineering, and computing. The bibliography at the end of each chapter is broken down by topic, listing a few of the many books I believe would be useful for immediate further study. It is my sincere hope that the reader will use the recommended books to begin a further study of the myriad important details and applications that quantum mechanics possesses. Because this monograph is pedagogical in nature and designed to be concise, many of the relevant research papers in quantum mechanics have not been listed. I apologize to the giants of physics, both living and dead, whose research, insights, and hard work laid the foundations of quantum mechanics and continue to strengthen it. However, a thorough list of all the relevant papers would be extremely long and would not substantially contribute to the intended pedagogical goal of this monograph. In that regard, some concepts presented have become so widely used that they are known simply by the name of their originator, and this form of attribution is often followed.

I have chosen to use certain notation conventions throughout the monograph. Spatial vectors in higher than one dimension are designated with bold font, so that $\boldsymbol{x}$ is understood as a position vector, while superscripts such as x^i refer to the vector components and $|\boldsymbol{x}|^2 = x^2$ is its magnitude squared. Three-dimensional Cartesian coordinates are commonly denoted $x^1 = x$, $x^2 = y$, and $x^3 = z$. Functions of vector position are written $f(\boldsymbol{x})$ and, where there is no chance of confusion, their arguments are sometimes suppressed and they are written simply as f. Integration over a volume of n-dimensional space is designated as $\int d^n x$, where the domain of integration is typically not indicated unless it is critical to the outcome of the integration. A single summation is indicated by Σ_j, where the limits on j in the sum may be explicitly displayed, $\Sigma_{j=1}^{N}$, or implicit in the context of the sum, as in the case that j represents an infinite sum over the powers of a variable. Multiple summations are written $\Sigma_{j,k}$. The symbol '$\equiv$' means 'is defined as', while the symbol '$\rightarrow$' is to be read 'becomes'. The expression '$x \in (a, b)$' is understood as 'x is contained in the interval a to b'. The occasional appearance of '$\Longrightarrow$' is meant to be read as 'implies', '$a \leftrightarrow b$' is understood as 'a and b are exchanged', '$a \gg b$' means 'a is much greater than b', and '$a \approx b$' is understood as 'a is approximately equal to b'.

I would like to thank my editor, John Navas, and his team at IOP for their assistance in bringing the second edition of this book to completion.

Author biography

Mark S Swanson

Mark S Swanson received his PhD in physics from the University of Missouri at Columbia in 1976 under the supervision of Justin Huang. After postdoctoral appointments at the University of Alberta and the University of Connecticut, he held a faculty appointment at the University of Connecticut at Stamford from 1983 to 2014, during which time he spent six years in university administrative roles. He is the author of 25 research articles and four monographs, with an emphasis on field theory and path integral techniques. He is currently Emeritus Professor of Physics at the University of Connecticut and lives in Monroe, Connecticut.

IOP Publishing

Mark S Swanson

Chapter 1

Classical mechanics and electromagnetism

The mechanics of motion formulated by Newton and contemporaries in the 17th century is the foundation of contemporary physics. It is immensely successful, explaining and quantifying a vast array of phenomena, ranging from planetary motion to the flow of fluids. By the latter part of the 19th century the study of electric and magnetic fields and their inclusion in the framework of Newtonian physics appeared near completion. Since motion and electromagnetism are central to the failures of Newtonian physics in the atomic realm, it is useful to review the essential ideas, concepts, and the implicit assumptions of what is now referred to as *classical physics*. Many of the results presented in this chapter are assumed to be familiar to the reader, but some concepts are more thoroughly presented because of their relevance to quantum mechanics.

1.1 Newtonian mechanics

For simplicity initial consideration is limited to a single *point particle* of inertial mass m. A point particle is treated as infinitesimally small and indivisible. It is possible to formulate Newtonian mechanics in terms of continuous matter distributions, referred to as *continuum mechanics*, but these will not be considered in this text. In that regard, point particles are often used as an approximation to the constituents of real matter in Newtonian mechanics.

A point particle's position therefore coincides with a point in space at every given moment of time, parameterized with t, and can be described mathematically by a vector $\boldsymbol{x}$ from the choice of origin to that point. As the point particle moves its position is given by the time-dependent *trajectory* $\boldsymbol{x}(t)$. An important assumption of Newtonian mechanics is that *all observers* are equipped with identical meter sticks and synchronized clocks, and they all agree upon the current position at the time t, although the choice of each observer's origin is *arbitrary*. The Cartesian components of the vector trajectory are denoted $x_i(t)$, where the subscript i runs over the dimensions of the space under consideration. In one dimension the subscript is

doi:10.1088/978-0-7503-5663-3ch1

dropped and the trajectory is written $x(t)$. As the point particle moves it has a velocity $v(t)$ and an acceleration $\boldsymbol{a}(t)$, both vector quantities and both obtained from the trajectory by differentiation,

$$v(t) = \frac{\mathrm{d}\boldsymbol{x}(t)}{\mathrm{d}t} \equiv \dot{\boldsymbol{x}}(t), \quad \boldsymbol{a}(t) = \frac{\mathrm{d}v(t)}{\mathrm{d}t} = \frac{\mathrm{d}^2\boldsymbol{x}(t)}{\mathrm{d}t^2} \equiv \ddot{\boldsymbol{x}}(t). \tag{1.1}$$

The definitions of (1.1) determine the velocity and acceleration when the particle is at the position $\boldsymbol{x}(t)$ since the time t is a parameter common to all three. It is important to note that the velocity and acceleration are *independent* of the Newtonian observer's choice of origin and will be agreed upon by all observers.

1.1.1 Newton's laws of motion

As the particle moves, it can be subject to a force, $\boldsymbol{F}(\boldsymbol{x}, t)$, which is also a vector with the same number of components as $\boldsymbol{x}$. The force $\boldsymbol{F}$ may depend explicitly on the time t as well as implicitly through the particle's current position $\boldsymbol{x}(t)$. Newton's first law of motion defines zero force by stating that zero force results in zero acceleration for the point mass and therefore the mass moves at a constant velocity. In what follows, the *scalar product* of two spatial vectors is designated as $\boldsymbol{x} \cdot \boldsymbol{y} = \sum_j x_j y_j$, and can be shown to be the same as $\boldsymbol{x} \cdot \boldsymbol{y} = |\boldsymbol{x}||\boldsymbol{y}|\cos\theta$, where θ is the angle between the two vectors.

Newton's second law of motion states that the acceleration of an inertial mass is proportional to the force and inversely proportional to its inertial mass. Using the definitions of (1.1) allows Newton's second law to be written

$$\boldsymbol{F}(\boldsymbol{x}, t) = m\boldsymbol{a}(t) = m\frac{\mathrm{d}^2\boldsymbol{x}(t)}{\mathrm{d}t^2}. \tag{1.2}$$

Since Newton's second law (1.2) is expressed as a second order differential equation, a *unique solution* requires two boundary or initial conditions for each spatial component to be specified. For example, specifying the particle's initial position and velocity is sufficient for a *unique solution* to (1.2). Once these boundary conditions are specified, the trajectory of the particle at all subsequent and previous times is determined solely from $\boldsymbol{F}$. The trajectory may be extremely sensitive to the initial conditions imposed, but at each moment the position is unique and therefore entirely *deterministic* in nature. *Determinism* is a cornerstone of Newtonian physics and its mechanical view of the world. It will be seen that determinism is incompatible with quantum mechanics. In what follows, the solution of Newton's second law (1.2) is referred to as the *classical trajectory* of the particle and is designated $\boldsymbol{x}_c(t)$.

Newton's second law (1.2) can be written

$$\boldsymbol{F}(\boldsymbol{x}, t) = m\frac{\mathrm{d}v(t)}{\mathrm{d}t} = \frac{\mathrm{d}(mv(t))}{\mathrm{d}t} \equiv \frac{\mathrm{d}\boldsymbol{p}(t)}{\mathrm{d}t}, \tag{1.3}$$

where $\boldsymbol{p} = mv$ is the *linear momentum* of the particle. Relation (1.3) shows that position $\boldsymbol{x}$ and momentum $\boldsymbol{p}$ are the two *canonically conjugate* variables used in the Newtonian description of motion. To begin the transition to quantum mechanics, position and momentum will be referred to as *observables*, which means that they are measurable aspects of motion. Newtonian mechanics uses these two observables to derive other important aspects of motion or observables, including energy and angular momentum. The same will remain true in quantum mechanics.

The Newtonian kinetic energy K for a point mass m as it moves along the classical trajectory $\boldsymbol{x}_c(t)$ is defined in terms of its velocity along the trajectory $\boldsymbol{v}_c(t)$ as

$$K = \frac{1}{2}mv_c^2 = \frac{1}{2}m\boldsymbol{v}_c \cdot \boldsymbol{v}_c = \frac{1}{2}m\dot{\boldsymbol{x}}_c \cdot \dot{\boldsymbol{x}}_c. \tag{1.4}$$

The rate at which K is changing along the trajectory is then given by

$$\frac{\mathrm{d}K}{\mathrm{d}t} = m\boldsymbol{v}_c \cdot \frac{\mathrm{d}\boldsymbol{v}_c}{\mathrm{d}t} = \boldsymbol{v}_c \cdot m\ddot{\boldsymbol{x}}_c = \boldsymbol{v}_c \cdot \mathbf{F}(\boldsymbol{x}_c) = \frac{\mathrm{d}\boldsymbol{x}_c}{\mathrm{d}t} \cdot \mathbf{F}(x_c). \tag{1.5}$$

By assumption, a *conservative force* acting on a mass m particle has no explicit time-dependence, and its components can be written

$$F_i(\boldsymbol{x}) = -\frac{\partial V(\boldsymbol{x})}{\partial x_i} \equiv -\partial_i V(\boldsymbol{x}) \implies m\ddot{x}_i = -\partial_i V(\boldsymbol{x}), \tag{1.6}$$

where $V(\boldsymbol{x})$ is a scalar function of position known as the *potential energy* and the components of the force are designated using a subscript. The combination of the kinetic and potential energy along the classical trajectory,

$$E = \frac{1}{2}mv_c^2 + V(\boldsymbol{x}_c), \tag{1.7}$$

is called the *total mechanical energy*. If all the forces on a particle are conservative, then the total mechanical energy E is time-independent or *conserved* as the particle moves along the classical trajectory that solves (1.2). Finding its time derivative via the chain rule and evaluating it along the classical trajectory gives

$$\left.\frac{\mathrm{d}E}{\mathrm{d}t}\right|_{x=x_c} = \sum_i \left(m\ddot{x}_i + \frac{\partial V(\boldsymbol{x})}{\partial x_i}\right)\dot{x}_i\Bigg|_{x=x_c} = \sum_i (m\ddot{x}_{ic} - F_i(\boldsymbol{x}_c))\dot{x}_{ic} = 0, \tag{1.8}$$

which vanishes since $\boldsymbol{x}_c$ solves (1.2). For the case of conservative forces, the conservation of total mechanical energy gives

$$\mathrm{d}K = -\mathrm{d}V, \tag{1.9}$$

so that the change in kinetic energy is equal and opposite to the change in potential energy. The concept of total energy is at the core of theoretical physics and plays a central role in formulating quantum mechanics. However, it should be noted that energy is defined *only up to an arbitrary constant*. This is because the potential energy can be modified by adding an arbitrary constant V_o,

$$V(\boldsymbol{x}) \to V(\boldsymbol{x}) + V_o, \tag{1.10}$$

without altering the conservation law (1.8) or Newton's equation of motion (1.2).

As a final note, Newton's third law of motion defines a true force by stating that the action of one body on another body is accompanied by the reaction of the other body on the first. This differentiates a true force as being a member of an action–reaction pair of forces. This distinguishes true forces from *fictitious forces* which can arise when a system is viewed in an accelerated or rotating coordinate system.

1.1.2 The variational formulation of Newtonian mechanics

In its modern formulation Newtonian mechanics is stated in terms of a *variational* or *action principle.* The variational version of (1.2) starts by considering an *arbitrary* trajectory $\boldsymbol{x}(t)$, assumed to be a differentiable function of t. This arbitrary trajectory is used to evaluate the *classical action* $S[\boldsymbol{x}, \dot{\boldsymbol{x}}]$ for the time interval T, which is given by

$$S[\boldsymbol{x}, \dot{\boldsymbol{x}}, T] = \int_0^T \mathrm{d}t\, (K(\boldsymbol{x}, \dot{\boldsymbol{x}}, t) - V(\boldsymbol{x}, \dot{\boldsymbol{x}}, t)) \equiv \int_0^T \mathrm{d}t\, \mathcal{L}(\dot{\boldsymbol{x}}, \boldsymbol{x}, t), \tag{1.11}$$

where both the kinetic energy K and the potential V may depend explicitly on $\boldsymbol{x}$, $\dot{\boldsymbol{x}}$, and t. The function $\mathcal{L}(\dot{\boldsymbol{x}}, \boldsymbol{x}, t)$ is referred to as the *Lagrangian density*. The value of the action $S[\boldsymbol{x}, \dot{\boldsymbol{x}}, T]$ is assumed to depend on the arbitrary trajectory $\boldsymbol{x}(t)$ and its first derivative, and for that reason the action is referred to as a *functional.*

The calculus of variations determines the form of the trajectory that *extremizes* the value of the functional S, consistent with boundary conditions, by determining the differential equation that this trajectory must solve. This differential equation is found by treating $\boldsymbol{x}$ and $\dot{\boldsymbol{x}}$ as independent degrees of freedom and finding the conditions under which the action is an *extremum*. The form (1.11) for S has been chosen to ensure that the extremal trajectory and the classical trajectory, $\boldsymbol{x}_c(t)$, coincide for simple Newtonian systems. Uniquely specifying the extremal trajectory requires two boundary conditions. As an example, *Dirichlet boundary conditions* specify $\boldsymbol{x}_c(0) = \boldsymbol{x}_o$ and $\boldsymbol{x}_c(T) = \boldsymbol{x}_f$. Other boundary conditions involving derivatives, such as *Neumann* and *Cauchy boundary conditions*, are possible. A trajectory that is close to the extremal trajectory can be written $\boldsymbol{x}(t) = \boldsymbol{x}_c(t) + \delta\boldsymbol{x}(t)$. The function $\delta\boldsymbol{x}(t)$ is an *infinitesimal* deviation from the extremal trajectory that must satisfy the boundary conditions $\delta\boldsymbol{x}(0) = \delta\boldsymbol{x}(T) = 0$. Since $\delta\boldsymbol{x}$ is infinitesimal, a Taylor series expansion around the extremal solution need only retain terms of order $O(\delta\boldsymbol{x}(t))$. Suppressing the time-dependence of $\boldsymbol{x}_c(t)$ and $\delta\boldsymbol{x}(t)$, the Taylor series gives

$$\begin{aligned} \mathcal{L}(\dot{\boldsymbol{x}}_c + \delta\dot{\boldsymbol{x}}, \boldsymbol{x}_c + \delta\boldsymbol{x}, t) &= \mathcal{L}(\dot{\boldsymbol{x}}_c, \boldsymbol{x}_c, t) + \sum_i \left(\delta\dot{x}_i \frac{\partial\mathcal{L}}{\partial\dot{x}_i}\bigg|_{x=x_c} + \delta x_i \frac{\partial\mathcal{L}}{\partial x_i}\bigg|_{x=x_c} \right) \\ &= \mathcal{L}(\dot{\boldsymbol{x}}_c, \boldsymbol{x}_c, t) + \sum_i \left\{ \frac{\mathrm{d}}{\mathrm{d}t}\left(\delta x_i \frac{\partial\mathcal{L}}{\partial\dot{x}_c^i}\bigg|_{x=x_c} \right) + \delta x_i \left(\frac{\partial\mathcal{L}}{\partial x_{ic}} - \frac{\mathrm{d}}{\mathrm{d}t}\frac{\partial\mathcal{L}}{\partial\dot{x}_{ic}} \right)\bigg|_{x=x_c} \right\}. \end{aligned} \tag{1.12}$$

Returning (1.12) to the action (1.11) shows that the total time derivative will vanish since $\delta\boldsymbol{x}(0) = \delta\boldsymbol{x}(T) = 0$. The *variation of the action* around the extremal trajectory becomes

$$\delta S[\boldsymbol{x}_c\, \dot{\boldsymbol{x}}_c, T] = S[\boldsymbol{x}_c + \delta\boldsymbol{x}, \dot{\boldsymbol{x}}_c + \delta\dot{\boldsymbol{x}}, T] - S[\boldsymbol{x}_c, \dot{\boldsymbol{x}}_c, T]$$
$$= \int_0^T \mathrm{d}t \sum_i \delta x_i(t) \left(\frac{\partial \mathcal{L}(\boldsymbol{x}, \dot{\boldsymbol{x}})}{\partial x_i} - \frac{\mathrm{d}}{\mathrm{d}t} \frac{\partial \mathcal{L}(\boldsymbol{x}, \dot{\boldsymbol{x}})}{\partial \dot{x}_i} \right) \bigg|_{\boldsymbol{x}=\boldsymbol{x}_c(t)}. \tag{1.13}$$

Similar to an extremum point of a function, an extremal trajectory must satisfy $\delta S = 0$. Since $\delta \boldsymbol{x}$ is arbitrary, the function $\boldsymbol{x}_c(t)$ must therefore satisfy the *Euler–Lagrange equation*, given by

$$\left(\frac{\partial \mathcal{L}(\boldsymbol{x}, \dot{\boldsymbol{x}})}{\partial x_i} - \frac{\mathrm{d}}{\mathrm{d}t} \frac{\partial \mathcal{L}(\boldsymbol{x}, \dot{\boldsymbol{x}})}{\partial \dot{x}_i} \right) \bigg|_{\boldsymbol{x}=\boldsymbol{x}_c(t)} = 0. \tag{1.14}$$

Using form (1.11) in the Euler–Lagrange equation (1.14) immediately reproduces the usual form of Newton's second law of motion (1.2) for a conservative force. It is worth noting that adding an arbitrary constant to the potential $V(\boldsymbol{x})$ appearing in $\mathcal{L}(\boldsymbol{x}, \dot{\boldsymbol{x}})$ does not alter (1.14).

The action formulation allows the generalization of both Newtonian mechanics and the concepts of momentum and energy. For example, the component of momentum p_i that is *canonically conjugate* to the component of position x_i can be defined as

$$p_i = \frac{\partial \mathcal{L}(\boldsymbol{x}, \dot{\boldsymbol{x}})}{\partial \dot{x}_i}. \tag{1.15}$$

It is assumed that (1.15) can be inverted to find the solution for $\dot{x}_i$ as a function of $\boldsymbol{p}$ and $\boldsymbol{x}$, denoted $\dot{x}_i(\boldsymbol{p}, \boldsymbol{x})$. This process simply introduces a new variable $\boldsymbol{p}$ that replaces the variable $\dot{\boldsymbol{x}}$. Using a *Legendre transformation*, the generalization of the total mechanical energy, called *the Hamiltonian* and denoted $H(\boldsymbol{p}, \boldsymbol{x})$, is defined as

$$H(\boldsymbol{p}, \boldsymbol{x}, t) = \boldsymbol{p} \cdot \dot{\boldsymbol{x}}(\boldsymbol{p}, \boldsymbol{x}) - \mathcal{L}(\dot{\boldsymbol{x}}(\boldsymbol{p}, \boldsymbol{x}), \boldsymbol{x}, t). \tag{1.16}$$

For the case that the Lagrangian density has no *explicit* time-dependence, it follows that the time derivative of the Hamiltonian along the classical trajectory is given by

$$\frac{\mathrm{d}H}{\mathrm{d}t} = \sum_i \left(\dot{p}_i \dot{x}_i + p_i \ddot{x}_i - \frac{\partial \mathcal{L}}{\partial \dot{x}_i} \ddot{x}_i - \frac{\partial \mathcal{L}}{\partial x_i} \dot{x}_i \right) \bigg|_{\boldsymbol{x}=\boldsymbol{x}_c} = \sum_i \left(\dot{p}_i - \frac{\partial \mathcal{L}}{\partial x_i} \right) \dot{x}_i \bigg|_{\boldsymbol{x}=\boldsymbol{x}_c}, \tag{1.17}$$

where (1.15) was used to cancel the second and third terms Once again, this vanishes along the classical trajectory since the combination of (1.15) with the Euler–Lagrange equation (1.14) gives

$$\dot{p}_i \big|_{\boldsymbol{x}=\boldsymbol{x}_c(t)} = \frac{\mathrm{d}}{\mathrm{d}t} \frac{\partial \mathcal{L}(\boldsymbol{x}, \dot{\boldsymbol{x}})}{\partial \dot{x}_i} \bigg|_{\boldsymbol{x}=\boldsymbol{x}_c(t)} = \frac{\partial \mathcal{L}(\boldsymbol{x}, \dot{\boldsymbol{x}})}{\partial x_i} \bigg|_{\boldsymbol{x}=\boldsymbol{x}_c(t)}. \tag{1.18}$$

The freedom to add an arbitrary constant to the Lagrangian $\mathcal{L}$ reflects the earlier observation that the energy is defined only up to an arbitrary constant. This means that only *differences* in energy are relevant to motion in Newtonian physics. It follows that

$$\mathcal{L} = \sum_i \frac{1}{2} m\dot{x}_i\dot{x}_i - V(\boldsymbol{x}) \implies p_i = m\dot{x}_i, \;\; H = \frac{1}{2m}\boldsymbol{p}_c \cdot \boldsymbol{p}_c + V(\boldsymbol{x}_c), \tag{1.19}$$

showing that the familiar results are obtained from the Lagrangian density of (1.19). Newton's law of motion then follows from (1.18), which gives

$$ma_{ic} = m\ddot{x}_{ic} = \dot{p}_{ic} = \frac{\mathrm{d}}{\mathrm{d}t}\frac{\partial \mathcal{L}(\dot{\boldsymbol{x}}_c, \boldsymbol{x}_c)}{\partial \dot{x}_{ic}} = \frac{\partial \mathcal{L}(\dot{\boldsymbol{x}}_c, \boldsymbol{x}_c)}{\partial x_{ic}} = -\frac{\partial V(\boldsymbol{x}_c)}{\partial x_{ic}} = F_i(\boldsymbol{x}_c). \tag{1.20}$$

It is the canonical momentum and position as well as the Hamiltonian generalization of energy that are used in formulating quantum mechanics.

A second means of implementing the variational principle (1.13) starts by using the Hamiltonian (1.16) to express the Lagrangian density as

$$\mathcal{L} = \boldsymbol{p} \cdot \dot{\boldsymbol{x}} - H(\boldsymbol{p}, \boldsymbol{x}, t), \tag{1.21}$$

where $\boldsymbol{p} = \boldsymbol{p}(\boldsymbol{x}, \dot{\boldsymbol{x}})$ is understood to relate $\dot{\boldsymbol{x}}$ to $\boldsymbol{p}$ and $\boldsymbol{x}$. The variation of the action is then given by varying both the momentum and the position according to $x_i = x_{ic} + \delta x_i$ and $p_i = p_{ic} + \delta p_i$, where both δp_i and δx_i vanish at $t = 0$ and $t = T$. The result is

$$\begin{aligned}\delta S[\boldsymbol{x}_c, \dot{\boldsymbol{x}}_c, T] &= \int_0^T \mathrm{d}t \sum_i \left\{ \delta\dot{x}_i\, p_{ic} + \delta p_i \left(\dot{x}_{ic} - \left.\frac{\partial H}{\partial p_i}\right|_{x=x_c(t)} \right) - \delta x_i \left.\frac{\partial H}{\partial x_i}\right|_{x=x_c(t)} \right\} \\ &= \int_0^T \mathrm{d}t \sum_i \left\{ \delta p_i \left(\dot{x}_{ic} - \left.\frac{\partial H}{\partial p_i}\right|_{x=x_c(t)} \right) - \delta x_i \left(\dot{p}_{ic} + \left.\frac{\partial H}{\partial x_i}\right|_{x=x_c(t)} \right) \right\},\end{aligned} \tag{1.22}$$

where the second step follows from an integration by parts employing the boundary conditions on δx_i. Since the two variations are independent, (1.22) yields

$$\dot{x}_{ic} - \frac{\partial H(\boldsymbol{p}_c, \boldsymbol{x}_c, t)}{\partial p_{ic}} = 0 \quad \text{and} \quad \dot{p}_{ic} + \frac{\partial H(\boldsymbol{p}_c, \boldsymbol{x}_c, t)}{\partial x_{ic}} = 0, \tag{1.23}$$

which are referred to as *Hamilton's equations*.

Hamilton's equations can be used to state the time evolution of a mechanical quantity $\mathrm{O}(\boldsymbol{p}, \boldsymbol{x}, t)$, referred to as an *observable*, along the classical trajectory $\boldsymbol{x}_c$,

$$\frac{\mathrm{dO}}{\mathrm{d}t} = \frac{\partial \mathrm{O}}{\partial t} + \sum_i \left(\frac{\partial \mathrm{O}}{\partial x_{ic}} \dot{x}_{ic} + \frac{\partial \mathrm{O}}{\partial p_{ic}} \dot{p}_{ic} \right) = \frac{\partial \mathrm{O}}{\partial t} + \sum_i \left(\frac{\partial \mathrm{O}}{\partial x_{ic}} \frac{\partial H}{\partial p_{ic}} - \frac{\partial \mathrm{O}}{\partial p_{ic}} \frac{\partial H}{\partial x_{ic}} \right). \tag{1.24}$$

Using (1.24) the *Poisson bracket* of any two mechanical quantities $u(\boldsymbol{p}, \boldsymbol{x}, t)$ and $v(\boldsymbol{p}, \boldsymbol{x}, t)$ is defined as

$$\{u, v\} \equiv \sum_i \left(\frac{\partial u}{\partial x_{ic}} \frac{\partial v}{\partial p_{ic}} - \frac{\partial u}{\partial p_{ic}} \frac{\partial v}{\partial x_{ic}} \right), \tag{1.25}$$

so that (1.24) can be written as

$$\frac{\mathrm{dO}}{\mathrm{d}t} = \frac{\partial \mathrm{O}}{\partial t} + \{\mathrm{O}, H\}. \tag{1.26}$$

The Poisson bracket has the property of antisymmetry, $\{u, v\} = -\{v, u\}$, and it obeys the *Jacobi identity*, $\{u, \{v, w\}\} + \{w, \{u, v\}\} + \{v, \{w, u\}\} = 0$, both of which are easily demonstrated by direct substitution using the definition (1.25). Antisymmetry guarantees that $\{u, u\} = 0$.

Notable Poisson bracket results include

$$\{x_i, p_j\} = \delta_{ij} \equiv \begin{cases} 1 & \text{if } i = j \\ 0 & \text{if } i \neq j \end{cases}, \tag{1.27}$$

where δ_{ij} is known as the *Kronecker delta*. A second important Poisson bracket involves the orbital angular momentum of a particle, $\boldsymbol{L} = \boldsymbol{r} \times \boldsymbol{p}$, where $\times$ denotes the *vector product*. The Cartesian components of the vector product in three dimensions are defined as

$$L_i = (\boldsymbol{r} \times \boldsymbol{p})_i = \sum_{j,k=1}^{3} \varepsilon_{ijk} x_j p_k, \tag{1.28}$$

where the *Levi–Civita symbol* is defined as

$$\varepsilon_{ijk} = \begin{cases} 1 & \text{if } ijk \text{ is an even permutation of 123} \\ -1 & \text{if } ijk \text{ is an odd permutation of 123} \\ 0 & \text{otherwise.} \end{cases} \tag{1.29}$$

It follows that the Cartesian x and y components of the vector product of $\boldsymbol{r}$ and $\boldsymbol{p}$ are given by

$$\begin{aligned} L_x &= yp_z - zp_y, \\ L_y &= zp_x - xp_z, \\ L_z &= xp_y - yp_x. \end{aligned} \tag{1.30}$$

Substituting these into the Poisson bracket gives

$$\{L_x, L_y\} = xp_y - yp_x = L_z, \tag{1.31}$$

with other *cyclic permutations* given by $\{L_y, L_z\} = L_x$ and $\{L_z, L_x\} = L_y$. The quantum mechanical versions of (1.27) and (1.31) are central to understanding atomic level behavior.

1.1.3 Newtonian mechanics and Galilean relativity

An often overlooked but key property of Newtonian mechanics is that it incorporates *Galilean relativity*. Two Newtonian observers, $\mathcal{O}$ and $\mathcal{O}'$, are equipped with identical meter sticks and clocks with which they measure the time and location of *events*, such as the collision of two objects or the flash of a light bulb. A Galilean transformation relates the time and position, t and $\boldsymbol{x}$, of an event measured in the $\mathcal{O}$ coordinate system or *frame* to the time and position, t' and $\boldsymbol{x}'$, for the *same event* measured in the $\mathcal{O}'$ frame. For simplicity, the two observers are assumed to synchronize their clocks so that their origins, located at $\boldsymbol{x} = \boldsymbol{x}' = 0$, coincide at the time $t = t' = 0$. For the case that the two observers are moving at the *constant relative velocity* $\boldsymbol{v}_o$ the Galilean coordinate transformations and their inverses are given by

$$\begin{aligned} \boldsymbol{x}' &= \boldsymbol{x} - \boldsymbol{v}_o t \implies \boldsymbol{x} = \boldsymbol{x}' + \boldsymbol{v}_o t', \\ t' &= t \implies t = t'. \end{aligned} \tag{1.32}$$

The origin in the $\mathcal{O}$ system at $\boldsymbol{x} = \mathbf{0}$ is at the location $\boldsymbol{x}' = -\boldsymbol{v}_o t'$ in the $\mathcal{O}'$ system at any later time. For the specific case that $\boldsymbol{v}$ is constant, the principle of Galilean relativity states that the equations of motion are *form invariant* under a *Galilean transformation*, so that two observers related by a Galilean transformation will agree on the laws of motion.

The transformations of (1.32) reflect the Newtonian belief in an absolute time that is the same for all observers and is unaffected by relative motion. The axiom of invariance under Galilean relativity arose from the observational work of Galileo, and so it bears his name. This axiom is reflected in Newton's law of motion (1.2), which is invariant under a Galilean transformation (1.32) as long as $\boldsymbol{v}$ is constant, $\mathrm{d}\boldsymbol{v}/\mathrm{d}t = 0$, since the acceleration transforms as

$$\boldsymbol{a}' = \frac{\mathrm{d}^2\boldsymbol{x}'}{\mathrm{d}t'^2} = \frac{\mathrm{d}^2\boldsymbol{x}'}{\mathrm{d}t^2} = \frac{\mathrm{d}}{\mathrm{d}t}\left(\frac{\mathrm{d}\boldsymbol{x}}{\mathrm{d}t} - \boldsymbol{v}_o\right) = \frac{\mathrm{d}^2\boldsymbol{x}}{\mathrm{d}t^2} = \boldsymbol{a}, \tag{1.33}$$

while the inertial mass m is invariant. If the force is static according to $\mathcal{O}$, so that $\mathbf{F} = \mathbf{F}(\boldsymbol{x})$, then according to $\mathcal{O}'$ the force is given by $\mathbf{F} = \mathbf{F}(\boldsymbol{x}' + \boldsymbol{v}_o t')$. This is consistent with the agent of the force appearing to be static according to $\mathcal{O}$, while it is in motion according to $\mathcal{O}'$.

It is possible to include *rotated reference frames* in Galilean relativity, and such rotated frames are discussed in chapter 7. It eventually became necessary to replace the Galilean transformations of (1.32) with the Lorentz transformations of Einstein's special relativity, which reduce to (1.32) for velocities small compared to that of light. However, for the purposes of this text the requirement for Einstein's special relativity will be waived and form invariance under Galilean transformations will be implemented in *nonrelativistic* quantum mechanics.

1.1.4 Determinism and classical systems

An often overlooked aspect of Newtonian physics is its deterministic nature. In the worldview adopted by Newton and contemporaries the physical world was *mechanical* in nature, so that the motion or dynamics of an object was completely predictable once the system of forces and objects to which it belonged was set in motion. In other words, the future behavior of an object, both its position and its velocity or momentum, was exactly determined from its initial situation.

This includes the idea that all the properties of an object, its mass, location, and velocity, could in principle be measured to arbitrary accuracy at any given moment. This assumption was unquestioned, despite the lack of accurate measuring devices available at that time. In addition, the act of measurement was taken for granted, so that any potential effect the measurement had on the aspect of the system being measured could be ignored. It was assumed that the system being observed and analysed could be reproduced to arbitrary accuracy in *all* its aspects by *all* observers and that *all* properly equipped observers would measure the same outcomes each time the experiment was performed. It was further assumed that the system being analysed was *isolated*, so that interactions between the objects in the system and the *environment of the system* could be ignored. For some cases this assumption was known to introduce error, such as the error induced by ignoring air resistance for a freely falling object. In other cases, this assumption was necessary due to the insoluble complexity of stating or including the effects of the environment. In many cases, the effects of the environment proved to be inconsequential.

These assumptions underlie the *ideal* laws of motion proposed by Newton and reflect the belief that a system, its environment, and its observers constitute separate *real and independent* objects. These assumptions, along with the laws of motion, constitute what is referred to as *classical physics*. With the advent of quantum mechanics these assumptions and their need for modification have become the subject of much debate. This text will draw attention to these modifications where relevant to the basic understanding of quantum mechanics.

1.2 Light and electromagnetism

One of the central goals for physics in the 19th century was understanding the dual phenomena of electricity and magnetism and their relationship to light. Because the interactions between light and matter were critical to probing atomic level phenomena, it is useful to review the structure that was in place at the end of the 19th century.

1.2.1 Fields and particles

The central postulate of electromagnetic theory is that *some but not all* matter is electrically charged. In the late 18th and early 19th century the atomic model of matter was in its infancy and it was not known which matter carried electric charge or if there was a fundamental element of charge. However, by the early 19th century Coulomb's law of electric force was experimentally established and used in applications of physics. It states that the electric force between two stationary electric *point charges*, q and Q, separated by the distance r has the magnitude in Gaussian or cgs units given by

$$F = \frac{|Qq|}{r^2}, \tag{1.34}$$

where the force F and distance r are measured in the usual cgs units of dynes (dy) and centimeters (cm), while the charges are measured in electrostatic units (esu = $\sqrt{\text{dy}}$ cm). The force is attractive if $Qq < 0$ and repulsive otherwise and acts along the line joining the charges. For two equal and *opposite* charges, q and $-q$, expression (1.34) is associated with a potential energy of separation given by

$$V(r) = -\frac{q^2}{r}. \tag{1.35}$$

These two results underlie the formal development of electromagnetic theory that took place during the first half of the 19th century.

The key step in formulating a consistent theory of electromagnetic forces is the introduction of the concept of the *field*. A charged point particle is visualized as accompanied by an *electric field* in the space around the particle. The electric field $\mathbf{E}(\boldsymbol{r}, t)$ is a vector valued quantity such that another point charge q at the position $\boldsymbol{r}$ at the time t experiences an *electric force* given by

$$\boldsymbol{F}_E(\boldsymbol{r}, t) = q\mathbf{E}(\boldsymbol{r}, t). \tag{1.36}$$

If the charge is in motion it creates a vector valued *magnetic field* $\mathbf{B}(\boldsymbol{r}, t)$ in the space around it, which creates the *Lorentz force* on a charge q moving with the relative velocity $\boldsymbol{v}$, given by

$$\boldsymbol{F}_\text{L}(\boldsymbol{r}, t) = \frac{q}{c}\boldsymbol{v} \times \mathbf{B}(\boldsymbol{r}, t), \tag{1.37}$$

where c is the speed of light. In the Gaussian system both the electric field and magnetic field have the units of dy/esu = $\sqrt{\text{dy}}$/cm.

It should be apparent that these two fields are fundamentally different from a point particle since the electric and magnetic fields are not localized at a point. Instead, they exist over a region of space and may be created or disappear as a result of interaction with charged matter.

1.2.2 Vectors and gradients

Both electromagnetic theory and quantum mechanics make use of *scalar* and *vector* functions. In simple terms, the scalar functions associate a number with a position and time, while the vector functions associate a vector with a position and time. The *gradient operator* is central to working with such functions, and is used in the formulation of electromagnetism as well as quantum mechanics. The gradient operator ∇ is a *vector differential operator*. In Cartesian coordinates it is written in vector notation as $\nabla = \sum_j \boldsymbol{e}_j \partial/\partial x_j$, where the $\boldsymbol{e}_j$ are the Cartesian coordinate *unit vectors*, so that

$$\nabla = \boldsymbol{e}_x\frac{\partial}{\partial x} + \boldsymbol{e}_y\frac{\partial}{\partial y} + \boldsymbol{e}_z\frac{\partial}{\partial z}. \tag{1.38}$$

The Cartesian unit vectors are *orthonormal*, which is to say their *scalar product* satisfies $\boldsymbol{e}_i \cdot \boldsymbol{e}_j = \delta_{ij}$, where δ_{ij} is the Kronecker delta defined earlier. Their vector product is defined to be consistent with (1.30),

$$\boldsymbol{e}_x \times \boldsymbol{e}_y = \boldsymbol{e}_z, \quad \boldsymbol{e}_y \times \boldsymbol{e}_z = \boldsymbol{e}_x, \quad \boldsymbol{e}_z \times \boldsymbol{e}_x = \boldsymbol{e}_y, \tag{1.39}$$

with all other vector products vanishing. When the gradient is applied to a scalar function $\varphi(\boldsymbol{x})$, the vector given by $\nabla\varphi(\boldsymbol{x})$ points in the direction of *maximum increase* for the function at the point $\boldsymbol{x}$.

Of particular importance in physics is the quantity $\nabla \cdot \nabla = \nabla^2$, referred to as the *Laplacian*. In Cartesian coordinates the Laplacian takes the straightforward form

$$\nabla^2 = \frac{\partial^2}{\partial x^2} + \frac{\partial^2}{\partial y^2} + \frac{\partial^2}{\partial z^2}. \tag{1.40}$$

In other coordinate systems the gradient and the Laplacian will take a different form. An illustrative example is the relationship between Cartesian coordinates (x, y, z) and cylindrical coordinates (ρ, ϕ, z), given by the expressions $x = \rho \cos\phi$, $y = \rho \sin\phi$, and $z = z$, along with their inverses, $\rho = \sqrt{x^2 + y^2}$, $\phi = \arctan y/x$, and $z = z$. It is important to remember that both the partial derivatives *and* the unit vectors change. Using the chain rule with these expressions gives

$$\frac{\partial}{\partial\rho} = \cos\phi\,\frac{\partial}{\partial x} + \sin\phi\,\frac{\partial}{\partial y}, \quad \frac{1}{\rho}\frac{\partial}{\partial\phi} = -\sin\phi\,\frac{\partial}{\partial x} + \cos\phi\,\frac{\partial}{\partial y}, \quad \frac{\partial}{\partial z} = \frac{\partial}{\partial z}. \tag{1.41}$$

The unit vectors in cylindrical coordinates in terms of Cartesian unit vectors can be read off from (1.41) by using the association $\partial_i \rightarrow \boldsymbol{e}_i$. This gives

$$\boldsymbol{e}_\rho = \cos\phi\,\boldsymbol{e}_x + \sin\phi\,\boldsymbol{e}_y, \quad \boldsymbol{e}_\phi = -\sin\phi\,\boldsymbol{e}_x + \cos\phi\,\boldsymbol{e}_y, \quad \boldsymbol{e}_z = \boldsymbol{e}_z. \tag{1.42}$$

This correspondence holds because the partial derivatives give the rate of change in the direction of the respective unit vectors. Using the orthonormality of the Cartesian unit vectors shows that the three unit vectors of (1.42) are also orthonormal, but $\boldsymbol{e}_\rho$ and $\boldsymbol{e}_\phi$ change orientation at different points around the z-axis, with $\boldsymbol{e}_\rho$ always perpendicular to the z-axis and $\boldsymbol{e}_\phi$ always tangent to circles around the z-axis. The vector product of these unit vectors is determined by using the vector products of the Cartesian unit vectors given by (1.39). For example,

$$\begin{aligned}\boldsymbol{e}_\phi \times \boldsymbol{e}_z &= (-\sin\phi\,\boldsymbol{e}_x + \cos\phi\,\boldsymbol{e}_y) \times \boldsymbol{e}_z = -\sin\phi\,(\boldsymbol{e}_x \times \boldsymbol{e}_z) + \cos\phi\,(\boldsymbol{e}_y \times \boldsymbol{e}_z) \\ &= \sin\phi\,\boldsymbol{e}_y + \cos\phi\,\boldsymbol{e}_x = \boldsymbol{e}_\rho,\end{aligned} \tag{1.43}$$

with $\boldsymbol{e}_z \times \boldsymbol{e}_\rho = \boldsymbol{e}_\phi$ and $\boldsymbol{e}_\rho \times \boldsymbol{e}_\phi = \boldsymbol{e}_z$. The relations of (1.42) are easily inverted by projecting the Cartesian unit vectors onto the cylindrical unit vectors,

$$\begin{aligned}\boldsymbol{e}_x &= (\boldsymbol{e}_x \cdot \boldsymbol{e}_\rho)\boldsymbol{e}_\rho + (\boldsymbol{e}_x \cdot \boldsymbol{e}_\phi)\boldsymbol{e}_\phi + (\boldsymbol{e}_x \cdot \boldsymbol{e}_z)\boldsymbol{e}_z = \cos\phi\,\boldsymbol{e}_\rho - \sin\phi\,\boldsymbol{e}_\phi, \\ \boldsymbol{e}_y &= (\boldsymbol{e}_y \cdot \boldsymbol{e}_\rho)\boldsymbol{e}_\rho + (\boldsymbol{e}_y \cdot \boldsymbol{e}_\phi)\boldsymbol{e}_\phi + (\boldsymbol{e}_y \cdot \boldsymbol{e}_z)\boldsymbol{e}_z = \sin\phi\,\boldsymbol{e}_\rho + \cos\phi\,\boldsymbol{e}_\phi, \\ \boldsymbol{e}_z &= (\boldsymbol{e}_z \cdot \boldsymbol{e}_\rho)\boldsymbol{e}_\rho + (\boldsymbol{e}_z \cdot \boldsymbol{e}_\phi)\boldsymbol{e}_\phi + (\boldsymbol{e}_z \cdot \boldsymbol{e}_z)\boldsymbol{e}_z = \boldsymbol{e}_z.\end{aligned} \tag{1.44}$$

The gradient in cylindrical coordinates is given by

$$\nabla = \boldsymbol{e}_\rho\,\frac{\partial}{\partial\rho} + \boldsymbol{e}_\phi\,\frac{1}{\rho}\frac{\partial}{\partial\phi} + \boldsymbol{e}_z\,\frac{\partial}{\partial z}. \tag{1.45}$$

Verifying (1.45) follows from substituting (1.41) and (1.42) into (1.45) to show that it reproduces the Cartesian gradient (1.38). In cylindrical coordinates the Laplacian is given by

$$\nabla^2 = \frac{\partial^2}{\partial \rho^2} + \frac{1}{\rho}\frac{\partial}{\partial \rho} + \frac{1}{\rho^2}\frac{\partial^2}{\partial \phi^2} + \frac{\partial^2}{\partial z^2}. \tag{1.46}$$

This follows by using (1.41) in (1.46) to show that it becomes (1.40).

Similarly, spherical coordinates (r, θ, ϕ) are related to Cartesian coordinates by $x = r\sin\theta\cos\phi$, $y = r\sin\theta\sin\phi$, and $z = r\cos\theta$. The chain rule gives

$$\begin{aligned}
\frac{\partial}{\partial r} &= \sin\theta\cos\phi\,\frac{\partial}{\partial x} + \sin\theta\sin\phi\,\frac{\partial}{\partial y} + \cos\theta\frac{\partial}{\partial z},\\
\frac{1}{r}\frac{\partial}{\partial \theta} &= \cos\theta\cos\phi\,\frac{\partial}{\partial x} + \cos\theta\sin\phi\,\frac{\partial}{\partial y} - \sin\theta\,\frac{\partial}{\partial z},\\
\frac{1}{r\sin\theta}\frac{\partial}{\partial \phi} &= -\sin\phi\frac{\partial}{\partial x} + \cos\phi\frac{\partial}{\partial y}.
\end{aligned} \tag{1.47}$$

The unit vectors in spherical coordinates in terms of the Cartesian unit vectors are read off from (1.47),

$$\begin{aligned}
\boldsymbol{e}_r &= \sin\theta\cos\phi\,\boldsymbol{e}_x + \sin\theta\sin\phi\,\boldsymbol{e}_y + \cos\theta\,\boldsymbol{e}_z,\\
\boldsymbol{e}_\theta &= \cos\theta\cos\phi\,\boldsymbol{e}_x + \cos\theta\sin\phi\,\boldsymbol{e}_y - \sin\theta\,\boldsymbol{e}_z,\\
\boldsymbol{e}_\phi &= -\sin\phi\,\boldsymbol{e}_x + \cos\phi\,\boldsymbol{e}_y.
\end{aligned} \tag{1.48}$$

Using the orthonormality of the Cartesian unit vectors shows that the spherical coordinate unit vectors are also orthonormal. Using the vector product of the Cartesian unit vectors shows that their vector products are given by

$$\boldsymbol{e}_r \times \boldsymbol{e}_\theta = \boldsymbol{e}_\phi, \quad \boldsymbol{e}_\theta \times \boldsymbol{e}_\phi = \boldsymbol{e}_r, \quad \boldsymbol{e}_\phi \times \boldsymbol{e}_r = \boldsymbol{e}_\theta. \tag{1.49}$$

The relations of (1.48) are also easily inverted using a vector projection similar to (1.44),

$$\begin{aligned}
\boldsymbol{e}_x &= (\boldsymbol{e}_x\cdot\boldsymbol{e}_r)\boldsymbol{e}_r + (\boldsymbol{e}_x\cdot\boldsymbol{e}_\theta)\boldsymbol{e}_\theta + (\boldsymbol{e}_x\cdot\boldsymbol{e}_\phi)\boldsymbol{e}_\phi\\
&= \sin\theta\cos\phi\,\boldsymbol{e}_r + \cos\theta\cos\phi\,\boldsymbol{e}_\theta - \sin\phi\,\boldsymbol{e}_\phi,\\
\boldsymbol{e}_y &= (\boldsymbol{e}_y\cdot\boldsymbol{e}_r)\boldsymbol{e}_r + (\boldsymbol{e}_y\cdot\boldsymbol{e}_\theta)\boldsymbol{e}_\theta + (\boldsymbol{e}_y\cdot\boldsymbol{e}_\phi)\boldsymbol{e}_\phi\\
&= \sin\theta\sin\phi\,\boldsymbol{e}_r + \cos\theta\sin\phi\,\boldsymbol{e}_\theta + \cos\phi\,\boldsymbol{e}_\phi,\\
\boldsymbol{e}_z &= (\boldsymbol{e}_z\cdot\boldsymbol{e}_r)\boldsymbol{e}_r + (\boldsymbol{e}_z\cdot\boldsymbol{e}_\theta)\boldsymbol{e}_\theta + (\boldsymbol{e}_z\cdot\boldsymbol{e}_\phi)\boldsymbol{e}_\phi = \cos\theta\,\boldsymbol{e}_r - \sin\theta\,\boldsymbol{e}_\theta.
\end{aligned} \tag{1.50}$$

Using these results shows that the gradient in spherical coordinates is given by

$$\nabla = \boldsymbol{e}_r\,\frac{\partial}{\partial r} + \boldsymbol{e}_\theta\,\frac{1}{r}\frac{\partial}{\partial \theta} + \boldsymbol{e}_\phi\,\frac{1}{r\sin\theta}\frac{\partial}{\partial \phi}, \tag{1.51}$$

while in spherical coordinates the Laplacian is given by

$$\nabla^2 = \frac{\partial^2}{\partial r^2} + \frac{2}{r}\frac{\partial}{\partial r} + \frac{\cot\theta}{r^2}\frac{\partial}{\partial\theta} + \frac{1}{r^2}\frac{\partial^2}{\partial\theta^2} + \frac{1}{r^2\sin^2\theta}\frac{\partial^2}{\partial\phi^2}. \tag{1.52}$$

As with cylindrical coordinates, these formulas are verified by using (1.47) and (1.48) to show that the gradient (1.51) and the Laplacian (1.52) reproduce the Cartesian versions (1.38) and (1.40).

The gradient can also be used combined with a general vector valued function, $\mathbf{A}(\boldsymbol{x}, t)$, to define the *divergence* of $\mathbf{A}(\boldsymbol{x}, t)$ in Cartesian coordinates as

$$\nabla \cdot \mathbf{A}(\boldsymbol{x}, t) = \frac{\partial A_x(\boldsymbol{x}, t)}{\partial x} + \frac{\partial A_y(\boldsymbol{x}, t)}{\partial y} + \frac{\partial A_z(\boldsymbol{x}, t)}{\partial z}. \tag{1.53}$$

This definition appears in the *divergence theorem*, which states that in three spatial dimensions

$$\int_V \mathrm{d}^3x\, \nabla \cdot \mathbf{A}(\boldsymbol{x}, t) = \oint_{S(V)} \mathrm{d}\mathbf{S} \cdot \mathbf{A}(\boldsymbol{x}, t), \tag{1.54}$$

where V is an arbitrary three-dimensional volume and $S(V)$ is the two-dimensional surface S forming the boundary of the three-dimensional volume V. The quantity $\mathrm{d}\mathbf{S}$ is an infinitesimal vector perpendicular to the two-dimensional surface S everywhere, with a magnitude equal to the infinitesimal surface patch to which it is perpendicular. The right-hand side of (1.54) is sometimes referred to as the *flux* of the vector quantity represented by $\mathbf{A}$ through the surface $\mathbf{S}$. If the divergence of the vector function $\mathbf{A}$ yields a positive result for the left-hand side of (1.54), then the overall flux of the quantity $\mathbf{A}$ is directed outward through the surface bounding that volume. The divergence theorem plays a central role in stating conservation laws in physics.

Another important occurrence of the gradient is in the *curl* of the vector function $\mathbf{A}(\boldsymbol{x})$, which is written $\nabla \times \mathbf{A}(\boldsymbol{x})$. The Cartesian components of the curl are given by

$$(\nabla \times \mathbf{A})_x = \frac{\partial A_z}{\partial y} - \frac{\partial A_y}{\partial z}, \quad (\nabla \times \mathbf{A})_y = \frac{\partial A_x}{\partial z} - \frac{\partial A_z}{\partial x}, \quad (\nabla \times \mathbf{A})_z = \frac{\partial A_y}{\partial x} - \frac{\partial A_x}{\partial y}. \tag{1.55}$$

The curl of the vector function appears in *Stokes' theorem*, which states that the line integral of the vector function around a closed path is given by the curl of the function through the surface $S(P)$ defined by the closed path P,

$$\oint_P \mathrm{d}\boldsymbol{\ell} \cdot \mathbf{A}(\boldsymbol{x}) = \int_{S(P)} \mathrm{d}\mathbf{S} \cdot (\nabla \times \mathbf{A}(\boldsymbol{x})), \tag{1.56}$$

where $\mathrm{d}\boldsymbol{\ell}$ is an infinitesimal vector tangent to the path at all points.

1.2.3 Maxwell's equations

The two physical quantities of electromagnetism are the *electric field* $\mathbf{E}(\boldsymbol{x}, t)$ and the *magnetic field* $\mathbf{B}(\boldsymbol{x}, t)$ introduced earlier. Both of these fields are vector valued functions of the position vector $\boldsymbol{x}$ as well as the time t. The charged matter is described by a charge density $\rho(\boldsymbol{x}, t)$, a charge per unit volume, and the vector charge current density $\mathbf{J}(\boldsymbol{x}, t)$, a rate of charge flow per unit area. It has been experimentally determined that electric charge is a *conserved quantity*, and this is stated mathematically by the relation

$$\nabla \cdot \mathbf{J}(\boldsymbol{x}, t) + \frac{\partial}{\partial t}\rho(\boldsymbol{x}, t) = 0. \tag{1.57}$$

Relation (1.57) can be combined with the divergence theorem (1.54) to show that the total electric charge Q_{T} in the three-dimensional volume V obeys the relation

$$\frac{\partial Q_{\mathrm{T}}}{\partial t} = \frac{\partial}{\partial t}\int_V \mathrm{d}^3x\, \rho(\boldsymbol{x}, t) = -\int_V \mathrm{d}^3x\, \nabla \cdot \mathbf{J}(\boldsymbol{x}, t) = -\oint_{S(V)} \mathrm{d}\boldsymbol{S} \cdot \mathbf{J}(\boldsymbol{x}, t), \tag{1.58}$$

where $S(V)$ is the surface enclosing the volume V. Result (1.58) states that the charge lost or gained in a volume V of space per unit time is identical to the charge flowing out of or into that volume of space through its surface $S(V)$ per unit time. In the presence of electric and magnetic fields the charged current undergoes a force per unit volume denoted $\boldsymbol{f}$ and known as the *Lorentz force density*. It is given by

$$\boldsymbol{f} = \rho\mathbf{E} + \frac{1}{c}\mathbf{J} \times \mathbf{B}, \tag{1.59}$$

generalizing the two previous results for the force on point charges, (1.36) and (1.37).

It is important to note that the classical current relations (1.57) and (1.59) make no statement regarding the physical nature of the constituents of electric charge. The charged matter may be treated as a collection of particles or a continuous distribution of mass. It was not until 1897 that Thomson identified the *electron* as a low mass point-like particle carrying electric charge. In effect, Thomson identified the first *elementary particle*. The electron continues to be considered *elementary* in that it cannot be subdivided into smaller constituents. Its electric charge is therefore the fundamental unit of all *observed* electric charges, which are *integer multiples* of its charge. At the time, the electron was treated as a massive point particle and therefore assumed to be governed by Newtonian mechanics.

There are four fundamental laws, referred to as *Maxwell's equations*, that govern electric and magnetic fields. In the absence of dielectric materials and using Gaussian units, they are given in differential form by

Gauss's law $$\nabla \cdot \mathbf{E} = 4\pi\rho, \tag{1.60}$$

Ampere's law $$\nabla \times \mathbf{B} - \frac{1}{c}\frac{\partial \mathbf{E}}{\partial t} = \frac{4\pi}{c}\mathbf{J}, \tag{1.61}$$

Faraday's law $$\nabla \times \mathbf{E} + \frac{1}{c}\frac{\partial \mathbf{B}}{\partial t} = 0, \tag{1.62}$$

No magnetic monopoles $$\nabla \cdot \mathbf{B} = 0, \tag{1.63}$$

where c is the speed of light. By the mid-19th century these laws were experimentally well-established and subsequently codified by Maxwell. It is possible to derive another useful formula using Ampere's law. This relates the magnetic field to the electric current I present,

$$\boldsymbol{B} = \frac{1}{c}\oint \frac{I\mathrm{d}\ell \times \boldsymbol{e}_r}{r^2}, \tag{1.64}$$

which is known as the *Biot–Savart law*.

1.2.4 Potentials and gauge invariance

Equations (1.62) and (1.63) are referred to as the *homogeneous equations* since they have no reference to charge. They can be solved by introducing the vector and scalar potentials, $\mathbf{A}(\boldsymbol{x}, t)$ and $\varphi(\boldsymbol{x}, t)$, such that

$$\begin{aligned} \mathbf{E}(\boldsymbol{x}, t) &= -\nabla\varphi(\boldsymbol{x}, t) - \frac{1}{c}\frac{\partial \mathbf{A}(\boldsymbol{x}, t)}{\partial t}, \\ \mathbf{B}(\boldsymbol{x}, t) &= \nabla \times \mathbf{A}(\boldsymbol{x}, t), \end{aligned} \tag{1.65}$$

where the Gaussian units of $\varphi(\boldsymbol{x}, t)$ and $\mathbf{A}(\boldsymbol{x}, t)$ are $\sqrt{\mathrm{dy}}$. Demonstrating that this satisfies the homogeneous equations uses the identities $\nabla \times \nabla\varphi(\boldsymbol{x}, t) = 0$ and $\nabla \cdot (\nabla \times \mathbf{A}(\boldsymbol{x}, t)) = 0$ for nonsingular potential functions. An important property of the $\mathbf{E}(\boldsymbol{x}, t)$ and $\mathbf{B}(\boldsymbol{x}, t)$ fields in terms of the potentials defined in (1.65) is that they are left unchanged by the simultaneous redefinitions

$$\mathbf{A}(\boldsymbol{x}, t) \to \mathbf{A}(\boldsymbol{x}, t) + \nabla\Lambda(\boldsymbol{x}, t), \qquad \varphi(\boldsymbol{x}, t) \to \varphi(\boldsymbol{x}, t) - \frac{1}{c}\frac{\partial \Lambda(\boldsymbol{x}, t)}{\partial t}, \tag{1.66}$$

where $\Lambda(\boldsymbol{x}, t)$ is an *arbitrary* nonsingular function. This freedom to redefine the potentials is known as *gauge invariance*. While the use of potentials automatically satisfies two of Maxwell's equation, the two potentials are not unique and therefore *not experimentally observable*. They require another equation, referred to as a *gauge condition*, to specify them uniquely. A common and useful choice is the *Lorentz condition*, which is given by

$$\nabla \cdot \mathbf{A}(\boldsymbol{x}, t) + \frac{1}{c}\frac{\partial \varphi(\boldsymbol{x}, t)}{\partial t} = 0. \tag{1.67}$$

Another common choice is the *Coulomb condition*, $\nabla \cdot \mathbf{A}(\boldsymbol{x}, t) = 0$.

A very useful relationship follows by combining Newtonian total mechanical energy with the electric force written in terms of the scalar potential. For a point charge q in the absence of a magnetic field, the definition (1.65) shows that the electric force can be written

$$\mathbf{F}_E = q\mathbf{E} = -q\nabla\varphi. \tag{1.68}$$

This shows that the total mechanical energy is conserved and given by

$$E = K + q\varphi \implies \mathrm{d}K = -q\,\mathrm{d}\varphi, \tag{1.69}$$

so that the change in the kinetic energy of the charge is equal and opposite to the change in the scalar potential.

1.2.5 Electromagnetic waves and light interference

Using (1.65) and (1.67) with the two inhomogeneous Maxwell equations reduces them to the following forms,

$$\left(\frac{1}{c^2}\frac{\partial^2}{\partial t^2} - \nabla^2\right)\varphi \equiv \Box\varphi = 4\pi\rho \quad \text{and} \quad \left(\frac{1}{c^2}\frac{\partial^2}{\partial t^2} - \nabla^2\right)\mathbf{A} \equiv \Box\mathbf{A} = \frac{4\pi}{c}\mathbf{J}, \tag{1.70}$$

where the differential operator $\Box$, defined as

$$\Box = \frac{1}{c^2}\frac{\partial^2}{\partial t^2} - \nabla^2, \tag{1.71}$$

is referred to as the *d'Alembertian operator*. In free space, where ρ and $\mathbf{J}$ vanish, relations (1.70) reduce to the *wave equation* for the propagation of electromagnetic fields. This electromagnetic wave is identified as *light*. Setting $\varphi = 0$, a *simple harmonic plane wave* solution to (1.70) and (1.67) takes the form

$$\mathbf{A}(\boldsymbol{x}, t) = \mathbf{A}_0(\boldsymbol{k})\sin(\boldsymbol{k}\cdot\boldsymbol{x} \pm \omega t + \phi), \tag{1.72}$$

where the *magnitude vector* $\mathbf{A}_0(\boldsymbol{k})$, the *wave vector* $\boldsymbol{k}$, the *angular frequency* ω, and the *phase* ϕ are all independent of $\boldsymbol{x}$ and t. The magnitude k of the wave vector $\boldsymbol{k}$ and the angular frequency ω must also satisfy $c^2k^2 - \omega^2 = 0$ in order for (1.72) to satisfy the wave equation (1.70). This allows the solution for (1.72) to be written as

$$\mathbf{A}(\boldsymbol{x}, t) = \mathbf{A}_0(\boldsymbol{k})\sin\left(k\left(\frac{\boldsymbol{k}\cdot\boldsymbol{x}}{k} \pm \frac{\omega t}{k}\right) + \phi\right) = \mathbf{A}_0(\boldsymbol{k})\sin\left(k(\boldsymbol{n}\cdot\boldsymbol{x} \pm ct) + \phi\right), \tag{1.73}$$

where $\boldsymbol{n} = \boldsymbol{k}/k$ is a unit vector, $\boldsymbol{n}\cdot\boldsymbol{n} = 1$, in the direction of the wave vector $\boldsymbol{k}$. This shows that the solution is a wave traveling at the speed of light c, justifying the identification of light as an electromagnetic wave. The two constants $\boldsymbol{k}$ and ω are often written as $\boldsymbol{k} = \boldsymbol{n}/2\pi\lambda$ and $\omega = 2\pi f$, where λ and f are referred to as the *wavelength* and *frequency* and obey the more commonly used expression

$$\lambda f = c. \tag{1.74}$$

In order to satisfy the Lorentz condition (1.67), which reduces to the Coulomb condition $\nabla\cdot\mathbf{A} = 0$ for $\varphi = 0$, the magnitude $\mathbf{A}_0(\boldsymbol{k})$ must be *transverse* to the direction of propagation, given by the wave vector $\boldsymbol{k}/k$, so that

$$\boldsymbol{k}\cdot\mathbf{A}_0(\boldsymbol{k}) = 0. \tag{1.75}$$

Because it must be perpendicular to $\boldsymbol{k}$ the vector $\mathbf{A}_0(\boldsymbol{k})$ must lie in the two-dimensional plane that is perpendicular to the direction of propagation for the wave. There are therefore *two independent polarizations* available to the light wave of (1.72), representing the two remaining degrees of freedom for **E** and **B**.

An extremely important property of the free space versions of (1.70) is that they define a *linear equation*. This means that if φ_1 and φ_2 are *any* two solutions to the equation $\Box\varphi = 0$, then so is the *linear superposition*, $\varphi = a\varphi_1 + b\varphi_2$, where a and b are *any* two real constants. This property of electromagnetic waves in vacuum gives rise to one of the important distinctions between wave phenomena and point particle behavior, which is the appearance of *interference* when two light waves overlap. These interference patterns can be explained using the *principle of superposition*, which is valid since light waves in vacuum obey a linear equation.

For example, if two light waves with zero phase, equal amplitude and frequency, and *opposite directions* overlap in vacuum, they are simply superposed to find the resultant wave. The trigonometric identity $\sin(\alpha \pm \beta) = \sin\alpha\cos\beta \pm \cos\alpha\sin\beta$ gives

$$\mathbf{A}_r(\boldsymbol{x}, t) = \mathbf{A}_0\sin(\boldsymbol{k}\cdot\boldsymbol{x} - \omega t) + \mathbf{A}_0\sin(\boldsymbol{k}\cdot\boldsymbol{x} + \omega t) = 2\mathbf{A}_0\sin(\boldsymbol{k}\cdot\boldsymbol{x})\cos(\omega t). \quad (1.76)$$

This is known as a *standing wave* since its *nodes*, those locations where $\sin(\boldsymbol{k}\cdot\boldsymbol{x}) = 0$, do not move.

As another example of interference, if two light waves with equal amplitudes, wavelengths, directions, but *opposite phases*, overlap in vacuum, the resultant wave is

$$\mathbf{A}_r = \mathbf{A}_0\sin(\boldsymbol{k}\cdot\boldsymbol{x} \pm \omega t + \frac{1}{2}\phi) + \mathbf{A}_0\sin(\boldsymbol{k}\cdot\boldsymbol{x} \pm \omega t - \frac{1}{2}\phi). \quad (1.77)$$

Applying the same trigonometric identity as before gives

$$\mathbf{A}_r = 2\mathbf{A}_0\cos\left(\frac{1}{2}\phi\right)\sin(\boldsymbol{k}\cdot\boldsymbol{x} \pm \omega t). \quad (1.78)$$

For the case that the two waves are *in phase* and $\phi = 2n\pi$, where n is an integer, the resultant wave will have maximum amplitude, referred to as *constructive interference*. For the case that the two waves are *out of phase* and $\phi = (2n + 1)\pi$, where n is an integer, the resultant wave will have zero amplitude, referred to as *destructive interference*. It must be stressed that this possibility is purely a property of wave phenomena and has no counterpart in the Newtonian mechanics of point particles.

An extremely important example of (1.78) occurs in the case of monochromatic coherent light passing through two slits separated by the distance d. The term *coherent light* is taken to mean that the light is exactly in phase as it passes through both slits. For such a case the light is assumed to leave the two slits at all angles and the light emanating from the two slits will therefore interfere when it recombines on a remote screen. If θ is the angle measured from a normal to the two-slit device, then the path difference between the two slits is $d\sin\theta$, so that the phase difference between the light from the two slits is $\phi = 2\pi d\sin\theta/\lambda$. As a result, the criterion for constructive interference, $\phi = 2\pi n$, yields a formula for the angular position of the *bright fringes* formed on a screen a long distance away compared to the wavelength,

$$d \sin \theta = n\lambda, \quad \text{where } n = 1, 2, 3.... \tag{1.79}$$

Of course, the number of bright fringes formed, if any, depends on both the slit separation d and the wavelength λ of the light.

A second important optical phenomenon is referred to as *diffraction*, and occurs when coherent light of wavelength λ is passed through a *single slit* of width a. The light is spread subsequent to passing through the slit, and bright and dark bands are formed on the screen behind the slit. The *dark bands* form at the angles θ that solve

$$a \sin \theta = n\lambda, \quad \text{where } n = 1, 2, 3.... \tag{1.80}$$

The origin of this formula is the interference between light passing through *different parts* of the slit as opposed to different slits in a grating. It is important to note that the angular spread of the central $n = 1$ peak, which occurs at $\sin \theta = \lambda / a$, is negligible when λ is small compared to the slit width a. In such a case the light wave passes through the slit with no discernible deviation, much as if the light was a stream of Newtonian particles passing through the slit. However, there is no real solution to (1.80) if $\lambda > a$, so that a detailed analysis using Maxwell's equations is required.

1.2.6 Electromagnetic waves and Galilean relativity

It is important for the development of quantum mechanics to understand the implications of Galilean relativity for electromagnetic waves. In Newtonian physics the wavelength of light is *unchanged* by the Galilean transformation of (1.32). This is because lengths are *absolute quantities in Newtonian physics*. The interference pattern formed on a screen by light passing through a diffraction grating is therefore unaffected by relative motion with respect to the diffraction grating since only the wavelength λ enters the interference formula (1.79).

To make the details of this clear, in the case of a light wave described by a harmonic wave of the form (1.72), a Galilean transformation to an observer in constant relative motion leaves the wavelength unchanged, so that $\lambda' = \lambda$. This means that the wave number vector is also unchanged, so that $\boldsymbol{k}' = \boldsymbol{k}$. In order for the wave to transform as a *Galilean scalar*, a Galilean transformation to a moving observer requires the frequency to be altered by a Galilean transformation. This follows by using the Galilean transformation (1.32) on the sinusoidal form of the wave, where invariance under the Galilean transformation requires that

$$\sin(\boldsymbol{k}' \cdot \boldsymbol{x}' - \omega' t') = \sin(\boldsymbol{k} \cdot \boldsymbol{x} - (\omega' + \boldsymbol{k} \cdot \boldsymbol{v}_o)t) = \sin(\boldsymbol{k} \cdot \boldsymbol{x} - \omega t), \tag{1.81}$$

where $\boldsymbol{v}_o$ is the relative velocity of the moving observer. Invariance of the sine function argument therefore yields the frequency observed by the moving observer,

$$\omega' = \omega - \boldsymbol{k} \cdot \boldsymbol{v}_o. \tag{1.82}$$

Using the velocity of propagation of light, $c = f\lambda$, and picking the observer's velocity parallel to the light wave, so that $\boldsymbol{k} \cdot \boldsymbol{v}_o = kv_o$, immediately gives

$$\omega' = \omega - \boldsymbol{k} \cdot v_o \implies 2\pi f' = 2\pi f - \frac{2\pi v_o}{\lambda} \implies f' = f\left(1 - \frac{v_o}{c}\right), \tag{1.83}$$

which is the Galilean version of the *Doppler effect*. The reader should bear in mind that this version of the Doppler effect for light is valid only if $v_o \ll c$ and breaks down as $v_o \to c$, where it must be amended by special relativity. This result also shows that the speed of light for the observer moving parallel to the light wave is predicted to be

$$c' = \frac{\omega'}{k'} = \frac{\omega}{k} - \frac{\boldsymbol{k} \cdot v_o}{k} = c - v_o = c\left(1 - \frac{v_o}{c}\right), \tag{1.84}$$

which is consistent with the Newtonian idea that the moving observer will claim a different speed of propagation for a wave moving through a medium.

From a historical perspective, in the latter 19th century the Michelson–Morley experiment demonstrated conclusively that the speed of light is constant for all observers regardless of relative motion, thereby exposing the failure of Galilean relativity when applied to light waves and Maxwell's equations in general. Resolving this dilemma led to Einstein's special theory of relativity, which lies outside the scope of this text. Instead, this text will use Galilean relativity in the development of *nonrelativistic* quantum mechanics with the understanding that its validity requires $v_o/c \ll 1$. As a result, some of the expressions derived using Galilean relativity, like (1.84), will be inconsistent with special relativity. The dual revolutions of special relativity and quantum mechanics occurred during the same period of physics development, and so are in many ways entangled.

1.2.7 Electromagnetic energy and momentum

The Lorentz force density (1.59) causes the energy and momentum of charged matter and its currents to be changed through interaction with electromagnetic fields and waves. The electromagnetic wave must therefore transport energy and momentum in order to deliver them to electrical charges. In the case of electric and magnetic fields, the *energy per unit volume* u ($\mathrm{erg/cm^3}$) in free space is given by

$$u = \frac{1}{8\pi}(\mathbf{E}^2 + \mathbf{B}^2). \tag{1.85}$$

Result (1.85) can be demonstrated from elementary arguments by calculating the work required to charge a capacitor and establish a current in a solenoid. The *flow of energy* in the electric and magnetic fields is given by the current density known as the *Poynting vector* **S**,

$$\mathbf{S} = \frac{c}{4\pi}\mathbf{E} \times \mathbf{B}, \tag{1.86}$$

which has the units of $\mathrm{erg/cm^2\ s}$. The definitions (1.85) and (1.86) can be combined with Maxwell's equations to give the work–energy relationship

$$\nabla \cdot \mathbf{S} + \frac{\partial u}{\partial t} = -\mathbf{E} \cdot \mathbf{J}. \tag{1.87}$$

The right-hand side is the rate at which mechanical work is done by the electric field on the charged matter current. The divergence theorem shows that the Poynting vector is the rate at which electromagnetic energy is flowing through the surface of a volume of space, matching the rate of change in electromagnetic energy plus the mechanical work done in the interior of the volume.

The *power* W of an electromagnetic wave moving in the z-direction is defined by the rate at which energy crosses the surface element $\mathrm{d}A_z$ per unit time. Since the wave travels at speed c, in the time interval $\mathrm{d}t$ the wave will flow into the infinitesimal volume $\mathrm{d}^3x = \mathrm{d}A_z\, c\, \mathrm{d}t$. This gives $\mathrm{d}E = u\, \mathrm{d}^3x = u\, \mathrm{d}A_z\, c\, \mathrm{d}t$, so that $W = \mathrm{d}E/(\mathrm{d}A_z\, \mathrm{d}t) = c\, u$. Similarly, the *momentum density* $\pi(\boldsymbol{x}, t)$ of an electromagnetic waves can be found by considering the force an electromagnetic wave creates. If f_z is the force density delivered in the wave, then the force must be generated by a loss in the momentum of the wave, so that Newton's second law gives $f_z\, \mathrm{d}^3x = -(\partial\pi/\partial t)\, \mathrm{d}^3x = -c(\partial\pi/\partial t)\, \mathrm{d}A_z\, \mathrm{d}t$. At the same time, this force must be given by the gradient of the energy density of the wave, which acts as a potential. Therefore $f_z\, \mathrm{d}^3x = -(\partial u/\partial z)\, \mathrm{d}^3x = -(\partial u/\partial z)\, \mathrm{d}A_z\, \mathrm{d}z = -(\partial u/\partial t)\, \mathrm{d}A_z\, \mathrm{d}t$, where the chain rule $\mathrm{d}z\, \partial/\partial z = \mathrm{d}t\, \partial/\partial t$ is valid since $\mathrm{d}z = c\, \mathrm{d}t$. Equating the two forms for the force gives $c\, \partial\pi/\partial t = \partial u/\partial t$. Therefore, the magnitude of the momentum density and the energy density of an electromagnetic wave are related by

$$c\, \pi(\boldsymbol{x}, t) = u(\boldsymbol{x}, t) \implies u^2(\boldsymbol{x}, t) - c^2\pi^2(\boldsymbol{x}, t) = 0, \tag{1.88}$$

where the absence of energy, $u(\boldsymbol{x}, t) = 0$, has been chosen to correspond to the absence of a wave. Result (1.88) is consistent with the relativistic energy–momentum expression for a *massless particle*,

$$E^2 - c^2p^2 = m^2c^4 = 0 \implies E = cp. \tag{1.89}$$

This relationship will be important in the next chapter when the concept of the photon is introduced.

The energy–momentum properties of light waves play a critical role in understanding quantum mechanics. To begin the analysis the positive moving simple harmonic wave (1.72) is used in (1.65), and this gives

$$\begin{aligned}
\mathbf{B}(\boldsymbol{x}, t) &= \nabla \times \mathbf{A}(\boldsymbol{x}, t) = (\boldsymbol{k} \times \mathbf{A}_0(\boldsymbol{k}))\sin(\boldsymbol{k} \cdot \boldsymbol{x} - \omega t + \phi) \\
&\equiv \mathbf{B}_0(\boldsymbol{k})\sin(\boldsymbol{k} \cdot \boldsymbol{x} - \omega t + \phi), \\
\mathbf{E}(\boldsymbol{x}, t) &= -\frac{1}{c}\frac{\partial \mathbf{A}(\boldsymbol{x}, t)}{\partial t} = \left(\frac{\omega \mathbf{A}_0(\boldsymbol{k})}{c}\right)\sin(\boldsymbol{k} \cdot \boldsymbol{x} - \omega t + \phi) \\
&\equiv \mathbf{E}_0(\boldsymbol{k})\sin(\boldsymbol{k} \cdot \boldsymbol{x} - \omega t + \phi),
\end{aligned} \tag{1.90}$$

so that $\mathbf{E}_0(\boldsymbol{k}) = \omega\mathbf{A}_0(\boldsymbol{k})/c$ and $\mathbf{B}_0(\boldsymbol{k}) = \boldsymbol{k} \times \mathbf{A}_0(\boldsymbol{k})$. The energy density (1.85) and the Poynting vector (1.86) can now be calculated using (1.90). Using the Coulomb condition $\boldsymbol{k} \cdot \mathbf{A}_0(\boldsymbol{k}) = 0$ and the identity

$$B_0^2(\boldsymbol{k}) = (\boldsymbol{k} \times \mathbf{A}(\boldsymbol{k})) \cdot (\boldsymbol{k} \times \mathbf{A}(\boldsymbol{k})) = k^2 A_0^2(\boldsymbol{k}) - (\boldsymbol{k} \cdot \mathbf{A}_0(\boldsymbol{k}))^2 = k^2 A_0^2(\boldsymbol{k}), \tag{1.91}$$

gives the magnitudes of the electric and magnetic fields in the wave,

$$\begin{aligned} E_0(\boldsymbol{k}) &= \frac{\omega}{c} A_0(\boldsymbol{k}), \\ B_0(\boldsymbol{k}) &= k A_0(\boldsymbol{k}). \end{aligned} \tag{1.92}$$

Using $\omega = ck$ shows that the magnitudes of the electric and magnetic fields in a light wave are equal, so that $E_0 = B_0$, and by virtue of the Coulomb condition they are perpendicular vectors. Using (1.92) gives the energy density of the wave,

$$\begin{aligned} u(\boldsymbol{x}, t) &= \frac{1}{8\pi}(E^2(\boldsymbol{x}, t) + B^2(\boldsymbol{x}, t)) \\ &= \frac{1}{8\pi}(E_0^2(\boldsymbol{k}) + B_0^2(\boldsymbol{k}))\sin^2(\boldsymbol{k} \cdot \boldsymbol{x} - \omega t + \phi) \\ &= \frac{1}{8\pi}\left(\frac{\omega^2}{c^2} + k^2\right)A_0^2(\boldsymbol{k})\sin^2(\boldsymbol{k} \cdot \boldsymbol{x} - \omega t + \phi) \\ &= \frac{k^2 A_0^2(\boldsymbol{k})}{4\pi}\sin^2(\boldsymbol{k} \cdot \boldsymbol{x} - \omega t + \phi). \end{aligned} \tag{1.93}$$

Similarly, using the Coulomb condition and the identity,

$$\begin{aligned} \mathbf{E}_0(\boldsymbol{k}) \times \mathbf{B}_0(\boldsymbol{k}) &= \frac{\omega}{c}\mathbf{A}_0(\boldsymbol{k}) \times (\boldsymbol{k} \times \mathbf{A}_0(\boldsymbol{k})) \\ &= \frac{\omega}{c}\boldsymbol{k}\, A_0{}^2(\boldsymbol{k}) - \frac{\omega}{c}(\boldsymbol{k} \cdot \mathbf{A}_0(\boldsymbol{k}))\, \mathbf{A}_0(\boldsymbol{k}) = \frac{\omega}{c}\boldsymbol{k} A_0^2(\boldsymbol{k}), \end{aligned} \tag{1.94}$$

gives the *instantaneous* Poynting vector of the wave,

$$\begin{aligned} \mathbf{S}(\boldsymbol{x}, t) &= \frac{c}{4\pi}\mathbf{E}(\boldsymbol{x}, t) \times \mathbf{B}(\boldsymbol{x}, t) = \left(\frac{\omega A_0{}^2(\boldsymbol{k})}{4\pi}\right)\boldsymbol{k}\sin^2(\boldsymbol{k} \cdot \boldsymbol{x} - \omega t + \phi) \\ &= \hat{\boldsymbol{n}}_k\left(\frac{\omega k A_0{}^2(\boldsymbol{k})}{4\pi}\right)\sin^2(\boldsymbol{k} \cdot \boldsymbol{x} - \omega t + \phi), \end{aligned} \tag{1.95}$$

where $\boldsymbol{n}_k = \boldsymbol{k}/k$ is a unit vector in the direction of propagation. It is straightforward to show that (1.93) and (1.95) satisfy (1.87) for the vacuum case $\mathbf{J} = 0$. Using $\omega = ck$ and (1.93) shows that the magnitude of the Poynting vector for the electromagnetic wave satisfies

$$S(\boldsymbol{x}, t) = cu(\boldsymbol{x}, t) = \frac{c}{4\pi}E(\boldsymbol{x}, t)B(\boldsymbol{x}, t) = \frac{c}{4\pi}E^2(\boldsymbol{x}, t). \tag{1.96}$$

Averaging $\sin^2(\boldsymbol{k} \cdot \boldsymbol{x} - \omega t + \phi)$ over a period $T = 2\pi/\omega$ yields

$$\langle\sin^2(\boldsymbol{k} \cdot \boldsymbol{x} - \omega t + \phi)\rangle \equiv \frac{1}{T}\int_0^T \mathrm{d}t \sin^2(\boldsymbol{k} \cdot \boldsymbol{x} - \omega t + \phi) = \frac{1}{2}. \tag{1.97}$$

Using (1.97) and $\omega = ck$ shows that the simple harmonic wave (1.72) has the average energy intensity

$$\langle \mathbf{S} \rangle = \left(\frac{\boldsymbol{k}}{k}\right)\frac{ck^2}{8\pi}A_0^2(\boldsymbol{k}) = \hat{\boldsymbol{n}}_k c\langle u \rangle, \tag{1.98}$$

where $\hat{\boldsymbol{n}}_k = \boldsymbol{k}/k$ is a unit vector directed parallel to the wave vector $\boldsymbol{k}$ and $\langle u \rangle$ is the time averaged energy density. Expression (1.98) can be rewritten using the relation $\omega = ck$ and (1.92) to give

$$\langle \mathbf{S} \rangle = \hat{\boldsymbol{n}}_k \frac{\omega k}{8\pi}A_0^2(\boldsymbol{k}) = \hat{\boldsymbol{n}}_k \frac{c}{8\pi}E_0(\boldsymbol{k})B_0(\boldsymbol{k}). \tag{1.99}$$

Expression (1.99) shows that the average energy delivered to a surface area per unit time by a light wave is proportional to the magnitudes of its electric and magnetic fields. The *average* intensity is *independent* of the light wave's frequency since $\boldsymbol{k}$ only serves to determine the *direction* of $E_0(\boldsymbol{k})$ and $B_0(\boldsymbol{k})$. The instantaneous Poynting vector (1.95) of the light wave retains frequency dependence.

Interference is a very important wave phenomena, and is characterized by the resultant amplitude A_r given by (1.78). Understanding its relation to energy intensity is extremely useful. Substituting A_r^2 into (1.95) and using $\omega = ck$ gives

$$\mathbf{S}(\boldsymbol{x}, t) = \hat{\boldsymbol{n}}_k \frac{\omega^2}{\pi c}\cos^2(\frac{1}{2}\phi)A_0^2(\boldsymbol{k})\sin^2(\boldsymbol{k}\cdot\boldsymbol{x} - \omega t). \tag{1.100}$$

This gives the intuitive result that the bright fringes formed by two-slit interference coincide with the maximum energy intensity positions determined by $\cos^2(\frac{1}{2}\phi) = 1$. It is useful to note that the energy intensity of both the monochromatic plane wave and the wave resulting from interference was found by *squaring the amplitude of the wave*. This insures that the energy intensity of the wave is a positive-definite quantity.

By the latter part of the 19th century the framework of Maxwell's equations had provided a wealth of understanding regarding electromagnetic phenomena and it appeared that electromagnetism, like gravitation, had been added to the list of natural forces to which Newtonian physics could be applied. A result of great importance was the understanding of the production of electromagnetic waves by *accelerated charges*. For the nonrelativistic case the *Larmor radiation formula* gives the power P (erg s^{-1}) radiated by a point charge q undergoing the instantaneous acceleraction a as

$$P = \frac{2}{3}\frac{q^2a^2}{c^3}. \tag{1.101}$$

For example, the transmission of radio waves is created by oscillating charges in a broadcast tower. However, Maxwell's equations predict that electromagnetic energy–momentum is radiated by any accelerated charge.

Electromagnetic field interactions with charged particles play a central role in probing the atomic world, and so it will be valuable to have the classical description of such phenomena for comparison with experiment. The interaction of a particle of mass m and electric charge q with the electromagnetic field is described by the Lagrangian density

$$\mathcal{L} = \frac{1}{2}m\dot{\boldsymbol{x}}^2 + \frac{q}{c}\mathbf{A}(\boldsymbol{x}, t) \cdot \dot{\boldsymbol{x}} - q\,\varphi(\boldsymbol{x}, t). \tag{1.102}$$

Applying the Euler–Lagrange equation (1.14) to (1.102) for the ith direction yields

$$\frac{q}{c}\sum_j \dot{x}_j(\partial_i A_j - \partial_j A_i) - q\left(\partial_i \varphi + \frac{1}{c}\frac{\partial A_i}{\partial t}\right) - m\ddot{r}_i = 0. \tag{1.103}$$

Using the forms (1.65) for the electromagnetic potentials and the vector identity

$$\sum_j \dot{x}_j(\partial_i A_j - \partial_j A_i) = (\dot{\boldsymbol{x}} \times (\nabla \times \mathbf{A}))_i = (\boldsymbol{v} \times \mathbf{B})_i, \tag{1.104}$$

result (1.103) becomes the sum of the electric and magnetic forces (1.36) and (1.37),

$$\boldsymbol{F} = q\mathbf{E} + \frac{q}{c}\boldsymbol{v} \times \mathbf{B}, \tag{1.105}$$

which justifies the choice of the Lagrangian density (1.102). Under the gauge transformation of (1.66) the Lagrangian density becomes

$$\mathcal{L} \to \mathcal{L} + \frac{q}{c}(\nabla\Lambda \cdot \dot{\boldsymbol{x}} + \dot{\Lambda}) = \mathcal{L} + \frac{q}{c}\mathrm{d}\Lambda/\mathrm{d}t. \tag{1.106}$$

The additional term does not contribute to the variation of the action (1.11) since it is an exact time derivative, so that it becomes $(q/c)[\Lambda(\boldsymbol{x}(T), T) - \Lambda(\boldsymbol{x}(0), 0)]$ after integration in the action. Therefore the action is gauge invariant, a hallmark of electromagnetism.

The Lagrangian identifies the canonical momentum of the particle as

$$p_i = \frac{\partial \mathcal{L}}{\partial \dot{x}_i} = m\dot{x}_i + \frac{q}{c}A_i \implies \dot{\boldsymbol{x}} = \frac{1}{m}\left(\boldsymbol{p} - \frac{q}{c}\mathbf{A}\right). \tag{1.107}$$

Using this in the definition of the Hamiltonian for the charged particle gives

$$H(\boldsymbol{p}, \boldsymbol{x}, t) = \boldsymbol{p} \cdot \dot{\boldsymbol{x}} - \mathcal{L}(\boldsymbol{x}, \dot{\boldsymbol{x}}) = \frac{1}{2m}\left(\boldsymbol{p} - \frac{q}{c}\mathbf{A}(\boldsymbol{x}, t)\right)^2 + q\,\varphi(\boldsymbol{x}, t). \tag{1.108}$$

The potentials $\mathbf{A}(\boldsymbol{x}, t)$ and $\varphi(\boldsymbol{x}, t)$ may be time-dependent if they are the result of an externally applied field. For such a case the Hamiltonian (1.108) for the charged particle may not be conserved.

1.3 Newtonian point particle solutions

There are several aspects of Newtonian mechanics that will play a critical role in its eventual failure in the atomic realm. In order to recognize these failures it is worth reviewing briefly the Newtonian mechanics for two important physical systems, the one-dimensional harmonic oscillator and the inverse square force. These two systems will prove to be significantly different quantum mechanically than their classical solutions,

1.3.1 The simple harmonic oscillator

In the case of the one-dimensional harmonic oscillator the potential energy is $V(x) = \frac{1}{2}kx^2$, where k is a real and positive constant. Equation (1.14) yields $m\ddot{x} = -kx$. This differential equation has the solution

$$x = A\sin(\omega t + \delta), \tag{1.109}$$

where $\omega = \sqrt{k/m}$. The values of the two constants A, the amplitude of the oscillation, and δ, the phase of the oscillation, are determined by combining the expressions for the initial position $x_o = A\sin\delta$ and the initial velocity $v_o = A\omega\cos\delta$. The solutions are given by

$$\begin{aligned} A &= \sqrt{x_o^2 + \frac{v_o^2}{\omega^2}}, \\ \delta &= \arctan\left(\frac{x_o}{\omega v_o}\right). \end{aligned} \tag{1.110}$$

There are well-defined for *all* values of x_o and v_o.

The total mechanical energy is constant and given by

$$E = \frac{1}{2}m\dot{x}^2 + \frac{1}{2}kx^2 = \frac{1}{2}kA^2 = \frac{1}{2}kx_o^2 + \frac{1}{2}mv_o^2. \tag{1.111}$$

Although the frequency of oscillations, $f = \omega/2\pi$, is constant, the energy of (1.111) can have any positive value since x_o and v_o are arbitrary. In addition, the instantaneous acceleration of the oscillating particle is given by

$$a = \frac{\mathrm{d}^2x}{\mathrm{d}t^2} = -A\omega^2\sin(\omega t + \delta). \tag{1.112}$$

As a result, (1.101) predicts that an oscillating electric charge will radiate energy–momentum in the form of electromagnetic waves.

1.3.2 The inverse square law

In the case of the standard inverse square laws, such as the Coulomb or Newtonian gravitation cases, a point particle with mass m is assumed to be moving in the presence of a fixed attractive potential given by $V(r) = -\beta/r$, where r is the radial position of the point particle and λ is real and positive. While (1.14) can be solved for both the bound and scattering behavior of the point particle, it is possible to use very basic physics to investigate the nature of circular orbits. The conditions for a circular orbit of radius r follow from the requirement for the orbital tangential velocity v to satisfy the centripetal force law,

$$\frac{mv^2}{r} = \frac{\beta}{r^2} \Longrightarrow a_c = \frac{v^2}{r} = \frac{\beta}{mr^2}, \tag{1.113}$$

so that the kinetic energy is $\frac{1}{2}mv^2 = \frac{1}{2}\beta/r$. Combining this with the potential energy gives the total mechanical energy E of a circular orbit with radius r,

$$E = -\frac{\beta}{2r}. \tag{1.114}$$

The negative value for the total energy reflects the fact that the point particle is, by assumption, bound and will therefore require an increase in energy to break free. However, once again the value of E can be any of a continuum of negative values since r is an arbitrary positive value. In addition, the orbital angular momentum L of the point particle, given by

$$L = mvr = m\sqrt{\frac{\beta}{mr}}\, r = \sqrt{m\beta r}, \tag{1.115}$$

is also arbitrary and continuous since r is arbitrary.

For the case that the mass m is in a circular orbit of radius r created by the Coulomb force between two opposite charges, q and $-q$, the constant β is given by $\beta = q^2$. The centripetal acceleration of (1.113) can be used in the Larmor radiation formula (1.101). This gives the rate of energy loss by the orbiting particle, so that the rate of energy radiation must be equal and opposite to the rate at which the total mechanical energy (1.114) is changing. This gives

$$P = -\frac{\mathrm{d}E}{\mathrm{d}t} \Longrightarrow \frac{2}{3}\frac{q^6}{m^2c^3r^4} = -\frac{q^2}{2r^2}\frac{\mathrm{d}r}{\mathrm{d}t} \tag{1.116}$$

$$\Longrightarrow r^2\frac{\mathrm{d}r}{\mathrm{d}t} = -\frac{4}{3}\frac{q^4}{m^2c^3}. \tag{1.117}$$

This formula yields the time T it takes for the charged particle to spiral to a zero radius from an initial value of R. This follows by integration,

$$\int_R^0 \mathrm{d}r\, r^2 = -\int_0^T \mathrm{d}t\, \frac{4}{3}\frac{q^4}{m^2c^3} \Longrightarrow T = \frac{m^2c^3R^3}{4q^4}. \tag{1.118}$$

This shows that a circular orbit created by the Coulomb force is not stable according to the *classical* laws of motion and electromagnetism.

1.3.3 Warning signs for classical physics

In a previous section the deterministic nature of Newtonian mechanics was discussed. Incorporating electromagnetism into classical physics initially indicated no obvious difficulty in maintaining the deterministic nature of the motion of charged particles. The first problem to surface was that, unlike Newton's laws of motion, Maxwell's equations are not invariant under a Galilean transformation. A solution was put forth that the form of Maxwell's equations presented earlier held only for observers at rest with respect to the putative medium of electromagnetic waves, referred to as the *ether*, giving rise to a privileged observer at rest with respect to the ether. This explanation eventually failed, and the resolution of the question raised by Maxwell's equations led to Einstein's special theory of relativity, which supplanted Galilean relativity for velocities near that of light.

A full development of the special theory of relativity lies beyond the scope of this text, but there is a result that will be of importance in quantum mechanics. Special relativity amends the Newtonian expression for the kinetic energy of a mass m free particle in terms of its momentum p to read

$$E = \sqrt{m^2c^4 + p^2c^2}, \tag{1.119}$$

so that the *rest mass energy* of the particle for $p = 0$ is mc^2. For values of p small compared to mc, this expression reduces to

$$E = mc^2\sqrt{1 + \frac{p^2}{m^2c^2}} \approx mc^2 + \frac{p^2}{2m}, \tag{1.120}$$

which includes the Newtonian expression $p^2/2m$. For values of p large compared to mc the expression becomes $E \approx cp$, which is consistent with the energy–momentum relation (1.88) for electromagnetic waves.

A second and unforeseen problem emerged from the experimental pursuit of the fundamental building blocks of matter, a theoretical concept dating back to Democritus. Newtonian mechanics does not possess a natural *length scale* that would indicate under what circumstances it would not be applicable. There was therefore no *a priori* reason to expect that Newtonian mechanics would fail to govern such fundamental entities, even if they were tiny compared to everyday objects. As the atomic model gained acceptance it was incumbent upon physics to investigate the experimental behavior of various atomic systems in the context of Newtonian mechanics. The next chapter details the failures of Newtonian mechanics to explain the observed atomic behavior and the first ad hoc steps in formulating quantum mechanics. These failures were in stark contrast to the successes Newtonian mechanics provides in describing the dynamical behavior of macroscopic objects. It follows that any alterations made to Newtonian mechanics *and* Newtonian determinism must be done with extreme care to preserve its obvious successes, while remedying its failures in the microscopic realm. This issue will form a central theme in the text.

References and recommended further reading

Monographs on Newtonian mechanics that present its variational formulation include

- Goldstein H, Poole C and Safko J 2000 *Classical Mechanics* 3rd edn (San Francisco: Addison-Wesley)
- Thornton S and Marion J 2022 *Classical Dynamics of Particles and Systems 6th edn* (Boston: Cengage Learning)

Monographs presenting mathematical aspects of the calculus of variations include

- Lanczos C 1970 *The Variational Principles of Mechanics* 4th edn (New York: Dover)
- Lovelock D and Rund H 1989 *Tensors, Differential Forms, and Variational Principles* (New York: Dover)

Monographs focusing on the myriad aspects and applications of classical electrodynamics include

- Eyges W 1980 *The Classical Electromagnetic Field* (New York: Dover)
- Westgard J 1997 *Electrodynamics: A Concise Introduction* (Berlin: Springer)
- Greiner W 1998 *Classical Electrodynamics* (Berlin: Springer)
- Jackson J 1999 *Classical Electrodynamics* 3rd edn (Hoboken, NJ: Wiley)

Monographs presenting the mathematical basis of classical electrodynamics include

- Bamberg P and Sternberg S 1988 *A Course in Mathematics for Students of Physics* vol I (Cambridge: Cambridge University Press)
- Dennery P and Krzywicki A 1997 *Mathematics For Physicists* (New York: Dover)
- Kusse B and Westwig E 1998 *Mathematical Physics* (New York: Wiley)
- Arfken G, Weber H, and Harris F 2013 *Mathematical Methods for Physicists* 7th edn (Amsterdam: Elsevier)

Introductory level solutions to Newton's laws can be found in

- Halliday P, Resnick R and Walker J 2021 *Fundamentals of Physics* 12th edn (New York: Wiley)

IOP Publishing

A Concise Introduction to Quantum Mechanics (Second Edition)

Mark S Swanson

Chapter 2

The origins of quantum mechanics

In the 19th century the science of modern chemistry emerged as a powerful tool for understanding the properties and behavior of matter. A key component of modern chemistry is its use of the atomic model, which explains the vast variety of matter occurring in nature in terms of compounds built from a much smaller set of more fundamental constituents, referred to generically as *atoms*. This approach to understanding matter and its behavior in terms of a simpler set of building blocks falls under the general rubric of *reductionism*, a concept that has dominated much of the research in physics over the last century. As the atomic model of matter gained acceptance it became important for physics to investigate the properties of atoms, in particular their size, mass, and internal structure, as well as their interaction with light. It was the inability of Newtonian mechanics to explain the experimentally observed structure and behavior of atoms that forced the development of quantum mechanics.

Of course, it is possible to present formal quantum mechanics without discussing the specific details of how Newtonian physics failed to explain atomic behavior. However, such an approach overlooks how quantum mechanics evolved logically from remedying the shortcomings of Newtonian physics while retaining its important and successful aspects. As a result, this book will analyse several of the prominent failures of Newtonian physics with the hope that the reader will gain a better understanding of how quantum mechanics itself works. The process of presenting these failures will also begin building the relationship of quantum mechanical processes to electromagnetic processes.

2.1 Blackbody radiation and Planck's constant

Before the advent of quantum mechanics, physicists in the late 19th century had already begun to deal with the ramifications of the atomic nature of matter. In particular, through the work of Boltzmann and contemporaries, *statistical mechanics* was developed to recast thermodynamics in terms of the statistical behavior of a large collection of atoms. This monograph makes no assumption of any familiarity

doi:10.1088/978-0-7503-5663-3ch2

by the reader with statistical mechanics. Moreover, since there are some structural similarities between quantum mechanics and statistical mechanics, particularly the use of probability, it is useful to present a brief digression into how statistical mechanics is formulated. Using the key concepts of *probability distributions* and *statistical averages*, statistical mechanics is able to bypass the impossible problem of solving classical Newtonian mechanics for a *system* containing typically 10^{23} particles.

2.1.1 Statistical mechanics

The distinguishing feature of statistical mechanics is its use of probability distributions to analyse the properties of a system in thermal equilibrium with its environment. The system's environment is viewed as a vast reservoir of thermal energy at a fixed absolute temperature T. The system under consideration exchanges energy with its environment until the system reaches equilibrium at the same temperature as its environment, sometimes referred to as the *zeroth law of thermodynamics.* In statistical mechanics the *state* of the system refers to the way that a system manifests a given total mechanical energy E in terms of macroscopic quantities such as volume, density, and pressure. For a system in thermal equilibrium with its environment the *relative probability* that it has the energy E is a function of both E and the temperature T and is denoted $P(E)$. By relative probability it is meant that the ratio $P(E_1)/P(E_2)$ has a well-defined probabilistic interpretation, so that the relative probability for the system having the total energy E_1 compared to having the total energy E_2 is given by the ratio.

The basic mathematical nature of $P(E)$ is revealed by using the earlier observation (1.10), which shows that only energy differences are physically meaningful. Therefore, the relative probability *must be independent of the freedom to add an arbitrary constant to the energy*. For E small compared to E_1 and E_2 this means that

$$\begin{aligned}\frac{P(E_1)}{P(E_2)} = \frac{P(E_1+E)}{P(E_2+E)} &\approx \frac{P(E_1)}{P(E_2)}\left(1 + \frac{E}{P(E_1)}\frac{\mathrm{d}P(E_1)}{\mathrm{d}E_1} - \frac{E}{P(E_2)}\frac{\mathrm{d}P(E_2)}{\mathrm{d}E_2}\right) \\ &= \frac{P(E_1)}{P(E_2)}\left(1 + E\left(\frac{\mathrm{d}\ln P(E_1)}{\mathrm{d}E_1} - \frac{\mathrm{d}\ln P(E_2)}{\mathrm{d}E_2}\right)\right),\end{aligned} \tag{2.1}$$

where a Taylor series expansion to $O(E)$ was used in (2.1). In order for the last two terms of result (2.1) to cancel it must be that $\mathrm{d}\ln P(E)/\mathrm{d}E$ is a constant independent of E. This constant is denoted $-\beta$, and this gives

$$P(E) \propto e^{-\beta E}. \tag{2.2}$$

The constant β is identified with the inverse temperature of the environment, $\beta = 1/k_B T$, where k_B is Boltzmann's constant. Boltzmann's constant is a fundamental constant, found by dividing the perfect gas constant R by Avogadro's number.

Because probability has been introduced, it is necessary to *normalize* $P(E)$ so that the sum of all possible probabilities is unity, i.e. $\sum_E P(E) = 1$. To accomplish this the *partition function* Q_β is defined as

$$Q_\beta = \sum_E e^{-\beta E}, \tag{2.3}$$

where the sum is over all possible energies available to the system as it exchanges energy with its environment. The *normalized probability* of the system having the energy E is then given by $P(E) = e^{-\beta E}/Q_\beta$. The *average value* $\langle \mathrm{O} \rangle$ of a measurable quantity or *observable* $\mathrm{O}(E)$ that depends on the energy is given by the *statistical* or *thermal average*,

$$\langle \mathrm{O} \rangle = \sum_E \mathrm{O}(E)\, P(E) = \frac{1}{Q_\beta} \sum_E \mathrm{O}(E)\, e^{-\beta E}. \tag{2.4}$$

This is *interpreted* as the value that the quantity $\mathrm{O}(E)$ will have when averaged over *many measurements*. Using (2.4) the *average energy* can be written as

$$\langle E \rangle = \frac{1}{Q_\beta} \sum_E E\, e^{-\beta E} = -\frac{1}{Q_\beta} \frac{\partial}{\partial \beta} \sum_E e^{-\beta E} = -\frac{1}{Q_\beta} \frac{\partial Q_\beta}{\partial \beta} = -\frac{\partial}{\partial \beta} \ln Q_\beta. \tag{2.5}$$

In the case of Newtonian physics, the sum in (2.3) is over all possible energies available to the classical system, which form a continuous set of values found from the Hamiltonian H that generalizes E. In the case of a single one-dimensional particle in contact with the reservoir that is governed by the Hamiltonian $H(p, x)$, the sum becomes an integral over *phase space*, which includes both p and x. As an example, for the case of a harmonic oscillator, where $H = p^2/2\,m + kx^2/2$, (2.3) yields

$$Q_\beta = \sum_E e^{-\beta E} \to \int_{-\infty}^{\infty} \frac{\mathrm{d}p\, \mathrm{d}x}{2\pi\hbar} \exp\left\{-\beta\left(\frac{p^2}{2m} + \frac{1}{2}kx^2\right)\right\} = \frac{1}{\beta\hbar\omega}, \tag{2.6}$$

where, for the moment, $\hbar$ is an arbitrary constant with cgs units of erg · second which renders the integral dimensionless, while $\omega = \sqrt{k/m}$ is the natural angular frequency of the harmonic oscillator. Using (2.6) in (2.5) immediately gives

$$\langle E \rangle = -\frac{\partial \ln(\beta\hbar\omega)^{-1}}{\partial \beta} = \frac{\partial \ln(\beta\hbar\omega)}{\partial \beta} = \frac{1}{\beta} = k_{\mathrm{B}}T, \tag{2.7}$$

so that the average thermal energy of a *classical* harmonic oscillator is independent of $\hbar$ and ω and varies continuously with temperature T.

2.1.2 Standing waves and blackbody radiation

Toward the end of the 19th century the thermal behavior of *blackbody radiation* was investigated experimentally. In this case the thermodynamic system is the electromagnetic radiation in a hollow cavity, a cube of side L, inside a large heat reservoir. The walls of the cavity consist of a perfect emitter and absorber of electromagnetic radiation, referred to as a *blackbody*. Understanding the classical analysis of blackbody radiation is somewhat complicated, but it presents great insights regarding the problems that classical physics was to experience in the atomic realm. It is also the genesis of quantum mechanics, and so it will be presented in some detail.

The first step is to understand the thermal electromagnetic energy present in the cavity. Since the walls of the cavity can both emit and absorb radiation, assumed to occur through the interaction of light with the atoms in the wall, thermal equilibrium requires equal rates of emission and absorption. These conditions create a *standing electromagnetic wave* inside the cavity. The standing wave solution (1.76) can be adapted to model the standing wave present in the cavity by writing the scalar potential in Cartesian coordinates for the cubical cavity as

$$\varphi(\boldsymbol{x}, t) = \varphi_0(k_1, k_2, k_3) \sin(k_1 x_1) \sin(k_2 x_2) \sin(k_3 x_3) \cos(\omega t), \tag{2.8}$$

where φ_0 is a function of the components k_1, k_2, and k_3 of the wave vector $\boldsymbol{k}$. There is an important subtlety with (2.8). For the case that k_i is negative, the sine function can be written $\sin(k_i x_i) = \sin(-|k_i| x_i) = -\sin(|k_i| x_i)$. This allows (2.8) to be written

$$\varphi(\boldsymbol{x}, t) = \tilde{\varphi}_0(k_1, k_2, k_3) \sin(|k_1| x_1) \sin(|k_2| x_2) \sin(|k_3| x_3) \cos(\omega t), \tag{2.9}$$

where $\tilde{\varphi}_0$ is obtained from φ_0 by breaking the three-dimensional space defined by $\boldsymbol{k}$ into eight octants. In the first octant, where the components k_i are all positive, the function $\tilde{\varphi}_0$ is given by $\varphi_0(|k_1|, |k_2|, |k_3|)$, while in the eighth octant, where the k_i are all negative, $\tilde{\varphi}_0$ is given by $-\varphi_0(-|k_1|, -|k_2|, -|k_3|)$. The other six octants are obtained similarly, with the sign of the amplitude determined by the number of negative k components. Once the amplitude in the first octant is specified, the other seven octant amplitudes are automatically determined. As a result, for the standing wave of (2.9) the only independent amplitude is the function $\varphi_0(k_1, k_2, k_3)$ in the *first octant*, which comprises *one-eighth* of the $\boldsymbol{k}$-space.

The associated vector potential $\mathbf{A}$ is determined from (2.8) so that the Lorentz condition (1.67) is satisfied. This requires

$$\begin{aligned} A_1(\boldsymbol{x}, t) &= A_{01} \cos(k_1 x_1) \sin(k_2 x_2) \sin(k_3 x_3) \sin(\omega t), \\ A_2(\boldsymbol{x}, t) &= A_{02} \sin(k_1 x_1) \cos(k_2 x_2) \sin(k_3 x_3) \sin(\omega t), \\ A_3(\boldsymbol{x}, t) &= A_{03} \sin(k_1 x_1) \sin(k_2 x_2) \cos(k_3 x_3) \sin(\omega t), \end{aligned} \tag{2.10}$$

where the A_{0j} are the components of the constant vector $\mathbf{A}_0$, which is a function of the k_i. Substituting (2.8) and (2.10) into the Lorentz condition (1.67) shows that the constants $\mathbf{A}_0$ and φ_0 are related,

$$\nabla \cdot \mathbf{A} + \frac{1}{c}\frac{\partial \varphi}{\partial t} = \frac{\partial A_1}{\partial x_1} + \frac{\partial A_2}{\partial x_2} + \frac{\partial A_3}{\partial x_3} + \frac{1}{c}\frac{\partial \varphi}{\partial t} = 0 \Longrightarrow \boldsymbol{k} \cdot \mathbf{A}_0 = -\frac{\omega}{c}\varphi_0. \tag{2.11}$$

Since there is no charge within the cavity, Gauss's law for $\rho = 0$ must be satisfied. Combining this with the Lorentz condition gives

$$\begin{aligned} \nabla \cdot \mathbf{E} = \nabla \cdot \left(-\nabla\varphi - \frac{1}{c}\frac{\partial \mathbf{A}}{\partial t}\right) &= -\nabla^2\varphi - \frac{1}{c}\frac{\partial}{\partial t}\nabla \cdot \mathbf{A} \\ = \left(-\nabla^2\varphi + \frac{1}{c^2}\frac{\partial^2\varphi}{\partial t^2}\right) &= \Box\varphi = 0. \end{aligned} \tag{2.12}$$

Using the form (2.8) shows that (2.12) is satisfied if

$$\omega^2 = c^2(k_1^2 + k_2^2 + k_3^2) = c^2\boldsymbol{k} \cdot \boldsymbol{k} = c^2k^2. \tag{2.13}$$

The standing wave must also satisfy Ampere's law for $\mathbf{J} = 0$. Combining the vector identity $\mathbf{A} \times (\mathbf{A} \times \mathbf{B}) = \mathbf{A}(\mathbf{A} \cdot \mathbf{B}) - (\mathbf{A} \cdot \mathbf{A})\mathbf{B}$ with the Lorentz condition gives the requirement

$$\begin{aligned} \nabla \times \mathbf{B} - \frac{1}{c}\frac{\partial \mathbf{E}}{\partial t} &= \nabla \times (\nabla \times \mathbf{A}) - \frac{1}{c}\frac{\partial}{\partial t}\left(-\nabla\varphi - \frac{1}{c}\frac{\partial \mathbf{A}}{\partial t}\right) \\ &= \nabla\left(\nabla \cdot \mathbf{A} + \frac{1}{c}\frac{\partial \varphi}{\partial t}\right) + \left(\frac{1}{c^2}\frac{\partial^2}{\partial t^2} - \nabla^2\right)\mathbf{A} = \Box\mathbf{A} = 0. \end{aligned} \tag{2.14}$$

Using (2.10) shows that Ampere's law (2.14) is also satisfied as long as (2.13) holds.

The walls of the cavity are assumed to be grounded, so that the wave numbers k_i of the three respective directions must be chosen so that $\varphi = 0$ at the walls, for convenience located at $x_i = 0$ and $x_i = L$. This requirement is met if $k_i = n_i\pi/L$, where n_i is a positive integer in order for k_i to be positive. These integers must therefore satisfy

$$n^2 \equiv n_1^2 + n_2^2 + n_3^2 = \frac{\omega^2L^2}{\pi^2c^2} \implies n = \frac{\omega L}{\pi c}. \tag{2.15}$$

Because the first octant of $\boldsymbol{k}$-space determines the scalar and vector potential, it is only the positive values of the n_i or the first octant of $\boldsymbol{n}$-space that are independent variables. Each possible set of values for the positive integers n_i that satisfies (2.15) for a given ω therefore constitutes a *mode* of the electromagnetic standing wave in the cavity.

The next step is to determine the thermal energy of a mode. This requires finding the total electromagnetic energy in the cavity and using it in the partition function. This begins by using the mode defined by the potentials (2.8) and (2.10) in the definitions (1.65) to find the associated forms for $\mathbf{E}$ and $\mathbf{B}$, and from them the total energy U of the mode is obtained by integrating the energy density (1.85) over the cubic volume V of the cavity. After an integration by parts that drops the surface terms, as well as combining the Lorentz condition with the equation of motion for φ and $\mathbf{A}$, the result for the $\mathbf{E}$ field reduces to

$$\begin{aligned} \frac{1}{8\pi}\int_V \mathrm{d}^3x\, \mathbf{E} \cdot \mathbf{E} &= \frac{1}{8\pi}\int_V \mathrm{d}^3x \left(\nabla\varphi + \frac{1}{c}\frac{\partial \mathbf{A}}{\partial t}\right) \cdot \left(\nabla\varphi + \frac{1}{c}\frac{\partial \mathbf{A}}{\partial t}\right) \\ &= \frac{1}{8\pi}\int_V \mathrm{d}^3x \left(-\varphi\nabla^2\varphi - \frac{2}{c}\varphi\nabla \cdot \frac{\partial \mathbf{A}}{\partial t} + \frac{1}{c^2}\frac{\partial \mathbf{A}}{\partial t} \cdot \frac{\partial \mathbf{A}}{\partial t}\right) \\ &= \frac{1}{8\pi c^2}\int_V \mathrm{d}^3x \left(\varphi\frac{\partial^2\varphi}{\partial t^2} + \frac{\partial \mathbf{A}}{\partial t} \cdot \frac{\partial \mathbf{A}}{\partial t}\right) \\ &= \frac{\omega^2L^3}{64\pi c^2}(A_0^2 - \varphi_0^2)\cos^2(\omega t) \\ &= \frac{L^3}{64\pi}(k^2A_0^2 - (\boldsymbol{k} \cdot \mathbf{A}_0)^2)\cos^2(\omega t), \end{aligned} \tag{2.16}$$

where $\omega = ck$ was used. Result (2.16) also used the spatial integrals

$$\int_0^L \mathrm{d}x \sin^2(|k_1|x) = \int_0^L \mathrm{d}x \cos^2(|k_1|x) = \frac{1}{2}L, \tag{2.17}$$

with identical results for the y and z integrations. Similarly, the contribution of the magnetic field is found by adapting the identity

$$(\mathbf{A} \times \mathbf{B}) \cdot (\mathbf{A} \times \mathbf{B}) = (\mathbf{A} \cdot \mathbf{A})(\mathbf{B} \cdot \mathbf{B}) - (\mathbf{A} \cdot \mathbf{B})^2. \tag{2.18}$$

After an integration by parts that drops the surface terms this gives

$$\begin{aligned}
\frac{1}{8\pi}\int_V \mathrm{d}^3x\, \mathbf{B} \cdot \mathbf{B} &= \frac{1}{8\pi}\int_V \mathrm{d}^3x\, (\nabla \times \mathbf{A}) \cdot (\nabla \times \mathbf{A}) \\
&= -\frac{1}{8\pi}\int_V \mathrm{d}^3x\, (\mathbf{A} \cdot \nabla^2\mathbf{A} + (\nabla \cdot \mathbf{A})^2) \\
&= -\frac{1}{8\pi c^2}\int_V \mathrm{d}^3x \left(\mathbf{A} \cdot \frac{\partial^2 \mathbf{A}}{\partial t^2} + \left(\frac{\partial \varphi}{\partial t}\right)^2\right) \\
&= \frac{\omega^2 L^3}{64\pi c^2}(A_0^2 - \varphi_0^2)\sin^2(\omega t) \\
&= \frac{L^3}{64\pi}(k^2 A_0^2 - (\boldsymbol{k} \cdot \mathbf{A}_0)^2)\sin^2(\omega t),
\end{aligned} \tag{2.19}$$

where $\omega = ck$ and the previous identities (2.17) were used.

The final result for the total electromagnetic energy U of the mode is the sum of (2.16) and (2.19),

$$U = \frac{L^3}{64\pi}(k^2 A_0^2 - (\boldsymbol{k} \cdot \mathbf{A}_0)^2). \tag{2.20}$$

The vector $\mathbf{A}_0$ can be written $\mathbf{A}_0 = \mathbf{A}_\perp + \mathbf{A}_\parallel$, where $\mathbf{A}_\perp$ is a vector in the plane perpendicular to $\boldsymbol{k}$, satisfying $\boldsymbol{k} \cdot \mathbf{A}_\perp = 0$, while $\boldsymbol{k} \cdot \mathbf{A}_\parallel = kA_\parallel$, so that $\mathbf{A}_\perp \cdot \mathbf{A}_\parallel = 0$. This gives

$$k^2 A_0^2 = k^2(\mathbf{A}_\parallel + \mathbf{A}_\perp) \cdot (\mathbf{A}_\parallel + \mathbf{A}_\perp) = k^2 A_\parallel^2 + k^2 \mathbf{A}_\perp \cdot \mathbf{A}_\perp, \tag{2.21}$$

$$(\boldsymbol{k} \cdot \mathbf{A}_0)^2 = (\boldsymbol{k} \cdot (\mathbf{A}_\parallel + \mathbf{A}_\perp))^2 = k^2 A_\parallel^2. \tag{2.22}$$

Using these results and expressing the two-dimensional vector $\mathbf{A}_\perp$ in terms of two perpendicular components, $\mathbf{A}_{\perp 1}$ and $\mathbf{A}_{\perp 2}$, gives

$$U = \frac{L^3}{64\pi}k^2 \mathbf{A}_\perp \cdot \mathbf{A}_\perp = \frac{L^3 k^2}{64\pi}[(A_{\perp 1})^2 + (A_{\perp 2})^2], \tag{2.23}$$

so that the energy of the mode has two quadratic degrees of freedom, $A_{\perp 1}$ and $A_{\perp 2}$. The energy (2.23) can written in a way that is independent of ω by noting that the magnetic and electric fields of the mode can be written as $B_\perp = kA_{\perp 1}$ and $E_\perp = \omega A_{\perp 2}/c = kA_{\perp 2}$, so that

$$U = \frac{L^3}{64\pi}(E_\perp^2 + B_\perp^2) = \frac{L^3}{32\pi}E_\perp B_\perp. \tag{2.24}$$

This is consistent with the earlier result (1.99) that showed the *average intensity* of a monochromatic electromagnetic wave is independent of its frequency when written in terms of its electric and magnetic fields.

The partition function is found by summing over all possible energy configurations. In the case of blackbody radiation this is accomplished by integrating over the two degrees of freedom in $\mathbf{A}_\perp$, so that the partition function is given by

$$Q_\beta = \sum_E e^{-\beta U} = \int \frac{\mathrm{d}A_{\perp 1}\,\mathrm{d}A_{\perp 2}}{2\pi\hbar\omega k} \exp\left(-\frac{\beta L^3 k^2}{64\pi}[(A_{\perp 1})^2 + (A_{\perp 2})^2]\right) = \frac{32\pi}{\beta\hbar L^3 \omega k^3}, \tag{2.25}$$

where the factor $2\pi\hbar\omega k$ is present to render the integration dimensionless. Using the partition function (2.25) immediately yields the average thermal energy of each standing wave mode,

$$\langle E \rangle = -\frac{\partial}{\partial\beta}\ln Q_\beta = \frac{1}{\beta} = k_B T, \tag{2.26}$$

so that the radiation energy is thermally equivalent to the harmonic oscillator. This will be an important factor in Planck's treatment of blackbody radiation.

Result (1.114) shows that each mode of the *classical blackbody radiation* has a thermal energy that is independent of its frequency ω, and therefore each mode is expected to contribute the same average thermal energy $\langle E \rangle$ found in (1.114). The basic assumption of statistical mechanics is that *all modes* are equally likely to occur, and will therefore be present in the blackbody radiation, sharing equally in the thermal energy exchanged with the reservoir. Therefore, the number of modes that correspond to a given frequency ω will determine the energy distribution as a *function of frequency* in the blackbody radiation.

The process of counting the modes associated with a given frequency ω begins by noting that (2.15) has the form of a radius n squared in *number space*, where the integers form a three-dimensional lattice. As ω changes an additional range of integers becomes available found by differentiating (2.15), so that

$$\mathrm{d}n = \frac{L}{\pi c}\,\mathrm{d}\omega. \tag{2.27}$$

In complete analogy with spherical coordinates, where $\mathrm{d}V = 4\pi R^2\,\mathrm{d}R$, the volume of numbers contained in the shell associated with $\mathrm{d}n$ is

$$\mathrm{d}N = 4\pi n^2\,\mathrm{d}n = \frac{4L^3\omega^2\,\mathrm{d}\omega}{\pi^2 c^3}. \tag{2.28}$$

It was shown earlier that the integers n_i must be positive, so that only the first octant of the spherical shell should be counted. As a result, the volume of numbers found in (2.28) must be divided by 8. Exchanging $\cos(\omega t) \leftrightarrow \sin(\omega t)$ in both (2.8) and $\mathbf{A}(\boldsymbol{x}, t)$ and then setting $\mathbf{A}_0 \to -\mathbf{A}_0$ gives a second solution with a *different polarization*, $\mathbf{B} \to -\mathbf{B}$ and $\mathbf{E} \to \mathbf{E}$, but with an identical relation for ω. As a result, the volume of modes must be multiplied by 2 to reflect this. Noting that $L^3 = V$ is the volume of the cavity, the final result is that

$$\frac{\mathrm{d}N}{V} \equiv \mathrm{d}\rho = \frac{\omega^2}{\pi^2 c^3}\,\mathrm{d}\omega, \tag{2.29}$$

where $\rho(\omega)$ is the number of wave modes per unit volume as a function of frequency. Multiplying (2.29) by $k_\mathrm{B}T$ then yields the energy per unit volume for the blackbody radiation lying in the frequency range $\mathrm{d}\omega$, giving

$$\langle E\rangle\,\mathrm{d}\rho = \frac{\omega^2\langle E\rangle}{\pi^2 c^3}\,\mathrm{d}\omega = \frac{\omega^2 k_\mathrm{B}T}{\pi^2 c^3}\,\mathrm{d}\omega = \frac{\omega^2}{\pi^2 c^3 \beta}\,\mathrm{d}\omega \equiv \mathrm{d}u(\omega), \tag{2.30}$$

where $u(\omega)$ is the total energy per unit volume, referred to as the *energy density*, of the cavity's blackbody radiation as a function of ω.

Relation (2.30) is the energy version of the classical result known as the *Rayleigh–Jeans radiation law*. Unfortunately for Newtonian physics, result (2.30) was found to be experimentally incorrect. Although (2.30) is reasonably accurate at low frequencies, it fails dramatically at high frequencies. This failure was known as the *ultraviolet catastrophe* since it becomes pronounced once ultraviolet frequencies are encountered. Worse yet, (2.30) predicts the energy density present in the blackbody radiation should grow as ω^3, which follows from

$$\mathrm{d}u(\omega) = \frac{\omega^2}{\pi^2 c^3 \beta}\,\mathrm{d}\omega \implies u(\omega) = \frac{\omega^3}{3\pi^2 c^3 \beta}. \tag{2.31}$$

This expression diverges as $\omega \to \infty$ and is therefore a completely unphysical result. In that regard, Stefan's law had been experimentally established, showing that the total energy W radiated per unit area per unit time from a blackbody is proportional to T^4. In an important observation, Wien deduced from experimental data that the energy spectrum $\mathrm{d}u/\mathrm{d}\omega$ of blackbody radiation is proportional to $\omega^3 e^{-\hbar\beta\omega}$ for large ω, where $\hbar$ is a very small constant with the cgs units of erg s.

2.1.3 Planck's quantization postulate

In the process of attempting to resolve the theoretical dilemma surrounding blackbody radiation, in 1900 Planck put forward the idea of modifying the classical expression (1.114) for the thermal energy average, replacing it with the frequency-dependent expression

$$\langle E\rangle = \frac{\hbar\omega}{e^{\hbar\beta\omega} - 1} = \frac{\hbar\omega\, e^{-\hbar\beta\omega}}{1 - e^{-\hbar\beta\omega}}, \tag{2.32}$$

where the constant $\hbar \approx 1.06 \times 10^{-27}$ erg · s is now known as *Planck's constant*. For ω small the approximation $e^{\hbar\beta\omega} \approx 1 + \hbar\beta\omega$ shows that (2.32) reduces to the classical result (1.114), $\langle E\rangle \approx 1/\beta$, recapturing the Rayleigh–Jeans formula. When (2.32) is used in (2.30), the result is

$$\mathrm{d}u = \langle E\rangle\, \mathrm{d}\rho = \frac{\hbar\omega^3\, e^{-\hbar\beta\omega}}{\pi^2 c^3(1 - e^{-\hbar\beta\omega})}\, \mathrm{d}\omega = \frac{\mathrm{d}u}{\mathrm{d}\omega}\, \mathrm{d}\omega. \tag{2.33}$$

For high frequencies (2.33) becomes $\mathrm{d}u/\mathrm{d}\omega \approx \hbar\omega^3\, e^{-\hbar\beta\omega}/\pi^2 c^3$, consistent with Wien's observation regarding blackbody radiation. The total energy per unit volume U contained in the blackbody radiation is therefore given by

$$\begin{aligned} U &= \int_0^U \mathrm{d}u = \int_0^\infty \frac{\mathrm{d}u}{\mathrm{d}\omega}\, \mathrm{d}\omega = \int_0^\infty \mathrm{d}\omega\, \frac{\hbar\omega^3\, e^{-\hbar\beta\omega}}{\pi^2 c^3(1 - e^{-\hbar\beta\omega})} \\ &= \frac{\pi^2}{15\hbar^3 c^3 \beta^4} = \frac{\pi^2 k_\mathrm{B}^4}{15\hbar^3 c^3} T^4. \end{aligned} \tag{2.34}$$

Result (2.34) yields *Stefan's law* for the power per unit area W radiated from the surface of a blackbody follows from (2.34) by using $W = c\ U/4$. The factor of 1/4 compensates for the random direction of the blackbody radiation inside the cavity, where half is traveling in the wrong direction to be emitted and the remaining half is randomly oriented to the surface of emission, yielding a second factor of 1/2 by averaging $\sin\theta\, \mathrm{d}\Omega$ over the hemisphere, as in (1.97). The result is $W = \sigma T^4$, where Stefan's constant σ is given by

$$\sigma = \frac{\pi^2 k_\mathrm{B}^4}{60\hbar^3 c^2}. \tag{2.35}$$

Stefan's constant was experimentally determined and using it in (2.35) gave a value for $\hbar$ that matched well with the value Wien had estimated. Since it appears with k_B and c, this also strongly indicated that $\hbar$ is a *fundamental constant of nature*.

Planck took the logical step of determining the form that the partition function (2.3) must take as a function of frequency in order to obtain the result (2.32). The solution is

$$Q_\beta(\omega) = \frac{1}{1 - e^{-\hbar\beta\omega}}, \tag{2.36}$$

which immediately yields (2.32) from (2.5). A series expansion of (2.36) gives

$$Q_\beta(\omega) = \frac{1}{1 - e^{-\hbar\beta\omega}} = \sum_{n=0}^{\infty} e^{-\beta n\hbar\omega} = \sum_{n=0}^{\infty} e^{-\beta E_n}, \tag{2.37}$$

where the values of the energies have been identified for consistency with the statistical mechanics form (2.3) for the partition function. The energy values available to the cavity radiation at the frequency ω are apparently given by $E_n = n\hbar\omega$, which are *integer multiples* of the base energy $E = \hbar\omega$. This result is clearly at odds with the classical result that the energy of each mode was independent of ω, but it was also in contradiction to the continuous nature of energy available to both Newtonian particles and waves.

This result was further complicated by the need to reconcile it with the idea that the source of the blackbody radiation is the atoms of the metal enclosing the cavity. It was widely believed that the atoms were radiating and absorbing light as oscillators. Since an oscillator is characterized by its natural frequency ω, Planck postulated that the atomic oscillators of the blackbody cavity wall with the natural frequency ω could emit and absorb radiation only by discrete amounts of energy, all of which were a multiple of $\hbar\omega$. Planck referred to these as a *quantum* of energy from the Latin for *amount*. Planck's solution required that both the blackbody radiation and the atomic oscillators have energies that are discontinuous. Planck's solution was initially unpopular with the physics community, but it gathered support as more experimental results reflecting atomic level behavior became available.

Planck's solution to the blackbody problem exhibits many of the important features that characterize the quantum mechanical description of atomic level phenomena. The identification of the new physical constant $\hbar$ established a quantitative tool to determine when Newtonian physics will break down and quantum mechanical behavior will manifest itself. It should be noted that expression (2.36) reduces to the Newtonian form (2.6) in the limit $\hbar \to 0$. Newtonian or *classical physics* is therefore *apparently* regained in the limit that corrections of $O(\hbar)$ are negligible, often referred to as the *classical limit*. Current research indicates that classical behavior can emerge from the quantum world even for the case that $\hbar$ is not zero. This will be discussed later in the text. Nevertheless, the remarkable successes of classical physics are therefore not abandoned, but instead are *modified* in the atomic realm to reflect the nature of atomic level phenomena. In the case just discussed, this meant that Planck could use the general structure of statistical mechanics for both atomic and classical phenomena. Finally, Planck's solution gave birth to the idea that the properties of both discrete quanta of energy and waves were entangled. While this interpretation was controversial at the time, the idea follows from the presence of only discrete particle-like energies, i.e. multiples of $\hbar\omega$, present in the frequency ω standing waves of the blackbody radiation. This strongly hints at the existence of *particles of light* making up the standing wave, each carrying the energy $\hbar\omega$. Because physics is a science that requires empirical verification, Planck's radical set of steps would have been rejected had these ideas not been found essential to explaining a wide range of emerging experimental results.

2.2 Light and photons

The particle nature of electromagnetic radiation appeared early in the 20th century in a second important experimental phenomenon.

2.2.1 The photoelectric effect and frequency

When a sliver of metal is exposed to monochromatic electromagnetic radiation, it is possible to strip electrons from it and create an electric current through the space around it, referred to as the *photoelectric effect*. Clearly, the light is supplying the energy to dislodge the electrons and create the *photocurrent*, and so the nature of the photocurrent reveals critical aspects of the interaction of light with the fundamental charge carrier in metals, the electron. There were a number of different photoelectric experiments beginning with those of Hertz. In particular, Lenard showed that the photocurrent yielded a surprising result: the amount of the photocurrent was highly dependent on the frequency of the monochromatic light used. This was a surprise because the energy delivered by the incident light is quantified by its average intensity, which the classical result (1.99) shows is *independent of frequency*. If the photocurrent was simply proportional to the energy delivered by the light, it should grow as the intensity of the incident light was increased.

Instead, the following results were obtained. A photocurrent occurred *only if the angular frequency of the incident light exceeded a critical value* ω_c. If the light's angular frequency was below the critical value there was virtually no photocurrent, regardless of available light intensities, and only a heating of the metal occurred. The critical frequency depended upon the particular metal being irradiated. Once the critical frequency was exceeded the kinetic energy of the electrons in the photocurrent grew with increasing frequency. For a fixed frequency exceeding the critical frequency the kinetic energy of the electrons in the photocurrent remained the same *regardless of intensity*. However, the *number* of ejected electrons grew with the light's intensity once the critical frequency was exceeded. These results could not be explained using the classical concept that each of the electrons in the metal received energy from the light wave in direct proportion to the intensity (1.99) of the wave, and that the frequency simply determined how long it took for the electron to accumulate enough energy to be ejected.

2.2.2 Einstein and photons

In 1905 Einstein resolved the paradox of the photoelectric effect by adapting Planck's blackbody solution to the case of the incident light wave. First, he postulated that each metal could be characterized by the amount of energy required to free an electron, which is referred to as the *work function* W of the metal. Next, he postulated that the incident light wave was composed of particles or *quanta of light*, referred to as *photons*, whose individual energy was given by the Planck formula $E = \hbar\omega$, where ω is the angular frequency of the *incident monochromatic light wave*. In this picture the light's intensity or power is determined from both the *density* of photons in the incident light wave and *the individual energy of each photon*. This seminal idea begins the process of merging the twin concepts of particle and wave, and will become the cornerstone of formal quantum mechanics and eventually quantum field theory. Finally, Einstein postulated that the electrons in the metal interact with the light *one photon at a time*. Only when the energy of these individual photons exceeds the particular metal's work function is the electron freed from the metal, so that the criterion for a photocurrent is $\hbar\omega - W > 0$. This gives a critical frequency, $\omega_c = W/\hbar$, that must be exceeded in order to create the photocurrent. This does *not* mean the photon is absorbed by the electron. Rather, it simply means that such a photon is energetic enough to free the electron through its interaction with it. The complete details of how individual photons interact with electrons and why it occurs almost exclusively one photon at a time would require the development of *quantum electrodynamics* in the late 1940s. Nevertheless, any excess energy in the photon was expected to be manifested in an increased kinetic energy K for the ejected electron. As a result, for $\omega > \omega_c$, Einstein argued that the photoelectric effect was governed by the simple energy relation

$$K = \hbar\omega - W. \tag{2.38}$$

Result (2.38) could be verified experimentally by determining the voltage V_s necessary to stop the electron from being ejected and therefore for zero current to occur, referred to as the *stopping voltage*. If q is the electron's charge, then this simply required that $qV_s = K$. Combining this with (2.38) yields a linear relationship between V_s and ω,

$$V_s = \frac{\hbar}{q}\omega - \frac{W}{q}, \tag{2.39}$$

which was experimentally observed. The slope of this linear relationship also gave a second means for measuring Planck's constant since the charge q of the electron had been experimentally established.

2.2.3 Photons and the Compton effect

As shorter wavelength light became available experiments focussed on the effect that such light had on charged particles. In particular, Compton examined the effects of scattering x-rays from electrons. The classical analysis of the process, referred to as Thomson scattering, does not include a change in the wavelength of the light. However, Compton observed a change in wavelength for the scattered light that grew larger with the angle of scattering.

The explanation of the *Compton effect* requires treating the scattering of the incident *light wave* in terms of the scattering of individual photons from electrons, treating both as *particles*. In so doing, it is assumed that *conservation of momentum and energy* occurs throughout the process. Although it seems like an obvious assumption, it is worth considering what evidence there is for it. If the light wave is considered as a stream of particles, the failure of energy–momentum conservation at the particle level would manifest itself at the macroscopic level if the light involved a sufficient number of photons. This would contradict the energy and momentum conservation relations obtained from Maxwell's equations, which are experimentally well established. As a result, photon scattering processes must respect energy and momentum conservation.

In order to understand Compton scattering it is therefore necessary to extend the concept of momentum to the photon. This follows from (1.88) and (1.89), which relate the energy E of the *massless* photon to its momentum p as $E = cp$. It follows that

$$p = \frac{E}{c} = \frac{\hbar\omega}{c} = \frac{2\pi\hbar f}{c} = \frac{hf}{c} = \frac{h}{\lambda}, \tag{2.40}$$

where λ is the wavelength of the associated electromagnetic wave and $h = 2\pi\hbar$ is a commonly used variant of Planck's constant. The momentum of the photon is elevated to a vector $\boldsymbol{p}$ by associating its direction with the direction in which the electromagnetic wave is propagating. Treating a light wave as simultaneously composed of particles requires extending the property of momentum to massless particles, a result not possible in Newtonian mechanics but required in quantum mechanics in order to be consistent with the electromagnetic wave energy–momentum relation (1.88).

For simplicity the electron is treated as initially at rest, so its initial momentum is zero. Denoting $\boldsymbol{p}$ and $\boldsymbol{p}'$ as the initial and final momenta of the photon and $\boldsymbol{p}_e$ as the final momentum of the electron, the conservation of momentum gives

$$\boldsymbol{p} = \boldsymbol{p}' + \boldsymbol{p}_e \implies \boldsymbol{p}_e = \boldsymbol{p} - \boldsymbol{p}', \tag{2.41}$$

so that squaring (2.41) gives

$$p_e^2 = \boldsymbol{p}_e \cdot \boldsymbol{p}_e = \boldsymbol{p} \cdot \boldsymbol{p} + \boldsymbol{p}' \cdot \boldsymbol{p}' - 2\boldsymbol{p} \cdot \boldsymbol{p}' = p^2 + p'^2 - 2pp' \cos\theta, \tag{2.42}$$

where θ is the angle between the incident and scattered photon. Conservation of energy is best expressed using the relativistic expression (1.119). For the case of a massless particle, this expression gives the energy of the incident and scattered photon as $E = cp$ and $E' = cp'$. The initial and final energy of the mass m electron are given by $E_i = mc^2$ and $E_f = \sqrt{m^2c^4 + p_e^2c^2}$. Conservation of energy then gives

$$E + E_i = E' + E_f \implies cp - cp' + mc^2 = \sqrt{m^2c^4 + p_e^2c^2}. \tag{2.43}$$

Squaring (2.43) and rearranging the result gives

$$p_e^2 = 2\,mc(p - p') + (p - p')^2. \tag{2.44}$$

Combining (2.42) and (2.44) gives

$$pp'(1 - \cos\theta) = mc(p - p') \implies 1 - \cos\theta = mc\left(\frac{1}{p'} - \frac{1}{p}\right). \tag{2.45}$$

Result (2.45) can now be combined with (2.40) to obtain the *Compton scattering formula*,

$$\lambda' - \lambda = \frac{h}{mc}(1 - \cos\theta), \tag{2.46}$$

which gives the observed change in wavelength for the scattered light as a function of the scattering angle θ. This experimental result validated both the particle nature of light and the conservation of energy and momentum in photon–electron scattering processes.

The quantity $\lambda_e = h/m_e c$ is referred to as the *Compton wavelength* of the electron since it corresponds to the approximate change in the wavelength of the light scattered from the electron. For the electron $\lambda_e \approx 2.4 \times 10^{-10}$ cm, so that the Compton effect becomes more pronounced for x-rays, $\lambda \approx 10^{-8}$ cm, as opposed to visible light, $\lambda \approx 6 \times 10^{-5}$ cm. Although Compton's original experiment focussed on the interaction of light and electrons, the Compton effect formula can be applied to the scattering of light from any charged particle. As a result, every massive charged particle has a Compton wavelength inversely proportional to its mass.

2.2.4 Light: wave and particle

In Newton's time and throughout the 18th century the nature of light was a topic of debate. Some physicists, such as Newton, subscribed to the idea that light was a stream of particles or *corpuscles*, while others, such as Huygens, believed that light was a wave composed of *wave fronts*. The two models had both strengths and weaknesses in explaining optical phenomena. However, by the time of Einstein and Compton it was almost universally accepted among physicists that light was a wave phenomenon composed of electric and magnetic fields and governed by Maxwell's equations. Introducing the concept of the photon as a particle of light resurrected many questions that physicists of the time had considered answered. Einstein's explanation of the photoelectric effect won the Nobel Prize, but it begged for a resolution of how light could simultaneously be described mathematically as both a stream of particles *and* an electromagnetic wave. In particular, how and when does a stream of particles behave as a wave or, equivalently, how and when does a wave behave as a stream of particles? If the photon is a fundamental particle, what is the spatial size of a photon?

A full explanation of wave–particle duality for light required the development of relativistic quantum field theory, which was pioneered in the 1940s by Feynman, Schwinger, and Tomonaga. However, a simple particle model represents a monochromatic light wave as having a *photon density* of $\rho(\boldsymbol{x}, t)$ with each photon possessing an *individual energy* of $\hbar\omega$. The energy density of the wave in such a picture is then given by

$$u(\boldsymbol{x}, t) = \hbar\omega\rho(\boldsymbol{x}, t). \tag{2.47}$$

Combining (2.47) and (1.97) gives the magnitude S of the Poynting vector in the direction of the wave,

$$S(\boldsymbol{x}, t) = cu(\boldsymbol{x}, t) = c\hbar\omega\rho(\boldsymbol{x}, t). \tag{2.48}$$

Comparing (2.48) to (1.98) gives

$$\rho(\boldsymbol{x}, t) = \left(\frac{kA_0^2(\boldsymbol{k})}{4\pi\hbar c}\right)\sin^2(\boldsymbol{k}\cdot\boldsymbol{x} - \omega t + \phi), \tag{2.49}$$

which has the requisite units of inverse volume. For the case of a light wave result (1.96) shows that magnitude of the Poynting vector is given by

$$S(\boldsymbol{x}, t) = \frac{c}{4\pi} E^2(\boldsymbol{x}, t). \tag{2.50}$$

Equating this result with (2.48) shows that the electric field of a light wave has the magnitude

$$E(\boldsymbol{x}, t) = \sqrt{4\pi\hbar\omega\rho(\boldsymbol{x}, t)}. \tag{2.51}$$

While this result depends on the energy of the photons associated with the wave, it is *not* the electric field of a photon. Rather, it is the electric field of the *light wave* associated with the photon.

This simple model can be used to obtain a sense of photon energies and their numbers in a typical light wave, and it is informative to analyse sunlight, which delivers in cgs units $\approx 1.4 \times 10^6$ erg cm^{-2} s^{-1} to the Earth's surface. Using its average frequency as that of yellow light, $f \approx 5.0 \times 10^{14}$ Hz, each photon in sunlight has the average energy $E = \hbar\omega = 2\pi\hbar f \approx 3.3 \times 10^{-12}$ erg. As a result, there are $\approx 1.4 \times 10^6/(3.3 \times 10^{-12}) \approx 4.2 \times 10^{17}$ photons of sunlight striking a square centimeter every second. The average density of photons in sunlight is given by

$$\rho = \frac{S}{\hbar\omega c} \approx 1.4 \times 10^7 \text{ photons/cm}^3. \tag{2.52}$$

Using (2.51) gives the average magnitude of the electric field in sunlight as $E \approx 2.4 \times 10^{-2}$ dy esu^{-1} ≈ 720 V m^{-1}, the observed value.

The energy of sunlight photons is typical of the energies associated with the atomic processes that create them. For that reason, it is much more convenient to measure atomic level energies in *electron-volts* (eV), which is the energy that an electron gains when moving through a rise of one volt. In SI units one eV is equivalent to $E = qV = 1.6 \times 10^{-19} \times 1 = 1.6 \times 10^{-19}$ J, so that in cgs units

$$1 \text{ eV} = 1.6 \times 10^{-12} \text{ erg}. \tag{2.53}$$

A photon of yellow light has an energy of 3.3×10^{-12} erg ≈ 2.1 eV.

As a specific example of the growing breakdown of classical physics brought about by the introduction of photons, the interference effect described by (1.78) occurs when light is shone through a two-slit interferometer. It is easily explained using the principle of superposition as in (1.78), a property of classical light waves in vacuum, since the two slits create two distinct sources of monochromatic light. However, if a light wave is treated as a stream of *fundamental particles*, then Newtonian logic dictates that each photon in the light wave must have gone through one slit or the other. Interference between the photons striking the screen and combining to form the distinctive interference pattern would require that each photon retained some quantifiable aspect of which slit it had gone through. This phenomenon is the case in the wave model, where the phase difference between the two waves is created by passing through one of the two slits and recombining at the screen. However, there was nothing *obvious* about a classical particle or its trajectory that yields a similar property. Even more troublesome was the idea that interference might occur for a *single photon*. Conceptually, a crisis was brewing in physics around the dichotomy in behavior between waves and Newtonian particles, and the possibility that the two very different concepts are somehow entangled.

2.3 Electron diffraction and the de Broglie wavelength

Following Thomson's identification of the electron in 1897, its properties were investigated experimentally. When an energetic electron strikes a suitable chemical photoplate, it leaves a very small image as it deposits its kinetic energy. This led to the conclusion that the electron, if not a true point particle, is an extremely compact object. As such, it was expected to be governed by Newtonian mechanics.

2.3.1 Electron diffraction and probability

Several of the prominent early experiments involved passing a stream of electrons through crystalline structures that act as multislit gratings. In the simplest of all possible cases, a stream of reasonably slow or low energy electrons is passed through a two-slit optical interferometer. Once the electrons passed through the crystalline structure or the optical interferometer, they struck a photoplate that recorded their impact location as a small image. Remarkably, as more and more electrons struck the photoplate they arrived *preferentially at specific locations.* Although each individual electron was largely unpredictable in its impact location, the aggregate of *many electrons* formed a histogram on the photoplate, a number of hits per angle, identifying the *most probable* impact angles for the electrons. The histogram or *probability distribution* that formed on the photoplate consisted of *alternating bands of maximum and minimum probability* for electron impact, and was identical in structure to the *intensity pattern* formed by optical light passing through the same device.

In the case of light the intensity pattern could be explained by using *wave interference* and the principle of superposition. The pattern on the screen was formed by the simultaneous arrival of light waves from the two sources that were out of phase due to the slit separation. As a result, the similar phenomenon for electrons became known as *electron diffraction.* Given the point-like nature of each electron's impact, this came as a surprise to the physicists of the time, who had expected the electrons simply to pass through one slit or the other and strike the photoplate directly behind each slit, like balls thrown through a door.

The previous conceptual problem of explaining how a light wave could be viewed as a stream of photons had returned in mirror form. In this case, a stream of what were assumed to be point-like particles, electrons, was now exhibiting a behavior identical to the interference of waves, at least when applied to the *probable outcome* of the experiment. It was natural at the time to consider this using statistical mechanics, since probability distributions were already in use in that discipline. However, the problem ran much deeper since statistical mechanics is applicable only to large *thermodynamic collections* of particles and explains their average behavior using concepts of Newtonian physics. It must be stressed that even the arrival of *individual* electrons falls along one of the histogram's probability peaks and *fails to strike the spaces in between the probable peaks.* This phenomenon was therefore present even in the behavior of *a single electron*, even though the impact image of the electron was identical to that expected of a point particle.

In that regard, a similarity to light wave interference can be drawn by treating the light intensity interference pattern as the most probable impact locations for the individual photons in the light wave. Viewed in this manner, there is a similarity between the light interference intensity pattern and the electron beam's probability pattern, and this indicates a wave interference effect that is occurring *in both cases.* However, it is to be stressed that photons and electrons are very different in their properties. Photons travel at the constant speed of light while electrons can have any speed less than that of light. In addition, photons are freely absorbed and reflected by almost all matter, and so their number is not constant. On the other hand, the charge of an electron is *conserved* in the processes in which it participates.

However, in both the case of light or electrons, Newtonian logic asserts that each individual particle, either photon or electron, must have gone through one or the other of the slits. After all, a similar experiment using everyday objects, throwing balls through two open doors side by side, does not result in a diffraction pattern formed by the impact of the balls on the wall behind the doors. Instead, if the ball goes cleanly through a door, it simply hits the wall directly behind the door it went through. The presence of a wave interference probability pattern in multislit diffraction experiments indicates a profound problem with the Newtonian concept of a particle when it is applied to the case of electrons or photons.

In that regard, it is a crucial observation that the photons of visible light and the electrons in the beam are extremely low energy objects compared to the balls being thrown through the two doors. The electron mass is extremely small, $m_e \approx 9.1 \times 10^{-28}$ g. Even if its velocity is 10^8 cm s^{-1}, it still has only ≈ 3 eV of Newtonian kinetic energy, comparable to the energy of a photon of yellow light. As a result, both the diffraction of light and electrons involve extremely low energy particles. In addition, the width of the doors through which the balls are thrown is many orders of magnitude greater than that of the slits in the interferometer. It became apparent to physicists that the small distances and energies present in the atomic realm result in behavior far different from that predicted by Newtonian physics.

2.3.2 de Broglie and matter waves

In 1924 de Broglie took the critical step in resolving the conceptual problems besetting physics. He postulated that massive point particles, such as electrons, *also have wave properties*, which is the counterpart to light waves exhibiting particle properties. This idea is referred to as *wave–particle duality* and is arguably the core concept of quantum mechanics. de Broglie simply adapted Einstein's ideas regarding photon particles in a light wave to find the wavelength for a particle in a particle stream. In the case of the photon, Einstein used Planck's postulate and treated the photon's energy as $E = \hbar\omega = \hbar ck$, where k is the magnitude of the wave vector and is given by $k = \omega/c$ for a *monochromatic light wave.* In order to explain Compton scattering this led to (2.40), which identifies the momentum of a photon from its wavelength, $p = h/\lambda$. de Broglie argued that this relationship could be reversed for *all particles*, in effect identifying the *wavelength* λ of a particle from the magnitude of its momentum p as

$$\lambda = \frac{h}{p}. \tag{2.54}$$

In the case of a nonrelativistic massive particle, the momentum is given by the familiar Newtonian expression $p = mv$, so that

$$\lambda = \frac{h}{mv}. \tag{2.55}$$

Relation (2.55) is referred to as the *de Broglie wavelength* of a massive particle with the magnitude of momentum $p = mv$.

The core assumption of wave–particle duality and quantum mechanics is the association of a *matter wave* with *each Newtonian particle*. It is the matter wave and its associated interference effects that determine the *probability* of observing the particle and measuring its physical properties. This statistical interpretation of the wave stems from the work of Born, who referred to the matter wave as a *guiding field*. It is important to note that the de Broglie wavelength of a massive particle is therefore fundamentally different from the wavelength associated with light since the matter wave is *not directly observable*. The electric and magnetic fields associated with light waves can have their strengths probed throughout the space in which they exist. However, the existence of an associated matter wave in a volume of space does *not* mean that an electron and its charge have expanded to fill that volume of space with a measurable field. Instead, the presence of the electron can be probed at any position in the space in which the matter wave exists, and the matter wave at that position determines the *probability* of observing the electron at that position. However, *when observed*, the electron behaves as a *point particle* with a radius experimentally determined to be less than $\approx 10^{-20}$ cm. The remainder of this text will develop both the nature of the matter wave and the methods for finding it mathematically, as well as many of the ramifications of the probabilistic interpretation of the matter wave. In that regard, the details of *observation* depend on the experimental apparatus being used, and in that sense the observer and the object being observed are no longer independent of each other. This is fundamentally different from Newtonian physics, and has led to many efforts to reconcile classical intuitions with quantum behavior.

As the concept of the matter wave was further refined and developed it became known as the *wave function*. In the examples discussed so far for photons and electrons, it is assumed that there is a wave that passes through *both slits* of the two-slit interferometer, subsequently interfering with itself through the principle of superposition to form the pattern observed on the screen. Where the interference pattern is constructive there is a greater chance of observing the particle's impact, either electron or photon. In that regard, the final step is to interpret the intensity of the resultant matter wave as providing *the probability of observing the impact at a specific angle.* When this method is used to analyse the experimental results of electron diffraction, the locations of maximum probability for the electron's impact occur precisely at those angles that satisfy (1.79) with *λ given by the de Broglie wavelength* (2.55). This success solidified the idea of wave–particle duality.

de Broglie's wave postulate forms the conceptual foundation of formal quantum mechanics. Neither he nor anyone since has been able to explain the existence of the matter wave. To that end, most physicists simply take the operational point of view that the wave function is *a postulate necessary to explain atomic level behavior.* However, a very serious objection to (2.54) was its ad hoc nature, since de Broglie could not present an equation which the wave function satisfied. As a result, he could not explain how the wave function was to be modified by either a potential energy acting on the particle or by a Galilean transformation to a moving frame of reference. The latter problem is discussed in a later subsection. The ad hoc nature of the wave function was resolved in 1926 by Schrödinger and will be discussed in detail in the coming chapters. However, prior to Schrödinger's work, de Broglie's idea provided an initial explanation of both light and electron diffraction by introducing two central concepts: *wave–particle duality and probability.* While some aspects of wave–particle duality could be interpreted as a problem in understanding the particle content of a wave, the idea that the behavior of a single point-like particle is governed by probability rather than determinism presented a profound and often controversial departure from Newtonian mechanics.

2.3.3 The scale of quantum effects

It is important to examine the scale at which the matter wave or wave function, and therefore probability, becomes the critical factor in the behavior of a point-like object possessing inertial mass. Using (2.55) shows that a one kilogram ball thrown at $25\,\mathrm{m\ s^{-1}}$ has a de Broglie wavelength of $\approx 2 \times 10^{-35}$ m. If the ball is thrown through a door of width one meter, the diffraction formula (1.80) shows that the diffraction minima occur at

$$a \sin\theta = n\lambda \implies \sin\theta = n\frac{\lambda}{a} \approx n \cdot (2 \times 10^{-35}). \tag{2.56}$$

If there is a wall ten meters behind the door the first diffraction minimum will occur $\approx 2 \times 10^{-34}$ m from the *classical* impact point. Treating the ball as a quantum mechanical object results in a set of diffraction peaks that are spaced apart by a distance nineteen orders of magnitude smaller than a typical nucleus. In this case, such a tiny distance results in the expected continuous Newtonian pattern directly across from the door. Carrying the observer and observed analysis further, it is the act of striking the screen that constitutes a measurement of the ball's position. If the ball leaves a mark on the wall with a width of $\approx 10^{-4}$ m, the mark would correspond to a possible $\approx 10^{20}$ diffraction maxima. The point to be made is that applying quantum mechanical considerations to an everyday object has given no indication that quantum mechanical effects will be either observable or relevant to its classical motion.

On the other hand, a relatively slow electron traveling at $\approx 10^5$ cm s^{-1} has a de Broglie wavelength of $\approx 7 \times 10^{-5}$ cm or about 700 nanometers, corresponding to visible red light. If a beam of such slow electrons is passed through a slit a micrometer in width, the electron beam undergoes diffraction minima at the angles predicted by the light formula,

$$a \sin \theta = n\lambda \implies \sin \theta = n\frac{\lambda}{a} \approx n \cdot 0.7, \tag{2.57}$$

resulting in a central peak with an angular size of $\approx 40°$. Such electrons are diffracted just as optical light is, spreading through a measurable angle. It is worth noting that in order to have a wavelength comparable to the electron the macroscopic ball would need to be moving no faster than $\approx 10^{-27}$ m s^{-1}. At such an absurdly slow speed it would take $\approx 10^{18}$ seconds, roughly twice the current age of the Universe, to move one nanometer. Such a slow speed is beyond experimental observation, giving another reason the de Broglie matter wave is irrelevant to the behavior of everyday objects.

Comparing the size of the de Broglie wavelength to the spatial size of the associated system is a commonly used method to determine if quantum effects are relevant. This is at the heart of *Bohr's correspondence principle*, which states that classical behavior occurs for macroscopic objects since macroscopic masses and lengths are many orders of magnitude larger than the associated de Broglie wavelength. If the object or the system is many orders of magnitude larger than the de Broglie wavelength then quantum mechanical effects can be considered irrelevant. This was referred to as *Bohr's magic wand* by Sommerfeld. This is consistent with the idea that classical behavior should emerge in the limit $\hbar \to 0$, since the de Broglie wavelength goes to zero in that limit. Historically, the phenomenon of electron diffraction was observed *after* de Broglie first proposed (2.55). However, its observation experimentally validated the concept of the wave function, as well as the specific de Broglie wavelength formula. Since it was first proposed, the de Broglie wavelength formula has been made consistent with special relativity and applied to all atomic and subatomic level particles, providing ample experimental evidence for the universal nature of the matter wave.

2.3.4 Galilean relativity and the de Broglie wavelength

It is important to understand how the photon energy and momentum are changed by the Galilean transformation (1.32) to an observer in relative motion. The energy of the photon is changed by a Galilean transformation since the frequency of the light wave changes according to (1.83), so that an observer moving in the same direction as the light wave will claim that the energy of an associated photon is given by

$$E' = \hbar\omega' = \hbar\omega\left(1 - \frac{v_o}{c}\right) = E\left(1 - \frac{v_o}{c}\right). \tag{2.58}$$

On the other hand, the momentum of the photon is unchanged,

$$p' = \frac{E'}{c'} = \hbar\omega = p, \tag{2.59}$$

which reflects the assumption that the wavelength of the photon is unchanged as are *all lengths* in Newtonian mechanics.

However, the effect is quite different for the case of a particle of mass m. If such a particle has the Newtonian momentum $\boldsymbol{p}$ according to one observer, a Galilean transformation to an observer moving at the constant velocity $\boldsymbol{v}_o$ relative to this observer results in the Newtonian momentum $\boldsymbol{p}' = \boldsymbol{p} - m\boldsymbol{v}_o$ when the velocity of the particle is measured with respect to the second observer. Therefore, a particle with inertial mass will have a different momentum when measured *with respect to an observer in constant relative motion*. Using these different momenta in the de Broglie wavelength formula (2.54) results in *different* de Broglie wavelengths for the same particle according to the different observers. This is unlike the case of the photon, where the wavelength and momentum were the same for both observers. If these different wavelengths are used in the interference formula (1.79), it results in different predictions for the angular interference pattern, in clear contradiction of the experimental outcome. This indicates that the matter wave of a massive point particle is different from a light wave, where Galilean relativity assumed that the wavelength λ of the light was the same for all observers.

In order to resolve this problem, de Broglie assumed that the wavelength should be determined using the particle's velocity measured *with respect to the physical apparatus interacting with the particle*. As an example, in the case of electron diffraction the velocity of the electrons passing through the interferometer will be measured with respect to the interferometer. This ad hoc solution will be replaced in the next chapter where the correct result of a Galilean transformation on the matter wave is developed.

2.4 Bohr and the atom

As the atomic model received more experimental scrutiny there were yet more issues with classical physics. These led Bohr to model the atom in a non-Newtonian way.

2.4.1 Atomic spectra and the Rydberg constant

The lightest atomic element is identified as hydrogen. A dilute hydrogen gas can be easily stimulated to emit light by passing an electric current through it. All elements can be similarly stimulated to emit light, but hydrogen provides the simplest case for study with a set of very distinct and easily enumerated wavelengths. As such, it provides important insights into basic atomic structure.

In the 1880s Balmer and Rydberg found an unusual empirical formula which provides the wavelengths at which elements emit light. This formula is based on integers, and for hydrogen is given by

$$\begin{aligned}
\frac{1}{\lambda} &= R_y\left(\frac{1}{1^2} - \frac{1}{n^2}\right) \text{ where } n = 2,\ 3, \ldots \text{ (Lyman series)},\\
\frac{1}{\lambda} &= R_y\left(\frac{1}{2^2} - \frac{1}{n^2}\right) \text{ where } n = 3, 4, \ldots \text{ (Balmer series)},\\
\frac{1}{\lambda} &= R_y\left(\frac{1}{3^2} - \frac{1}{n^2}\right) \text{ where } n = 4, 5 \ldots \text{ (Paschen series)},\\
&\ldots,
\end{aligned} \tag{2.60}$$

where R_y is known as the *Rydberg constant* and is given by $\approx 1.1 \times 10^5$ cm^{-1}. Only the Balmer series of wavelengths is visible, while the Lyman series lies in the ultraviolet and the Paschen and higher series lie in the infrared. Considerable theoretical effort was then directed at explaining these *discrete wavelengths* and their relationship to sets of integers.

2.4.2 Keplerian atoms and classical instability

From early experiments it was known that atoms could become ionized, breaking into equal amounts of positive and negative charges, the latter composed of point-like electrons that were much less massive than the atom. It was natural to posit that the neutral atom itself is not fundamental, but rather formed from some kind of composite of these positive and negative charges. However, early models of the electrically neutral atom struggled to explain why the atom was stable. An obvious model of the atom is similar to the Kepler model of the solar system, with the light negatively charged electrons undergoing the attractive Coulomb force (1.34) and moving in orbits around a compact and heavier positive nucleus. However, in such a Keplerian model the Larmor radiation formula (1.118) predicts that the classical electron should collapse into the nucleus through emission of radiation. For an initial orbit of radius R the collapse time in Gaussian units is given by $T = m^2R^3c^3/4q^4$. The mass of the electron $m \approx 9 \times 10^{-28}$ g, the charge $q = 4.8 \times 10^{-10}$ esu, and the experimentally observed size of an atom, $R \approx 5 \times 10^{-9}$ cm, yields $T \approx 10^{-11}$ s. In effect, the Keplerian model of the atom is *extremely unstable* according to the combination of Newtonian physics and electromagnetism, which predicts the electron will *rapidly* spiral into the nucleus, something that is clearly not the case.

This problem led Thomson to develop an alternative model of the atom in which the point-like electrons were embedded in a tenuous extended positively charged cloud, like raisins in a pudding. While stable, such a model had problems explaining the particular wavelengths of light emitted by energetically excited hydrogen. Worse yet, subsequent experiments by Rutherford, where positively charged alpha particles were passed through a thin gold foil, demonstrated that the atom indeed had a *dense positive nucleus* surrounded by a *tenuous cloud of negatively charged electrons*, the opposite of Thomson's model. This simply added to the crisis in theoretical physics, as it now had to explain how the Keplerian atom is stable.

2.4.3 Bohr's model of the atom

Bohr attempted to understand the electron orbits in the hydrogen atom using a method that can be cast in terms of wave–particle duality. Bohr's method predated de Broglie's hypothesis, but in what follows Bohr's results will be found using the de Broglie wavelength. His approach was equivalent to assuming the presence of a wave associated with each possible orbit. For simplicity, only circular orbits will be considered. A circular orbit has the circumference $2\pi R$, and fitting a simple harmonic wave with wavelength λ *smoothly* around the circumference requires that the allowed wavelengths have the form $n\lambda = 2\pi R$, so that

$$\lambda = \frac{2\pi R}{n}, \tag{2.61}$$

where n is an integer greater than zero. Equating this with the de Broglie wavelength, $\lambda = h/mv$, where m is the mass of the electron and v is its orbital velocity, immediately yields the orbital angular momentum L of the electron,

$$\lambda = \frac{2\pi R}{n} = \frac{h}{mv} \implies mvR = L = n\frac{h}{2\pi} = n\hbar, \quad (n = 1, 2, \ldots). \tag{2.62}$$

In this model of the atom the orbital angular momentum of the electron is therefore an *integer multiple* of $\hbar$, which is the essence of Bohr's original assumption. Borrowing Planck's terminology, the orbital angular momentum of the electron is said to be *quantized* in units of $\hbar$. Connecting this result to classical physics is made by using the Newtonian result (1.115) for the case that $\beta = q^2$, where q is the charge of the nucleus and $-q$ is the charge of the electron. The result is

$$L = n\hbar = \sqrt{mq^2 R} \implies R_n = \frac{n^2\hbar^2}{mq^2}, \quad (n = 1, 2, \ldots). \tag{2.63}$$

The allowed radii R_n for the electron are determined by the integer n and are referred to as the *Bohr orbits*. The smallest allowed Bohr orbit occurs for $n = 1$, and is given by

$$a_0 = \frac{\hbar^2}{mq^2} = 5.3 \times 10^{-9}\ \text{cm}, \tag{2.64}$$

which is the experimentally inferred size of a hydrogen atom. The smallest orbital distance is referred to as the *Bohr radius*, while the larger radii are referred to as *outer shells*.

Equally important, these radii can be combined with the classical expression (1.114) for the energy to obtain the energy associated with each orbit, given by

$$E_n = -\frac{1}{2}\frac{q^2}{R_n} = -\frac{mq^4}{2\hbar^2 n^2} = -\frac{E_o}{n^2}. \tag{2.65}$$

The resulting energies available to the electron are *not continuous* as in classical physics. Instead, they are *discrete* and indexed by the positive integer n, which is referred to as a *quantum number*. The energy is therefore said to be *quantized*. The lowest or most negative energy occurs when $n = 1$, and is given by

$$E_1 = -E_o = -\frac{mq^4}{2\hbar^2} \approx -2.18 \times 10^{-11}\ \text{erg} = -13.6\ \text{eV}. \tag{2.66}$$

This is referred to as the *ground state energy* of the electron in the hydrogen atom, since there are no lower energies available to the electron *in this approach*. This result presented a reason why the Keplerian atom is stable by arguing that there was a ground state energy below which the electron could not go. However, this method did not explain why the electron in the ground state could not radiate away its remaining energy since it is still undergoing a centripetal acceleration. In addition, the assumption of a simple circular orbit ignored the possibility of more complex orbits.

Energy levels associated with integers greater than one are referred to as *excited states* since they correspond to energies higher than the ground state. Since energy must still be conserved, result (2.65) means that the electron can change energy levels only through the absorption or emission of a *discrete quantum of energy*, borrowing Planck's terminology again. Such a discrete change in energy is often referred to as a *quantum jump*. In the case of the atom, the absorption or emission of light can be responsible for changing the energy level of the electron. In such a case, Einstein's postulates for the photoelectric effect require that the energy of the absorbed or emitted photon, given by $\hbar\omega$, must account for the difference between the two energy levels. If the electron is in the initial energy level associated with n_i and transits to the final energy level associated with n_f, then for the case of an initial excited state, $n_\text{i} > n_\text{f}$, the energy of the emitted photon is given by

$$E = \hbar\omega = \frac{2\pi\hbar c}{\lambda} = E_{n_\text{i}} - E_{n_\text{f}} = \frac{mq^4}{2\hbar^2}\left(\frac{1}{{n_\text{f}}^2} - \frac{1}{{n_\text{i}}^2}\right), \tag{2.67}$$

so that the wavelengths of light emitted from a Hydrogen atom are given by the formula

$$\frac{1}{\lambda} = \left(\frac{mq^4}{4\pi\hbar^3 c}\right)\left(\frac{1}{{n_\text{f}}^2} - \frac{1}{{n_\text{i}}^2}\right) = R_y\left(\frac{1}{{n_\text{f}}^2} - \frac{1}{{n_\text{i}}^2}\right), \tag{2.68}$$

where

$$R_y = \frac{mq^4}{4\pi\hbar^3 c} \approx 1.1 \times 10^5\ \text{cm}^{-1}. \tag{2.69}$$

Results (2.68) and (2.69) exactly match the empirical formula (2.60) as well as the observed value of Rydberg's constant R_y. This was considered a powerful verification of the emerging nature of quantum mechanics.

2.5 The need for further development

The derivation of Bohr's results (2.68) and (2.69) using wave–particle duality and the concept of *quantized angular momentum* (2.62) constituted one of the important successes of early quantum mechanics. Result (2.68) shows that only a discrete set of frequencies accompanies the emission of light from the electron, and this set of frequencies is therefore referred to as an *emission spectrum*. It follows that these frequencies are also the same ones that will be absorbed as the electron goes from a lower to a higher energy level, and so (2.68) also describes the *absorption spectrum* of the atom. Formulas of the form (2.68) involving integers are typical of quantum mechanical spectra. In particular, the experimentally observed *discrete emission and absorption spectra* of atoms had defied explanation prior to the advent of quantum mechanics.

However, there was a growing list of observed atomic properties that Bohr's simple model was unable to explain. For example, the change in the wavelengths of emitted light that occurs when excited hydrogen is placed in a strong magnetic field, referred to as the Zeeman effect, had no immediate explanation. Furthermore, the results obtained so far, ranging from the photoelectric effect to atomic spectra, have come in a haphazard manner by merging particle and wave properties within a classical framework. Although Bohr's model helped confirm the utility of de Broglie's matter wave hypothesis, by the mid-1920s a more formal development of quantum mechanics and the concept of the matter wave was necessary to consolidate and further develop these early successes.

References and recommended further reading

Classical statistical mechanics is presented in

- Greiner W, Neise L, and Stöcker H 1995 *Thermodynamics and Statistical Mechanics* (New York: Springer)
- McQuarrie D 1973 *Statistical Thermodynamics* (Mill Valley: University Science Books)

Planck's original papers on blackbody radiation are available in

- Planck M 1972 *Original Papers in Quantum Physics* (London: Taylor and Francis)

Blackbody radiation and Planck's quantization condition are also presented in

- Tipler P and Llewellyn R 2012 *Modern Physics* 6th edn (New York: Freeman)

Lenard's experimental work on the photoelectric effect was presented in

- Lenard P 1902 Ueber die lichtelektrische Wirkung *Ann. Phys.* **313 5** (https://doi.org/10.1002/andp.19023130510)

Einstein's original work on the photoelectric effect is found in

- Einstein A 1905 Über einen die Erzeugung und Verwandlung des Lichtes betreffenden heuristischen Gesichtspunkt *Ann. Phys.* **17 132** (https://doi.org/10.1002/andp.19053220607)

The Compton effect was first reported in

- Compton A 1923 A quantum theory of the scattering of x-rays by light elements *Phys. Rev.* **21 483** (https://doi.org/10.1103/PhysRev.21.483)

The matter wavelength formula was formally presented by de Broglie in his doctoral thesis,

- de Broglie L 1924 *Researches on the Quantum Theory* (Paris: Paris University)

The guiding field interpretation of the matter wave was first presented in

- Born M 1926 Quantenmechanik der Stoßvorgänge *Zs. Phys.* **38 803** (https://doi.org/10.1007/BF01397184)

Rydberg presented his results on atomic spectra in

- Rydberg J 1889 Investigations of the composition of the emission spectra of chemical elements *Proceedings of the Royal Swedish Academy of Science* **23 1**

Rutherford presented his results on atomic scattering in

- Rutherford E 1911 LXXIX. The scattering of α and β particles by matter and the structure of the atom *Lon. Edin. Dub. Phil. Mag. J. Sci.* **21 669** (https://doi.org/10.1080/14786440508637080)

Bohr presented his theory of atomic structure in

- Bohr N 1913 I. On the constitution of atoms and molecules *Phil. Mag.* Ser. 1 **26 1** (https://doi.org/10.1080/14786441308634955)

Bohr presented the correspondence principle in

- Bohr N 1923 Über die Anwendung der Quantentheorie auf den Atombau *Zs. Phys.* **13 117** (https://doi.org/10.1007/BF01328209)

IOP Publishing

A Concise Introduction to Quantum Mechanics (Second Edition)

Mark S Swanson

Chapter 3

The wave function and observable quantities

Initially, it was hoped that adapting simple matter wave techniques, such as those used for Bohr's atom, might be sufficient to explain the growing catalog of quantum phenomena. In this regard Bohr and Sommerfeld developed a predecessor to modern quantum mechanics based on the postulate that the action-angle variables of classical mechanics are integer multiples of $\hbar$, often referred to as the *old quantum theory*. Despite some successes, this approach did not provide a consistent version of quantum mechanics, and it was unable to explain such phenomena as the Zeeman effect in atomic spectroscopy. It became evident that wave–particle duality required a more formal approach.

In the development of formal quantum mechanics it was necessary to understand the role and the meaning of the wave function as well as its relationship to experimentally measurable quantities. Such quantities or aspects of a system are known generically as *observables*. In this chapter specific properties of the wave function and its associated observables will be developed in the context of the *Copenhagen interpretation* of quantum mechanics. This interpretation is historically understood as stemming from a dialog between Bohr and his student Heisenberg in the 1920s, but did not become part of the lexicon of physics until after the Second World War. The Copenhagen interpretation has controversial aspects, leading some physicists to declare it incomplete and to advocate for alternative interpretations. Nevertheless, the Copenhagen interpretation is the most commonly used basis for understanding the nature of quantum phenomena, and so it will be presented in this chapter. Some of its controversial aspects will be sketched briefly but in no way resolved in this text.

3.1 Basic properties of the wave function

Quantum mechanics abandons Newtonian determinism and introduces the use of probability to predict the outcome of quantum mechanical processes and measurements. The relationship of probability to the wave function is key to the formulation

doi:10.1088/978-0-7503-5663-3ch3

of quantum mechanics. When the wave function is expressed in terms of position, this version of quantum mechanics is often referred to as *wave mechanics*. In that regard, the nature of probability and its connection to the Galilean observers of Newtonian physics places profound restrictions on the structure of wave mechanics.

3.1.1 Wave mechanics and probability

The first goal of wave mechanics is to find the *probability density* $\mathcal{P}(\boldsymbol{x}, t)$ for observing a particle at the point $\boldsymbol{x}$ at the time t. The probability density is found from the wave function $\Psi(\boldsymbol{x}, t)$ for the particle. It is therefore critical to understand how the wave function determines the probability density, and the first step in this process is to obtain an understanding of the nature of the wave function $\Psi(\boldsymbol{x}, t)$, including its general mathematical properties.

This starts by defining what a probability density means. The general physical system under consideration is that of a point-like particle located in some spatial volume. It is often the case that quantum mechanics is applied to models that are one-dimensional rather than three-dimensional. As a result, this spatial volume will be considered to be n-dimensional. If a particle detector of infinitesimally small n-dimensional volume $\mathrm{d}^n x$ is switched on at the time t at some point $\boldsymbol{x}$ in the volume, then the infinitesimal probability $\mathrm{d}P(\boldsymbol{x}, t)$ of observing the particle in the detector volume $\mathrm{d}^n x(\boldsymbol{x})$ at the time t is given by

$$\mathrm{d}P(\boldsymbol{x}, t) = \mathrm{d}^n x\, \mathcal{P}(\boldsymbol{x}, t). \tag{3.1}$$

It is assumed that the reader is familiar with the basic aspects of probability. If the particle is stable and is not destroyed in some physical process, then this probability density must also have the feature that summing up *all* the infinitesimal probabilities of the particle being *somewhere* in the volume *must result in unity at all times* as long as the particle remains in that volume. Mathematically, this means that

$$\int \mathrm{d}P(\boldsymbol{x}, t) = \int_V \mathrm{d}^n x\, \mathcal{P}(\boldsymbol{x}, t) = 1, \tag{3.2}$$

where the integral is over the entirity of the spatial volume V available to the particle. Requirement (3.2) is referred to as *normalization*, and it is similar to the normalization of the probabilities described by the partition function (2.3) in statistical mechanics. Statement (3.2) is simply the requirement that the probability of finding the particle in the volume V is unity. In what follows the subscript V on the integral will be implicit. In order to satisfy (3.2) the probability density must be an integrable function.

In order to be interpreted as a probability density, the function $\mathcal{P}(\boldsymbol{x}, t)$ *must be positive-definite at all locations at all times* since negative probability has no meaning. The probability density therefore must have the property that

$$\mathcal{P}(\boldsymbol{x}, t) \geqslant 0 \quad \text{for all } \boldsymbol{x} \text{ and } t, \tag{3.3}$$

so that $\mathcal{P}(\boldsymbol{x}, t)$ is a *positive-definite function*. In addition, in order for quantum mechanics to be logically consistent with Newtonian mechanics, it must be that *all observers related by the Galilean transformation of* (1.32) *will agree upon the probability of observing the particle at a specific location.*

3.1.2 The wave function and Galilean relativity

The constraint that all Galilean observers agree upon the probability density for observing a particle can be stated mathematically as the requirement that the probability density function is a *Galilean scalar*. In addition to depending on position and time, it is possible that the probability density depends on the momentum $\boldsymbol{p}$ of the massive particle which it is describing, so that $\mathcal{P} = \mathcal{P}(\boldsymbol{p}, \boldsymbol{x}, t)$. The requirement that the probability density is a Galilean scalar means that the probability density in the $\mathcal{O}'$ frame, which is moving at the relative velocity $\boldsymbol{v}_o$ with respect to the $\mathcal{O}$ frame, is found from the probability density $\mathcal{P}(\boldsymbol{p}, \boldsymbol{x}, t)$ in the $\mathcal{O}$ frame according to

$$\mathcal{P}'(\boldsymbol{p}', \boldsymbol{x}', t') = \mathcal{P}(\boldsymbol{p} - m\boldsymbol{v}_o, \boldsymbol{x} - \boldsymbol{v}_o t, t) = \mathcal{P}(\boldsymbol{p}, \boldsymbol{x}, t). \tag{3.4}$$

This insures that both observers agree upon the probability density at all points in their respective coordinate systems and that the momentum of a massive particle is no longer required to be measured in the frame of the experimental apparatus. The restriction to a Galilean scalar expressed by (3.4) places a significant constraint on how the probability density for a massive particle is found from the wave function.

3.1.3 The analogy to light interference

In order to identify the general nature of the wave function for a massive point particle it is very useful first to apply wave–particle duality to the case of light interference, which was analysed in (1.100) for the case that a monochromatic light wave is passed through a two-slit device. In the nascent wave–particle duality introduced in the previous chapter, the photoelectric effect was explained by treating a frequency f electromagnetic wave as composed of photons, each with the energy $E = hf = \hbar\omega$ and momentum $p = h/\lambda$. It is instructive to express the monochromatic electric and magnetic waves in terms of the momentum and energy of the individual photons comprising them. This begins by noting that the vector versions of the momentum $\boldsymbol{p}$ of the photons and the wave number $\boldsymbol{k}$ of the wave are related through the de Broglie relation,

$$p = \frac{h}{\lambda} = \frac{2\pi\hbar}{\lambda} = \hbar k \implies \boldsymbol{k} = \frac{\boldsymbol{p}}{\hbar}. \tag{3.5}$$

Similarly, the energy and the angular frequency of the photon are related,

$$E_{\mathrm{p}} = \hbar\omega \implies \omega = \frac{E_{\mathrm{p}}}{\hbar}. \tag{3.6}$$

As a result, the positive-moving version of the electromagnetic wave (1.72) can be written

$$\mathbf{A}(\boldsymbol{x}, t) = \mathbf{A}_0(\boldsymbol{k})\sin\left(\boldsymbol{k}\cdot\boldsymbol{x} - \omega t + \frac{1}{2}\phi\right) = \mathbf{A}_0(\boldsymbol{p})\sin\left((\boldsymbol{p}\cdot\boldsymbol{x} - E_{\mathrm{p}}t + \frac{1}{2}\hbar\phi)\Big/\hbar\right). \quad (3.7)$$

Light arriving at the screen from the two separated slits is out of phase, so that (1.78) gives the resultant wave striking the screen,

$$\begin{aligned}\mathbf{A}_{\mathrm{r}} &= \mathbf{A}_0\sin\left((\boldsymbol{p}\cdot\boldsymbol{x} - E_{\mathrm{p}}t + \frac{1}{2}\hbar\phi)\Big/\hbar\right) + \mathbf{A}_0\sin\left((\boldsymbol{p}\cdot\boldsymbol{x} - E_{\mathrm{p}}t - \frac{1}{2}\hbar\phi)\Big/\hbar\right)\\ &= 2\mathbf{A}_0\cos(\frac{1}{2}\phi)\sin\left((\boldsymbol{p}\cdot\boldsymbol{x} - E_{\mathrm{p}}t)/\hbar\right).\end{aligned} \quad (3.8)$$

The resultant light wave (3.8) forms an *intensity pattern* on the screen, which is quantified by the Poynting vector (1.86). Using the amplitude of (3.8) in (1.100) gives the *instantaneous* intensity of the resultant light wave,

$$\mathbf{S} = \hat{\boldsymbol{n}}_p\left(\frac{cp^2A_0^{\,2}(\boldsymbol{p})}{\pi\hbar^2}\right)\cos^2(\frac{1}{2}\phi)\sin^2\left((\boldsymbol{p}\cdot\boldsymbol{x} - E_{\mathrm{p}}t)/\hbar\right), \quad (3.9)$$

where $\hat{\boldsymbol{n}}_{\mathrm{p}} = \boldsymbol{p}/p$. This is positive-definite, as both a probability distribution and an energy pattern must be. Using (1.81) shows that it is also a Galilean scalar. This follows by noting that photon momentum $p = h/\lambda$ is invariant, while the Galilean transformed energy $E_{\mathrm{p}}' = c'p' = (1 - v_o/c)cp = (1 - v_o/c)E_{\mathrm{p}}$ undergoes the Doppler shift (1.83) for the case of an observer moving in the same direction as the light wave. As a result, *all Galilean observers* will agree on the intensity pattern formed on the screen by the interference pattern.

The next step in developing quantum mechanics is to examine the light pattern *from the particle perspective*. This allows the intensity of the light wave striking the screen given by (3.9) to be understood as quantifying the *number of photons per unit time per unit area* arriving at a particular location on the screen. Wave–particle duality therefore allows viewing the intensity pattern of the light wave both in terms of photons striking the screen at the most probable angles *and* in terms of the energy delivered by the resultant wave subsequent to interference. In effect, (3.9) serves as the *probability pattern for photons striking the screen*. The key observation is that this probability pattern is proportional to the square of the resultant electromagnetic wave associated with the photons.

3.1.4 The need for a complex valued wave function

It was argued in the previous chapter that wave–particle duality can also be used to explain electron diffraction, where charged point particles are associated with a wave via the de Broglie wavelength formula (2.54). The impact pattern of the electrons is the result of the matter wave of each electron passing through the diffraction grating and forming an interference pattern. The most probable impact angles on the screen are determined by using the de Broglie wavelength in the *same* interference formula (1.79) that was used for light. The physical interpretation of the interference pattern created by the matter wave is that it provides the *probability* for the electron striking the photoplate at a given angle. Where there is destructive interference, there is a low probability of the electron striking the photoplate and leaving an image, while constructive interference yields a high probability of striking the photoplate and leaving an image. In that regard, both the energy of the diffracted light wave and the probability of the electron impact distribution on the screen are required to be *positive-definite*.

It is important to note that the tentative explanation of the electron diffraction pattern using the analogy to light interference presented in the previous chapter implicitly borrows the properties of electromagnetic energy (1.85), where the *wave squared* was used to find the light intensity. The act of squaring the sum of the electron waves from both slits was implicit in the explanation, since, like the case with light, squaring the resultant wave results in a probability density that is positive-definite at all points *as long as the wave is represented by a real function.* In the case of light interference the electromagnetic wave used to find the energy density of the light wave was represented by the *real function* (3.7) since both the electric and magnetic fields are *observable* in other experimental contexts and must therefore correspond to a *real quantity.*

Following the example of light wave and photon duality, the wave function of the electron could tentatively be defined in terms of an identical recipe, so that the probability density is found from a real valued wave function,

$$\mathcal{P}(\boldsymbol{p}, \boldsymbol{x}, t) \propto \Psi^2(\boldsymbol{p}, \boldsymbol{x}, t). \tag{3.10}$$

Choosing the matter wave function to be similar to that for light,

$$\Psi(\boldsymbol{p}, \boldsymbol{x}, t) = \Psi_0 \sin\left((\boldsymbol{p} \cdot \boldsymbol{x} - E_{\mathrm{p}}t)/\hbar\right), \tag{3.11}$$

immediately yields

$$\mathcal{P}(\boldsymbol{p}, \boldsymbol{x}, t) \propto \Psi_0^2 \sin^2\left((\boldsymbol{p} \cdot \boldsymbol{x} - E_{\mathrm{p}}t)/\hbar\right). \tag{3.12}$$

In order for the argument of the sine function to be dimensionless $\boldsymbol{p}$ must have the units of momentum and E_{p} must have the units of energy. It follows that the choice for the case of a massive Newtonian particle must be $E_{\mathrm{p}} = p^2/2m$. Expression (3.12) then yields a positive-definite probability density as well as allowing the required probability interference effects by superimposing two waves that are out of phase with each other.

However, there is a fatal flaw with the recipe (3.12) for massive Newtonian particles. Because nonrelativistic quantum mechanics must be consistent with Newtonian mechanics, its predictions for the probability must also be invariant under the Galilean transformation (1.32). This requirement is implemented in expression (3.7) for the light wave since the photon momentum $\boldsymbol{p}$ is invariant for all observers and the energy of the photon undergoes the Dopper effect (1.83). For the case of (3.12) a Galilean transformation gives

$$\begin{aligned} \mathcal{P}'(\boldsymbol{p}', \boldsymbol{x}', t') &= \mathcal{P}(\boldsymbol{p} - m\boldsymbol{v}_o, \boldsymbol{x} - \boldsymbol{v}_o t, t) \\ &= \Psi_0^2 \sin^2\left((\boldsymbol{p} \cdot \boldsymbol{x} - E_{\mathrm{p}}t - m\boldsymbol{v}_o \cdot \boldsymbol{x} + \frac{1}{2}mv_o^2)\Big/\hbar\right) \neq \mathcal{P}(\boldsymbol{p}, \boldsymbol{x}, t). \end{aligned} \tag{3.13}$$

It is not possible to alter the expression for E_p or the argument of the sine function in such a way that eliminates the extra terms generated by the Galilean transformation. The wave function associated with a massive particle must therefore be different from an electromagnetic wave. The eventual solution is to allow the wave function Ψ to be *complex valued.* The next two sections briefly present the relevant properties of complex variables and the critically important nature of *function spaces.*

3.2 A review of complex variables

Complex variables play an extremely important role in quantum mechanics. Complex analysis is a rich subject and the following brief review only sketches the results immediately relevant to quantum mechanics.

3.2.1 Basic definitions

A *complex number*, denoted $z = x + iy$, consists of two real numbers, x and y, and the imaginary number i, which satisfies $i^2 = -1$. A complex number z can be visualized as a point in the *complex plane*, given by the two values x and y, which are known, respectively, as the real and the imaginary part of z. Two complex numbers are equal *if and only if both their real and imaginary parts are identical.* The *complex conjugate* of z is denoted z^* and is given by reversing the sign of i, so that if $z = x + iy$ then $z^* = x - iy$. Complex conjugation has the obvious property that $z^{**} = z$. It follows that

$$z^*z = (x - iy) \cdot (x + iy) = x^2 - i^2y^2 = x^2 + y^2, \tag{3.14}$$

which is a positive-definite expression. For that reason it is often written $z^*z = |z|^2$, where $|z| = +\sqrt{x^2 + y^2}$ is known as the *modulus* of z.

An arbitrary complex number, $z = x + iy$, can be expressed as

$$z = r\cos\theta + ir\sin\theta. \tag{3.15}$$

Equivalence requires that

$$x = r\cos\theta, \quad y = r\sin\theta, \tag{3.16}$$

which have the solutions

$$\begin{aligned} r^2 = r^2\sin^2\theta + r^2\cos^2\theta = x^2 + y^2 = |z|^2 \quad &\Longrightarrow \quad r = |z|, \\ \tan\theta = \frac{r\sin\theta}{r\cos\theta} = \frac{y}{x} \quad &\Longrightarrow \theta = \tan^{-1}\frac{y}{x}. \end{aligned} \tag{3.17}$$

The solution for θ is not unique since $\theta \to \theta + 2\pi n$, where n is an integer, yields $\tan(\theta + 2\pi n) = \tan\theta$ and corresponds to the same complex number.

Complex numbers with the exponential form $\phi = e^{i\theta}$, where θ is a real number, play a critical role in quantum mechanics and are often referred to as a *phase factor* in physics. Using the property of the exponential shows that the modulus squared of the phase is given by

$$|\phi|^2 = |e^{i\theta}|^2 = e^{-i\theta}e^{i\theta} = e^0 = 1. \tag{3.18}$$

This can be understood by using the Taylor series expansion for $e^{i\theta}$ and separating the real and imaginary parts. The imaginary parts occur for odd powers of i, while the real parts are given by the even powers of i, resulting in two power series that sum to $\cos\theta$ and $\sin\theta$,

$$e^{i\theta} = \sum_{n=0}^{\infty}\frac{(i\theta)^n}{n!} = \sum_{n=0}^{\infty}\frac{(-1)^n\theta^{2n}}{(2n)!} + i\sum_{n=0}^{\infty}\frac{(-1)^n\theta^{2n+1}}{(2n+1)!} = \cos\theta + i\sin\theta, \tag{3.19}$$

which is known as the *Euler relation*. It is important to note that for n an integer relation (3.19) gives

$$e^{i(\theta+2\pi n)} = e^{i\theta}e^{2\pi in} = e^{i\theta}(\cos(2\pi n) + i\sin(2\pi n)) = e^{i\theta}, \tag{3.20}$$

so that $e^{i2\pi n} = 1$, as well as the famous relation $e^{i\pi} + 1 = \cos\pi + 1 = 0$.

Result (3.19) allows an arbitrary complex number to be expressed in *polar form*,

$$z = re^{i\theta}, \tag{3.21}$$

where r and θ are real numbers. It follows that $z^*z = re^{-i\theta}re^{i\theta} = r^2 = x^2 + y^2$, so that $r = +\sqrt{x^2 + y^2}$ is revealed as the modulus of the complex number. The Euler relation gives the two identifications previously derived, $r\cos\theta = x$ and $r\sin\theta = y$, and the quotient $\sin\theta/\cos\theta = \tan\theta = y/x$. If the polar form is known, then the relations can be reversed to find $x = r\cos\theta$ and $y = r\sin\theta$.

3.2.2 Complex functions, analyticity, and the Cauchy–Riemann conditions

A complex function $f(z) = f(x + iy)$ maps a complex number into another complex number, which is another point in the complex plane. As a result, the function can be written

$$f(z) = u(x, y) + iv(x, y), \tag{3.22}$$

where $u(x, y)$ and $v(x, y)$ are real valued functions.

Similarly to complex numbers, a complex function $f(z)$ can also be expressed in polar form. If $f(z) = f(x + iy) = u(x, y) + iv(x, y)$, then the polar form is given by

$$f(z) = \zeta(x, y)e^{i\varphi(x, y)}, \tag{3.23}$$

where

$$\zeta(x, y) = \sqrt{u^2(x, y) + v^2(x, y)}, \quad \varphi(x, y) = \tan\left(\frac{v(x, y)}{u(x, y)}\right). \tag{3.24}$$

The demonstration of this is identical to the one for the complex variable. If the polar form of the function is known then $u(x, y) = \zeta(x, y)\cos\varphi(x, y)$ and $v(x, y) = \zeta(x, y)\sin\varphi(x, y)$.

Defining the derivative of $f(z)$ requires picking an infinitesimal complex number $\epsilon = \epsilon_R + i\epsilon_I$ and writing

$$\frac{\mathrm{d}f(z)}{\mathrm{d}z} = \lim_{\epsilon\to 0}\frac{f(z+\epsilon)-f(z)}{\epsilon}. \tag{3.25}$$

If (3.25) exists and *is unique*, it must be independent of how the limit $\epsilon \to 0$ is taken. Starting with $\epsilon_{\mathrm{I}} = 0$ gives

$$\begin{aligned}\frac{\mathrm{d}f(z)}{\mathrm{d}z} &= \lim_{\epsilon_{\mathrm{R}}\to 0}\frac{u(x+\epsilon_{\mathrm{R}}, y) + iv(x+\epsilon_{\mathrm{R}}, y) - u(x, y) - iv(x, y)}{\epsilon_{\mathrm{R}}} \\ &= \frac{\partial u}{\partial x} + i\frac{\partial v}{\partial x}.\end{aligned} \tag{3.26}$$

Starting with $\epsilon_{\mathrm{R}} = 0$ gives

$$\begin{aligned}\frac{\mathrm{d}f(z)}{\mathrm{d}z} &= \lim_{\epsilon_{\mathrm{I}}\to 0}\frac{u(x, y+\epsilon_{\mathrm{I}}) + iv(x, y+\epsilon_{\mathrm{I}}) - u(x, y) - iv(x, y)}{i\epsilon_{\mathrm{I}}} \\ &= -i\frac{\partial u}{\partial y} + \frac{\partial v}{\partial y}.\end{aligned} \tag{3.27}$$

If (3.26) and (3.27) are to be identical, it must be that

$$\frac{\partial u}{\partial x} = \frac{\partial v}{\partial y} \quad \text{and} \quad \frac{\partial u}{\partial y} = -\frac{\partial v}{\partial x}, \tag{3.28}$$

which are known as the *Cauchy–Riemann conditions.* If a complex function satisfies the Cauchy–Riemann conditions at a point in the complex plane and at every point in a neighborhood of that point, the function is said to be *analytic* at that point since it possesses a unique derivative. For example, the function e^{iz} is analytic everywhere, but the singular function $1/z$ is not analytic at $z = 0$.

3.2.3 Cauchy's theorem

It is often important to evaluate the integral of a complex function along a path in the complex plane, which is referred to as a *contour*. This is similar to integrating a vector function along a path in the real plane as in Stokes' theorem (1.55). The elements of the contour C are parameterized as $\mathrm{d}z = \mathrm{d}x + i\,\mathrm{d}y$, and this gives $f(z)\,\mathrm{d}z = (u + iv)(\mathrm{d}x + i\,\mathrm{d}y) = (u\,\mathrm{d}x - v\,\mathrm{d}y) + i(u\,\mathrm{d}y + v\,\mathrm{d}x)$. This can be visualized as a pair of separate line integrals in a real two-dimensional plane. In the case of the real term, a two-dimensional vector function $\boldsymbol{A}$ with the components $A_1(x, y) = u(x, y)$ and $A_2(x, y) = -v(x, y)$ is identified. In the imaginary term, a vector function $\boldsymbol{B}$ is defined with the components $B_1(x, y) = v(x, y)$ and $B_2(x, y) = u(x, y)$. A *closed contour* integral can now be analysed by treating it as two line integrals in the x–y plane. Denoting the surface of the disk formed by the closed contour as S and using Stokes' theorem (1.55) gives

$$\begin{aligned}\oint_C \mathrm{d}z\, f(z) &= \oint_C (u\,\mathrm{d}x - v\,\mathrm{d}y) + i(u\,\mathrm{d}y + v\,\mathrm{d}x) = \oint_C \boldsymbol{A}\cdot \mathrm{d}\boldsymbol{\ell} + i\oint_C \boldsymbol{B}\cdot \mathrm{d}\boldsymbol{\ell} \\ &= \int_S \mathrm{d}\boldsymbol{S}\cdot \nabla\times(\boldsymbol{A} + i\boldsymbol{B}) = \int_S \left\{-\left(\frac{\partial u}{\partial y} + \frac{\partial v}{\partial x}\right) + i\left(\frac{\partial u}{\partial x} - \frac{\partial v}{\partial y}\right)\right\}\mathrm{d}x\,\mathrm{d}y,\end{aligned} \tag{3.29}$$

where $\nabla \times \boldsymbol{A}$ has the component perpendicular to the disk given by $\partial A_2/\partial x - \partial A_1/\partial y = -(\partial v/\partial x + \partial u/\partial y)$. The result for $\boldsymbol{B}$ gives $\partial B_2/\partial x - \partial B_1/\partial y = \partial v/\partial x - \partial u/\partial y$. Result (3.29) vanishes by virtue of the Cauchy–Riemann conditions (3.28) *if the function is analytic everywhere* in the disk S. As a result, the closed contour integral of an analytic function is zero regardless of the contour chosen.

If $f(z)$ is analytic everywhere inside a closed contour C, then the function $g(z) = f(z)/(z - z_o)$ is said to have a *simple pole* at $z = z_o$. For a counter-clockwise or right-handed circular contour around z_o it follows that

$$\oint_C \mathrm{d}z\, g(z) = \oint_C \mathrm{d}z\, \frac{f(z)}{(z - z_o)} = 2\pi i f(z_o), \tag{3.30}$$

regardless of the radius of the circular contour. This is demonstrated by using the fact that $f(z)$ is analytic to expand it in a Taylor series around the singular point. This starts by writing $z = z_o + \delta z$, where the complex number δz is written in polar form $\delta z = re^{i\theta}$ using result (3.21). The Taylor series for $f(z)$ is then given by

$$f(z) = f(z_o + re^{i\theta}) = \sum_{n=0}^{\infty} \frac{1}{n!} r^n e^{in\theta} \frac{\mathrm{d}^n f(z_o)}{\mathrm{d}z_o{}^n}. \tag{3.31}$$

The denominator of the contour integral is given by $1/(z - z_o) = e^{-i\theta}/r$ and, since the contour is a circle of constant radius r, $\mathrm{d}z = ire^{i\theta}\,\mathrm{d}\theta$. This gives $\mathrm{d}z/(z - z_o) = i\,\mathrm{d}\theta$. The circular contour integral around z_o is therefore given by

$$\oint_C \mathrm{d}z\, \frac{f(z)}{z - z_o} = i \sum_{n=0}^{\infty} \frac{1}{n!} r^n \frac{\mathrm{d}^n f(z_o)}{\mathrm{d}z_o{}^n} \int_0^{2\pi} \mathrm{d}\theta\, e^{in\theta} = 2\pi i f(z_o). \tag{3.32}$$

The final step occurs since the integral over θ vanishes unless $n = 0$. Result (3.32) can be extended to any contour C that encloses the simple pole at $z = z_o$ and is known as *Cauchy's integral formula*. Result (3.32) was derived for a right-handed or counter-clockwise contour around the pole. If the contour is left-handed or clockwise around the pole, the result changes sign to $-2\pi i f(z_o)$. For the case that there are several different simple poles enclosed by a counter-clockwise contour Cauchy's integral theorem gives

$$\oint_C \mathrm{d}z\, \frac{f(z)}{(z - z_o)(z - z_1)} = 2\pi i \left(\frac{f(z_o) - f(z_1)}{z_o - z_1} \right). \tag{3.33}$$

It is often the case that an integral along of a function along the real line can be continued to a contour integral in the complex plane to allow (3.32) to be used. This is especially useful if a closed contour can be chosen for (3.32) such that the integrand vanishes off the real axis. If the integrand $g(z)$ appearing in (3.30) can be written

$$g(z) = e^{iaz} f(z), \tag{3.34}$$

where a is a positive parameter, then *Jordan's lemma* states that a semicircular contour C_R in the upper half plane, parameterized by $z = Re^{i\theta}$ where R is real and $0 \leqslant \theta \leqslant \pi$, is bounded by

$$\left| \int_{C_R} dz \, g(z) \right| \leqslant \frac{\pi}{a} M_R, \tag{3.35}$$

where M_R is the maximum value of $|f(Re^{i\theta})|$ along the semicircular contour. As a result, if

$$\lim_{R\to\infty} f(Re^{i\theta}) = 0, \tag{3.36}$$

then

$$\lim_{R\to\infty} \int_{C_R} dz \, e^{iaz} f(z) = 0. \tag{3.37}$$

For such a case, the integral along the real axis is reduced to identifying the simple poles of the integrand and invoking Cauchy's integral formula (3.32). If a is a negative number, then Jordan's lemma remains valid as long as the semicircular contour is in the *lower half plane*. The Jordan lemma is consistent with the observation that

$$\lim_{z_I\to\infty} e^{iaz} = 0 \tag{3.38}$$

as long as $a > 0$, while

$$\lim_{z_I\to-\infty} e^{-iaz} = 0 \tag{3.39}$$

as long as $a > 0$. This is often used as the reason for how the contour is closed, but the Jordan lemma is a more rigorous explanation. The Jordan lemma can be adapted to quarter circle contours as long as the radius tends to infinity and the argument of the function being integrated tends to zero in that limit.

3.3 Fourier analysis and function spaces

Fourier analysis is an indispensable mathematical tool used throughout physics and quantum mechanics in particular. The associated Fourier series and transforms are also the prototype for a very general and important mathematical structure known as a *function space*. Since the *wave function* $\Psi(\boldsymbol{x}, t)$ is the central feature of wave mechanics, it is critical to recognize and incorporate the properties of function space into the formal structure of quantum mechanics. The Fourier series representation of an arbitrary function is therefore a very useful place to begin.

3.3.1 The inner product of complex vectors

The first step in understanding a function space is to generalize the concept of the scalar product, familiar from working with position vectors, to the case of complex vectors. A complex vector $\boldsymbol{x}$ is simply a vector whose components x_j are complex numbers. The *inner product* of two n-dimensional complex vectors $\boldsymbol{x}$ and $\boldsymbol{y}$ is denoted $(\boldsymbol{x}, \boldsymbol{y})$ and is defined in a manner that reduces to the familiar scalar product for the case of real vectors, but also in a way that manifests the complex nature of the two vectors,

$$\boldsymbol{x} \cdot \boldsymbol{y} \equiv (\boldsymbol{x}, \boldsymbol{y}) = \sum_{j=1}^{n} x_j^* y_j. \tag{3.40}$$

In general, the inner product (3.40) is also *complex number*, but from its definition it satisfies

$$(\boldsymbol{x}, \boldsymbol{y})^* = \left(\sum_{j=1}^{n} x_j^* y_j \right)^* = \sum_{j=1}^{n} y_j^* x_j = (\boldsymbol{y}, \boldsymbol{x}). \tag{3.41}$$

Since the order of the vectors is critically important, the notation $(\boldsymbol{x}, \boldsymbol{y})$ for the inner product far more effectively represents its nature, and therefore will be adapted to the inner product of complex functions in the next subsection and to Dirac notation, which is introduced in chapter 6. For real vectors (3.40) reduces to the usual scalar product. However, even for complex vectors, it has the important property that

$$(\boldsymbol{x}, \boldsymbol{x}) = \sum_{j=1}^{n} |x_j|^2 \equiv |\boldsymbol{x}|^2 \geqslant 0, \tag{3.42}$$

so that $(\boldsymbol{x}, \boldsymbol{x})$ is a positive-definite real number. As a result, the quantity $|\boldsymbol{x}| = \sqrt{(\boldsymbol{x}, \boldsymbol{x})}$ can viewed as the *magnitude* or *norm* of the complex vector, just as $|\boldsymbol{x}| = \sqrt{\boldsymbol{x} \cdot \boldsymbol{x}}$ is the magnitude of a three-dimensional position vector. Definition (3.40) also inherits the *linear* nature of the inner product of vectors, so that

$$(\boldsymbol{x}, \alpha\, \boldsymbol{y} + \beta\, \boldsymbol{z}) = \alpha\, (\boldsymbol{x}, \boldsymbol{y}) + \beta\, (\boldsymbol{x}, \boldsymbol{z}), \tag{3.43}$$

where α and β are general complex numbers. Like the position vectors, two complex vectors are said to be *orthogonal* if their inner product vanishes.

The definition (3.40) of the inner product satisfies the very useful *Cauchy–Schwarz inequality*,

$$|\boldsymbol{x}|^2 |\boldsymbol{y}|^2 \geqslant |(\boldsymbol{x}, \boldsymbol{y})|^2. \tag{3.44}$$

This relation is proved by first defining the vector $\boldsymbol{w} = \boldsymbol{x} - \lambda \boldsymbol{y}$, where λ is a complex number. Using the fact that the magnitude of this vector is positive-definite, $(\boldsymbol{w}, \boldsymbol{w}) = |\boldsymbol{w}|^2 = |\boldsymbol{x} - \lambda \boldsymbol{y}|^2 \geqslant 0$, along with the linearity of the inner product, it follows that

$$(\boldsymbol{w}, \boldsymbol{w}) = |\boldsymbol{x}|^2 - \lambda\, (\boldsymbol{x}, \boldsymbol{y}) - \lambda^*\, (\boldsymbol{y}, \boldsymbol{x}) + \lambda^* \lambda\, |\boldsymbol{y}|^2 \geqslant 0. \tag{3.45}$$

The left-hand side is minimized by demanding that its derivatives with respect to λ and λ^* vanish, which gives

$$\lambda|y|^2 - (y, x) = 0 \implies \lambda = \frac{(y, x)}{|y|^2}, \tag{3.46}$$

along with its complex conjugate

$$\lambda^*|y|^2 - (x, y) = 0 \implies \lambda^* = \frac{(x, y)}{|y|^2}. \tag{3.47}$$

Substituting these expressions for λ and λ^* into (3.45) immediately gives the Cauchy–Schwarz inequality,

$$|x|^2|y|^2 \geqslant (x, y)(y, x) = |(x, y)|^2. \tag{3.48}$$

This inequality will provide important insights into function spaces and quantum mechanics.

3.3.2 The inner product of functions and the Cauchy–Schwarz inequality

The definition (3.40) can be adapted to define the *inner product of two complex valued functions*. For simplicity the two functions are denoted $f(x)$ and $g(x)$, where x is a single real variable. If the interval where the two complex functions are well behaved and are of interest is $(-L, L)$, then the infinitesimal subinterval $\Delta x = L/N$ can be defined, where N is an arbitrarily large integer. For an integer j, such that $-N \leqslant j \leqslant N$, the value of the complex function f at $x = j\ \Delta x$ is used to define the complex vector component $f_j = \alpha f(j\ \Delta x)$, where α is a constant of proportionality that will be chosen to regulate the inner product. This quantity becomes the jth component of the $(2N + 1)$-*dimensional complex vector* f, with a similar procedure performed for a second complex function g. The inner product of these two functions, denoted (f, g), is defined identically to (3.40), so that

$$(f, g) = \sum_{j=-N}^{N} f_j^* g_j. \tag{3.49}$$

Setting $|\alpha|^2 = L/N = \Delta x$ as the constant of proportionality and taking the limit $N \to \infty$ allows the identification $\Delta x \to \mathrm{d}x$, and the inner product of the two functions becomes a Riemann sum or *definite integral*,

$$(f, g) = \lim_{N\to\infty} \sum_{j=-N}^{N} \Delta x\, f^*(j\ \Delta x)\, g(j\Delta x) = \int_{-L}^{L} \mathrm{d}x\, f^*(x)\, g(x). \tag{3.50}$$

An important property of (3.50) is that the limit $N \to \infty$ creates an *infinite-dimensional* complex vector space to represent the functions. The definition (3.50) requires that the functions under consideration are *square integrable* over the interval, which simply means that (f, f) is both well-defined and results in a *finite and positive value*.

For a square integrable nonzero function f, (3.50) also allows the definition of a positive real valued *norm* of the function f, denoted $|f|$, from the expression

$|f|^2 = (f, f)$. Since the norm is a well-defined finite real positive number, the inner product (3.50) allows a statement of the Cauchy–Schwarz inequality *for functions,*

$$|f|^2 |g|^2 \geqslant |(f, g)|^2. \tag{3.51}$$

It is worth noting that the equality in (3.51) occurs for the case that $f(x)$ is proportional to $g(x)$, so that $f(x) = \alpha g(x)$, where α is a complex number.

The set of all *square integrable functions* form a *linear space of functions,* referred to as L^2, that plays a key role in quantum mechanics since these functions are *normalizable.* It is a space of functions in the sense that a linear combination *or superposition* of members of this set is also a member of the set. This is stated mathematically in the following way. If $f(x)$ and $g(x)$ are two members of L^2, then the function $h(x) = f(x) + g(x)$ is also a member of L^2. The proof starts by writing the inner product $(f, g) = x + iy$, where x and y are *real numbers,* so that both x^2 and y^2 are positive-definite. The Cauchy–Schwarz inequality (3.51) then gives

$$|(f, g)|^2 = x^2 + y^2 \leqslant |f|^2 |g|^2 \implies x^2 \leqslant |f|^2 |g|^2 - y^2, \tag{3.52}$$

which shows that $0 \leqslant y^2 \leqslant |f|^2 |g|^2$ since the right-hand side of the inequality must be positive-definite. Using this in the inequality for x^2 in (3.52) shows that

$$0 \leqslant x^2 \leqslant |f|^2 |g|^2 \implies -|f||g| \leqslant x \leqslant |f||g|. \tag{3.53}$$

The norm of the function $h(x)$ is given by

$$|h|^2 = (f + g, f + g) = |f|^2 + |g|^2 + (f, g) + (g, f) = |f|^2 + |g|^2 + 2x. \tag{3.54}$$

Using the inequality (3.53) for x shows that

$$\begin{aligned} &(|f| - |g|)^2 = |f|^2 + |g|^2 - 2|f||g| \leqslant |h|^2 \leqslant |f|^2 + |g|^2 + 2|f||g| = (|f| + |g|)^2 \\ &\implies 0 \leqslant |h| \leqslant |f| + |g|, \end{aligned} \tag{3.55}$$

where the second line is known as the *triangle inequality*. Since $|f|$ and $|g|$ are bounded by virtue of the square integrability of $f(x)$ and $g(x)$, this shows that $h(x)$ is also a square integrable function. As a result, the set of square integrable functions forms a *space of functions* since it is closed under addition. This is identical to a *vector space*, where the member vectors are closed under addition.

A *normalized* function $\hat{f}(x)$ is obtained from $f(x)$ by simply dividing it by its norm, so that $\hat{f}(x) = f(x)/|f|$ has a norm of unity. This is identical to dividing a vector by its magnitude to create a *unit vector*. For normalized functions the Cauchy–Schwarz inequality becomes

$$|(\hat{f}, \hat{g})|^2 \leqslant 1. \tag{3.56}$$

Generalizing these definitions to functions on a space of higher dimensions is straightforward by simply increasing the space of integration in the inner product (3.50).

3.3.3 The Kronecker delta

In the description of vector and function spaces, it is very useful to use the *Kronecker delta* δ_{jk}, which was defined earlier in (1.27) as

$$\delta_{jk} \equiv \begin{cases} 1 \text{ if } j = k \\ 0 \text{ if } j \neq k \end{cases}, \tag{3.57}$$

where both j and k are integers. The definition (3.57) has the easily demonstrated properties that, for N an integer such that $-N \leqslant k \leqslant N$,

$$\sum_{j=-N}^{N} \delta_{jk} = 1 \quad \text{and} \quad \sum_{j=-N}^{N} \delta_{jk}\, x_j = x_k. \tag{3.58}$$

If $k < -N$ or $k > N$ both expressions in (3.58) are zero. The Kronecker delta also appears in the context of the matrix formulation of quantum mechanics presented in chapter 6.

3.3.4 Orthonormal functions

Similarly to vectors, if two functions f and g have an inner product that is zero, $(f, g) = 0$, then they are said to be *orthogonal*. In effect, the term orthogonal extends the concept of perpendicularity to a multi-dimensional space. For the case that the interval of integration is $(-L, L)$, it is straightforward to verify that $\sin(j\pi x/L)$ and $\cos(j\pi x/L)$, where j is an arbitrary integer that includes zero for the cosine functions, form a set of orthogonal functions. As an example, the sine and cosine functions yield the inner product

$$\int_{-L}^{L} \mathrm{d}x \cos\left(\frac{j\pi x}{L}\right)\cos\left(\frac{k\pi x}{L}\right) = \int_{-L}^{L} \mathrm{d}x \sin\left(\frac{j\pi x}{L}\right)\sin\left(\frac{k\pi x}{L}\right) = L\,\delta_{jk}, \tag{3.59}$$

so that the norm of these functions is $\sqrt{L}$. Dividing these functions by $\sqrt{L}$ results in a set of *orthonormal functions*, which means they are both orthogonal and their inner product is *normalized* to unity. This is similar in structure to the unit vectors in three-dimensional space. The complex combination of these trigonometric functions,

$$f_j(x) = \frac{1}{\sqrt{2L}} \exp\left(\frac{i\pi j x}{L}\right) = \frac{1}{\sqrt{2L}} \cos\left(\frac{j\pi x}{L}\right) + \frac{i}{\sqrt{2L}} \sin\left(\frac{j\pi x}{L}\right), \tag{3.60}$$

results in a set of orthonormal complex functions indexed by the integer j. Using the result that

$$\exp(\pm i\pi(j-k)) = (-1)^{(j-k)}, \tag{3.61}$$

the orthonormality of the functions of (3.60) is demonstrated by their inner product,

$$(f_j, f_k) = \int_{-L}^{L} \mathrm{d}x \frac{1}{\sqrt{2L}} \exp\left(-\frac{i\pi j x}{L}\right)\frac{1}{\sqrt{2L}} \exp\left(\frac{i\pi k x}{L}\right) = \begin{cases} 1 & \text{if } j = k \\ 0 & \text{if } j \neq k \end{cases} = \delta_{jk}. \tag{3.62}$$

3.3.5 Fourier series and completeness

It is the central result of Fourier analysis that the orthonormal functions of (3.60) provide a *complete set of basis functions* to expand *any piecewise continuous member* of L^2 that is periodic over the interval $(-L, L)$. The *Fourier series* representation of a piecewise continuous function $g(x)$ contained in L^2 is given by an expansion using the functions of (3.60),

$$g(x) = \sqrt{\frac{1}{2L}} \sum_{j=-\infty}^{\infty} a_j \exp\left(\frac{i\pi jx}{L}\right) = \sqrt{\frac{1}{2L}} \sum_{j=0}^{\infty} \left\{ A_j \cos\left(\frac{j\pi x}{L}\right) + B_j \sin\left(\frac{j\pi x}{L}\right) \right\}, \quad (3.63)$$

where $A_0 = a_0$ and $B_0 = 0$. For $j > 0$ the coefficients are given by $A_j = a_j + a_{-j}$, and $B_j = i(a_j - a_{-j})$. The coefficients a_j appearing in (3.63) and in the definitions of A_j and B_j are obtained by integrating g against f_j^* and using the orthonormality of the f_j as expressed in (3.62). The result can be written

$$\begin{aligned} a_j &= \sum_{k=-\infty}^{\infty} a_k \delta_{jk} \\ &= \sum_{k=-\infty}^{\infty} a_k (f_j, f_k) = (f_j, g) = \sqrt{\frac{1}{2L}} \int_{-L}^{L} \mathrm{d}x \, g(x) \exp\left(-\frac{i\pi jx}{L}\right). \end{aligned} \quad (3.64)$$

The generalizations of (3.63) and (3.64) to multivariable functions is straightforward, with the basis functions becoming products of the one-dimensional forms, $f_{jk\ell}(x, y, z) = f_j(x) f_k(y) f_\ell(z)$, and the integrations occurring over appropriate higher-dimensional volumes, such as $\mathrm{d}^3x = \mathrm{d}x\,\mathrm{d}y\,\mathrm{d}z$.

Completeness is a crucial property of the basis functions $f_j(x)$. The mathematical statement of completeness for the function set $\{f_j(x)\}$ is given by

$$\lim_{n\to\infty} \left(g(x) - \sum_{j=-n}^{n} a_j f_j(x) \right) = 0, \quad (3.65)$$

where $g(x)$ is a member of L^2 that is periodic over the interval $(-L, L)$ and the Fourier coefficient is given by $a_j = (f_j, g)$. An important result in proving completeness is Bessel's inequality. Using the linearity of the inner product (3.43) and the positive-definite norm of a function, Bessel's inequality states that

$$\begin{aligned} \left(g - \sum_{j=-n}^{n} a_j f_j, \; g - \sum_{k=-n}^{n} a_k f_k \right) &= (g, g) - \sum_{j=-n}^{n} a_j^* (f_j, g) - \sum_{k=-n}^{n} a_k (g, f_k) \\ &\quad + \sum_{k,j=-n}^{n} a_j^* a_k (f_j, f_k) \\ &= (g, g) - \sum_{j=-n}^{n} a_j^* a_j \geqslant 0, \end{aligned} \quad (3.66)$$

where the orthonormality $(f_j, f_k) = \delta_{jk}$ and the definition $(g, f_k) = (f_k, g)^* = a_k^*$ were used. Because $g(x)$ is a member of L^2 the norm (g, g) is well-defined. As a result, demonstrating that the equality holds in the limit $n \to \infty$ shows that the functions are complete. This equality is straightforward to demonstrate for *specific* simple periodic functions. For example, choosing $g(x) = \sin(\ell\pi x/L)$, where ℓ is an integer, results in (3.50) giving

$$(g, g) = L = \int_{-L}^{L} \mathrm{d}x \, \sin^2\left(\frac{\ell\pi x}{L}\right) = L. \tag{3.67}$$

Similarly, (3.64) gives

$$\begin{aligned} a_j &= \frac{1}{\sqrt{2L}} \int_{-L}^{L} \mathrm{d}x \, \sin\left(\frac{\ell\pi x}{L}\right) \exp\left(-\frac{ij\pi x}{L}\right) \\ &= -i\sqrt{\frac{L}{2}}\left(\delta_{\ell j} - \delta_{-\ell, j}\right) \implies \lim_{n\to\infty} \sum_{j=-n}^{n} a_j^* a_j = L, \end{aligned} \tag{3.68}$$

demonstrating that the equality is met. A general proof is available in the references for this chapter.

The function space of L^2 has a further similarity to a vector space, since there exists a *complete set* of orthonormal basis functions f_j that play the same role for L^2 that a complete set of orthonormal basis vectors $\boldsymbol{e}_j$ does for a vector space. A complete set of basis vectors for a vector space allows an arbitrary vector $\boldsymbol{x}$ in that vector space to be written as *a linear superposition*, $\boldsymbol{x} = \sum_j x_j \boldsymbol{e}_j$, where the x_j are the components of the vector along each of the basis vectors and are found by using the inner product, $x_j = \boldsymbol{e}_j \cdot \boldsymbol{x} = (\boldsymbol{e}_j, \boldsymbol{x})$. This is identical to the method of (3.64) to find the coefficients of the Fourier series. The addition of two vectors gives the vector $\boldsymbol{x} + \boldsymbol{y} = \sum_j (x_j + y_j)\boldsymbol{e}_j$. Similarly, the addition of two functions in L^2 can be written

$$f(x) + g(x) = \sum_j a_j f_j(x) + \sum_j b_j f_j(x) = \sum_j (a_j + b_j) f_j(x). \tag{3.69}$$

It was shown earlier that this sum is square integrable. It is instructive to note that expression (3.63) is identical in form to a vector, with the two major differences that the inner product of (3.64) for the function components is now an integral and the set of basis functions is *infinite-dimensional.*

3.3.6 The Dirac delta

It is common to state the completeness of a set of functions using the *Dirac delta*, which generalizes the Kronecker delta to the case of continuous indices. Because the Dirac delta appears in many physical applications, it is important to define it and catalog its properties. This process begins by introducing the *Heaviside step function*, $\theta(x)$, which has the values

$$\theta(x) = \begin{cases} 1 & \text{if } x > 0 \\ 0 & \text{if } x < 0 \end{cases}. \tag{3.70}$$

From its definition, the step function has an integral given by

$$\int_c^b \mathrm{d}x\, \theta(x - a) = \int_a^b \mathrm{d}x = b - a, \tag{3.71}$$

as long as $b > a$ and $a > c$. Defining $\theta(0) = \frac{1}{2}\lim_{\epsilon\to 0}(\theta(\epsilon) + \theta(-\epsilon))$ gives $\theta(0) = 1/2$. Such a function has a derivative that, for $\epsilon > 0$, formally obeys

$$\lim_{\epsilon\to 0}\int_{a-\epsilon}^{a+\epsilon} \mathrm{d}x\, \frac{\mathrm{d}}{\mathrm{d}x}\theta(x - a) = \lim_{\epsilon\to 0}(\theta(\epsilon) - \theta(-\epsilon)) = \lim_{\epsilon\to 0}(1 - 0) = 1. \tag{3.72}$$

A continuous function would not exhibit this behavior and it is the step function's discontinuity at the argument of zero that creates this result. The derivative of the step function $\theta(x - a)$ is called the *Dirac delta*, $\delta(x - a)$,

$$\delta(x - a) = \frac{\mathrm{d}}{\mathrm{d}x}\theta(x - a), \tag{3.73}$$

For the purposes of visualization, $\delta(x - a)$ can be treated as singular at $x = a$ and zero everywhere else.

Using (3.70) and the definition (3.73) gives the continuous version of the first sum in (3.58),

$$\int_a^c \mathrm{d}x\, \delta(x - b) = \begin{cases} 1 & \text{if } b \in (a, c) \\ 0 & \text{otherwise} \end{cases}. \tag{3.74}$$

For a non-singular function $f(x)$, definition (3.73), assuming $b \in (a, c)$ and integrating by parts, gives

$$\begin{aligned} \int_c^a \mathrm{d}x\, f(x)\, \delta(x - b) &= \int_c^a \mathrm{d}x\, f(x)\frac{\mathrm{d}}{\mathrm{d}x}\theta(x - b) \\ &= \int_c^a \mathrm{d}x \left\{\frac{\mathrm{d}}{\mathrm{d}x}[f(x)\theta(x - b)] - \theta(x - b)\frac{\mathrm{d}f}{\mathrm{d}x}\right\} \\ &= f(a) - \int_b^a \mathrm{d}x\, \frac{\mathrm{d}f}{\mathrm{d}x} = f(a) - f(a) + f(b) \\ &= f(b), \end{aligned} \tag{3.75}$$

which is a key property of the Dirac delta. If $f(x) = x - b$ then $(x - b)\delta(x - b) = 0$ when integrated, and so $(x - b)\, \delta(x - b)$ can be treated as zero. Result (3.74) can also be viewed as a singular version of the Kronecker delta by turning a sum into an integration,

$$1 = \sum_{n=0}^{\infty}\delta_{nm} = \lim_{\Delta x_n \to \mathrm{d}x_n}\sum_{n=0}^{\infty}\Delta x_n\, \frac{\delta_{nm}}{\Delta x_n} = \int_0^{\infty} \mathrm{d}x_n\, \delta(x_n - x_m), \tag{3.76}$$

where the limit $\Delta x_n \to 0$ creates both the integral and the singular nature of the Dirac delta via the expression $\delta(x_n - x_m) = \delta_{nm}/\Delta x_n$. This is consistent with viewing the Dirac delta as vanishing for all arguments other than zero, and there it is singular. These are not the required continuity properties of a function, and so the Dirac delta is not a function.

The Dirac delta has a number of important properties. Using the change of variables, $x = -y$, and assuming that $-b < a < b$, it follows that

$$\begin{aligned}\int_{-b}^{b} \mathrm{d}x\, f(x)\, \delta(-x + a) &= \int_{-b}^{b} \mathrm{d}y\, f(-y)\, \delta(y + a) \\ &= f(a) \implies \delta(-x + a) = \delta(x - a).\end{aligned} \tag{3.77}$$

A second useful property is $\delta(ax) = \delta(x)/|a|$, where $|a|$ is the absolute value of the constant a. This follows from the previous result and a change of variables,

$$\int_{-b}^{b} \mathrm{d}x\, \delta(ax) = \int_{-b}^{b} \mathrm{d}x\, \delta(|a|\, x) = \frac{1}{|a|} \int_{-b|a|}^{b|a|} \mathrm{d}y\, \delta(y) = \frac{1}{|a|}. \tag{3.78}$$

Using the visualization of the Dirac delta, $\delta(f(x))$ is zero everywhere except where the function vanishes. Therefore, if $f(x)$ has a zero at x_o, a Taylor series expansion around x_o gives

$$f(x) = f(x_o + x - x_o) \approx (x - x_o) \frac{\mathrm{d}f(x_o)}{\mathrm{d}x_o}. \tag{3.79}$$

Coupling this result with (3.78) gives

$$\delta(f(x)) = \delta(x - x_o) \left| \frac{\mathrm{d}f(x_o)}{\mathrm{d}x_o} \right|^{-1}, \tag{3.80}$$

where x is assumed to be in the neighborhood of x_o. If the function has more than one zero, the Dirac delta breaks into a sum of terms around each zero, each of which is identical to (3.80).

A useful expression of completeness for the set of basis functions appearing in (3.65) is obtained by combining the expression

$$f(x) = \sum_j a_j f_j(x), \tag{3.81}$$

with the form of the coefficients,

$$a_j = (f_j, f) = \int \mathrm{d}y\, f_j^*(y) f(y). \tag{3.82}$$

The limits on the sum and the integration are omitted to allow a variety of cases. Inserting (3.82) into (3.81) gives

$$f(x) = \sum_j \int \mathrm{d}y\, f_j^*(y) f(y) f_j(x) = \int \mathrm{d}y \left(\sum_j f_j^*(y) f_j(x) \right) f(y). \tag{3.83}$$

Result (3.83) can hold only if the sum gives a Dirac delta,

$$\sum_j f_j^*(y)f_j(x) = \delta(x - y). \tag{3.84}$$

Result (3.84) is often used to express completeness of the functions $f_j(x)$. In the event that the basis functions are indexed by a continuous variable the sum becomes an integral. This is discussed in the section on Fourier transforms.

The Dirac delta can be generalized to higher dimensions and is designated $\delta^n(\boldsymbol{x} - \boldsymbol{a})$ in n-dimensions. This comes with some mild complications. In Cartesian coordinates the three-dimensional version is given by

$$\delta^3(\boldsymbol{x} - \boldsymbol{b}) = \delta(x - b_x)\,\delta(y - b_y)\,\delta(z - b_z), \tag{3.85}$$

so that

$$\int_V \mathrm{d}^3x\, f(\boldsymbol{x})\,\delta^3(\boldsymbol{x} - \boldsymbol{b}) = f(\boldsymbol{b}), \tag{3.86}$$

where it is assumed that the volume V contains the position $\boldsymbol{b}$, but results in zero if it does not. The Dirac delta can be expressed in other coordinate systems. For example, in polar coordinates in two dimensions, where $x = r\cos\theta$ and $y = r\sin\theta$, the two-dimensional Dirac delta is given by

$$\delta^2(\boldsymbol{x} - \boldsymbol{a}) = \delta(x - a_x)\,\delta(y - a_y) = \frac{1}{r}\delta(r - a)\,\delta(\theta - \arctan(a_y/a_x)), \tag{3.87}$$

where $a = \sqrt{a_x{}^2 + a_y{}^2}$. In cylindrical coordinates, where $\boldsymbol{r} = (\rho, \phi, z)$, the coordinates are related to Cartesian coordinates according to $\rho = \sqrt{x^2 + y^2}$ and $\phi = \arctan(y/x)$. The Dirac delta centered on the point $\boldsymbol{r}_o = (\rho_o, \phi_o, z_o)$ is given by

$$\delta^3(\boldsymbol{r} - \boldsymbol{r}_o) = \frac{1}{\rho}\delta(\rho - \rho_o)\delta(\phi - \phi_o)\delta(z - z_o). \tag{3.88}$$

It is useful to note that $\delta(f)$ has the inverse units of f.

It is important to be aware that different definitions of the Dirac delta are possible, and the area of mathematics known as *distribution theory* provides a formal analysis of different representations. These alternative representations often occur in physics applications. For example, using a Taylor series representation of the function $f(x)$ around $x = a$, straightforward integration of the Taylor series shows that

$$\begin{aligned}\lim_{\alpha\to\infty}\sqrt{\frac{\alpha}{\pi}}\int_{-\infty}^{\infty}\mathrm{d}x\, f(x)\, e^{-\alpha(x-a)^2} = f(a) \implies \lim_{\alpha\to\infty}\sqrt{\frac{\alpha}{\pi}}\, e^{-\alpha(x-a)^2} \\ = \delta(x - a).\end{aligned} \tag{3.89}$$

Similarly, the Cauchy residue theorem (3.32) gives

$$\lim_{\alpha\to 0}\frac{1}{\pi}\int_{-\infty}^{\infty} dx\,\frac{\alpha f(x)}{(x-a)^2+\alpha^2} = f(a) \implies \lim_{\alpha\to 0}\frac{\alpha}{\pi((x-a)^2+\alpha^2)} = \delta(x-a). \tag{3.90}$$

In addition, the *sinc squared* representation of the Dirac delta is given by the limit

$$\delta(x-a) = \lim_{t\to\infty}\frac{\sin^2((x-a)t)}{t\pi(x-a)^2}. \tag{3.91}$$

In the case of (3.91), integration by parts gives

$$\int_{-\infty}^{\infty} dx\,\frac{\sin^2(tx)}{t\pi x^2} = \int_{-\infty}^{\infty} dx\,\frac{\sin(2tx)}{\pi x} = 1. \tag{3.92}$$

In the limit $t \to \infty$ only the neighborhood of $x = 0$ contributes to the integrand, demonstrating the emergence of the Dirac delta. The detailed proofs of these are available in the references. It is worth noting that all three representations given by (3.89), (3.90), and (3.91) are zero in their respective limits *unless* the argument $(x - a)$ is zero, and in that case all three are singular in the limit. This reinforces the visualization of the Dirac delta as zero if its argument is nonzero and singular if its argument vanishes.

3.3.7 The Fourier transform and completeness

The Dirac delta allows the Fourier series to be generalized to non-periodic functions. Inserting (3.64) into (3.63) and interchanging the order of sum and integration yields

$$g(x) = \int_{-L}^{L} dy\, g(y)\frac{1}{2L}\sum_{j=-\infty}^{\infty}\exp\left(i\frac{\pi j}{L}(x-y)\right). \tag{3.93}$$

Comparing (3.93) to (3.75) allows the identification

$$\delta(x-y) = \frac{1}{2L}\sum_{j=-\infty}^{\infty}\exp\left(i\frac{\pi j}{L}(x-y)\right) = \sum_{j=-\infty}^{\infty} f_j^*(y) f_j(x), \tag{3.94}$$

which gives a *functional representation* of the Dirac delta valid over the interval $(-L, L)$. This is consistent with the previous general statement of completeness given by (3.84).

If the function $g(x)$ is not periodic, it can still represented using a Fourier series by allowing the range of periodicity L to become infinite. For large L the new variable $k_j = j\pi/L$ is defined, so that $\Delta k_j = \Delta j\;\pi/L \implies \Delta j = \Delta k_j\; L/\pi$. Since $\Delta j = 1$, it follows that $\Delta k_j = \pi/L \equiv \Delta k$. As a result, $L\,\Delta k/\pi = 1$ can be inserted into (3.94) and the limit $L \to \infty$ taken, resulting in the integral representation of the one-dimensional non-periodic Dirac delta,

$$\delta(x-y) = \lim_{L\to\infty}\frac{1}{2L}\sum_{j=-\infty}^{\infty}\frac{\Delta k\, L}{\pi}\exp\left(ik_j(x-y)\right) = \int_{-\infty}^{\infty}\frac{dk}{2\pi}\, e^{ik(x-y)}. \tag{3.95}$$

The continuous variable k has the units of inverse length, just as the wave number k does, so that $\delta(x - y)$ has the required units of inverse length.

Results (3.94) and (3.95) are consistent with the *completeness* (3.84) of the functions being summed or integrated on the right-hand side. They can be used to create a series or integral representation of an arbitrary function in terms of a *linear superposition* of *basis functions.* For example, (3.95) can be used to give an integral representation of an arbitrary non-periodic function $g(x)$ in terms of the function e^{ikx},

$$\begin{aligned} g(x) &= \int_{-\infty}^{\infty} \mathrm{d}y\, g(y)\, \delta(x - y) = \int_{-\infty}^{\infty} \mathrm{d}y\, g(y) \int_{-\infty}^{\infty} \frac{\mathrm{d}k}{2\pi} e^{ik(x-y)} \\ &= \int_{-\infty}^{\infty} \frac{\mathrm{d}k}{2\pi}\, e^{ikx} \int_{-\infty}^{\infty} \mathrm{d}y\, g(y)\, e^{-iky} \equiv \int_{-\infty}^{\infty} \frac{\mathrm{d}k}{2\pi}\, \tilde{g}(k)\, e^{ikx}. \end{aligned} \tag{3.96}$$

The function $\tilde{g}(k)$ is called the *Fourier transform* of $g(x)$, and from (3.96) it is given by

$$\tilde{g}(k) = \int_{-\infty}^{\infty} \mathrm{d}y\, g(y)\, e^{-iky}. \tag{3.97}$$

This method of representing a function is often referred to as the *continuum limit* of the Fourier series, since the discrete values $k_j = j\pi/L$ of the Fourier series have become the *continuous variable k.* The form (3.96) can be obtained from (3.63) by inserting $\Delta k\; L/\pi = 1$ and absorbing the normalization $\sqrt{2L}$ into the coefficients a_j when $L \to \infty$,

$$\begin{aligned} g(x) &= \lim_{L\to\infty} \frac{1}{\sqrt{2L}} \sum_{j=-\infty}^{\infty} a_j\, e^{i\pi jx/L} \\ &= \lim_{L\to\infty} \sum_{j=-\infty}^{\infty} \frac{\Delta k}{2\pi} (a_j\sqrt{2L}) e^{ik_j x} = \int \frac{\mathrm{d}k}{2\pi}\, \tilde{g}(k) e^{ikx}. \end{aligned} \tag{3.98}$$

The identification of $\tilde{g}(k)$ holds since (3.64) shows that $a_j\sqrt{2L}$ remains finite in the limit $L \to \infty$. In higher spatial dimensions n, this result generalizes to $a_j\sqrt{(2L)^n} = a_j\sqrt{V} \to \tilde{g}(\boldsymbol{k})$, showing that $\tilde{g}(\boldsymbol{k})$ has the same units as $\sqrt{V}$.

It is important to note that the function $\exp(ikx)$ satisfies a *continuum form* of the Fourier orthonormality condition (3.62). The first step in demonstrating this is to use (3.76) to convert the Kronecker delta $\delta_{nn'}$ into a Dirac delta in the wave number, $\delta(k - k') = \lim_{\Delta k\to 0} (\delta_{nn'}/\Delta k)$, where $\Delta k = \Delta n\; \pi/L = \pi/L \to \mathrm{d}k$ in the $L \to \infty$ limit. Dividing both sides of (3.62) by $\Delta k = \pi/L$ and taking the $L \to \infty$ limit gives

$$\lim_{L\to\infty} \frac{\delta_{nn'}}{\Delta k} = \delta(k - k') = \int_{-\infty}^{\infty} \frac{\mathrm{d}x}{2\pi}\, e^{i(k-k')x}. \tag{3.99}$$

This gives an integral representation of the Dirac delta for the wave number k that is identical to the expression (3.95) for position when the interchange $x \leftrightarrow k$ is made.

Since the Dirac delta is zero if its argument is nonzero, result (3.99) is a statement of the orthonormality of the functions (3.60) *in the continuum limit* where $L \to \infty$.

It is often the case in quantum mechanical applications that the two forms of orthonormality are used interchangeably. The orthonormality of the discrete wave number functions,

$$\frac{1}{2L}\int_{-L}^{L} \mathrm{d}x\, e^{i(n-n')\pi x/L} = \delta_{nn'}, \tag{3.100}$$

is often replaced by the continuum statement

$$\int_{-\infty}^{\infty} \frac{\mathrm{d}x}{2\pi}\, e^{i(k-k')x} = \delta(k-k'). \tag{3.101}$$

However, it is important to note that the units of the two statements are different. Statement (3.100), referred to as *box normalization*, is dimensionless, while (3.101), referred to as *continuum normalization*, has the units of length. It is possible to use a hybrid version of the two normalizations, which is written

$$\frac{\pi}{L}\int_{-L}^{L} \frac{\mathrm{d}x}{2\pi}\, e^{i(k-k')x} = \delta_{kk'}, \tag{3.102}$$

where the limit $L \to \infty$ is implicit. In one-dimensional quantum mechanics the quantity $2L$ is treated as the *volume* V *of the system*, so that (3.102) can be written

$$\frac{1}{V}\int_{V} \mathrm{d}x\, e^{i(k-k')x} = \delta_{kk'}. \tag{3.103}$$

In that regard, the dimensionless Kronecker delta of (3.102) is loosely defined as

$$\delta_{kk'} = \frac{\pi}{L}\delta(k-k') = \frac{\delta(k-k')}{\delta(0)}. \tag{3.104}$$

Such an identification ignores potential issues with rationalizing infinite quantities. The identification (3.104) is understood in the continuum limit,

$$\sum_k \delta_{kk'} = 1 \;\to\; \int_{-\infty}^{\infty} \mathrm{d}k\; \delta(k-k') = 1, \tag{3.105}$$

so that the sum over the dimensionless Kronecker delta can be replaced by an integral over the Dirac delta,

$$\sum_k \delta_{kk'} = \sum_k \frac{\pi}{L}\delta(k-k') = \sum_k \frac{2\pi}{V}\delta(k-k') = \int_{-\infty}^{\infty} \mathrm{d}k\; \delta(k-k'). \tag{3.106}$$

Use of the hybrid normalization where $V = 2L$ in one dimension therefore uses the identification

$$\sum_k \leftrightarrow \frac{V}{2\pi}\int_{-\infty}^{\infty} \mathrm{d}k, \tag{3.107}$$

where the limit $V \to \infty$ is taken subsequent to all calculations. This identification preserves the dimensionless nature of the sum.

3.4 The complex valued wave function

Choosing the wave function to be complex will allow Galilean relativity to be implemented in a manner that preserves the probability density for different Galilean observers *and* will allow the energy scale to be shifted by an arbitrary amount as in (2.1). The demonstration of the first property is the subject of this section.

3.4.1 The first postulate of quantum mechanics

It is useful to enumerate the assumptions that are made to formulate quantum mechanics. The reader should bear in mind that, unlike Euclidean geometry, the list of postulates presented in this text are not universally agreed upon by all physicists. Various interpretations of quantum mechanics often have differing sets of assumptions, and the sets of assumptions have varying levels of mathematical abstraction. The first postulate of this chapter deals with the fundamental properties of the wave function expressed as a function of $\boldsymbol{x}$ and t. Such a restriction to position and time is referred to as *wave mechanics*. In chapter 6 this restriction is relaxed when a more general version of quantum mechanics is introduced.

Postulate one: Each massive Newtonian point particle is associated with a complex wave function $\Psi(\boldsymbol{x}, t)$ that is a member of the space of square integrable functions L^2. The probability density $\mathcal{P}(\boldsymbol{x}, t)$ of observing the particle at the point $\boldsymbol{x}$ at the time t is given by the *positive-definite real quadratic quantity*

$$\mathcal{P}(\boldsymbol{x}, t) = \Psi^*(\boldsymbol{x}, t)\Psi(\boldsymbol{x}, t) = |\Psi(\boldsymbol{x}, t)|^2. \tag{3.108}$$

The wave function is required to satisfy the normalization condition for the probability density,

$$\int \mathrm{d}^n x \, \mathcal{P}(\boldsymbol{x}, t) = \int \mathrm{d}^n x \, |\Psi(\boldsymbol{x}, t)|^2 = 1, \tag{3.109}$$

where the integration is over the spatial volume available to the particle.

It is important to clarify the use of the phrase *observing the particle* in the first assumption. This is understood to refer to a *measurement* made on the position of the particle in a chosen coordinate system. In this context, *particle* refers to a very compact object such as an electron. Of course, measuring the position of an object can be done using a variety of techniques, and each one may have a level of *experimental error* associated with the measurement. However, once a measurement is made on the location of the particle, there will be only *one unique result* reported by the *observer*. The observer *will not* report the particle at two simultaneous locations that are separated by more than the size of the particle. In the Copenhagen interpretation of quantum mechanics, the wave function and its associated probability density give the *possible* locations of the particle at the time of measurement along with *predicting* the probability that the particle is at that *specific location*. It is important to note that this does *not* claim that the particle is at all possible locations at once. In that sense these predictions only apply *prior to the measurement*, since after the measurement a specific location has been obtained within experimental error. The Copenhagen interpretation asserts that the wave function encapsulates *all the possible outcomes of a measurement* for the system under consideration. This will include measurements on quantities other than position, such as momentum and angular momentum. In that sense, the Copenhagen interpretation is *epistemic*, since it speaks to a lack of knowledge regarding the particle *prior to a measurement*. This is not the only interpretation of quantum mechanics and what the wave function means, but it is the interpretation that this text will follow. It is also often the case that the complete information property of the wave function is included in the postulate just made regarding the wave function. Because there are other possible interpretations of quantum mechanics and the wave function, this text will not include this in the first postulate. Instead, it will be listed as part of the Copenhagen interpretation.

The first postulate can be expressed in more general terms by introducing the concept of an *observable*, which refers to an aspect of an object that can be measured. Clealy, position is a measurable property of an object, as are other quantities associated with motion such as momentum, energy differences, and angular momentum. In that sense, there is a collection of observables associated with an object, and understanding their relation to the wave function is at the core of quantum mechanics. In that sense, the first postulate can be expressed in the following way: the probability density $\mathcal{P}(\boldsymbol{x}, t) = |\Psi(\boldsymbol{x}, t)|^2$ gives the probability density of measuring the observable known as position with the value $\boldsymbol{x}$ at the time t. Statement (3.1) holds, so that $\mathrm{d}P = \mathcal{P}(\boldsymbol{x}, t)\,\mathrm{d}^n x$ is the infinitesimal probability of measuring the position of the particle to be in the infinitesimal volume $\mathrm{d}^n x$ containing the position $\boldsymbol{x}$ at the time t. Finding the probability density for other observables from $\Psi(\boldsymbol{x}, t)$ will be an important result of the remainder of this chapter.

The first postulate asserts that the particle and the associated wave function are *inseparable aspects* of an *object*, in this case a point-like particle. This clearly clashes with the classical treatment of a material object, where it is given no wave properties whatsoever, and comes into profound conflict with human intuitions formed from a lifetime of experience with macroscopic objects. The objects of the everyday world simply do not behave as some inseparable combination of particle and wave function. However, the merger of particle and wave is necessary for quantum mechanics to have any *predictive power* regarding the processes that occur at the *atomic level* of matter. It is a serious and important question as to how large aggregates of atoms cease to behave quantum mechanically and transit into behaving classically. This is an ongoing area of research and will be briefly discussed later in this text.

The first postulate also borrows the quadratic nature of the Poynting vector in the definition of the probability density for massive particles and is consistent with treating the Poynting vector as describing the energy flux of *light particles* or photons. Allowing the wave function to be complex will allow Galilean relativity to be implemented. Using the notation of function space the normalization requirement (3.109) can be written

$$\int d^n x\, \Psi^*(\boldsymbol{x}, t)\, \Psi(\boldsymbol{x}, t) = \int d^n x\, |\Psi(\boldsymbol{x}, t)|^2 = (\Psi, \Psi) = 1, \tag{3.110}$$

where the spatial volume of integration coincides with the volume where the particle may be located. By virtue of belonging to L^2 it is always possible to normalize $\Psi(\boldsymbol{x}, t)$.

Implementing (3.110) starts by noting that (3.22) allows a complex wave function to be written

$$\Psi(\boldsymbol{x}, t) = \zeta(\boldsymbol{x}, t)\, e^{i\varphi(\boldsymbol{x}, t)}, \tag{3.111}$$

where ζ and φ are both *real valued functions*. Normalization of the probability density therefore places a condition *only on the modulus of* Ψ since $\Psi^*(\boldsymbol{x}, t)\, \Psi(\boldsymbol{x}, t) = \zeta^2(\boldsymbol{x}, t)$. If the modulus of the wave function satisfies

$$\int d^n x\, \zeta^2(\boldsymbol{x}, t) = 1, \tag{3.112}$$

then the normalization requirement for the probability (3.2) is automatically satisfied *regardless of the phase function* $\varphi(\boldsymbol{x}, t)$. This also requires $\zeta(\boldsymbol{x}, t)$ or $|\Psi(\boldsymbol{x}, t)|$ to be a *real valued member* of L^2. As a result, the properties of this general space of functions discussed in the previous section play a critical role in quantum mechanics.

3.4.2 Phase transformation symmetry

A complex wave function has the critically important freedom to transform with a multiplicative factor, *as long as the multiplicative factor takes the form of a phase.* This is because changing the *phase* of the wave function has no effect on the probability density for the observable position,

$$\Psi(\boldsymbol{x}, t) \to e^{i\theta(\boldsymbol{x}, t)}\Psi(\boldsymbol{x}, t) \implies \mathcal{P}(\boldsymbol{x}, t) \to \mathcal{P}(\boldsymbol{x}, t), \tag{3.113}$$

where $\theta(\boldsymbol{x}, t)$ is *an arbitrary phase angle.* The two wave functions $\Psi(\boldsymbol{x}, t)$ and $e^{i\theta(\boldsymbol{x}, t)}\Psi(\boldsymbol{x}, t)$ are *experimentally indistinguishable* since they yield identical probability densities. This is similar to the electromagnetic potentials $\mathbf{A}(\boldsymbol{x}, t)$ and $\varphi(\boldsymbol{x}, t)$, which can undergo the gauge transformation (1.66) without affecting the *observable* electric and magnetic fields. However, in the case of electromagnetism the freedom to perform gauge transformations removes two degrees of freedom from the electric and magnetic fields. This freedom to redefine or transform the fields of quantum mechanics and electromagnetism without affecting physical observables is referred to as a *symmetry*. Symmetries are associated with *conservation laws*, and in the case of quantum mechanical phase symmetry the conserved quantity is probability.

3.4.3 Phase factors and Galilean relativity

In the case of quantum mechanics, the freedom to change the phase of the wave function allows Galilean relativity to be implemented. In order to show how this is possible, the wave function of a massive Newtonian particle will be written $\Psi(\boldsymbol{p}, \boldsymbol{x}, t)$. This shows its dependence on position $\boldsymbol{x}$ and time t, as well as its possible dependence on the particle's momentum $\boldsymbol{p}$. This was the case with the electromagnetic wave (3.7). In order to implement Galilean relativity the wave function according to observer $\mathcal{O}$ will be transformed under a Galilean transformation to the wave function according to $\mathcal{O}'$, given by

$$\Psi'(\boldsymbol{p}', \boldsymbol{x}', t') = \Psi(\boldsymbol{p} - m\boldsymbol{v}_o, \boldsymbol{x} - \boldsymbol{v}_o t, t) = e^{-i\alpha(\boldsymbol{p}, \boldsymbol{v}_o, \boldsymbol{x}, t)}\Psi(\boldsymbol{p}, \boldsymbol{x}, t), \tag{3.114}$$

where the phase angle α is assumed to depend on the relative velocity $\boldsymbol{v}_o$ of the two observers as well as the position $\boldsymbol{x}$ and time t under consideration. The momentum $\boldsymbol{p}$ has been included as one of the possible wave function variables. The transformation property (3.114) also assumes that transforming the wave function with the phase angle $\alpha(\boldsymbol{p}, \boldsymbol{v}_o, \boldsymbol{x}, t)$ is consistent with the equation that governs the wave function. This requirement is central to determining the correct form of the Schrödinger equation and will be presented in detail in the next chapter. If the transformation property (3.114) holds, then the phase factor drops out of the definition (3.108), and the probability densities for the two observers obey the desired equality,

$$\mathcal{P}'(\boldsymbol{p}', \boldsymbol{x}', t') = |\Psi'(\boldsymbol{p}', \boldsymbol{x}', t')|^2 = |\Psi(\boldsymbol{p}, \boldsymbol{x}, t)|^2 = \mathcal{P}(\boldsymbol{p}, \boldsymbol{x}, t). \tag{3.115}$$

It is important to remember that $\boldsymbol{x}' = \boldsymbol{x} - \boldsymbol{v}_o t$ and $\boldsymbol{x} = \boldsymbol{x}' + \boldsymbol{v}_o t$ are the same point in space, but described in coordinate systems that are in relative motion. Result (3.115) means that the probability density is a *true scalar* for Galilean observers, and the probability density obtained by the first observer will coincide with the probability density obtained by the second observer even though they use *different values* of momentum. This alleviates the previous problem associated with the de Broglie wavelength changing for a moving observer. Again, (3.114) means that the wave function is itself *not a scalar*. Since an *arbitrary* complex phase factor may result from a Galilean transformation, the wave function Ψ *is not directly observable*.

The next step is to determine the form for $\alpha(\boldsymbol{p}, \boldsymbol{v}_o, \boldsymbol{x}, t)$ that corresponds to a Galilean transformation. This is found from considering how two phase angles for different velocities compose to make a single phase angle. This begins by writing the transformed wave function of (3.114) as

$$\begin{aligned}\Psi'(\boldsymbol{p} - m\boldsymbol{v}_o, \boldsymbol{x} - \boldsymbol{v}_o t, t) &= \Psi'(\boldsymbol{p} - m(\boldsymbol{v}_o - \boldsymbol{v}_o') - m\boldsymbol{v}_o', \boldsymbol{x} - (\boldsymbol{v}_o - \boldsymbol{v}_o')t - \boldsymbol{v}_o' t, t)\\ &= \Psi'(\boldsymbol{p}' - m\boldsymbol{v}_o', \boldsymbol{x}' - \boldsymbol{v}_o' t, t),\end{aligned} \tag{3.116}$$

where $\boldsymbol{p}' = \boldsymbol{p} - m(\boldsymbol{v}_o - \boldsymbol{v}_o')$ and $\boldsymbol{x}' = \boldsymbol{x} - (\boldsymbol{v}_o - \boldsymbol{v}_o')t$. Applying the rules of the transformation defined by (3.114) twice in succession gives

$$
\begin{aligned}
\Psi'(\boldsymbol{p} - m\boldsymbol{v}_o, \boldsymbol{x} - \boldsymbol{v}_o t, t) &= \Psi'(\boldsymbol{p}' - m\boldsymbol{v}_o', \boldsymbol{x}' - \boldsymbol{v}_o' t, t) \\
&= e^{-i\alpha(\boldsymbol{p}', \boldsymbol{v}_o', \boldsymbol{x}', t)}\Psi'(\boldsymbol{p}', \boldsymbol{x}', t) \qquad (3.117) \\
&= e^{-i\alpha(\boldsymbol{p}', \boldsymbol{v}_o', \boldsymbol{x}', t)} e^{-i\alpha(\boldsymbol{p}, \boldsymbol{v}_o - \boldsymbol{v}_o', \boldsymbol{x}, t)}\Psi(\boldsymbol{p}, \boldsymbol{x}, t).
\end{aligned}
$$

Since (3.114) and (3.117) must result in identically transformed wave functions, the definitions of $\boldsymbol{p}'$ and $\boldsymbol{x}'$ shows that the phase angle function must satisfy

$$\alpha(\boldsymbol{p} - m(\boldsymbol{v}_o - \boldsymbol{v}_o'), \boldsymbol{v}_o', \boldsymbol{x} - (\boldsymbol{v}_o - \boldsymbol{v}_o')t, t) + \alpha(\boldsymbol{p}, \boldsymbol{v}_o - \boldsymbol{v}_o', \boldsymbol{x}, t) = \alpha(\boldsymbol{p}, \boldsymbol{v}_o, \boldsymbol{x}, t). \qquad (3.118)$$

Setting $\boldsymbol{v}_o' = \boldsymbol{v}_o$ shows that $\alpha(\boldsymbol{p}, 0, \boldsymbol{x}, t)$ must vanish. The requirement (3.118) has a simple solution for α that satisfies this property, given by

$$\alpha(\boldsymbol{p}, \boldsymbol{v}_o, \boldsymbol{x}, t) = \left(m\boldsymbol{v}_o \cdot \boldsymbol{x} - \frac{1}{2}m v_o^2 t\right)\Big/\hbar, \qquad (3.119)$$

where the factor $\hbar^{-1}$ insures the requirement that the phase angle is dimensionless. This solution is independent of any value of $\boldsymbol{p}$ which may appear in the wave function and depends only on the position $\boldsymbol{x}$, the time t, and the relative velocity $\boldsymbol{v}_o$ of the two observers. This is in keeping with the definition of the Galilean transformation originally made in (1.32), which has no reference to the momentum of any particle that the second observer may measure.

The demonstration that the required relation (3.118) is satisfied is straightforward. Using the functional form for α given by (3.119) it follows that

$$\alpha(\boldsymbol{p} - m(\boldsymbol{v}_o - \boldsymbol{v}_o'), \boldsymbol{v}_o', \boldsymbol{x} - (\boldsymbol{v}_o - \boldsymbol{v}_o')t, t) = \left(m\boldsymbol{v}_o' \cdot \boldsymbol{x} - m\boldsymbol{v}_o' \cdot \boldsymbol{v}_o t + \frac{1}{2}m v_o'^2\right)\Big/\hbar \qquad (3.120)$$

and

$$\alpha(\boldsymbol{p}, \boldsymbol{v}_o - \boldsymbol{v}_o', \boldsymbol{x}, t) = \left(m\boldsymbol{v}_o \cdot \boldsymbol{x} - m\boldsymbol{v}_o' \cdot \boldsymbol{x} - \frac{1}{2}m v_o^2 t + m\boldsymbol{v}_o \cdot \boldsymbol{v}_o' t - \frac{1}{2}m v_o'^2 t\right)\Big/\hbar, \qquad (3.121)$$

which sum to the required expression (3.119). The properties of α will play a significant role in identifying the form of the Schrödinger equation.

A second important property of Newtonian mechanics and statistical mechanics is that only energy *differences* are of importance in determining the dynamics of a particle. This property should be present in quantum mechanics as well for the case that the wave function depends on the energy E, so that $\Psi = \Psi(E, \boldsymbol{x}, t)$. In that regard, the probability density (3.108) becomes independent of the overall scale chosen for the energy E *if the wave function obeys the condition*

$$\Psi(E + \delta E, \boldsymbol{x}, t) = e^{i\,\delta\alpha}\Psi(E, \boldsymbol{x}, t), \qquad (3.122)$$

where $\delta\alpha$ is *a phase factor* that depends on δE. Arbitrarily changing the scale of energy then leaves the probability density unchanged since such phase factors will once again cancel in the expression $|\Psi(\boldsymbol{x}, t)|^2$. Additional properties of the wave function, such as conservation of probability, will emerge when the Schrödinger equation is developed and analysed in the next chapter.

3.4.4 The free particle wave function

As an instructive starting point, a possible wave function for a particle confined to the spatial volume V is given by

$$\Psi(\boldsymbol{x}, t) = \frac{1}{\sqrt{V}} e^{i\varphi(\boldsymbol{p}, \boldsymbol{x}, t)/\hbar}, \tag{3.123}$$

where $\varphi(\boldsymbol{p}, \boldsymbol{x}, t)$ is a real valued function and $\boldsymbol{p}$ is the momentum of the particle according to a specific observer $\mathcal{O}$. This gives a probability density that satisfies the normalization condition (3.2),

$$\mathcal{P}(\boldsymbol{x}, t) = \Psi^*(\boldsymbol{x}, t)\Psi(\boldsymbol{x}, t) = \frac{1}{V} \Longrightarrow \int \mathrm{d}^n x\, \mathcal{P}(\boldsymbol{x}, t) = \frac{1}{V}\int \mathrm{d}^n x = 1, \tag{3.124}$$

and shows that a wave function of the general form (3.123) describes a particle with a probability density that is the same *everywhere* in the volume V. The probability density is also *independent of* a specific form for $\varphi(\boldsymbol{x}, t)$. For the case that (3.123) is the particle's wave function, the particle would have an equal probability to be at *any location* in the spatial volume V, and so the wave function (3.123) describes a situation where the particle is in a completely *delocalized state* in the sense that no location in the volume V is more probable than another. In what follows, the word *state* will be used to refer to the wave function of the system under consideration. As formal quantum mechanics is developed, it will be seen that such a wave function is valid only in the limit that $V \to \infty$. In what follows this limit will be understood.

It is extremely instructive to consider the case that the phase function $\varphi(\boldsymbol{x}, t)$ for observer $\mathcal{O}$ becomes zero everywhere according to an observer $\mathcal{O}'$ moving with the velocity v_o relative to observer $\mathcal{O}$. This means that the wave function according to $\mathcal{O}'$ is given by

$$\Psi_{\mathcal{O}'}(\boldsymbol{x}, t) = \frac{1}{\sqrt{V}}. \tag{3.125}$$

Using (3.114) and (3.119) shows that an observer $\mathcal{O}'$ moving with the velocity v_o relative to $\mathcal{O}$ obtains the wave function given by

$$\begin{aligned}\Psi_{\mathcal{O}'}(\boldsymbol{x}, t) = \frac{1}{\sqrt{V}} = \Psi'_{\mathcal{O}}(\boldsymbol{x}', t') &= e^{-i\alpha(v_o, \boldsymbol{x}, t)}\Psi_{\mathcal{O}}(\boldsymbol{x}, t) = \frac{1}{\sqrt{V}} e^{i(\varphi(\boldsymbol{p}, \boldsymbol{x}, t) - \hbar\alpha(v_o, \boldsymbol{x}, t))/\hbar} \\ &= \frac{1}{\sqrt{V}} \exp\left(\frac{i}{\hbar}\varphi(\boldsymbol{p}, \boldsymbol{x}, t) - \frac{i}{\hbar}\left(\boldsymbol{p}\cdot\boldsymbol{x} - \frac{p^2}{2m}t\right)\right),\end{aligned} \tag{3.126}$$

where $v_o = \boldsymbol{p}/m$ has been used to express $\alpha(v_o, \boldsymbol{x}, t)$ in terms of a momentum $\boldsymbol{p}$. In order to satisfy (3.126) the phase function according to observer $\mathcal{O}$ must be given by

$$\Psi(\boldsymbol{p}, \boldsymbol{x}, t) = \frac{1}{\sqrt{V}} \exp\left(\frac{i}{\hbar}\varphi(\boldsymbol{p}, \boldsymbol{x}, t)\right) = \frac{1}{\sqrt{V}} \exp\left(\frac{i}{\hbar}(\boldsymbol{p}\cdot\boldsymbol{x} - E_{\mathrm{p}}t)\right), \tag{3.127}$$

where $E_{\rm p} = p^2/2m$ is the Newtonian kinetic energy associated with the momentum $\boldsymbol{p}$. It follows directly that the phase angle of (3.127) has the Galilean transformation property

$$\begin{aligned}\varphi'(\boldsymbol{p}', \boldsymbol{x}', t) &= \varphi(\boldsymbol{p} - m\boldsymbol{v}_o, \boldsymbol{x} - \boldsymbol{v}_o t, t) = (\boldsymbol{p} - m\boldsymbol{v}_o)\cdot(x - \boldsymbol{v}_o t) - (\boldsymbol{p} - m\boldsymbol{v}_o)\cdot(\boldsymbol{p} - m\boldsymbol{v}_o)t/2\,m \\ &= \boldsymbol{p}\cdot\boldsymbol{x} - E_{\rm p}t - (m\boldsymbol{v}_o\cdot\boldsymbol{x} - \frac{1}{2}mv_o^2 t) \\ &= \varphi(\boldsymbol{p}, \boldsymbol{x}, t) - \hbar\alpha(\boldsymbol{v}_o, \boldsymbol{x}, t),\end{aligned} \tag{3.128}$$

where $\alpha(\boldsymbol{v}_o, \boldsymbol{x}, t)$ is precisely the Galilean transformation property phase factor predicted by (3.119). A Galilean transformation will therefore *not* change the probability density $1/V$. As a result, all Galilean observers will agree that $\mathcal{P}(\boldsymbol{x}, t) = 1/V$, so that the wave function of (3.127) is associated with a probability density that is *the same everywhere in the volume* according to all observers. However, for an observer traveling at the relative velocity $\boldsymbol{v}_o = \boldsymbol{p}/m$, the particle will appear to have zero momentum.

The complex wave function (3.127) for a specific momentum $\boldsymbol{p}$ has been identified from the requirement of Galilean relativity and the probabilistic interpretation of the wave function. Because $\boldsymbol{p}$ is constant at all times, it is the case that the wave function (3.127) corresponds to the absence of force on the particle and therefore is the wave function of a *free particle*. The resultant quantum mechanical description of a free particle with momentum $\boldsymbol{p}$ is in stark contrast to the results of Newtonian mechanics, where the free particle has both a definite momentum *and* a definite position *at all times* after its initial position is fixed. Instead, the quantum mechanical free particle with a definite momentum has an indeterminate position at all times. Although a form for the matter wave of a free particle with a definite momentum has been derived by indirect means, an equation for the wave function which (3.127) satisfies has not yet been identified. This leaves the question of how *boundary conditions* and the presence of a *potential energy* will modify these results. While Galilean relativity and the basic concept of the complex wave function have led to (3.127), it is clear that the further development of quantum mechanics is required to answer these questions.

3.4.5 A first look at wave function preparation, measurement, and collapse

It is natural to ask how the *object*, the particle and wave function, could happen to be in the particular *quantum state* characterized by (3.127). In the Copenhagen interpretation of quantum mechanics the *initial state* of the system is the result of *preparation* by an experimental apparatus *or* by a previous physical process. For the case of the free particle the preparation of the wave function reduces to selecting a specific momentum for an otherwise free particle.

An idealization of this preparation process is depicted in figure 3.1, where a beam of electrons with random velocities is passed through a slit into a region of space where there is a known uniform magnetic field **B**. Such a beam is available from a filament of wire heated by an electric current. This gives a means to select a specific momentum particle from the beam since *classically* a charged particle like the

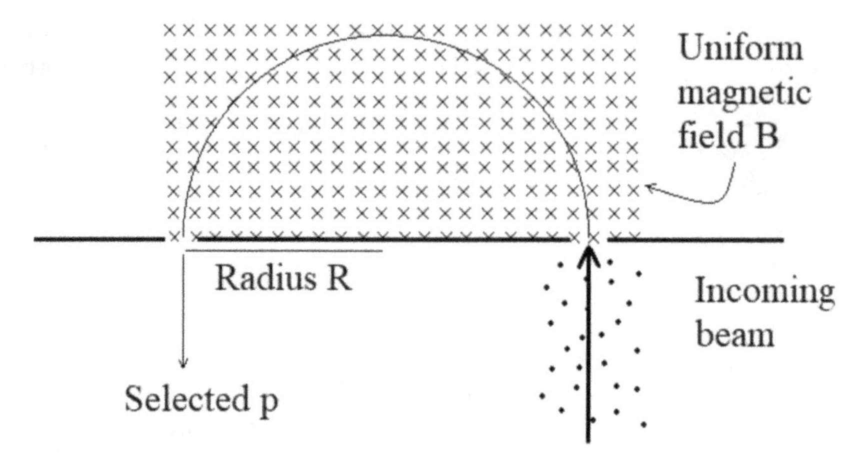

Figure 3.1. Free particle wave function preparation.

electron will move in a circle of radius R. This radius is found by equating the Lorentz force qpB/mc of the magnetic field to the centripetal force p^2/mR required for circular motion, yielding the magnitude of momentum, $p = qBR/c$, from the radius of motion. Placing a second slit at a distance $2R$ from the first slit will allow these electrons to be selected. It should be pointed out that there is an *uncertainty* in the momentum brought about by the slit size. If the slit size is d, the measured radius has an uncertainty of d, which gives an experimental uncertainty in p given by $\Delta p = qBd/c$. It is further possible to imagine this electron directed at a *random angle* into an ideal cubic box with perfectly reflecting walls and volume $V = L^3$, so that no energy is lost and the particle will retain the magnitude of $\boldsymbol{p}$. As a final touch, it is possible to imagine that this identical process has been followed to *initially prepare* a large number of containers for an experimentalist to examine. Although this situation is an *ideal thought experiment*, it captures many of the aspects of quantum preparation since it selects for *observation* only those particles with the wave function given by (3.127).

However, an important *expected property* of the wave function is revealed from this thought experiment. Once the wave function is prepared at some initial time, then it is assumed to evolve in time in a *unique way*. This means that the equation which the wave function obeys must be *linear* in the time derivative or, in the parlance of differential equations, it must be *first order in time*. Only an equation that is first order in the time derivative will provide a unique function at later times from a single initial function. This will be an important feature of the Schrödinger equation, which governs the wave function.

It is then natural to ask what would occur if these identical systems are each subjected to an experiment to *measure* the location of the *particle*. A particle detector can be placed in the container and moved from location to location, until it eventually detects the electron at *the current location of the detector*. Of course, even the smallest of modern electron detectors are not infinitesimal, although some are on the order of 10 micrometers in dimension. If this experiment were repeatedly performed for the many identically prepared boxes, the probabilistic interpretation of the wave function states that the eventual detection of the particle could occur anywhere in the volume V with *an equal probability*. For an infinitesimal detector, examining each box would result in detecting the *particle* at a specific location, but there would be no favored location because each location in the box has the probability density $1/V$ of being the current location of the particle. This probability density will remain true even if the electron undergoes instantaneous collisions with the perfect reflectors of the container walls since only the direction of $\boldsymbol{p}$ will change.

In the Copenhagen interpretation of quantum mechanics this means that a *statistical analysis* of the individual experimental results obtained subsequent to this preparation would exhibit no favored location. Each individual location has the probability $\mathcal{P}(\boldsymbol{x}, t)\,\mathrm{d}^n x$ of being the location of the particle. As a result, the *statistical average* after many measurements is known as the *expectation value of* $\boldsymbol{x}$, denoted $\langle \boldsymbol{x} \rangle$, and is given by

$$\begin{aligned} \langle \boldsymbol{x} \rangle &= \int \mathrm{d}^n x \, \boldsymbol{x} \, \mathcal{P}(\boldsymbol{x}, t) = \int \mathrm{d}^n x \, \boldsymbol{x} \, |\Psi(\boldsymbol{x}, t)|^2 \\ &= \int \mathrm{d}^n x \, \Psi^*(\boldsymbol{x}, t) \, \boldsymbol{x} \, \Psi(\boldsymbol{x}, t), \end{aligned} \tag{3.129}$$

where the integration is over the spatial volume where the wave function is nonzero.

A familiar example explains this definition. A six-sided die has an equal probability of 1/6 for landing on any one of its six sides. The sides are marked with numbers N from one to six. After many tosses, the average of the results of the number on the side landed upon, or expectation value, is given by

$$\langle N \rangle = 1 \times \frac{1}{6} + 2 \times \frac{1}{6} + 3 \times \frac{1}{6} + 4 \times \frac{1}{6} + 5 \times \frac{1}{6} + 6 \times \frac{1}{6} = \frac{21}{6} = 3.5. \tag{3.130}$$

During no toss of the die will the number 3.5 occur, and yet the average of many die rolls will be 3.5.

In the quantum mechanical context, for a cubical box of side L the *statistical average* of the measurements of the particle's x coordinate, denoted $\langle x \rangle$, is predicted by the probability density $1/V = 1/L^3$ to be given by

$$\begin{aligned} \langle x \rangle &= \int \mathrm{d}^3 x \, x \mathcal{P}(\boldsymbol{x}, t) = \int \mathrm{d}^3 x \, \Psi^*(\boldsymbol{x}, t) \, x \, \Psi(\boldsymbol{x}, t) \\ &= \frac{1}{L^3} \int_0^L \mathrm{d}x \, x \int_0^L \mathrm{d}y \int_0^L \mathrm{d}z = \frac{1}{L^3} \frac{1}{2} L^4 = \frac{1}{2} L. \end{aligned} \tag{3.131}$$

The same holds for the y and z coordinates. This statistical average of many position measurements will yield this expectation value for the position as long as the probability density remains the same. This does *not* mean that the particle is observed midway in the box during each measurement, just as the roll of a die will not show a side marked 3.5. Instead, in the Copenhagen interpretation of quantum mechanics the expectation value of the particle's position is understood as the statistical average of many measurements of the particle's position for identically prepared systems. For the particular wave function (3.127) the expectation value of position is *predicted* to be the center of the box. In this sense, the wave function does not predict a well-defined position for the particle, as does Newtonian mechanics. Instead, it predicts the probability for a specific outcome of a measurement of position. The statistical average of measuring *any observable quantity* for identically prepared systems will be generalized to the expectation value of the *observable* using the wave function for these systems. However, it will be seen that the probability of an *individual measurement* of an observable quantity is also given by the wave function. This is presented in the next section where the nature of quantum mechanical observables is developed.

However, at the moment $t_{\rm d}$ the detection occurs in each measurement, the wave function (3.127) would no longer be valid since the probability of the particle being at the detector's location *has become unity*. In other words, if the particle is detected at the position $\boldsymbol{a}$ at the time $t_{\rm d}$, the observer will claim that the probability density representing the particle underwent a collapse onto the position $\boldsymbol{a}$, so that

$$\mathcal{P}(\boldsymbol{x},\, t_{\rm d}) \approx \delta^3(\boldsymbol{x} - \boldsymbol{a}). \tag{3.132}$$

The successful observation process has therefore *altered* the particle's wave function *at that moment* from (3.127) to a form that is peaked around the detector's position. In the Copenhagen interpretation, successfully observing the particle's location is said to have triggered *wave function collapse*, sometimes referred to as *wave function update*. In the case under consideration, this phrase is understood to mean that the successful observation or measurement of the particle's position has collapsed the wave function onto *one of the particle's possible positions*, which is a position that had a nonzero probability according to the wave function describing the state of the particle. In the case under consideration, this means that the particle's position has become localized, within the spatial resolution $\mathrm{d}^n x$ of the detector, at a specific position within the volume V of all *possible locations*.

In the case of the free particle in three dimensions each possible location initially has the *same* nonzero probability d^3x/V to contain the particle, and so the particle is initially exhibiting delocalized *wave-like* behavior. However, the measurement forces *particle behavior* by measuring a point-like position, at least within the resolution $\mathrm{d}^n x$ of the detector. Bohr generalized this to what is known as the *complementarity principle*, which states that a quantum mechanical system can exhibit either particle properties or wave properties, but not both simultaneously. The use of *wave* versus *particle* here refers to *uncertainty* versus *certainty*. In the case under consideration, the momentum is known to great precision, while the position is completely uncertain. It will be seen that measuring the position to great precision, so that position is like that of a particle, renders the subsequent measurement of momentum uncertain, so that momentum is delocalized in momentum space. It is in this sense that Bohr referred to these two *observables* as complementary. This will be clarified when the nature of observables is examined in the coming section. However, at its core, Bohr's principle of complementarity states that the outcome of a measurement must include the circumstances or methods used to perform the measurement. In that sense, the nature of what is measured is also dependent upon how it is measured.

The concept of wave function collapse or update as the result of measurement has been and remains a controversial aspect of the Copenhagen interpretation. There is no widely accepted mathematical model for wave function collapse as the result of interaction with a measuring apparatus, but it is typically envisioned as an extremely sudden process during which all the other possible outcomes of measurement simply disappear. A large part of this problem stems from the fact that the particle is no longer isolated, and is instead interacting with the experimental apparatus during the act of measurement. Numerous physicists have argued that wave function collapse indicates that the current form of quantum mechanics is an *incomplete theory* since it cannot account for the measurement induced disappearance of possible outcomes, which constitutes a *loss of information* about the state of the system that existed *prior to measurement*. Bohm's *hidden variable theory*, Everett's *many worlds theory*, and the group of *objective-collapse theories* of quantum mechanics were developed in large part to deal with this conceptual paradox. Experimental measurements have not favored any of these alternatives to wave function collapse. In particular, Bell's theorem shows that hidden variable theories are incompatible with experimental results. Informal polls suggest that roughly half of physicists adhere to the Copenhagen interpretation of quantum mechanics that includes wave function collapse, while most of the rest are undecided. Questions surrounding the interpretation of quantum mechanics will be discussed further where relevant to the content being developed. There are further references at the end of this chapter.

3.5 Observables in wave mechanics

The next step in developing formal quantum mechanics is to understand the relationship of the wave function to *observable quantities*. This term is intended to include any of the observable aspects of motion in Newtonian physics. So far attention has been limited to position and momentum, which are the most basic aspects of motion. However, quantum mechanical observables should also include aspects of motion such as mechanical energy and angular momentum. The argument for their inclusion is based on their presence in the motion of macroscopic objects where Newtonian physics is valid. If such aspects of motion were not available for observation in the microscopic world, it is difficult to see how they could emerge in the macroscopic world.

3.5.1 The momentum operator

In the previous case of a free particle in a volume V, the momentum was selected to be identical within the limitations of preparation for each copy of the system. Momentum measurement can be modeled ideally as the application of momentum conservation to the collision of the particle with the detector. Since the wave function has been prepared for a particle with specific momentum $\boldsymbol{p}$, this means that each initial measurement of momentum in identically prepared systems must result in $\boldsymbol{p}$. As a result, the expectation value or statistical average for the momentum will also be $\boldsymbol{p}$. After the initial measurement, wave function collapse during the measurement may have altered the state of the particle, and the system may no longer be modeled by the free particle wave function. However, it is a critical step in the development of quantum mechanics to understand how the same formal statement (3.131) used to find the expectation value of position for a free particle can be adapted to find the expectation value of the momentum for a free particle. This expression must hold even though the probability density computed from the wave function gives the probability of finding the particle at the position $\boldsymbol{x}$ at the time t, rather than the probability of observing the particle momentum $\boldsymbol{p}$ at the time t.

The key step in this process is to note that the free particle wave function, given explicitly by (3.127) and written $\Psi(\boldsymbol{p}, \boldsymbol{x}, t)$ to emphasize its dependence on the momentum $\boldsymbol{p}$, has the mathematical property that

$$-i\hbar\nabla\Psi(\boldsymbol{p}, \boldsymbol{x}, t) = \boldsymbol{p}\,\Psi(\boldsymbol{p}, \boldsymbol{x}, t). \tag{3.133}$$

The quantity $-i\hbar\nabla$ has the requisite units of momentum and it *operates* on the wave function to produce the *vector valued* momentum $\boldsymbol{p}$ as a factor. For that reason, it is referred to as an *operator*, and in this specific case it is often referred to as a *differential operator* from the mathematical nature of the gradient. In the mathematical terminology of function space, the wave function $\Psi(\boldsymbol{p}, \boldsymbol{x}, t)$ is referred to as an *eigenfunction* of the *operator* $-i\hbar\nabla$ and $\boldsymbol{p}$ is the associated *eigenvalue*. The equation (3.133) is therefore referred to as an *eigenvalue equation*. It is important to note that the eigenvalue $\boldsymbol{p}$ must have no dependence on position in order that the eigenvalue equation of (3.133) has the property that

$$\begin{aligned}(-i\hbar\nabla)\cdot(-i\hbar\nabla)\,\Psi(\boldsymbol{p}, \boldsymbol{x}, t) &= -\,i\hbar\nabla\cdot(\boldsymbol{p}\,\Psi(\boldsymbol{p}, \boldsymbol{x}, t))\\ &= \boldsymbol{p}\cdot(-i\hbar\nabla\Psi(\boldsymbol{p}, \boldsymbol{x}, t)) = p^2\Psi(\boldsymbol{p}, \boldsymbol{x}, t).\end{aligned} \tag{3.134}$$

The action of the gradient is unaffected by the presence of the constant eigenvalue $\boldsymbol{p}$.

The differential operator $-i\hbar\nabla$ is therefore identified as the *momentum operator* in wave mechanics. Operators are designated with a caret, so that the momentum operator $\hat{\boldsymbol{p}}$ in *position space* is defined as

$$\hat{\boldsymbol{p}} = -i\hbar\nabla. \tag{3.135}$$

The eigenvalue equation of (3.133) is written

$$\hat{\boldsymbol{p}}\,\Psi(\boldsymbol{p}, \boldsymbol{x}, t) = \boldsymbol{p}\,\Psi(\boldsymbol{p}, \boldsymbol{x}, t), \tag{3.136}$$

which shows that the free particle wave function (3.127) is an *eigenfunction* of the momentum operator $\hat{\boldsymbol{p}}$ at all times t. Eigenvalue equations such as (3.136) characterize the nature of quantum mechanical observables. The restriction to *position space* is because (3.135) is the momentum operator that specifically operates on functions that depend on position. As quantum mechanics is developed it will be shown that the momentum operator can have different representations. Similarly, the position operator $\hat{\boldsymbol{x}}$ in *position space* will be defined by its action on a function $f(\boldsymbol{x}, t)$ of position and time, which is given by

$$\hat{\boldsymbol{x}}f(\boldsymbol{x}, t) = \boldsymbol{x}f(\boldsymbol{x}, t), \tag{3.137}$$

where $\boldsymbol{x}$ is the position argument of the function. Similarly, when applied to the function $f(\boldsymbol{y}, t)$ it gives

$$\hat{\boldsymbol{x}}f(\boldsymbol{y}, t) = \boldsymbol{y}f(\boldsymbol{y}, t), \tag{3.138}$$

At first glance, the position operator may seem trivial, but that is the result of choosing a position space representation for the wave function. This will change when the general formulation of quantum mechanics is developed in chapter 6.

Before proceeding, it is reasonable to consider if the identification (3.135) is limited to the specific case of a free particle. In that regard, the identification (3.135) can be further justified by considering the effect that the Galilean transformation (1.32) has on expression (3.133). Noting that the Galilean transformation can be writtent $\boldsymbol{x} = \boldsymbol{x}' + \boldsymbol{v}_o t'$, the chain rule shows that the gradient operator according to observer $\mathcal{O}'$ is related to the gradient operator according to observer $\mathcal{O}$ by

$$\frac{\partial}{\partial x_j'} = \sum_{k=1}^{3} \frac{\partial x_k}{\partial x_j'} \frac{\partial}{\partial x_k} + \frac{\partial t}{\partial x_j'} \frac{\partial}{\partial t} = \sum_{k=1}^{3} \delta_{jk} \frac{\partial}{\partial x_k} = \frac{\partial}{\partial x_j} \implies \nabla' = \nabla, \tag{3.139}$$

which follows since distances remain constant as a result of Galilean transformations. On the other hand, the partial time derivative becomes

$$\frac{\partial}{\partial t'} = \frac{\partial t}{\partial t'} \frac{\partial}{\partial t} + \sum_{k=1}^{3} \frac{\partial x_k}{\partial t'} \frac{\partial}{\partial x_k} = \frac{\partial}{\partial t} + \sum_{k=1}^{3} v_o^k \frac{\partial}{\partial x_k} = \frac{\partial}{\partial t} + \boldsymbol{v}_o \cdot \nabla, \tag{3.140}$$

which occurs because velocities, the *rate* of distance change, are not invariant under a Galilean transformation.

Applying the rule (3.115) for the Galilean transformation of the wave function to equation (3.133) and using the phase angle (3.119) shows that the transformed version of (3.133) is given by

$$\begin{aligned} -i\hbar\nabla'\Psi'(\boldsymbol{p}', \boldsymbol{x}', t') &= -i\hbar\nabla(e^{-i\alpha(\boldsymbol{v}_o, \boldsymbol{x}, t)}\Psi(\boldsymbol{p}, \boldsymbol{x}, t)) \\ &= e^{-i\alpha(\boldsymbol{v}_o, \boldsymbol{x}, t)}(-i\hbar\nabla\Psi(\boldsymbol{p}, \boldsymbol{x}, t) - \hbar\nabla\alpha(\boldsymbol{v}_o, \boldsymbol{x}, t)\Psi(\boldsymbol{p}, \boldsymbol{x}, t)) \\ &= (\boldsymbol{p} - m\boldsymbol{v}_o)e^{-i\alpha(\boldsymbol{v}_o, \boldsymbol{x}, t)}\Psi(\boldsymbol{p}, \boldsymbol{x}, t) \\ &= \boldsymbol{p}'\Psi'(\boldsymbol{p}', \boldsymbol{x}', t'), \end{aligned} \tag{3.141}$$

where $\boldsymbol{p}' = \boldsymbol{p} - m\boldsymbol{v}_o$ is the correct Newtonian momentum observed by a Galilean observer moving relative to the first observer. The eigenvalue equation (3.133) is said to be *covariant* under Galilean transformations, which means that all Galilean observers will agree on its form.

Result (3.141) strongly supports the idea that momentum is correctly represented by the differential operator $\hat{\boldsymbol{p}}$ of (3.135). However, this identification must hold even when dealing with wave functions that are *not* eigenfunctions of $\hat{\boldsymbol{p}}$. In order to verify this, the expectation value of momentum *for an arbitrary wave function* is defined in a manner identical to the expectation value of position given by (3.129), so that

$$\langle \hat{\boldsymbol{p}} \rangle = \int_V \mathrm{d}^n x \, \Psi^*(\boldsymbol{x}, t) \, (\hat{\boldsymbol{p}} \, \Psi(\boldsymbol{x}, t)) = -i\hbar \int_V \mathrm{d}^n x \, \Psi^*(\boldsymbol{x}, t) \nabla\Psi(\boldsymbol{x}, t). \tag{3.142}$$

The expectation value $\langle \boldsymbol{p} \rangle$ has the property that a Galilean transformation of the wave function and the momentum operator results in

$$\begin{aligned}\langle \hat{\boldsymbol{p}}' \rangle &= -i\hbar \int \mathrm{d}^n x\, \Psi'^*(\boldsymbol{x}', t')\nabla'\Psi'(\boldsymbol{x}', t') = -i\hbar \int \mathrm{d}^n x\, e^{ia}\Psi^*(\boldsymbol{x}, t)\nabla(e^{-ia}\Psi(\boldsymbol{x}, t)) \\ &= -i\hbar \int \mathrm{d}^n x\, \Psi^*(\boldsymbol{x}, t)\nabla\Psi(\boldsymbol{x}, t) - m\boldsymbol{v}_o \int_V \mathrm{d}^n x\, \Psi^*(\boldsymbol{x}, t)\Psi(\boldsymbol{x}, t) \\ &= \langle \hat{\boldsymbol{p}} \rangle - m\boldsymbol{v}_o,\end{aligned} \tag{3.143}$$

where the normalization of the wave function was used in the last step. This shows that, even in the case that the wave function is not an eigenfunction of the momentum operator, the expectation value of $\hat{\boldsymbol{p}}$ behaves as momentum under a Galilean transformation. The interpretation of $\langle \hat{\boldsymbol{p}} \rangle$ is analogous to $\langle \hat{\boldsymbol{x}} \rangle$. It is the statistical average of many measurements of the momentum in a set of systems that are identically prepared.

It is instructive to evaluate (3.142) for the case that the wave function is given by the free particle version (3.127) in a three-dimensional box, which is an eigenfunction of $\hat{\boldsymbol{p}}$. It follows that

$$\begin{aligned}\langle \hat{\boldsymbol{p}} \rangle &= -i\hbar \int \mathrm{d}^3 x\, \frac{1}{L^3} e^{-i(\boldsymbol{p}\cdot\boldsymbol{x} - p^2 t/2m)/\hbar} \nabla\, e^{i(\boldsymbol{p}\cdot\boldsymbol{x} - p^2 t/2m)/\hbar} \\ &= \frac{\boldsymbol{p}}{L^3} \int_0^L \mathrm{d}x \int_0^L \mathrm{d}y \int_0^L \mathrm{d}z = \boldsymbol{p},\end{aligned} \tag{3.144}$$

which is the anticipated result for the case that each copy of the system has the same value of $\boldsymbol{p}$. This further supports identifying the momentum operator as (3.135).

3.5.2 Quantum mechanical observables as linear operators

Using the position operator $\hat{\boldsymbol{x}}$ and the momentum operator $\hat{\boldsymbol{p}}$ allows the observables of Newtonian physics to be translated into the quantum mechanical format, where they become *linear operators* that act on the wave function. The term linear is derived from the action of the operators on linear combinations of functions. For example, the momentum operator has the property that

$$\hat{\boldsymbol{p}}(a\Psi_1(\boldsymbol{x}, t) + b\Psi_2(\boldsymbol{x}, t)) = a\,\hat{\boldsymbol{p}}\,\Psi_1(\boldsymbol{x}, t) + b\,\hat{\boldsymbol{p}}\,\Psi_2(\boldsymbol{x}, t), \tag{3.145}$$

where a and b are constants. This is the definition of a linear operator. By virtue of the linearity of the momentum and position operators, the observables of quantum mechanics formed from products of the two will also be linear. It is important to note that this definition excludes an operator $\hat{A}$ whose action is given by $\hat{A}f(\boldsymbol{x}, t) = f^2(\boldsymbol{x}, t)$ since

$$\hat{A}(a\Psi_1(\boldsymbol{x}, t) + b\Psi_2(\boldsymbol{x}, t)) = (a\Psi_1(\boldsymbol{x}, t) + b\Psi_2(\boldsymbol{x}, t))^2 \neq a\Psi_1^2(\boldsymbol{x}, t) + b\Psi_2^2(\boldsymbol{x}, t). \tag{3.146}$$

An operator $\hat{A}$ with such an action is known as a *nonlinear operator*.

As an example of a linear operator formed from products of the $\hat{\boldsymbol{x}}$ or $\hat{\boldsymbol{p}}$, the kinetic energy K of classical mechanics becomes the wave mechanical operator $\hat{K}$, given by the prescription

$$K = \frac{p^2}{2m} \rightarrow \hat{K} = \frac{\hat{\boldsymbol{p}} \cdot \hat{\boldsymbol{p}}}{2m} = -\frac{\hbar^2}{2m}\nabla \cdot \nabla = -\frac{\hbar^2}{2m}\nabla^2. \tag{3.147}$$

Another example is the angular momentum vector $\boldsymbol{L}$, which becomes the vector operator $\hat{\boldsymbol{L}}$ given by

$$\boldsymbol{L} = \boldsymbol{x} \times \boldsymbol{p} \rightarrow \hat{\boldsymbol{L}} = \hat{\boldsymbol{x}} \times \hat{\boldsymbol{p}} = -i\hbar\, \hat{\boldsymbol{x}} \times \nabla. \tag{3.148}$$

Using this prescription, classical observables are translated into linear operators that represent the associated *quantum mechanical observables.*

It is instructive to apply the two operators (3.147) and (3.148) to the free particle wave function (3.127). It follows that

$$\begin{aligned} \hat{K}\Psi(\boldsymbol{p}, \boldsymbol{x}, t) &= \left(-\frac{\hbar^2}{2m}\nabla^2\right)\left(\frac{1}{\sqrt{V}}\exp\left(i(\boldsymbol{p}\cdot\boldsymbol{x} - E_{\mathrm{p}}t)/\hbar\right)\right) = \left(\frac{p^2}{2m}\right)\Psi(\boldsymbol{p}, \boldsymbol{x}, t), \\ \hat{\boldsymbol{L}}\Psi(\boldsymbol{p}, \boldsymbol{x}, t) &= (-i\hbar\, \hat{\boldsymbol{x}} \times \nabla)\left(\frac{1}{\sqrt{V}}\exp\left(i(\boldsymbol{p}\cdot\boldsymbol{x} - E_{\mathrm{p}}t)/\hbar\right)\right) = (\boldsymbol{x} \times \boldsymbol{p})\Psi(\boldsymbol{p}, \boldsymbol{x}, t). \end{aligned} \tag{3.149}$$

This shows that the free particle wave function (3.127) is an eigenfunction of the kinetic energy operator $\hat{K}$ with the expected eigenvalue of $p^2/2m$. However, it is *not* an eigenfunction of the angular momentum operator $\hat{\boldsymbol{L}}$ since $\boldsymbol{x} \times \boldsymbol{p}$ is not independent of $\boldsymbol{x}$.

In the functional notation of (3.50) the position and momentum expectation values are written

$$\begin{aligned} \langle \hat{\boldsymbol{x}} \rangle &= (\Psi, \hat{\boldsymbol{x}}\, \Psi), \\ \langle \hat{\boldsymbol{p}} \rangle &= (\Psi, \hat{\boldsymbol{p}}\, \Psi). \end{aligned} \tag{3.150}$$

This result will now be generalized to an arbitrary linear operator $\hat{\mathrm{O}}$ derived using the prescription employed in (3.147) and (3.148). The expectation value of $\hat{\mathrm{O}}$ for the wave function $\Psi(\boldsymbol{x}, t)$ is given by an expression identical to (3.150),

$$\langle \hat{\mathrm{O}} \rangle = \int \mathrm{d}^n x\, \Psi^*(\boldsymbol{x}, t)\, (\hat{\mathrm{O}}\, \Psi(\boldsymbol{x}, t)) = (\Psi, \hat{\mathrm{O}}\Psi), \tag{3.151}$$

where the operator $\hat{\mathrm{O}}$ is acting on $\Psi(\boldsymbol{x}, t)$ in the inner product. It has the same Copenhagen interpretation discussed earlier, which is that it is the predicted statistical average of many measurements of the observable $\hat{\mathrm{O}}$ for identical preparations of the wave function of the system. It is important to note that the complex conjugate of (3.151) is given in the functional notation of (3.50) by

$$\begin{aligned} \langle \hat{\mathrm{O}} \rangle^* &= \left(\int_V \mathrm{d}^n x\, \Psi^*(\boldsymbol{x}, t)\, (\hat{\mathrm{O}}\, \Psi(\boldsymbol{x}, t))\right)^* \\ &= \int_V \mathrm{d}^n x\, (\hat{\mathrm{O}}\, \Psi(\boldsymbol{x}, t))^* \Psi(\boldsymbol{x}, t) = (\hat{\mathrm{O}}\Psi, \Psi). \end{aligned} \tag{3.152}$$

The inner product of different functions and operators have the property that

$$\begin{aligned}(\Psi_1, \hat{O}\Psi_2)^* &= \left(\int_V d^n x \, \Psi_1^*(\boldsymbol{x}, t)(\hat{O}\,\Psi_2(\boldsymbol{x}, t))\right)^* \\ &= \int_V d^n x \, (\hat{O}\,\Psi_2(\boldsymbol{x}, t))^* \Psi_1(\boldsymbol{x}, t) = (\hat{O}\Psi_2, \Psi_1).\end{aligned} \tag{3.153}$$

Because expectation values of observables must be real, this property will be of great importance.

3.5.3 Hermitian linear operators

Before proceeding to a formal postulate regarding quantum mechanical observables, it is important to outline additional properties and classifications of linear operators. This will be done in the context of wave mechanics, where the operators act upon functions in L^2. It is useful to define an operator related to $\hat{O}$, denoted $\hat{O}^\dagger$, that is referred to as *the adjoint of* $\hat{O}$. This operator is defined using two possible wave functions $\Psi_1(\boldsymbol{x}, t)$ and $\Psi_2(\boldsymbol{x}, t)$ by the equality

$$\begin{aligned}(\hat{O}^\dagger\Psi_1, \Psi_2) &= \int d^n x \left(\hat{O}^\dagger \Psi_1(\boldsymbol{x}, t)\right)^* \Psi_2(\boldsymbol{x}, t) \\ &= \int_V d^n x \, \Psi_1^*(\boldsymbol{x}, t)\, \hat{O}\Psi_2(\boldsymbol{x}, t) = (\Psi_1, \hat{O}\Psi_2).\end{aligned} \tag{3.154}$$

Exchanging Ψ_1 and Ψ_2 and then taking the complex conjugate of both sides of (3.154) and using (3.153) gives

$$\begin{aligned}(\Psi_2, \hat{O}\Psi_1) = (\hat{O}^\dagger\Psi_2, \Psi_1) &\implies (\Psi_2, \hat{O}\Psi_1)^* \\ = (\hat{O}^\dagger\Psi_2, \Psi_1)^* &\implies (\hat{O}\Psi_1, \Psi_2) = (\Psi_1, \hat{O}^\dagger\Psi_2).\end{aligned} \tag{3.155}$$

Using properties (3.154) and (3.155) sequentially gives

$$(\Psi_1, \hat{O}\Psi_2) = (\hat{O}^\dagger\Psi_1, \Psi_2) = (\Psi_1, \hat{O}^{\dagger\dagger}\Psi_2), \tag{3.156}$$

which immediately yields

$$\hat{O}^{\dagger\dagger} = \hat{O}. \tag{3.157}$$

The definition of the adjoint operator shows that the product of two operators, $\hat{A}$ and $\hat{B}$, has the adjoint determined by

$$((\hat{A}\hat{B})^\dagger\Psi_1, \Psi_2) = (\Psi_1, \hat{A}\hat{B}\Psi_2) = (\hat{A}^\dagger\Psi_1, \hat{B}\Psi_2) = (\hat{B}^\dagger\hat{A}^\dagger\Psi_1, \Psi_2), \tag{3.158}$$

which shows that

$$(\hat{A}\hat{B})^\dagger = \hat{B}^\dagger\hat{A}^\dagger. \tag{3.159}$$

This is the first indication that the order of operators is of great importance in quantum mechanics.

As a final important result, it is possible to consider the adjoint of a linear combination of operators, $\hat{C} = \alpha\hat{A} + \beta\hat{B}$, where α and β are complex numbers. This begins by noting that if α is a complex constant then

$$\begin{aligned}((\alpha\hat{A})\Psi_1, \Psi_2) &= \alpha \int d^n x(\alpha\hat{A}\Psi_1(\boldsymbol{x}, t))^* \Psi_2(\boldsymbol{x}, t) \\ &= \alpha^* \int d^n x\, (\hat{A}\Psi_1(\boldsymbol{x}, t))^* \Psi_2(\boldsymbol{x}, t) = \alpha^*(\hat{A}\Psi_1, \Psi_2).\end{aligned} \tag{3.160}$$

Using this result shows that the linear combination of operators has the property that

$$\begin{aligned}(\hat{C}^\dagger\Psi_1, \Psi_2) &= (\Psi_1, \hat{C}\Psi_2) = \alpha(\Psi_1, \hat{A}\Psi_2) + \beta(\Psi_1, \hat{B}\Psi_2) \\ &= \alpha(\hat{A}^\dagger\Psi_1, \Psi_2) + \beta(\hat{B}^\dagger\Psi_1, \Psi_2) \\ &= ((\alpha^*\hat{A}^\dagger + \beta^*\hat{B}^\dagger)\Psi_1, \Psi_2),\end{aligned} \tag{3.161}$$

from which it follows that

$$(\alpha\hat{A} + \beta\hat{B})^\dagger = \alpha^*\hat{A}^\dagger + \beta^*\hat{B}^\dagger. \tag{3.162}$$

An operator is said to be *self-adjoint* or *Hermitian* if $\hat{O}^\dagger = \hat{O}$. It is instructive to examine the two cases of $\hat{\boldsymbol{x}}$ and $\hat{\boldsymbol{p}}$. The wave mechanical position operator $\hat{\boldsymbol{x}}$ is trivial in the sense that it simply creates a factor of $\boldsymbol{x}$ when applied to a function of $\boldsymbol{x}$, and has no further action on the function. Furthermore, since the position $\boldsymbol{x}$ is real, so that $\boldsymbol{x}^*=\boldsymbol{x}$, it follows that the position operator is Hermitian,

$$\begin{aligned}(\hat{\boldsymbol{x}}\Psi_1, \Psi_2) = \int d^n x\, (\boldsymbol{x}\, \Psi_1(\boldsymbol{x}, t))^* \Psi_2(\boldsymbol{x}, t) &= \int d^n x\, \Psi_1^*(\boldsymbol{x}, t)\, \boldsymbol{x}\Psi_2(\boldsymbol{x}, t) \\ &= (\Psi_1, \hat{\boldsymbol{x}}\Psi_2).\end{aligned} \tag{3.163}$$

The wave mechanical momentum operator is also Hermitian, but the demonstration is more complicated since it requires an integration by parts,

$$\begin{aligned}(\hat{\boldsymbol{p}}\Psi_1, \Psi_2) &= \int d^n x\, (-i\hbar\nabla\Psi_1(\boldsymbol{x}, t))^* \Psi_2(\boldsymbol{x}, t) = \int d^n x\, (i\hbar\nabla\Psi_1^*(\boldsymbol{x}, t))\Psi_2(\boldsymbol{x}, t)) \\ &= i\hbar \int d^n x\, \nabla(\Psi_1^*(\boldsymbol{x}, t)\Psi_2(\boldsymbol{x}, t)) + \int d^n x\, \Psi_1^*(\boldsymbol{x}, t)(-i\hbar\nabla\Psi_2(\boldsymbol{x}, t)) \\ &= (\Psi_1, \hat{\boldsymbol{p}}\Psi_2),\end{aligned} \tag{3.164}$$

where the surface term generated by the first integral is assumed to vanish. For the case of expectation values the surface term will vanish if the probability density at the boundaries is zero or, as is the case with the free particle, the probability density is the same at the two boundaries of each surface term. However, this strongly indicates that boundary conditions for the probability density and therefore for the wave function will be important. For these restrictions it follows that $\hat{\boldsymbol{x}}^\dagger = \hat{\boldsymbol{x}}$ and $\hat{\boldsymbol{p}}^\dagger = \hat{\boldsymbol{p}}$. Result (3.159) shows that products of either $\hat{\boldsymbol{x}}$ or $\hat{\boldsymbol{p}}$ will also be Hermitian. However, operators constructed from the product of the two, such as angular momentum, will require further examination to determinine their *Hermiticity* due to result (3.159), which changes their order in the adjoint version of the operator.

A very important property of the self-adjoint or Hermitian operators acting on L^2 is that they have *orthonormal eigenfunctions with real eigenvalues.* Demonstrating the reality of the eigenvalues begins by assuming that the Hermitian operator $\hat{O}$ has a nonzero eigenfunction $f_\lambda(\boldsymbol{x})$ that satisfies

$$\hat{O}f_\lambda(\boldsymbol{x}) = \lambda f_\lambda(\boldsymbol{x}), \tag{3.165}$$

where λ is the associated eigenvalue. For the moment, λ is treated as a scalar. However, if $\hat{O}$ is a vector valued operator its eigenvalues will also be vector valued. Taking the complex conjugate of (3.165) gives

$$\left(\hat{O}f_\lambda(\boldsymbol{x})\right)^* = \left(\lambda f_\lambda(\boldsymbol{x})\right)^* = \lambda^* f_\lambda^*(\boldsymbol{x}). \tag{3.166}$$

Using (3.165) and (3.166) in the definition (3.154) of a self-adjoint operator immediately gives

$$(\hat{O}f_\lambda, f_\lambda) - (f_\lambda, \hat{O}f_\lambda) = (\lambda^* - \lambda)(f_\lambda, f_\lambda) = 0 \implies \lambda = \lambda^*, \tag{3.167}$$

where, by virtue of belonging to L^2 the inner product (f_λ, f_λ) is bounded and positive-definite. As a result, the eigenvalue λ is real. Next, the orthogonality of the eigenfunctions is demonstrated by using a second eigenfunction $f_{\lambda'}(\boldsymbol{x})$. Using the self-adjointness and the reality of the eigenvalues gives

$$(\hat{O}f_{\lambda'}, f_\lambda) - (f_{\lambda'}, \hat{O}f_\lambda) = (\lambda' - \lambda)(f_{\lambda'}, f_\lambda) = 0 \implies (f_{\lambda'}, f_\lambda) = 0, \tag{3.168}$$

which holds as long as the two eigenvalues satisfy $\lambda \neq \lambda'$.

The eigenvalue statement (3.165) assumes that $\hat{O}$ and its eigenfunctions have no *explicit* time-dependence. The conditions under which the wave function $\Psi(\lambda, \boldsymbol{x}, t)$ can be an eigenfunction of $\hat{O}$ for a range of times t requires understanding how both the wave function and the operator evolve in time. In other words, if the wave function of the system is an eigenfunction of $\hat{O}$ at $t = 0$, will it be an eigenfunction at a later time t,

$$\hat{O}\Psi(\lambda, \boldsymbol{x}, t = 0) = \lambda\Psi(\lambda, \boldsymbol{x}, t = 0) \overset{?}{\implies} \hat{O}\Psi(\lambda, \boldsymbol{x}, t) = \lambda\Psi(\lambda, \boldsymbol{x}, t). \tag{3.169}$$

Clearly, (3.169) holds for the case that $\hat{O} = \hat{\boldsymbol{p}}$ and the wave function is given by the free particle form (3.127). For now, (3.169) will remain an open question. This is revisited later in this chapter when Poisson brackets are discussed, but its full resolution is presented in chapter 4.

If $\lambda = \lambda'$ for two *distinct eigenfunctions*, then the eigenfunctions are said to be *degenerate* and the proof presented in (3.168) breaks down. However, even if the two eigenfunctions are not orthogonal, they can be combined linearly to obtain two orthogonal eigenfunctions. This is easily demonstrated for the case of two degenerate eigenfunctions $f(\boldsymbol{x})$ and $g(\boldsymbol{x})$. These are used to define two orthogonal eigenfunctions, $f_1(\boldsymbol{x})$ and $f_2(\boldsymbol{x})$, from $f(\boldsymbol{x})$ and $g(\boldsymbol{x})$ through the definitions

$$f_1(x) = \frac{f(x)}{|f|},$$
$$f_2(x) = g(x) - (g, f_1)f_1(x) = g(x) - \frac{1}{|f|^2}(g, f)f(x). \tag{3.170}$$

It follows that these two functions are both eigenfunctions since they are linear combinations of $f(x)$ and $g(x)$, and they are orthogonal,

$$(f_2, f_1) = (g, f_1) - (g, f_1)(f_1, f_1) = (g, f_1) - (g, f_1) = 0, \tag{3.171}$$

where the property $(f_1, f_1) = 1$ was used. The function $f_2(x)$ is not necessarily normalized, but using the form for $f_2(x)$ it follows that

$$(f_2, f_2) = (g, g) - |(f_1, g)|^2 = (g, g) - \frac{1}{|f|^2}|(f, g)|^2 = |f_2|^2, \tag{3.172}$$

where the property $(f_1, g)^* = (g, f_1)$ was used. This allows $f_2(x)$ to be normalized. This can be generalized to an arbitrary number of degenerate eigenfunctions and is known as the *Gram–Schmidt orthonormalization procedure*. Its details are available in the references.

Because these functions are members of L^2 they can be made into orthonormal functions by dividing by their norm. The L^2 eigenfunctions of a Hermitian operator are therefore both *orthonormal* and associated with *real eigenvalues*. As a preview of further developments, it is the eigenvalues of quantum mechanical observables that will represent the possible outcome of experiments to measure that observable, and so the reality of the eigenvalues is essential in order to be a measurable and therefore observable quantity.

It is important to examine the simplest orthonormal eigenfunctions of the operator $\hat{p}$, which are given in three spatial dimensions by

$$f_p(x) = \frac{e^{ip\cdot x/\hbar}}{\sqrt{(2\pi\hbar)^3}}. \tag{3.173}$$

These functions clearly satisfy the eigenvalue equation

$$\hat{p}f_p(x) = pf_p(x). \tag{3.174}$$

As predicted by (3.168) these eigenfunctions are orthonormal in the *continuum sense*, since result (3.99) shows that they satisfy

$$(f_{p'}, f_p) = \int \frac{d^3x}{(2\pi\hbar)^3} e^{i(p-p')\cdot(x)/\hbar} = \delta^3(p - p'). \tag{3.175}$$

The eigenfunctions of (3.173) are a *complete set* of orthonormal functions. This follows from combining the Fourier transform (3.95) with the identification $k = p/\hbar$, which shows that the three-dimensional Dirac delta can be written

$$\delta^3(\boldsymbol{x} - \boldsymbol{y}) = \int \frac{\mathrm{d}^3 p}{(2\pi\hbar)^3} e^{i\boldsymbol{p}\cdot(\boldsymbol{x}-\boldsymbol{y})/\hbar} = \int \mathrm{d}^3 p\, f_{\boldsymbol{p}}^*(\boldsymbol{y}) f_{\boldsymbol{p}}(\boldsymbol{x}). \tag{3.176}$$

This allows an *arbitrary* piecewise and square integrable function $f(\boldsymbol{x})$ to be expanded in terms of the momentum eigenfunctions by using (3.176) and interchanging the order of integration to obtain

$$\begin{aligned} f(\boldsymbol{x}) &= \int \mathrm{d}^3 y\, f(\boldsymbol{y})\, \delta^3(\boldsymbol{x} - \boldsymbol{y}) = \int \frac{\mathrm{d}^3 p}{(2\pi\hbar)^3} \left(\int \mathrm{d}^3 y\, f(\boldsymbol{y})\, e^{-i\boldsymbol{p}\cdot\boldsymbol{y}/\hbar} \right) e^{i\boldsymbol{p}\cdot\boldsymbol{x}/\hbar} \\ &= \int \frac{\mathrm{d}^3 p}{(2\pi\hbar)^3} \tilde{f}(\boldsymbol{p})\, e^{i\boldsymbol{p}\cdot\boldsymbol{x}/\hbar}, \end{aligned} \tag{3.177}$$

where

$$\tilde{f}(\boldsymbol{p}) = \int \mathrm{d}^3 y\, f(\boldsymbol{y})\, e^{-i\boldsymbol{p}\cdot\boldsymbol{y}/\hbar}. \tag{3.178}$$

Result (3.177) shows that an arbitrary function of position can be written as a *continuous superposition of orthonormal momentum eigenfunctions.*

This ignores possible problems with convergence of the superposition to the original function. This can be addressed by adapting the Bessel inequality for the convergence of the Fourier series derived in (3.66) to the one-dimensional version of the Fourier transform (3.177), which gives the inequality

$$(f, f) - \int_{-\infty}^{\infty} \frac{\mathrm{d}p}{2\pi\hbar} |\tilde{f}(p)|^2 \geqslant 0, \tag{3.179}$$

with the equality holding for the case that the integral representation (3.177) converges to the function $f(x)$. This inequality can be examined for the specific simple but nontrivial case $f(x) = \exp(-\frac{1}{2}\alpha x^2)$. It follows that

$$(f, f) = \int_{-\infty}^{\infty} \mathrm{d}x \exp(-\alpha x^2) = \sqrt{\frac{\pi}{\alpha}}. \tag{3.180}$$

Likewise, the coefficient (3.178) is given by

$$\tilde{f}(p) = \int_{-\infty}^{\infty} \mathrm{d}y \exp\left(-\frac{ipy}{\hbar}\right) \exp\left(-\frac{1}{2}\alpha y^2\right) = \sqrt{\frac{2\pi}{\alpha}} \exp\left(-\frac{p^2}{2\hbar^2\alpha}\right). \tag{3.181}$$

This gives the result

$$\int_{-\infty}^{\infty} \frac{\mathrm{d}p}{2\pi\hbar} |\tilde{f}(p)|^2 = \frac{2\pi}{\alpha} \int_{-\infty}^{\infty} \frac{\mathrm{d}p}{2\pi\hbar} \exp\left(-\frac{p^2}{\hbar^2\alpha}\right) = \sqrt{\frac{\pi}{\alpha}}, \tag{3.182}$$

which shows that the equality in (3.179) is met. As a result, there is exact convergence in this simple case. A general proof of completeness for the basis functions of (3.173) is given in the references.

3.5.4 Commutators and eigenfunctions

In classical mechanics position and momentum are the two fundamental observables. Other quantities, such as energy and angular momentum, are typically constructed from them. This will also be the case for quantum mechanical observables. However, there is an unusual aspect to the identification (3.135) that must be understood first, which is that the two quantum mechanical observables $\hat{\boldsymbol{x}}$ and $\hat{\boldsymbol{p}}$ no longer *commute*. This follows from the simple observation that, for an arbitrary wave function $\Psi(\boldsymbol{x}, t)$, the product of components of the two obeys the following relation,

$$(\hat{x}_j\hat{p}_k - \hat{p}_k\hat{x}_j)\Psi(\boldsymbol{x}, t) = -i\hbar x_j\frac{\partial\Psi(\boldsymbol{x}, t)}{\partial x_k} + i\hbar\frac{\partial}{\partial x_k}\big(x_j\Psi(\boldsymbol{x}, t)\big) = i\hbar\, \delta_{jk}\Psi(\boldsymbol{x}, t), \quad (3.183)$$

where δ_{jk} is the Kronecker delta (3.57). This leads to the definition of the *commutator*, which is written

$$[x_j, \hat{p}_k] = x_j\hat{p}_k - \hat{p}_k x_j = i\hbar\, \delta_{jk}. \quad (3.184)$$

The commutator has the obvious property $[x_j, \hat{p}_k] = -[\hat{p}_k, x_j]$.

It will be useful to note that the commutator of powers is defined by

$$[x_j^n, \hat{p}_k] = x_j^n\hat{p}_k - \hat{p}_k x_j^n. \quad (3.185)$$

For the case that $n = 2$ the commutator is given by

$$\begin{aligned}\left[x_j^2, \hat{p}_k\right] &= x_j^2\hat{p}_k - x_j\hat{p}_k x_j + x_j\hat{p}_k x_j - \hat{p}_k x_j^2 \\ &= x_j\big(x_j\hat{p}_k - \hat{p}_k x_j\big) + \big(x_j\hat{p}_k - \hat{p}_k x_j\big)x_j = 2i\hbar\, x_j\delta_{jk}.\end{aligned} \quad (3.186)$$

This generalizes by induction to give

$$[x_j^n, \hat{p}_k] = i\hbar n\, x_j^{n-1}\delta_{jk} = i\hbar\, \delta_{jk}\frac{\partial}{\partial x_j}\big(x_j^n\big). \quad (3.187)$$

Likewise, it follows that

$$[\hat{p}_k^n, \hat{x}_j] = -i\hbar n\, \hat{p}_k^{n-1}\delta_{jk} = -i\hbar\, \delta_{jk}\frac{\partial}{\partial \hat{p}_k}\big(\hat{p}_k^n\big). \quad (3.188)$$

For the case of a potential $V(x)$ that is solely a function of position, it follows that

$$[V(x), \hat{p}_k] = i\hbar\frac{\partial}{\partial x_j}V(x). \quad (3.189)$$

These results will be used again in chapter 6.

The failure of two observables to commute, which occurs if their commutator is nonzero, has profound implications for their simultaneous observability. To start the process of understanding this, two possible observables $\hat{\mathrm{A}}$ and $\hat{\mathrm{B}}$ are identified. The next step is to ask if there is a set of functions that serve as mutual eigenfunctions for both of them. Such an eigenfunction $f(a, b, \boldsymbol{x})$ would have the property

$$\hat{\mathrm{A}}f(a, b, \boldsymbol{x}) = af(a, b, \boldsymbol{x}), \quad \hat{\mathrm{B}}f(a, b, \boldsymbol{x}) = bf(a, b, \boldsymbol{x}), \tag{3.190}$$

where a and b are the respective eigenvalues of $\hat{\mathrm{A}}$ and $\hat{\mathrm{B}}$. Using the properties of linear operators given by (3.145), it follows that

$$\hat{\mathrm{A}}\hat{\mathrm{B}}f(a, b, \boldsymbol{x}) = \hat{\mathrm{A}}bf(a, b, \boldsymbol{x}) = b\hat{\mathrm{A}}f(a, b, \boldsymbol{x}) = baf(a, b, \boldsymbol{x}), \tag{3.191}$$

with a similar statement for the reversed order,

$$\hat{\mathrm{B}}\hat{\mathrm{A}}f(a, b, \boldsymbol{x}) = \hat{\mathrm{B}}af(a, b, \boldsymbol{x}) = a\hat{\mathrm{B}}f(a, b, \boldsymbol{x}) = abf(a, b, \boldsymbol{x}), \tag{3.192}$$

Subtracting (3.192) from (3.191) gives

$$\begin{aligned}(\hat{\mathrm{A}}\hat{\mathrm{B}} - \hat{\mathrm{B}}\hat{\mathrm{A}})f(a, b, \boldsymbol{x}) &= [\hat{\mathrm{A}}, \hat{\mathrm{B}}]f(a, b, \boldsymbol{x}) \\ &= (ba - ab)f(a, b, \boldsymbol{x}) = 0 \implies [\hat{\mathrm{A}}, \hat{\mathrm{B}}] = 0,\end{aligned} \tag{3.193}$$

where the fact that the eigenvalues commute was used. This gives the result that, for two operators to share a set of mutual eigenfunctions, the two operators *must commute.* Since $\hat{\boldsymbol{x}}$ and $\hat{\boldsymbol{p}}$ do not commute, it is impossible to find a set of mutual eigenfunctions. This is the first step toward understanding why the momentum eigenfunction (3.127) corresponds to a particle that is delocalized. It will lead to the uncertainty principle in the next chapter.

The next step is to consider the implications of the commutator $[\hat{\mathrm{A}}, \hat{\mathrm{B}}]$ vanishing. If $f_a(\boldsymbol{x})$ is an eigenfunction of the operator $\hat{\mathrm{A}}$, such that $\hat{\mathrm{A}}f_a(\boldsymbol{x}) = af_a(\boldsymbol{x})$, then it follows that

$$\hat{\mathrm{A}}\left(\hat{\mathrm{B}}f_a(\boldsymbol{x})\right) = \hat{\mathrm{B}}\left(\hat{\mathrm{A}}f_a(\boldsymbol{x})\right) + [\hat{\mathrm{A}}, \hat{\mathrm{B}}]f_a(\boldsymbol{x}) = \hat{\mathrm{B}}\left(af_a(\boldsymbol{x})\right) = a\left(\hat{\mathrm{B}}f_a(\boldsymbol{x})\right). \tag{3.194}$$

Ignoring the possibility that $\hat{\mathrm{B}}f_a(\boldsymbol{x}) = 0$, this can hold only if $\hat{\mathrm{B}}f_a(\boldsymbol{x}) \propto f_a(\boldsymbol{x})$, so that $\hat{\mathrm{B}}f_a(\boldsymbol{x}) = c\, f_a(\boldsymbol{x})$. The constant of proportionality c is therefore an eigenvalue of $\hat{\mathrm{B}}$, and so this shows that an eigenfunction $f_a(\boldsymbol{x})$ of the observable $\hat{\mathrm{A}}$ is also an eigenfunction of $\hat{\mathrm{B}}$ as long as $[\hat{\mathrm{A}}, \hat{\mathrm{B}}] = 0$.

3.5.5 Commutators and Poisson brackets

It is very useful to note that the commutator (3.184) is identical in form to the Poisson bracket (1.27) for the classical quantities x_j and p_k, although the commutator possesses an additional factor of $i\hbar$. Both the commutator and the Poisson bracket are antisymmetric. The one to one relation of Poisson brackets and commutators is therefore defined by the following recipe. Given two classical observables, $A(\boldsymbol{x}, \boldsymbol{p}, t)$ and $B(\boldsymbol{x}, \boldsymbol{p}, t)$, all classical Poisson bracket relationships can be replaced by the quantum mechanical commutator of the two associated operators, $\hat{\mathrm{A}}(\hat{\boldsymbol{x}}, \hat{\boldsymbol{p}}, t)$ and $\hat{\mathrm{B}}(\hat{\boldsymbol{x}}, \hat{\boldsymbol{p}}, t)$, with an additional factor of $-i/\hbar$. For example, this provides an important insight into the time-dependence of quantum mechanical observables by adapting the Poisson bracket result (1.26). This gives what is referred to as *Heisenberg's equation*,

$$\frac{\mathrm{d}\hat{O}}{\mathrm{d}t} = \frac{\partial \hat{O}}{\partial t} - \frac{i}{\hbar}[\hat{O}, \hat{H}] = \frac{\partial \hat{O}}{\partial t} + \frac{i}{\hbar}[\hat{H}, \hat{O}], \tag{3.195}$$

where $\hat{H} = \hat{H}(\hat{\boldsymbol{p}}, \hat{\boldsymbol{x}})$ is the operator obtained from the classical Hamiltonian by replacing $\boldsymbol{p}$ and $\boldsymbol{x}$ with the operators $\hat{\boldsymbol{p}}$ and $\hat{\boldsymbol{x}}$. This results in the Hamiltonian determining the time development of observables in the quantum mechanical formalism just as it did in classical mechanics. Because $\hat{\boldsymbol{x}}$ and $\hat{\boldsymbol{p}}$ no longer commute, ambiguities will occur if mixed products of $\hat{\boldsymbol{x}}$ and $\hat{\boldsymbol{p}}$ are present in the classical Hamiltonian.

It is instructive to examine a simple example of applying (3.195). For the case that the Hamiltonian is that of a free particle, $\hat{H} = \hat{p}^2/2m$, it follows that

$$\frac{\mathrm{d}\hat{\boldsymbol{p}}}{\mathrm{d}t} = \frac{\partial \hat{\boldsymbol{p}}}{\partial t} - \frac{i}{\hbar}[\hat{\boldsymbol{p}}, \hat{H}] = 0, \tag{3.196}$$

since $\partial\hat{\boldsymbol{p}}/\partial t = 0$ and $\hat{\boldsymbol{p}}$ commutes with itself. For a free quantum particle, the momentum expectation value $\langle\hat{\boldsymbol{p}}\rangle$ is therefore a *constant of motion*. This confirms earlier results. Analysing the time-dependence of $\hat{\boldsymbol{x}}$ is facilitated by the commutator identity

$$\begin{aligned}[\hat{A}, \hat{B}^2] &= \hat{A}\hat{B}^2 - \hat{B}^2\hat{A} \\ &= \hat{A}\hat{B}\hat{B} - \hat{B}\hat{A}\hat{B} + \hat{B}\hat{A}\hat{B} - \hat{B}\hat{B}\hat{A} = [\hat{A}, \hat{B}]\,\hat{B} + \hat{B}\,[\hat{A}, \hat{B}],\end{aligned} \tag{3.197}$$

where the last step follows from the definition of the commutator. Using this result for the case that $\hat{O} = \hat{\boldsymbol{x}}$, it follows that

$$[\hat{x}_k, \hat{p}^2] = \sum_j [\hat{x}_k, \hat{p}_j]\,\hat{p}_j + \sum_j \hat{p}_j\,[\hat{x}_k, \hat{p}_j] = 2i\hbar\sum_j \delta_{kj}\hat{p}_j = 2i\hbar\,\hat{p}_k. \tag{3.198}$$

Using $\partial\hat{\boldsymbol{x}}/\partial t = 0$ gives

$$\frac{\mathrm{d}\hat{\boldsymbol{x}}}{\mathrm{d}t} = -\frac{i}{\hbar}[\hat{\boldsymbol{x}}, \hat{H}] = -\frac{i}{2m\hbar}[\hat{\boldsymbol{x}}, \hat{p}^2] = \frac{\hat{\boldsymbol{p}}}{m}, \tag{3.199}$$

which is the standard relation between velocity and momentum that follows from classical mechanics when the Hamiltonian contains a term quadratic in the momentum and no other terms where both position and momentum occur. There is an important exception to this general result, and that occurs for the Hamiltonian (1.108) of a charged particle interacting with an electromagnetic field. This is discussed in a later chapter.

It is important to note that the Poisson brackets provide no direct information regarding the nature or time evolution of the wave function. However, the time dependency of the wave function must be consistent with (3.195). In that regard, the classical to quantum recipe of (3.195) begins to answer the question about the persistence of eigenvalues raised in (3.169). An observable $\hat{O}$ is time-independent if

$$\frac{\mathrm{d}\hat{O}}{\mathrm{d}t} = \frac{\partial \hat{O}}{\partial t} - \frac{i}{\hbar}[\hat{O}, \hat{H}] = 0. \tag{3.200}$$

This requires $\hat{O}$ to have no explicit time-dependence, so that $\partial\hat{O}/\partial t = 0$, *and it must also commute with the Hamiltonian*, so that $[\hat{O}, \hat{H}] = 0$. This is the case for the momentum operator $\hat{\boldsymbol{p}}$ and the free particle Hamiltonian $\hat{H} = \hat{p}^2/2m$. As a result, it is possible to find time-dependent wave functions that are also eigenfunctions of $\hat{\boldsymbol{p}}$ for the free particle. This is discussed in greater detail in the next chapter when the Schrödinger equation is developed and analysed.

3.5.6 The second postulate of quantum mechanics

The previous developments of this section can be summarized in the second postulate of quantum mechanics, which formally identifies quantum mechanical observables as Hermitian operators acting on wave functions in L^2.

Postulate two: The observables of the wave mechanical version of quantum mechanics correspond to Hermitian operators obtained from classical mechanics through the replacements

$$\begin{aligned} \boldsymbol{x} &\rightarrow \hat{\boldsymbol{x}}, \\ \boldsymbol{p} &\rightarrow \hat{\boldsymbol{p}} = -i\hbar\nabla, \end{aligned} \tag{3.201}$$

in the classical expression for the observable. Each of the observables obtained in this manner is assumed to possess a complete set of orthonormal eigenfunctions that span the function space L^2. The expectation value of the Hermitian observable $\hat{O}$ is obtained from the wave function $\Psi(\boldsymbol{x}, t)$ according to

$$\langle\hat{O}\rangle = (\Psi, \hat{O}\Psi) = \int \mathrm{d}^3x\, \Psi^*(\boldsymbol{x}, t)\hat{O}\Psi(\boldsymbol{x}, t) = (\hat{O}\Psi, \Psi) = \int \mathrm{d}^3x\, (\hat{O}\Psi(\boldsymbol{x}, t))^*\Psi(\boldsymbol{x}, t). \tag{3.202}$$

This postulate is consistent with the results obtained so far for the momentum operator $\hat{\boldsymbol{p}}$. The assumption of hermiticity ensures that each observable has eigenvalues that are real and eigenfunctions that are orthonormal. In the case of momentum the eigenfunctions were shown to be complete. However, the completeness of the eigenfunctions of other observables must be verified on a case by case basis, and so this is listed as an assumption. However, it is crucial to understand that the observables of quantum mechanics derive their values from the wave function. In the Copenhagen interpretation of quantum mechanics the wave function *encapsulates* all that can be known *or observed* about the system. The expectation value (3.202) is one example of this.

An additional complication arises since the results of the previous section show that among the set of all possible quantum mechanical observables there are pairs of observables that do not commute, e.g. $\hat{x}$ and $\hat{p}$. Result (3.193) then shows that these two observables do not share a common set of eigenfunctions. In the Copenhagen interpretation of quantum mechanics all measurements of observables must result in a real value. It will now be shown that two observables must commute in order for their product to yield a real expectation value, and so their product is *not* an observable unless they commute. This follows by first writing the product of two observables $\hat{A}$ and $\hat{B}$ as

$$\hat{A}\hat{B} = \frac{1}{2}(\hat{A}\hat{B} - \hat{B}\hat{A}) + \frac{1}{2}(\hat{A}\hat{B} + \hat{B}\hat{A}) \equiv \frac{1}{2}[\hat{A}, \hat{B}] + \frac{1}{2}\{\hat{A}, \hat{B}\}, \tag{3.203}$$

where $\{\hat{A}, \hat{B}\}$ now stands for the *anticommutator* rather than the Poisson bracket, and is defined by

$$\left\{\hat{A}, \hat{B}\right\} = \hat{A}\hat{B} + \hat{B}\hat{A}. \tag{3.204}$$

Using result (3.159), $(\hat{A}\hat{B})^{\dagger} = \hat{B}^{\dagger}\hat{A}^{\dagger}$, shows that the adjoint of the commutator of two Hermitian operators is given by

$$[\hat{A}, \hat{B}]^{\dagger} = (\hat{A}\hat{B})^{\dagger} - (\hat{B}\hat{A})^{\dagger} = \hat{B}\hat{A} - \hat{A}\hat{B} = [\hat{B}, \hat{A}] = -[\hat{A}, \hat{B}]. \tag{3.205}$$

This reveals the commutator of two Hermitian operators is *anti-Hermitian*, changing signs under Hermitian conjugation, whereas the anticommutator is Hermitian since

$$\left\{\hat{A}, \hat{B}\right\}^{\dagger} = (\hat{A}\hat{B})^{\dagger} + (\hat{B}\hat{A})^{\dagger} = \hat{B}\hat{A} + \hat{A}\hat{B} = \left\{\hat{A}, \hat{B}\right\}. \tag{3.206}$$

The expectation value of an anti-Hermitian operator $\hat{O}$, satisfying $\hat{O}^{\dagger} = -\hat{O}$, has the property that its complex conjugate is given by

$$\begin{aligned} \langle\hat{O}\rangle^{*} &= \left(\int_V \mathrm{d}^n x\, \Psi^{*}(\boldsymbol{x}, t)[\hat{O}\Psi(\boldsymbol{x}, t)]\right)^{*} = \int_V \mathrm{d}^n x\, [\hat{O}^{*}\Psi^{*}(\boldsymbol{x}, t)]\Psi(\boldsymbol{x}, t) \\ &= \int_V \mathrm{d}^n x\, \Psi^{*}(\boldsymbol{x}, t)\left[\hat{O}^{\dagger}\Psi(\boldsymbol{x}, t)\right] = -\int_V \mathrm{d}^n x\, \Psi^{*}(\boldsymbol{x}, t)[\hat{O}\,\Psi(\boldsymbol{x}, t)] = -\langle\hat{O}\rangle, \end{aligned} \tag{3.207}$$

showing that its expectation value is *pure imaginary*. The anticommutator of two Hermitian operators is Hermitian, so it has a real expectation value. It is useful to define the Hermitian operator $\hat{C}$ by factoring out the i,

$$\hat{C} = -i[\hat{A}, \hat{B}], \tag{3.208}$$

which clearly satisfies $\hat{C}^{\dagger} = \hat{C}$. Since $\hat{C}$ is Hermitian it will have real expectation values. Therefore, (3.203) shows that the product of two Hermitian operators has an expectation value that breaks into a real and an imaginary part,

$$\begin{aligned} \langle\hat{A}\hat{B}\rangle &= \frac{1}{2}\langle\{\hat{A}, \hat{B}\}\rangle + \frac{1}{2}\langle[\hat{A}, \hat{B}]\rangle \\ &= \frac{1}{2}\langle\{\hat{A}, \hat{B}\}\rangle + \frac{1}{2}i\langle\hat{C}\rangle = X + iY, \end{aligned} \tag{3.209}$$

where X and Y are real numbers. The imaginary part, $Y = \frac{1}{2}\langle\hat{C}\rangle$, is guaranteed to vanish *only if the commutator of the two operators vanishes.* In the particular case of $\hat{A} = \hat{x}_j$ and $\hat{B} = \hat{p}_j$ the commutator is a pure imaginary number, which renders the expectation value of their product unmeasurable since all physical measurements of observables must result in a real value. As a result, *all observables whose products can be measured in an experimental setting must commute with each other.*

In the Copenhagen interpretation of quantum mechanics it is the choice of what is to be measured that selects the set of observables that can be simultaneously measured. If the experimental apparatus measures values for the observable $\hat{A}$, then among the set of *all possible observables* there will be a *maximal set* of observables that commute with $\hat{A}$. This maximal set is referred to as a *complete set of commuting observables* (CSCO). In the Copenhagen interpretation it is the values of the CSCO that constitute *all* that the experiment can reveal about the *state of the particle.*

As an example, in the case of the free particle presented earlier, the experiment located the position $\boldsymbol{x}$ of the particle in the container. This renders a simultaneous measurement of the momentum observable $\hat{\boldsymbol{p}}$ or any of its powers impossible since these observables do not commute with $\hat{\boldsymbol{x}}$. This is the case even though the particle was prepared with a specific momentum $\boldsymbol{p}_o$. At first glance, this seems contradictory, since the experimentalist apparently knows the momentum at the moment of the location detection. However, this apparent contradiction is resolved by the assumption of wave function collapse as described by (3.132). For the case that the probability density collapses onto an approximate Dirac delta function of position $\delta^3(\boldsymbol{x} - \boldsymbol{a})$ at the moment of measurement, assumed to be $t = t_d$, the wave function created by the measurement can be modeled by

$$\Psi(\boldsymbol{x}, t_d) = \left(\frac{\alpha}{\pi}\right)^{3/4} e^{i\boldsymbol{p}_o\cdot\boldsymbol{x}/\hbar} e^{-\frac{1}{2}\alpha|\boldsymbol{x}-\boldsymbol{a}|^2}, \tag{3.210}$$

where α is a parameter characterizing the locality of the measuring device. For the case that α is large, the probability density at the moment of measurement is given by

$$\mathcal{P}(\boldsymbol{x}, t_d) = |\Psi(\boldsymbol{x}, t_d)|^2 = \left(\frac{\alpha}{\pi}\right)^{3/2} \exp\left(-\alpha|\boldsymbol{x} - \boldsymbol{a}|^2\right) \approx \delta^3(\boldsymbol{x} - \boldsymbol{a}), \tag{3.211}$$

where the probability density becomes the product of three Dirac delta functions given by (3.89). As a result, for α large, the probability density (3.211) describes the experimental localization of the particle. The constant α has the physical units of inverse length squared, and it can be visualized as characterizing the inverse size of the detector area. As this size becomes small the detection volume for the particle begins to approximate a Dirac delta.

At the moment of detection the wave function (3.210) yields an expectation value for $\hat{\boldsymbol{p}}$ given by

$$\begin{aligned}
\langle \hat{\boldsymbol{p}} \rangle &= -i\hbar \int \mathrm{d}^3x\, \Psi^*(\boldsymbol{x}, t_d)\nabla\Psi(\boldsymbol{x}, t_d) \\
&= \boldsymbol{p}_o \int \mathrm{d}^3x \left(\frac{\alpha}{\pi}\right)^{3/2} \exp\left(-\alpha|\boldsymbol{x} - \boldsymbol{a}|^2\right) \\
&\quad + \frac{1}{2}i\hbar\alpha \int \mathrm{d}^3x\, (\boldsymbol{x} - \boldsymbol{a})\left(\frac{\alpha}{\pi}\right)^{3/2} \exp\left(-\alpha|\boldsymbol{x} - \boldsymbol{a}|^2\right) \\
&= \boldsymbol{p}_o \int \mathrm{d}^3x\, \delta^3(\boldsymbol{x} - \boldsymbol{a}) + \frac{1}{2}i\hbar\alpha \int \mathrm{d}^3x\, (\boldsymbol{x} - \boldsymbol{a})\, \delta^3(\boldsymbol{x} - \boldsymbol{a}) = \boldsymbol{p}_o.
\end{aligned} \tag{3.212}$$

where α is assumed large enough to allow the representation of the Dirac delta (3.89) to hold. This can be interpreted as indicating the outcome of measuring $\hat{\boldsymbol{p}}$ at the same moment of detection the particle's position results in a *random* outcome, but that its statistical average is still given by the initial momentum $\boldsymbol{p}_o$. It will be shown in the next section that this is consistent with visualizing the act of localizing the particle as disturbing the particle's momentum in a *random manner*. This aspect of simultaneously measuring noncommuting observables will eventually lead to the *Heisenberg uncertainty principle*, which is presented in the next chapter. However, it raises the question of what, if anything, can be inferred from the wave function regarding the probability density for *single measurements* of observables other than position.

3.5.7 Probability densities and the characteristic function

In order to answer the question raised at the end of the last subsection, it is useful to recall that the first postulate of quantum mechanics *assumes* the wave function $\Psi(\boldsymbol{x}, t)$ provides the probability density $\mathcal{P}(\boldsymbol{x}, t)$ for observing the particle at the position $\boldsymbol{x}$ at the time t through the formula

$$\mathcal{P}(\boldsymbol{x}, t) = |\Psi(\boldsymbol{x}, t)|^2. \tag{3.213}$$

The expression $\mathcal{P}(\boldsymbol{x}, t)$ provides the probability density associated with of a *single measurement* of the particle's position. This is quite different from the expectation value of position $\langle \hat{\boldsymbol{x}} \rangle$ since $\langle \hat{\boldsymbol{x}} \rangle$ is interpreted as the statistical average of *many measurements* of the particle's position for the same wave function. The expectation value $\langle \hat{\boldsymbol{x}} \rangle$ was initially obtained from the probability density (3.213) using the basic definition (3.129). For that reason, the position probability density will be written $\mathcal{P}_{\hat{\boldsymbol{x}}}(\boldsymbol{x}, t) = |\Psi(\boldsymbol{x}, t)|^2$ to denote its relevance to the position observable.

While the probability density for position is defined from the wave function, it is natural to ask if a similar probability density can be found for other observables. Such a probability density for the observable $\hat{\mathrm{A}}$ is written $\mathcal{P}_{\hat{\mathrm{A}}}(a, t)$, and gives the probability density that the observable $\hat{\mathrm{A}}$ will be measured with value a at the time t.

For example, $\mathcal{P}_{\hat{p}}(\boldsymbol{p}, t)$ yields the probability density that the particle has the momentum $\boldsymbol{p}$ at the time t. This means that the probability $\mathrm{d}P_{\hat{p}}$ of observing the particle with the momentum $\boldsymbol{p}$ in the infinitesimal volume of momentum space d^3p around $\boldsymbol{p}$ at the time t is given by

$$\mathrm{d}P_{\hat{p}} = \mathcal{P}_{\hat{p}}(\boldsymbol{p}, t)\, \mathrm{d}^3p. \tag{3.214}$$

This is identical to the relationship of the probability density for position to the probability of observing the particle at the location $\boldsymbol{x}$ at the time t first introduced in (3.1). Such a probability density must also be subject to normalization, so that

$$\int \mathrm{d}P_{\hat{\mathrm{A}}} = \int \mathrm{d}a\, \mathcal{P}_{\hat{\mathrm{A}}}(a, t) = 1, \tag{3.215}$$

where the integral is over the full range of possible values for a. In addition, the expectation value of the observable $\hat{\mathrm{A}}$ is given by a formula identical to the one first defined for position by (3.131), so that

$$\langle \hat{\mathrm{A}} \rangle = \int \mathrm{d}a\, a\, \mathcal{P}_{\hat{\mathrm{A}}}(a, t). \tag{3.216}$$

The connection to the wave function $\Psi(\boldsymbol{x}, t)$ is made by demanding that (3.216) yields a result identical to the expectation value defined in terms of $\Psi(\boldsymbol{x}, t)$ by expression (3.202).

The statistical theory of random variables provides a method to find the probability density for any observable in such a way that (3.202) and (3.216) match. This starts by defining the *characteristic function* for the observable $\hat{\mathrm{A}}$,

$$\zeta(\lambda) = \langle e^{-i\lambda\hat{\mathrm{A}}} \rangle, \tag{3.217}$$

where λ is a real parameter. For the specific case of a quantum mechanical observable, the expectation value is found using the wave function and the recipe given by (3.202). The operator $e^{-i\lambda\hat{\mathrm{A}}}$ is defined using the power series expansion of the exponential, so that

$$\zeta(\lambda) = \langle e^{-i\lambda\hat{\mathrm{A}}} \rangle = \sum_{n=0}^{\infty} \frac{(-i\lambda)^n}{n!} \langle \hat{\mathrm{A}}^n \rangle. \tag{3.218}$$

The desired probability density for $\hat{\mathrm{A}}$ is then given by

$$\mathcal{P}_{\hat{\mathrm{A}}}(a, t) = \int_{-\infty}^{\infty} \frac{\mathrm{d}\lambda}{2\pi}\, e^{i\lambda a}\, \zeta(\lambda) = \int_{-\infty}^{\infty} \frac{\mathrm{d}\lambda}{2\pi}\, e^{i\lambda a}\, \langle e^{-i\lambda\hat{\mathrm{A}}} \rangle. \tag{3.219}$$

This gives the probability density for the observable $\hat{\mathrm{A}}$ to be measured with the value a at the time t.

The proof that (3.219) is the correct probability density consists of showing that the expectation value obtained from it is identical to the one obtained from the wave function using (3.202). This is demonstrated using an integration by parts,

$$
\begin{aligned}
\int_{-\infty}^{\infty} \mathrm{d}a\, a\, \mathcal{P}_{\hat{\mathrm{A}}}(a, t) &= \int_{-\infty}^{\infty} \mathrm{d}a \int_{-\infty}^{\infty} \frac{\mathrm{d}\lambda}{2\pi}\, a\, e^{i\lambda a}\langle e^{-i\lambda \hat{\mathrm{A}}}\rangle \\
&= -i \int_{-\infty}^{\infty} \mathrm{d}a \int_{-\infty}^{\infty} \frac{\mathrm{d}\lambda}{2\pi} \left(\frac{\partial}{\partial \lambda} e^{i\lambda a}\right)\langle e^{-i\lambda \hat{\mathrm{A}}}\rangle \\
&= -i \int_{-\infty}^{\infty} \mathrm{d}a \int_{-\infty}^{\infty} \frac{\mathrm{d}\lambda}{2\pi} \frac{\partial}{\partial \lambda}(e^{i\lambda a}\langle e^{-i\lambda \hat{\mathrm{A}}}\rangle) \\
&\quad + i \int_{-\infty}^{\infty} \mathrm{d}a \int_{-\infty}^{\infty} \frac{\mathrm{d}\lambda}{2\pi}\, e^{i\lambda a} \frac{\partial}{\partial \lambda}\langle e^{-i\lambda \hat{\mathrm{A}}}\rangle \\
&= \int_{-\infty}^{\infty} \mathrm{d}\lambda\, \langle \hat{\mathrm{A}} e^{-i\lambda \hat{\mathrm{A}}}\rangle \int_{-\infty}^{\infty} \frac{\mathrm{d}a}{2\pi} e^{i\lambda a} = \int_{-\infty}^{\infty} \mathrm{d}\lambda\, \langle \hat{\mathrm{A}} e^{-i\lambda \hat{\mathrm{A}}}\rangle\, \delta(\lambda) \\
&= \langle \hat{\mathrm{A}}\rangle,
\end{aligned}
\tag{3.220}
$$

where the expectation value $\langle \hat{\mathrm{A}}\rangle$ is found using the wave function recipe (3.202). The endpoint terms of the integration by parts in the third step vanish,

$$
\begin{aligned}
-i \lim_{\lambda\to\infty}\left(e^{i\lambda a}\langle e^{-i\lambda \hat{\mathrm{A}}}\rangle - e^{-i\lambda a}\langle e^{i\lambda \hat{\mathrm{A}}}\rangle\right) &= -i \lim_{\lambda\to\infty}\langle\left(e^{i\lambda(a-\hat{\mathrm{A}})} - e^{-i\lambda(a-\hat{\mathrm{A}})}\right)\rangle \\
&= \lim_{\lambda\to\infty} 2\langle \sin(\lambda(a - \hat{\mathrm{A}})\rangle = 0.
\end{aligned}
\tag{3.221}
$$

This follows by applying a variant of the Riemann–Lebesgue lemma,

$$
\begin{aligned}
\lim_{\lambda\to\infty} \sin(\lambda\omega) &= \int_{0}^{\infty} \mathrm{d}\lambda \frac{\partial}{\partial\lambda} \sin(\lambda\omega) \\
&= \frac{\omega}{2}\int_{-\infty}^{\infty} \mathrm{d}\lambda \cos(\lambda\omega) = \frac{\pi\omega}{2}(\delta(\omega) + \delta(-\omega)) = \pi\omega\, \delta(\omega),
\end{aligned}
\tag{3.222}
$$

which vanishes for the space of functions bounded at $\omega = 0$.

It is important to show that (3.219) is consistent with the first postulate of quantum mechanics. This requires extending (3.219) to a vector valued observable by making a vector of λ, from which it follows that

$$
\begin{aligned}
P_{\hat{x}}(\boldsymbol{x}, t) &= \int \frac{\mathrm{d}^3\lambda}{(2\pi)^3} e^{i\boldsymbol{\lambda}\cdot\boldsymbol{x}}\langle e^{-i\boldsymbol{\lambda}\cdot\hat{\boldsymbol{x}}}\rangle = \int_{-\infty}^{\infty} \frac{\mathrm{d}^3\lambda}{(2\pi)^3} e^{i\boldsymbol{\lambda}\cdot\boldsymbol{x}} \int \mathrm{d}^3y\, \Psi^*(\boldsymbol{y}, t) e^{-i\boldsymbol{\lambda}\cdot\hat{\boldsymbol{x}}}\Psi(\boldsymbol{y}, t) \\
&= \int \mathrm{d}^3y \int \frac{\mathrm{d}^3\lambda}{(2\pi)^3} e^{i\boldsymbol{\lambda}\cdot(\boldsymbol{x}-\boldsymbol{y})}|\Psi(\boldsymbol{y}, t)|^2 = \int \mathrm{d}^3y\, \delta^3(\boldsymbol{x} - \boldsymbol{y})|\Psi(\boldsymbol{y}, t)|^2 \\
&= |\Psi(\boldsymbol{x}, t)|^2,
\end{aligned}
\tag{3.223}
$$

where the integral representation (3.95) of the Dirac delta was used. Result (3.223) demonstrates that the definition (3.219) for the probability density is consistent with the first postulate of quantum mechanics.

The probability density for the momentum operator can now be derived by using the completeness of the momentum eigenfunctions and adapting (3.177) and (3.178) to express the wave function as

$$\Psi(\boldsymbol{x}, t) = \int \frac{\mathrm{d}^3p}{(2\pi\hbar)^3}\, \tilde{\Psi}(\boldsymbol{p}, t)\, e^{i\boldsymbol{p}\cdot\boldsymbol{x}/\hbar}, \tag{3.224}$$

where

$$\tilde{\Psi}(\boldsymbol{p}, t) = \int \mathrm{d}^3y\, \Psi(\boldsymbol{y}, t)\, e^{-i\boldsymbol{p}\cdot\boldsymbol{y}/\hbar}. \tag{3.225}$$

Combining the result

$$e^{-i\boldsymbol{\lambda}\cdot\hat{\boldsymbol{p}}}\, e^{i\boldsymbol{k}\cdot\boldsymbol{x}/\hbar} = \sum_{n=0}^{\infty} \frac{(-i\boldsymbol{\lambda}\cdot\hat{\boldsymbol{p}})^n}{n!} e^{i\boldsymbol{k}\cdot\boldsymbol{x}/\hbar} = \sum_{n=0}^{\infty} \frac{(-i\boldsymbol{\lambda}\cdot\boldsymbol{k})^n}{n!} e^{i\boldsymbol{k}\cdot\boldsymbol{x}/\hbar} = e^{-i\boldsymbol{\lambda}\cdot\boldsymbol{k}} e^{i\boldsymbol{k}\cdot\boldsymbol{x}/\hbar} \tag{3.226}$$

with (3.224), it follows that the characteristic function for the momentum operator is given by

$$\begin{aligned}
\chi(\boldsymbol{\lambda}) &= \int \mathrm{d}^3x\, \Psi^*(\boldsymbol{x}, t)\, e^{-i\boldsymbol{\lambda}\cdot\hat{\boldsymbol{p}}}\, \Psi(\boldsymbol{x}, t) \\
&= \int \mathrm{d}^3x \int \frac{\mathrm{d}^3p}{(2\pi\hbar)^3} \int \frac{\mathrm{d}^3k}{(2\pi\hbar)^3}\, \tilde{\Psi}^*(\boldsymbol{p}, t)\, e^{-i\boldsymbol{p}\cdot\boldsymbol{x}/\hbar}\, e^{-i\boldsymbol{\lambda}\cdot\boldsymbol{k}}\, \tilde{\Psi}(\boldsymbol{k}, t)\, e^{i\boldsymbol{k}\cdot\boldsymbol{x}/\hbar} \\
&= \int \frac{\mathrm{d}^3p}{(2\pi\hbar)^3} \int \mathrm{d}^3k\, \tilde{\Psi}^*(\boldsymbol{p}, t)\, \tilde{\Psi}^*(\boldsymbol{k}, t)\, e^{-i\boldsymbol{\lambda}\cdot\boldsymbol{k}}\, \delta^3(\boldsymbol{k} - \boldsymbol{p}) \\
&= \int \frac{\mathrm{d}^3k}{(2\pi\hbar)^3}\, e^{-i\boldsymbol{\lambda}\cdot\boldsymbol{k}}\, |\tilde{\Psi}(\boldsymbol{k}, t)|^2.
\end{aligned} \tag{3.227}$$

The probability density for the momentum operator is then given by

$$\begin{aligned}
\mathcal{P}_{\hat{\boldsymbol{p}}}(\boldsymbol{p}, t) &= \int \frac{\mathrm{d}^3\lambda}{(2\pi)^3} \int \frac{\mathrm{d}^3k}{(2\pi\hbar)^3}\, e^{i\boldsymbol{\lambda}\cdot(\boldsymbol{p}-\boldsymbol{k})}\, |\tilde{\Psi}(\boldsymbol{k}, t)|^2 \\
&= \int \frac{\mathrm{d}^3k}{(2\pi\hbar)^3}\, \delta^3(\boldsymbol{p} - \boldsymbol{k})\, |\tilde{\Psi}(\boldsymbol{k}, t)|^2 \\
&= \frac{1}{(2\pi\hbar)^3}\, |\tilde{\Psi}(\boldsymbol{p}, t)|^2.
\end{aligned} \tag{3.228}$$

This shows that projecting the wave function onto the eigenfunctions of the momentum operator using (3.225) yields the probability density for the momentum. The normalization of (3.228) follows from the assumed normalization of $\Psi(\boldsymbol{x}, t)$,

$$\begin{aligned}
\int \mathrm{d}^3p\, \mathcal{P}_{\hat{\boldsymbol{p}}}(\boldsymbol{p}, t) &= \int \frac{\mathrm{d}^3p}{(2\pi\hbar)^3}\, |\tilde{\Psi}(\boldsymbol{p}, t)|^2 \\
&= \int \frac{\mathrm{d}^3p}{(2\pi\hbar)^3} \int \mathrm{d}^3x \int \mathrm{d}^3y\, \Psi^*(\boldsymbol{x}, t) e^{i\boldsymbol{p}\cdot\boldsymbol{x}/\hbar} e^{-i\boldsymbol{p}\cdot\boldsymbol{y}/\hbar} \Psi(\boldsymbol{y}, t) \\
&= \int \mathrm{d}^3x \int \mathrm{d}^3y\, \delta^3(\boldsymbol{x} - \boldsymbol{y})\, \Psi^*(\boldsymbol{x}, t) \Psi(\boldsymbol{y}, t) \\
&= \int \mathrm{d}^3x\, |\Psi(\boldsymbol{x}, t)|^2 = 1.
\end{aligned} \tag{3.229}$$

It is insightful to apply this result to the wave function (3.210) used to model wave function collapse resulting from the free particle detection at the position a at the time $t_{\rm d}$. For simplicity, attention will be limited to one spatial dimension. The result for $\tilde{\Psi}(p, t_{\rm d})$ is given by

$$\begin{aligned}\tilde{\Psi}(p, t_{\rm d}) &= \int_{-\infty}^{\infty} {\rm d}y\; e^{-ipy/\hbar}\, \Psi(x, t_{\rm d}) = \left(\frac{\alpha}{\pi}\right)^{1/4} \int_{-\infty}^{\infty} {\rm d}y\; e^{-ipy/\hbar} e^{ip_o y/\hbar} e^{-\frac{1}{2}\alpha(y-a)^2} \\ &= \left(\frac{4\pi}{\alpha}\right)^{1/4} e^{-i(p-p_o)a/\hbar}\, e^{-(p-p_o)^2/2\hbar^2\alpha}.\end{aligned} \tag{3.230}$$

This gives the one-dimensional probability density for momentum at the moment of detection,

$$\mathcal{P}_{\hat{p}}(p, t_{\rm d}) = \frac{1}{2\pi\hbar}|\tilde{\Psi}(p, t_{\rm d})|^2 = \frac{1}{\sqrt{\pi\alpha\hbar^2}}\, e^{-(p-p_o)^2/\hbar^2\alpha}. \tag{3.231}$$

For the case that α is large, which corresponds to measuring the particle's position in a small volume, the momentum probability density is still peaked around p_o, but it has spread around p_o and is lowered to the peak value $1/\sqrt{\pi\alpha\hbar^2}$, which tends to zero for large α. On the other hand, for the case that $\alpha \approx 0$, the probability density becomes

$$\begin{aligned}\lim_{\alpha\to 0}\mathcal{P}_{\hat{p}}(p, t_{\rm d}) &= \frac{1}{\hbar}\lim_{\alpha\to 0}\frac{1}{\sqrt{\pi\alpha}}\, e^{-(p-p_o)^2/\hbar^2\alpha} \\ &\approx \frac{1}{\hbar}\,\delta\big((p-p_o)/\hbar\big) = \delta(p-p_o).\end{aligned} \tag{3.232}$$

This shows that a weak localization of the particle leaves its momentum probability density sharply peaked around the initial momentum p_o.

As a final step, it is important to consider the case that the Hermitian observable $\hat{A}$ has a *discrete set* of eigenvalues e_j indexed by an integer j, so that

$$\hat{A}\varphi_j(\boldsymbol{x}) = e_j\varphi_j(\boldsymbol{x}), \tag{3.233}$$

and that the eigenfunctions are orthonormal, so that

$$(\varphi_j, \varphi_k) = \int {\rm d}^3x\; \varphi_j^*(\boldsymbol{x})\, \varphi_k(\boldsymbol{x}) = \delta_{jk}. \tag{3.234}$$

Assuming these eigenfunctions are also complete in the same manner as the discrete functions of (3.63) gives

$$\sum_j \varphi_j^*(\boldsymbol{x})\, \varphi_j(\boldsymbol{y}) = \delta^3(\boldsymbol{x}-\boldsymbol{y}). \tag{3.235}$$

This allows an arbitrary wave function $\Psi(\boldsymbol{x}, t)$ to be projected onto these eigenfunctions, so that

$$\Psi(\boldsymbol{x}, t) = \sum_j \bar{\Psi}_j(t)\, \varphi_j(\boldsymbol{x}), \tag{3.236}$$

where

$$\bar{\Psi}_j(t) = (\varphi_j, \Psi) = \int \mathrm{d}^3y\, \varphi_j^*(\boldsymbol{y})\Psi(\boldsymbol{y}, t). \tag{3.237}$$

It is useful to note that $\bar{\Psi}_k(t)$ is a dimensionless quantity since both φ_j and Ψ have the units $1/\sqrt{V}$.

The representation of the wave function given by (3.236) gives the characteristic function

$$\begin{aligned}
\chi(\lambda) = \langle e^{-i\lambda\hat{\mathrm{A}}}\rangle &= \int \mathrm{d}^3x \sum_j \bar{\Psi}_j^*(t)\, \varphi_j^*(\boldsymbol{x})\, e^{-i\lambda\hat{\mathrm{A}}} \sum_k \bar{\Psi}_k(t)\, \varphi_k(\boldsymbol{x}) \\
&= \sum_{j,k} \bar{\Psi}_j^*(t)\bar{\Psi}_k e^{-i\lambda e_k} \int \mathrm{d}^3x\, \varphi_j^*(\boldsymbol{x})\, \varphi_k(\boldsymbol{x}) = \sum_{j,k} \bar{\Psi}_j^*(t)\bar{\Psi}_k e^{-i\lambda e_k}\delta_{jk} \\
&= \sum_j e^{-i\lambda e_j} |\bar{\Psi}_j(t)|^2.
\end{aligned} \tag{3.238}$$

This gives the probability density for the observable $\hat{\mathrm{A}}$ as

$$\mathcal{P}_{\hat{\mathrm{A}}}(a, t) = \int_{-\infty}^{\infty} \frac{\mathrm{d}\lambda}{2\pi} \sum_j e^{i\lambda(a-e_j)} |\bar{\Psi}_j(t)|^2 = \sum_j |\bar{\Psi}_j(t)|^2\, \delta(a - e_j). \tag{3.239}$$

The delta function shows that the probability density (3.239) is nonzero *only* for values of a that match one of the observable's eigenvalues. Like the momentum probability density (3.229), the normalization of (3.239) follows from the normalization of $\Psi(\boldsymbol{x}, t)$,

$$\begin{aligned}
\int \mathrm{d}a\, \mathcal{P}_{\hat{\mathrm{A}}}(a, t) &= \sum_j |\bar{\Psi}_j(t)|^2 \int \mathrm{d}a\, \delta(a - e_j) = \sum_j |\bar{\Psi}_j(t)|^2 \\
&= \sum_j \int \mathrm{d}^3x\, \Psi^*(\boldsymbol{x}, t)\varphi^*(\boldsymbol{x}) \int \mathrm{d}^3y\, \varphi(\boldsymbol{y})\Psi(\boldsymbol{y}, t) \\
&= \int \mathrm{d}^3x \int \mathrm{d}^3y\, \delta^3(\boldsymbol{x} - \boldsymbol{y})\Psi^*(\boldsymbol{x}, t)\Psi(\boldsymbol{y}, t) = \int \mathrm{d}^3x\, |\Psi(\boldsymbol{x}, t)|^2 = 1.
\end{aligned} \tag{3.240}$$

It is important to note that the formalism has ensured that the sum of the individual probabilities satisfies

$$\sum_j |\bar{\Psi}_j(t)|^2 = 1, \tag{3.241}$$

and this follows from wave function normalization. Even in the case where the outcomes of measurement are discrete eigenvalues, the normalization of the wave function ensures that the sum of all the probabilities for the observation of the individual eigenvalues is unity.

3.5.8 Born's rule

In complete analogy to the first postulate for the measurement of the particle's position, $\mathcal{P}_{\hat{A}}(a, t)$ is the probability density for obtaining the result a from a measurement of the observable $\hat{A}$ for the particle at the time t. Like the measurement of position, this can be combined with the assumption that each measurement of the observable $\hat{A}$ results in a *distinct value* for a within the experimental limits of the measurement and its apparatus. Such an assumption provides a restriction on the type of measurements that are the purview of quantum mechanics, and so it is not included in the first or second postulate. However, it then follows from result (3.239) that this probability is predicted to be zero *unless* the measured value a coincides with one of the eigenvalues of $\hat{A}$. In other words, a *successful* measurement of the observable $\hat{A}$ will result in *one* of the eigenvalues of $\hat{A}$. The infinitesimal probability of measuring the eigenvalue e_k for $\hat{A}$ is given by

$$\mathrm{d}P_{\hat{A}}(a) = \mathcal{P}_{\hat{A}}(a, t)\,\mathrm{d}a = |\bar{\Psi}_k(t)|^2\,\delta(a - e_k)\,\mathrm{d}a, \tag{3.242}$$

where the infinitesimal interval of possible values $\mathrm{d}a$ includes e_k. The probability of observing $\hat{A}$ with the value e_k is then given by integrating (3.242) over the interval around e_k, which gives

$$P_{\hat{A}}(e_k, t) = \lim_{\Delta\to 0}\int_{e_k-\Delta}^{e_k+\Delta} \mathrm{d}a\,|\bar{\Psi}_k(t)|^2\,\delta(a - e_k) = |\bar{\Psi}_k(t)|^2. \tag{3.243}$$

This assumes that the separation between e_k and its neighboring eigenvalues allows (3.243) to be defined. This assumption breaks down if there is a set of degenerate eigenvalues. In such a case, the result will be the total probability of observing the degenerate eigenvalue.

This result, that a single measurement of the observable $\hat{A}$ will result in an eigenvalue of $\hat{A}$ with the probability (3.243), constitutes what is known as *Born's rule*. Born's rule was originally considered a fundamental assumption of quantum mechanics, but in the analysis just presented it has followed from the first two postulates and the restriction to a single valued result for measurement of an observable. In that regard, the first postulate of quantum mechanics can be considered as a statement of Born's rule applied to position measurement. Viewed in that light this section has simply extended Born's rule to other observables. It should not be considered a proof of Born's rule since it begins by assuming Born's rule is valid for position measurements. The references list alternate axiomatic approaches to deriving Born's rule.

It was shown earlier that commuting observables share a mutual set of eigenfunctions. As a result, each measurement of the observables belonging to the particular CSCO chosen by the experimental situation will result in a distinct set of eigenvalues for these observables. These eigenvalues constitute everything that can be measured for the given choice of CSCO. It is in this sense that the wave function is said to encapsulate *all that can be known* about the system in terms of the probabilities for the measurements made on a complete set of commuting observables.

3.6 Superposition and mixed states

One of the driving factors in the development of quantum mechanics was the manifestation of interference effects in electron diffraction and its similarity to the interference observed with light waves. In the case of light, this effect is explained using the principle of superposition, which in turn is a property of the *linear nature* of the wave equation governing light waves in vacuum. The purpose of this section is to examine the probability densities obtained using superposition of wave functions.

3.6.1 Superposition in quantum mechanics

In order to explain electron diffraction the wave function is *assumed* to obey the principle of superposition, similar to the principle of superposition for light waves in vacuum. In the case of light this was the result of the vacuum wave equation (1.70), which is a linear differential equation. The principle of superposition has already been used quantum mechanically to express a wave function in terms of the orthonormal eigenfunctions of an observable, as in (3.177) and (3.236). However, in the case of eigenfunction superposition the time development of the wave function was subsumed into the coefficients of the expansion.

In what follows, the differential equation governing the time development of the wave function will also be assumed to be a *linear differential equation.* This assumption will be developed in detail in the next chapter and stated as the third postulate of quantum mechanics when the Schrödinger equation is presented. Quantum mechanical superposition can be stated mathematically by using two possible *wave functions* for the particle, $\Psi_1(\boldsymbol{x}, t)$ and $\Psi_2(\boldsymbol{x}, t)$, both of which are assumed to satisfy both the normalization condition (3.2) and the linear wave equation. The principle of superposition then allows the linear combination,

$$\Psi(\boldsymbol{x}, t) = a\Psi_1(\boldsymbol{x}, t) + b\Psi_2(\boldsymbol{x}, t), \tag{3.244}$$

where a and b are constants, to be another *possible* wave function or *state* for the particle, as long as the constants a and b are chosen so that the normalization condition (3.2) holds for Ψ. It was shown earlier in (3.55) that a linear combination of functions in $\mathbb{L}^2$ is also a function in $\mathbb{L}^2$ and therefore normalizable. The superposition defined by (3.244) is therefore a possible wave function for the system. The resulting wave function differs from the superposition of eigenfunctions used earlier in (3.177) and (3.236) since the two wave functions $\Psi_1(\boldsymbol{x}, t)$ and $\Psi_2(\boldsymbol{x}, t)$ are *not necessarily* eigenfunctions of any observable. To distinguish this aspect of the wave function of (3.244) it is referred to as a *mixed state*, where the term *state* is shorthand for *state of the system.* In what follows the wave functions $\Psi_1(\boldsymbol{x}, t)$ and $\Psi_2(\boldsymbol{x}, t)$ will be referred to as the *constituent states* of the mixed state.

3.6.2 Quantum interference

To understand the general properties of the mixed state (3.244), the probability density (3.108) associated with it will be examined. Since $\Psi^{*}=a^{*}\Psi_1^{*} + b^{*}\Psi_2^{*}$, the position probability density for this mixed state is given by

$$
\begin{aligned}
(\boldsymbol{x}, t) = |\Psi(\boldsymbol{x}, t)|^2 &= |a|^2|\Psi_1(\boldsymbol{x}, t)|^2 \\
&+ a^{*}b\Psi_1^{*}(\boldsymbol{x}, t)\Psi_2(\boldsymbol{x}, t) + b^{*}a\Psi_2^{*}(\boldsymbol{x}, t)\Psi_1(\boldsymbol{x}, t) + |b|^2|\Psi_2(\boldsymbol{x}, t)|^2.
\end{aligned} \tag{3.245}
$$

The cross terms between Ψ_1 and Ψ_2 in (3.245) are referred to as *quantum interference terms*. Their presence shows that quantum probabilities *do not add like classical probabilities*. It is important to clarify this statement. Relation (3.108) identifies the *constituent* probability densities associated with the *constituent* wave functions as

$$
\begin{aligned}
\mathcal{P}_1(\boldsymbol{x}, t) &= |a|^2|\Psi_1(\boldsymbol{x}, t)|^2, \\
\mathcal{P}_2(\boldsymbol{x}, t) &= |b|^2|\Psi_2(\boldsymbol{x}, t)|^2.
\end{aligned} \tag{3.246}
$$

Using these identifications, the *classical* probability density $\mathcal{P}_c(\boldsymbol{x}, t)$ of observing the particle at $\boldsymbol{x}$ at the time t is defined as

$$
\mathcal{P}_c(\boldsymbol{x}, t) = \mathcal{P}_1(\boldsymbol{x}, t) + \mathcal{P}_2(\boldsymbol{x}, t) = |a|^2|\Psi_1(\boldsymbol{x}, t)|^2 + |b|^2|\Psi_2(\boldsymbol{x}, t)|^2. \tag{3.247}
$$

If the two wave functions are normalized, the classical probability density is normalized by the condition

$$
\begin{aligned}
\int \mathrm{d}^n x\, \mathcal{P}_c(\boldsymbol{x}, t) &= |a|^2 \int \mathrm{d}^n x\, |\Psi_1(\boldsymbol{x}, t)|^2 + |b|^2 \int \mathrm{d}^n x\, |\Psi_2(\boldsymbol{x}, t)|^2 \\
&= |a|^2 + |b|^2 = 1.
\end{aligned} \tag{3.248}
$$

While this combination is present in (3.245), the quantum interference terms represent a significant deviation from classical behavior and give rise to uniquely quantum mechanical effects.

3.6.3 Normalization of the mixed state

Requiring that the normalization condition (3.110) holds for the mixed state $\Psi(\boldsymbol{x}, t)$ defined by (3.244) will also define a relationship between the two constants, a and b. Denoting the inner product of the two functions as the complex number λ and using the notation of (3.50), λ is given by

$$
\lambda = \int \mathrm{d}^n x\, \Psi_1^{*}(\boldsymbol{x}, t)\Psi_2(\boldsymbol{x}, t) = (\Psi_1, \Psi_2), \tag{3.249}
$$

the normalization condition is obtained,

$$
\begin{aligned}
(\Psi, \Psi) &= \int \mathrm{d}^n x\, |\Psi(\boldsymbol{x}, t)|^2 \\
&= \int \mathrm{d}^n x\, (a^* \Psi_1^*(\boldsymbol{x}, t) + b^* \Psi_2^*(\boldsymbol{x}, t))(a\Psi_1(\boldsymbol{x}, t) + b\Psi_2(\boldsymbol{x}, t)) \\
&= |a|^2 + \lambda a^* b + \lambda^* a b^* + |b|^2 = 1,
\end{aligned}
\tag{3.250}
$$

where the normalization of both Ψ_1 and Ψ_2 was used. It is important to note that if Ψ_1 and Ψ_2 are different eigenfunctions of a Hermitian observable, then they are orthogonal and the integral vanishes so that $\lambda = 0$. For such a case, the normalization condition (3.250) reduces to $|a|^2 + |b|^2 = 1$, which is identical to the classical normalization condition (3.248). This also occurs if $\lambda a^* b$ is pure imaginary, since then $\lambda a^* b + \lambda^* a b^* = 0$.

Solving (3.250) for the case that $\lambda \neq 0$ is most easily achieved using the polar representation (3.21) for a, b, and λ, so that

$$
a = |a|\, e^{i\alpha}, \quad b = |b|\, e^{i\beta}, \quad \lambda = |\lambda|\, e^{i\theta}. \tag{3.251}
$$

However, it is important to note that if $\Psi(\boldsymbol{x}, t)$ is normalized, then so is $e^{-i\beta}\Psi(\boldsymbol{x}, t)$, where β is the phase angle for b. As a result, there is only one degree of phase freedom in the two constants a and b, and β will be treated as zero, so that α is the *relative phase* between a and b. Using the Euler relation (3.19) reduces (3.250) to a quadratic equation for the two coefficients,

$$
(\Psi, \Psi) = |a|^2 + 2|\lambda||a||b|\cos(\theta - \alpha) + |b|^2 = 1. \tag{3.252}
$$

The Cauchy–Schwarz inequality for normalized functions (3.56) shows that $|\lambda|^2 \leqslant 1$. The properties of the cosine give $-1 \leqslant \cos\theta \leqslant 1$. Combining these results show that

$$
(|a| - |b|)^2 \leqslant (\Psi, \Psi) \leqslant (|a| + |b|)^2, \tag{3.253}
$$

which is simply a variant of the previously derived triangle inequality (3.55). The positive solution to the quadratic equation (3.252) for $|a|$ is given by

$$
|a| = -|\lambda||b|\cos(\theta - \alpha) + \sqrt{|\lambda|^2 |b|^2 \cos^2(\theta - \alpha) + 1 - |b|^2}\,. \tag{3.254}
$$

A particularly simple solution to (3.252) for $\lambda \neq 0$ is obtained by choosing the relative phase α such that

$$
\theta - \alpha = \pi/2. \tag{3.255}
$$

This is equivalent to fixing $\lambda a^* b$ to be pure imaginary since the Euler relation gives

$$
\lambda a^* b = |\lambda||a||b|\, e^{i(\theta - \alpha)} = |\lambda||a||b|\, e^{i\pi/2} = i|\lambda||a||b|. \tag{3.256}
$$

For that choice, it follows that

$$
|a|^2 + |b|^2 = 1, \tag{3.257}
$$

which has real solutions as long as the restriction $|a|, |b| \leqslant 1$ is met.

Results (3.255) and (3.257) determine a in terms of b, and so b can be viewed as an independent variable. This reflects the fact that the choice of the mixture coefficients,

a and b, can be adjusted to represent the *relative presence* of the two constituent states. For the case $a = 1$ and $b = 0$ the mixed state corresponds to the particle being entirely in the *normalized state* associated with Ψ_1. Similarly, if $a = 0$ and $b = 1$, then the particle is entirely in the *normalized state* associated with Ψ_2. If both are nonzero, the wave function represents a situation where the particle is described by a weighted mixture of the two constituent states, each of which correspond to different probability densities. Of course, a mixed state is not limited to just two states. The mixed state can have one or both of the constituent states replaced by yet another mixed state, and this process can proceed *ad infinitum*.

3.6.4 Electron diffraction and superposition

It is important to note that the *relative phase* between the two possible wave functions will manifest itself *only in the quantum interference terms*. This can be seen by considering the two constituent wave functions in the mixed state to have the general form given by

$$\begin{aligned} \Psi_1(\boldsymbol{x}, t) &= \psi_1(\boldsymbol{x}, t)e^{i\varphi_1(\boldsymbol{x}, t)}, \\ \Psi_2(\boldsymbol{x}, t) &= \psi_2(\boldsymbol{x}, t)e^{i\varphi_2(\boldsymbol{x}, t)}, \end{aligned} \tag{3.258}$$

where ψ_1, ψ_2, φ_1, and φ_2 are real valued functions. Using the Euler relation (3.19) and the previous polar representations of the constants a and b gives

$$\begin{aligned} \mathcal{P}_{\hat{x}}(\boldsymbol{x}, t) &= |\Psi(\boldsymbol{x}, t)|^2 \\ &= |a|^2\psi_1^2(\boldsymbol{x}, t) \\ &\quad + |b|^2\psi_2^2(\boldsymbol{x}, t) + 2|a||b|\, \psi_1(\boldsymbol{x}, t)\psi_2(\boldsymbol{x}, t)\cos(\alpha + \varphi_2(\boldsymbol{x}, t) - \varphi_1(\boldsymbol{x}, t)). \end{aligned} \tag{3.259}$$

Choosing $|a| = |b| = 1/\sqrt{2}$ for the normalization (3.257) and $\psi_1 = \psi_2$ along with the identity $1 + \cos\theta = 2\cos^2\frac{1}{2}\theta$ shows that (3.259) becomes

$$\mathcal{P}_{\hat{x}}(\boldsymbol{x}, t) = 2\psi_1^2(\boldsymbol{x}, t)\cos^2\left(\frac{1}{2}(\alpha + \varphi_1(\boldsymbol{x}, t) - \varphi_2(\boldsymbol{x}, t)\right). \tag{3.260}$$

Result (3.260) is identical in form to the light intensity (3.9) resulting from two-slit interference, and so it can be used to explain the probability pattern of electron diffraction that results from passing the electron beam through the two-slit interferometer. Treating the wave function as having passed through both slits results in a mixed state of the form (3.244), where Ψ_1 and Ψ_2 originate from the respective slits. These constituent wave functions have a phase difference induced by the different path lengths from the two slits to the photoplate in a manner similar to light waves. As a result, an equal mixture of the two equal amplitude wave functions results in an interference pattern that depends on the relative phase difference $\varphi_1 - \varphi_2$ between the two wave functions. The analysis of electron diffraction is revisited in much greater detail in chapter 6 where the relative phase functions are found.

For the case that $|a| = |b|$, the Copenhagen interpretation states that the phase difference *encapsulates* the effect of each electron passing through the interferometer with an equal probability of passing through either slit. It is important to note that if the two phase functions satisfy $\varphi_1(\boldsymbol{x}, t) = \varphi_2(\boldsymbol{x}, t)$, then the constant λ defined by (3.249) is real and (3.255) shows that the relative phase of the two constants a and b satisfies $\alpha = \pi/2$, with the result that

$$\mathcal{P}_{\hat{x}}(\boldsymbol{x}, t) = \psi_1^2(\boldsymbol{x}, t). \tag{3.261}$$

This shows that an absence of a phase difference between the constituent wave functions results in no quantum interference.

It is important to consider the difference between the previous case where the constituent wave functions are *not* orthonormal and the case where the constituent wave functions are orthonormal eigenfunctions. As an example of the latter, it is possible to prepare a wave function that is the superposition of two different values of the free particle wave function of (3.127), which are eigenfunctions of the momentum. A simple example of a such a state is given by

$$\begin{aligned}\Psi(\boldsymbol{x}, t) &= \frac{1}{\sqrt{2}}\Psi(\boldsymbol{p}, \boldsymbol{x}, t) + \frac{1}{\sqrt{2}}\Psi(\boldsymbol{p}', \boldsymbol{x}, t)\\ &= \frac{1}{\sqrt{2V}}\exp\left(\frac{i}{\hbar}(\boldsymbol{p}\cdot\boldsymbol{x} - E_{\mathrm{p}}t)\right) + \frac{1}{\sqrt{2V}}\exp\left(\frac{i}{\hbar}(\boldsymbol{p}'\cdot\boldsymbol{x} - E_{\mathrm{p}'}t)\right),\end{aligned} \tag{3.262}$$

where $\boldsymbol{p} \neq \boldsymbol{p}'$. The *preparation* of such a state would require a source of particles that have two *distinct* momenta that are equally likely to be produced. The wave function of (3.262) is correctly normalized. Demonstrating this uses the orthonormality of the two free particle wave functions,

$$\begin{aligned}&\int \mathrm{d}^3x \exp\left(\frac{i}{\hbar}\left(\boldsymbol{p} - \boldsymbol{p}'\right)\cdot\boldsymbol{x} - \frac{i}{\hbar}\left(E_{\mathrm{p}} - E_{\mathrm{p}'}\right)t\right)\\ &= (2\pi\hbar)^3\delta^3\left(\boldsymbol{p} - \boldsymbol{p}'\right)\exp\left(\frac{i}{\hbar}\left(E_{\mathrm{p}} - E_{\mathrm{p}'}\right)t\right) = 0,\end{aligned} \tag{3.263}$$

where the delta function vanishes since $\boldsymbol{p} \neq \boldsymbol{p}'$. For the case that $\boldsymbol{p} \neq \boldsymbol{p}'$ the state (3.262) yields the probability density

$$\begin{aligned}\mathcal{P}_{\hat{x}}(\boldsymbol{x}, t) &= \frac{1}{2V}\left(1 + \cos\left((\boldsymbol{p} - \boldsymbol{p}')\cdot\boldsymbol{x}/\hbar - (E_{\mathrm{p}} - E_{\mathrm{p}'})t/\hbar\right)\right)\\ &= \frac{1}{V}\cos^2\left(\frac{1}{2\hbar}\left((\boldsymbol{p} - \boldsymbol{p}')\cdot\boldsymbol{x} - (E_{\mathrm{p}} - E_{\mathrm{p}'})t\right)\right).\end{aligned} \tag{3.264}$$

The resulting probability density is no longer uniform throughout the volume V. Limiting consideration to $t = 0$ shows that the probability density is zero at those positions $\boldsymbol{x}$ that satisfy

$$(\boldsymbol{p} - \boldsymbol{p}')\cdot\boldsymbol{x} = (2n + 1)\pi\hbar, \tag{3.265}$$

where n is an arbitrary integer. In this case the mixed state yields a probability density that is *more localized* than either of the constituent wave functions. However, the state is a mixture of two momentum states, and so the state has become *delocalized in momentum space*. It will be seen that this is a general result that occurs whenever two observables do not commute. Simultaneously measuring both to absolute accuracy is impossible since they do not share a mutual set of eigenfunctions. This is detailed in the next chapter where the Heisenberg uncertainty principle is derived.

3.6.5 Interpreting the mixed state

It is natural to ask how such a mixed state could occur. In the Copenhagen interpretation the answer is that a mixed state is the result of the initial *preparation* of the physical system, which may involve the choice made by an *observer* for the experiment's design or the experimental apparatus used. It is the preparation of the system that determines the relative values of the coefficients of all the wave functions that may appear in the mixed state wave function. For example, if an electron beam is passed through a velocity selector, a range of electron momenta will be present afterwards, but this range of values will manifest itself for *each electron* that is prepared in this manner as long as the velocity selector behaves uniformly as time passes. The Copenhagen interpretation describes this by saying that the state of the electrons in the beam, i.e. their wave function, has been *prepared in a mixed momentum state* by the experimental situation. Bohr's complementarity principle argues that, unlike classical physics, it is possible to prepare and subsequently measure the system in ways that exhibit *either* localized particle properties *or* delocalized wave-like properties.

In the case of electron diffraction, the preparation allowed the electron beam to pass through *both slits* with equal probability, resulting in a mixed state that manifests the interference pattern (3.260) on the photoplate. Prior to striking the photoplate each electron is governed by the mixed state wave function that is consistent with passing through both slits. Striking the photoplate is a measurement of position that results in a particle property. This is distinctly different from Newtonian determinism, which *assumes* that the point particle must have passed through one slit or the other. If the diffraction experiment is designed to determine which of the slits each electron passes through *before* striking the photoplate, then the *quantum interference disappears*. The disappearance of quantum interference has been *experimentally* observed for a suitably configured interferometer. This experimental result is explained in the Copenhagen interpretation using wave function collapse. In this interpretation of quantum mechanics, determining the slit through which the electron passes causes the wave function to collapse onto the position of either slit one or two, which eliminates either Ψ_1 or Ψ_2 in the mixed state (3.245). The details of this are presented in chapter 7 where *decoherence*, the vanishing of the quantum interference terms, is shown to occur through *environmental interaction or detection*.

There is another aspect to mixed states that is important for their interpretation. A wave function can be expanded in terms of a complete set of the orthonormal eigenfunctions of an observable $\hat{\mathrm{O}}$, denoted $f_j(\boldsymbol{x}, t)$, where

$$\hat{\mathrm{O}}f_j(\boldsymbol{x}, t) = \lambda_j f_j(\boldsymbol{x}, t). \tag{3.266}$$

Such an expansion takes the general form

$$\Psi(\boldsymbol{x}, t) = \sum_j a_j f_j(\boldsymbol{x}, t). \tag{3.267}$$

If the system is prepared in the *same* state $\Psi(\boldsymbol{x}, t)$ prior to *each* measurement of the observable $\hat{\mathrm{O}}$, then Born's rule states that each measurement will result in an eigenvalue of $\hat{\mathrm{O}}$, but only those eigenvalues associated with the eigenfunctions actually present in the expansion (3.267). The observation of the eigenvalue λ_j during a measurement occurs with the Born probability given by $|a_j|^2$. However, if a_j is not present in the expansion (3.267) then the associated eigenvalue λ_j will *never be observed* since $a_j = 0$. After many measurements involving the same wave function, these probabilities will become experimentally determined and therefore the moduli of the coefficients a_j present in (3.267) will become known. The state (3.267) is often referred to as a *pure state* since it is purely quantum mechanical in nature.

However, it is possible for the experimental situation to prepare the system in two possible pure states, denoted $\Psi_1(\boldsymbol{x}, t)$ and $\Psi_2(\boldsymbol{x}, t)$. The wave function is then represented by a mixed state of the form (3.244). As an example, electrons may be ejected by two devices oriented in such a way that they arrive at the detector traveling in different directions that distinctly identify which of the two devices it was ejected from. The apparatus for measuring their momenta will be able to distinguish which device ejected the electron based on the direction of its momentum. For such a case, the two possible states can be distinguished from each other and the observer will know which device prepared the electron since the eigenvalue of momentum will have a distinct direction. On the other hand, if the electrons are ejected from the two devices in the same direction, there is no experiment based on measuring the direction of momentum for the electrons that can distinguish which device ejected the electron. Quantum mechanically, the two situations are characterized by whether or not the two constituent pure states of the mixed state have a subset of eigenfunctions that are present in both expansions. If they do, then the two states are experimentally indistinguishable since similar eigenvalues will be measured when observing both states. They will be distinguished using an experiment measuring $\hat{\mathrm{O}}$ only if the two states do not share any eigenfunctions of $\hat{\mathrm{O}}$. This means that the two states must satisfy

$$\begin{aligned} &\int \mathrm{d}^n x\, \Psi_2^*(\boldsymbol{x}, t)\Psi_1(\boldsymbol{x}, t) = 0, \\ &\int \mathrm{d}^n x\, \Psi_2^*(\boldsymbol{x}, t)\hat{\mathrm{O}}\Psi_1(\boldsymbol{x}, t) = 0. \end{aligned} \tag{3.268}$$

A vanishing inner product or orthogonality as well as the action of $\hat{O}$ on $\Psi_1(x, t)$ remaining orthogonal indicates the two states can be distinguished experimentally by measuring $\hat{O}$. In the event of a mixed state of the form (3.244), the orthogonality of the two states corresponds to the integral (3.249) vanishing and the normalization condition reduces to $|a|^2 + |b|^2 = 1$. This becomes important in understanding the outcome of experiments, and is discussed again in the last section of chapter 7.

To some physicists this interpretation raises the disquieting idea that it is the act of observation that determines *reality*, or at least the observer's *knowledge of reality*. As mentioned earlier, the Copenhagen interpretation is referred to as *epistemic* since it casts quantum mechanics as a theory of *possible* measurements or knowledge. To clarify what this means in the case of particle diffraction, it is the availability of two slits through which the incident electrons pass that results in the mixed state probability density of (3.259). As long as the observer does not obtain knowledge of which slit the electron passed through by *interacting* with the particle and *localizing* it at one of the slits prior to its arrival at the photoplate, then the observer must treat both slits as possible sources of the electron arriving at the screen. This is discussed in more detail in chapter 7 where *decoherence* by detection is examined. Much effort has been spent trying to reconcile this conceptual conundrum, but there is no consensus regarding an alternative interpretation of quantum mechanics as of yet. Recent work by Pusey, Barrett, and Rudolph argues that the wave function must be interpreted as representing a physical reality, leading to an *ontic* theory. The interested reader should consult the references.

As an important point, in order to allow the superposition of wave functions, the phase angle $\alpha(v, x, t)$ created by the Galilean transformation of (3.114) must be *identical* for both Ψ_1 and Ψ_2. Only if *all* the wave functions included in the linear superposition transform *identically* to (3.114) will the probability density (3.108) for a linear superposition remain invariant under a Galilean transformation. This is an important reason why the phase angle must be independent of the particle momentum, since the two wave functions may correspond to different momenta values and Galilean invariance would then break down. However, the principle of superposition can be invoked only if the wave function obeys a *linear equation*, like light waves in vacuum. It will be seen in the next chapter that the Schrödinger equation is indeed a *linear partial differential equation* and therefore allows the principle of superposition. However, because it is also a partial differential equation, it is necessary to specify appropriate *boundary and continuity conditions* in order to make the solutions to the Schrödinger equation unique. As an example, allowing the electrons to pass through either slit of the two-slit interferometer constitutes a boundary condition on their wave function, as it did with the light wave, and the principle of superposition is therefore crucial to matching such boundary conditions.

Although the results of this section have been developed without use of the Schrödinger equation, they demonstrate many of the fundamental properties of quantum mechanical measurement as well as the nature of the wave function. In the next chapter the formal structure of quantum mechanics that is consistent with these results will be developed and discussed. In that regard, the results of this chapter will serve as a guide.

References and recommended further reading

The old quantum theory was presented in

- Wilson W 1915 LXXXIII. The quantum-theory of radiation and line spectra *Phil. Mag.* **29** 795 (https://doi.org/10.1080/14786440608635362)
- Sommerfeld A 1916 Zur Quantentheorie der Spektrallinien *Ann. Phys.* **51** 1 (https://doi.org/10.1002/andp.19163561702)
- Bohr N 1925 Atomic theory and mechanics *Nature* **116** 845 (https://doi.org/10.1038/116845a0)

Excellent introductory textbooks on quantum mechanics that begin with the concept of the wave function include

- Greiner W 2001 *Quantum Mechanics: An Introduction* 4th edn (New York: Springer)
- Gasiorowicz S 2003 *Quantum Physics* 3rd edn (New York: Wiley)
- Messiah A 2014 *Quantum Mechanics* (New York: Dover)
- Cohen–Tannoudji C, Diu B and Laloë F 2019 *Quantum Mechanics* vol 1 2nd edn (New York: Wiley)
- Sakurai J and Napolitano J 2020 *Modern Quantum Mechanics* (New York: Cambridge University Press)

Bohr presented his complementarity principle and acknowledged the influence of Heisenberg in

- Bohr N 1928 The quantum postulate and the recent development of atomic theory *Nature* **121** 580 (https://doi.org/10.1038/121580a0)

Heisenberg discussed the development of quantum theory and the Copenhagen interpretation in

- Heisenberg W 1955 *Niels Bohr and the Development of Physics* ed W Pauli (London: Pergamon)
- Heisenberg W 1958 *Physics And Philosophy: the Revolution in Modern Science* (New York: Harper and Row)

The interested reader can find a thorough discussion of the philosophical ramifications of quantum mechanics in

- Faye J 2019 Copenhagen interpretation of quantum mechanics *The Stanford Encyclopedia of Philosophy* (Winter 2019 Edition) ed E Zalta (https://plato.stanford.edu/archives/win2019/entries/qm-copenhagen/)

Complex variables are developed in

- Brown J and Churchill R 2009 *Complex Variables and Applications* 9th edn (New York: McGraw–Hill)
- Gamelin T 2001 *Complex Analysis* (New York: Springer)

The mathematical methods used in quantum mechanics cover a broad range of concepts, including distribution theory, operator theory, the completeness of eigenfunctions, and the Gram–Schmidt orthonormalization procedure. The following books present most of these concepts, if not all, in a form that is accessible to physical scientists. In order of publication they include:

- Courant R and Hilbert D 1989 *Methods of Mathematical Physics* vols I and II (New York: Wiley)
- Byron F and Fuller R 1992 *Mathematics of Classical and Quantum Physics* (New York: Dover)
- Kusse B and Westwig E 1998 *Mathematical Physics, Applied Mathematics for Scientists and Engineers* (New York: Wiley Interscience)
- Hassani S 1999 *Mathematical Physics* (New York: Springer)
- Boas M 2006 *Mathematical Methods in the Physical Sciences* (Hoboken, NJ: Wiley)
- Arfken G, Weber H–J, and Harris F 2013 *Mathematical Methods for Physicists* 7th edn (Cambridge, MA: Elsevier)

Alternate interpretations of quantum mechanics have been presented in

- Bohm D 1952 A suggested interpretation of the quantum theory in terms of "hidden" variables. I *Phys. Rev.* **85** 166 (https://doi.org/10.1103/PhysRev.85.166)
- Everett H 1957 "Relative state" formulation of quantum mechanics *Rev. Mod. Phys.* **29** 454 (https://doi.org/10.1103/RevModPhys.29.454)
- Ghirardi G, Rimini A, and Weber T 1986 Unified dynamics for microscopic and macroscopic systems *Phys. Rev. D* **34** 470 (https://doi.org/10.1103/PhysRevD.34.470)

Galilean transformations of the wave function are presented in

- Ehrenfest P 1927 Bemerkung über die angenäherte Gültigkeit der klassischen Mechanik innerhalb der Quantenmechanik *Z. Phys.* **45** 455 (https://doi.org/10.1007/BF01329203)
- Kaempffer F 1965 *Concepts in Quantum Mechanics* (New York: Academic Press)
- Landau L and Lifshitz E 1977 *Quantum Mechanics: Non-Relativistic Theory* 3rd edn (New York: Pergamon)

The characteristic function of random variables is also discussed in

- Schwabl F 2007 *Quantum Mechanics* 4th edn (New York: Springer)

Born's rule was first presented in

- Born M 1926 Zur Quantenmechanik der Stoßvorgänge *Z. Phys.* **37** 863 (https://doi.org/10.1007/BF01397477)

The derivation of Born's rule from an axiomatic approach is discussed in

- Gleason A 1957 Measures on the closed subspaces of a Hilbert space *Indiana Univ. Math. J.* **6** 885 (http://www.iumj.indiana.edu/IUMJ/FULLTEXT/1957/6/56050)
- Landsman N 2008 *Compendium of Quantum Physics* ed F Weinert *et al* (New York: Springer)

The argument for the ontic interpretation of the wave function is found in

- Pusey M, Barrett J and Rudolph T 2012 On the reality of the quantum state *Nature Phys.* **8** 475 (https://doi.org/10.1038/nphys2309)

IOP Publishing

A Concise Introduction to Quantum Mechanics (Second Edition)

Mark S Swanson

Chapter 4

Formal wave mechanics

In 1926 Schrödinger proposed a partial differential equation, now known as the *Schrödinger equation*, that governs the wave function, and formal quantum mechanics was developed from it. Solving the Schrödinger equation and applying the resulting wave function is often referred to as *wave mechanics*. Understanding its basic implications for the wave function and quantum mechanics is the subject of this chapter.

4.1 The Schrödinger equation

In the previous chapter a number of the basic properties of the wave function were developed. This was done primarily using physical arguments such as superposition and the need for normalizing probabilities. It is useful to start by recapitulating the mathematical properties of the wave function before introducing the Schrödinger equation.

4.1.1 The requirements for the wave function equation

The properties required for the wave function listed in the previous chapter were derived from the probabilistic interpretation of the wave function and the need to explain quantum mechanical phenomena. These are the properties of linear superposition, membership in the space of square integrable functions L^2, the need to be consistent with the recipe (3.114) for a massive particle wave function undergoing a Galilean transformation of the form (1.32), an independence of the base value chosen for the energy, the ability to represent a large variety of spatial boundary conditions through a complete set of basis functions, and an equation that is linear in the time derivative. In addition, there is the need to include potential energies in determining the form of the wave function.

The most viable candidate for a wave equation that fulfills all these requirements is a *linear partial differential equation of the Sturm–Liouville type*, which typically

doi:10.1088/978-0-7503-5663-3ch4

generates complete sets of eigenfunctions. Such an equation would take the general form

$$\hat{D}\left(\nabla, \frac{\partial}{\partial t}, x, t\right)\Psi(x, t) = 0, \tag{4.1}$$

where $\hat{D}$ is a collection of the gradient, time derivative, position, and time which defines the action of the *linear and Hermitian operator* $\hat{D}$. Linear and Hermitian operators were defined in the previous chapter by the dual requirements of (3.145) and (3.154). The requirement of linearity allows the superposition of solutions to represent quantum interference effects, while the hermiticity insures that the eigenvalues of the operator are both real and the associated eigenfunctions are typically complete. The reality of the eigenvalues is necessary if they are to represent the outcome of physical measurements, while the completeness of the eigenfunctions allows an arbitrary wave function to be represented as a superposition of these eigenfunctions.

4.1.2 Motivating the Schrödinger equation

Examining the free particle wave function (3.127) found from the Galilean transformation property (3.126) in the previous chapter enables the general structure of the operator $\hat{D}$ to be deduced. Using the identification $\hat{\boldsymbol{p}} = -i\hbar\nabla$ found in the previous chapter, it follows that

$$\frac{\hat{p}^2}{2m}\Psi(\boldsymbol{p}, \boldsymbol{x}, t) = -\frac{\hbar^2}{2m}\nabla^2\left(\frac{1}{\sqrt{V}}\exp\left(\frac{i}{\hbar}(\boldsymbol{p}\cdot\boldsymbol{x} - E_{\mathrm{p}}t)\right)\right) = \frac{p^2}{2m}\Psi(\boldsymbol{p}, \boldsymbol{x}, t). \tag{4.2}$$

The operator $\hat{K} = \hat{p}^2/2m$ has the free particle wave function as an eigenfunction with the usual Newtonian expression $p^2/2m$ for the kinetic energy as an eigenvalue. It was shown in the previous chapter that the eigenfunctions of $\hat{K}$ are both *complete* and *orthonormal* in the continuum sense. Result (4.2) can be coupled with the result that

$$i\hbar\frac{\partial}{\partial t}\Psi(\boldsymbol{p}, \boldsymbol{x}, t) = i\hbar\frac{\partial}{\partial t}\left(\frac{1}{\sqrt{V}}\exp\left(\frac{i}{\hbar}(\boldsymbol{p}\cdot\boldsymbol{x} - E_{\mathrm{p}}t)\right)\right) = E_{\mathrm{p}}\Psi(\boldsymbol{p}, \boldsymbol{x}, t) = \frac{p^2}{2m}\Psi(\boldsymbol{p}, \boldsymbol{x}, t). \tag{4.3}$$

As a result, the free particle wave function satisfies

$$\left(-\frac{\hbar^2}{2m}\nabla^2 - i\hbar\frac{\partial}{\partial t}\right)\Psi(\boldsymbol{p}, \boldsymbol{x}, t) = \left(\hat{H}(\hat{\boldsymbol{p}}) - i\hbar\frac{\partial}{\partial t}\right)\Psi(\boldsymbol{p}, \boldsymbol{x}, t) = 0. \tag{4.4}$$

The partial differential equation of (4.4) serves as the motivation for the form of the Schrödinger equation.

4.1.3 Galilean covariance and the wave equation

In order to identify (4.4) as the correct wave equation for the case of a free particle, it is critical to show that it is consistent with the Galilean transformation of the wave function given by the phase (3.119), where the transformed wave function is given by

$$\Psi'(\boldsymbol{p}', \boldsymbol{x}', t') = e^{-i\alpha(\boldsymbol{v}_o, \boldsymbol{x}, t)}\Psi(\boldsymbol{p}, \boldsymbol{x}, t),$$
$$\alpha(\boldsymbol{v}_o, \boldsymbol{x}, t) = \frac{1}{\hbar}(m\boldsymbol{v}_o \cdot \boldsymbol{x} - \frac{1}{2}mv_o^2 t), \tag{4.5}$$

with $\boldsymbol{v}_o$ the relative velocity of the second observer. Consistency is demonstrated by using the Galilean transformed versions of the gradient and time derivative founder earlier in (3.139) and (3.140),

$$\nabla' = \nabla,$$
$$\frac{\partial}{\partial t'} = \frac{\partial}{\partial t} + \boldsymbol{v}_o \cdot \nabla. \tag{4.6}$$

Applying these forms to the equation (4.4) gives, after some rearrangement of the derivatives,

$$\left(-\frac{\hbar^2}{2m}\nabla'^2 - i\hbar\frac{\partial}{\partial t'}\right)\Psi'(\boldsymbol{p}', \boldsymbol{x}', t') = \left(-\frac{\hbar^2}{2m}\nabla^2 - i\hbar\left(\frac{\partial}{\partial t} + \boldsymbol{v}_o \cdot \nabla\right)\right)(e^{-i\alpha(\boldsymbol{v}_o, \boldsymbol{x}, t)}\Psi(\boldsymbol{p}, \boldsymbol{x}, t))$$
$$= \left(\frac{1}{2m}(-i\hbar\nabla + m\boldsymbol{v}_o)^2 - \left(i\hbar\frac{\partial}{\partial t} + \frac{1}{2}mv_o^2\right)\right)(e^{-i\alpha(\boldsymbol{v}_o, \boldsymbol{x}, t)}\Psi_p(\boldsymbol{x}, t)). \tag{4.7}$$

The appearance of the combination $(-i\hbar\nabla + m\boldsymbol{v}_o)$ in the transformed equation (4.7) is consistent with the identification of the momentum operator, $\hat{\boldsymbol{p}} = -i\hbar\nabla$, first made in (3.135). This identification is corroborated by using the form of the phase angle $\alpha(\boldsymbol{v}_o, \boldsymbol{x}, t) = (m\boldsymbol{v}_o \cdot \boldsymbol{x} - \frac{1}{2}mv_o^2 t)/\hbar$, which gives two useful results,

$$(-i\hbar\nabla + m\boldsymbol{v}_o)e^{-i\alpha(\boldsymbol{v}_o, \boldsymbol{x}, t)} = e^{-i\alpha(\boldsymbol{v}_o, \boldsymbol{x}, t)}(-\hbar\nabla\alpha(\boldsymbol{v}_o, \boldsymbol{x}, t) + m\boldsymbol{v}_o) = 0,$$
$$\left(i\hbar\frac{\partial}{\partial t} + \frac{1}{2}mv_o^2\right)e^{-i\alpha(\boldsymbol{v}_o, \boldsymbol{x}, t)} = e^{-i\alpha(\boldsymbol{v}_o, \boldsymbol{x}, t)}\left(\hbar\frac{\partial}{\partial t}\alpha(\boldsymbol{v}_o, \boldsymbol{x}, t) + \frac{1}{2}mv_o^2\right) = 0. \tag{4.8}$$

As a result the transformed equation (4.7) becomes

$$\left(-\frac{\hbar^2}{2m}\nabla'^2 - i\hbar\frac{\partial}{\partial t'}\right)\Psi'(\boldsymbol{p}', \boldsymbol{x}', t') = e^{-i\alpha(\boldsymbol{v}_o, \boldsymbol{x}, t)}\left(-\frac{\hbar^2}{2m}\nabla^2 - i\hbar\frac{\partial}{\partial t}\right)\Psi(\boldsymbol{p}, \boldsymbol{x}, t) = 0. \tag{4.9}$$

Up to the overall physically irrelevant phase factor, equation (4.4) takes the same form for *all Galilean observers regardless of relative motion.* Result (4.9) shows that the wave equation (4.4) is *covariant under a Galilean transformation.* In other words, all Galilean observers will solve the same equation using their coordinate system. As a result, (4.4) is the correct quantum mechanical *equation of motion* for the case of a free Newtonian particle. The linearity in the time derivative is critical to the Galilean invariance of (4.4) and is a requirement for the time evolution of the wave equation to determine a unique wave function at later times from its initial preparation.

4.1.4 The third postulate of quantum mechanics

Because the free particle equation (4.4) takes the form

$$\frac{\hat{p}^2}{2m}\Psi(\boldsymbol{x}, t) = i\hbar\frac{\partial}{\partial t}\Psi(\boldsymbol{x}, t), \tag{4.10}$$

the Hamiltonian for the free particle expressed in operator form has appeared. It is logical to extrapolate this form to the case that the Hamiltonian of the particle contains a potential energy $V(\boldsymbol{x}, t)$. Adding a potential energy to the Hamiltonian is consistent with Galilean relativity since the phase factor of (4.5) commutes with the potential as long as the potential is solely a function of $\boldsymbol{x}$ and t. This yields the Schrödinger equation for the wave function of a single massive particle in the presence of such a potential energy.

Postulate three: the wave function for a single particle moving in a potential $V(\boldsymbol{x}, t)$ obeys the partial differential equation known as Schrödinger's equation,

$$\begin{aligned} \hat{H}(\hat{\boldsymbol{p}}, \hat{\boldsymbol{x}}, t)\Psi(\boldsymbol{x}, t) &= \left(\frac{\hat{p}^2}{2m} + V(\hat{\boldsymbol{x}}, t)\right)\Psi(\boldsymbol{x}, t) \\ &= \left(-\frac{\hbar^2}{2m}\nabla^2 + V(\boldsymbol{x}, t)\right)\Psi(\boldsymbol{x}, t) = i\hbar\frac{\partial}{\partial t}\Psi(\boldsymbol{x}, t). \end{aligned} \tag{4.11}$$

The form of the Schrödinger equation (4.11) is tightly constrained by the interpretation of the wave function in probabilistic terms, as well as the requirement of Galilean relativity. It is straightforward to repeat the steps of (4.9) to show that the differential operator appearing in (4.11) transforms to

$$\begin{aligned} &\left(-\frac{\hbar^2}{2m}\nabla'^2 + V(\boldsymbol{x}', t') - i\hbar\frac{\partial}{\partial t'}\right)\Psi'(\boldsymbol{x}', t') \\ &= e^{-i\alpha(\boldsymbol{v}_o, \boldsymbol{x}, t)}\left(-\frac{\hbar^2}{2m}\nabla^2 + V(\boldsymbol{x} + \boldsymbol{v}_o t, t) - i\hbar\frac{\partial}{\partial t}\right)\Psi(\boldsymbol{x}, t). \end{aligned} \tag{4.12}$$

This shows that the moving observer will solve the *same differential equation*, but with the spatial shape of the potential moving at the relative velocity $\boldsymbol{v}_o$. This is identical to the classical result (1.33).

The Schrödinger equation also introduces the Hamiltonian as the generator of linear time development for the wave function. It will be seen that this is identical to the role of the Hamiltonian in classical physics, as expressed in the Poisson bracket formulation of (1.26). Of course, (4.11) must remain a postulate. There is no overarching postulate that can be used to justify the inclusion of the potential. In that regard, over the approximate century since its first appearance, there have been many variations of the Schrödinger equation presented and analysed. However, the solutions of the Schrödinger equation in the form given by (4.11) are supported by a myriad of experiments.

4.2 General properties of the Schrödinger equation solutions

The solutions to the Schrödinger equation (4.11) have general properties that are both critical to the interpretation of quantum mechanics and useful tools in solving the behavior of specific systems.

4.2.1 Invariance under changes in the energy scale

A critical feature of the Schrödinger equation is consistency with the classical result (1.10), which allows the potential to be altered by an arbitrary constant V_o. Quantum mechanically this means that a change in the potential energy scale,

$$V(\boldsymbol{x}, t) \rightarrow V(\boldsymbol{x}, t) + V_o, \tag{4.13}$$

must result in the *same probability density*, $\mathcal{P}(\boldsymbol{x}, t) = |\Psi(\boldsymbol{x}, t)|^2$. This property is demonstrated by using the phase symmetry of the probability density to define a simultaneous change in the wave function in the Schrödinger equation,

$$\Psi(\boldsymbol{x}, t) \rightarrow \Psi'(\boldsymbol{x}, t) = e^{-iV_o t/\hbar}\Psi(\boldsymbol{x}, t). \tag{4.14}$$

Because V_o is a constant, the transformed Schrödinger equation is given by

$$\begin{aligned} \hat{H}'\Psi'(\boldsymbol{x}, t) - i\hbar\frac{\partial}{\partial t}\Psi'(\boldsymbol{x}, t) &= \left(-\frac{\hbar^2}{2m}\nabla^2 + V(\boldsymbol{x}, t) + V_o\right)(e^{-iV_o t/\hbar}\Psi(\boldsymbol{x}, t)) \\ &\quad - i\hbar\frac{\partial}{\partial t}(e^{-iV_o t/\hbar}\Psi(\boldsymbol{x}, t)) \\ &= e^{-iV_o t/\hbar}\left(-\frac{\hbar^2}{2m}\nabla^2 + V(\boldsymbol{x}, t) + V_o\right)\Psi(\boldsymbol{x}, t) \\ &\quad - e^{-iV_o t/\hbar}\left(V_o + i\hbar\frac{\partial}{\partial t}\right)\Psi(\boldsymbol{x}, t) \\ &= e^{-iV_o t/\hbar}\left(\hat{H} - i\hbar\frac{\partial}{\partial t}\right)\Psi(\boldsymbol{x}, t) = 0. \end{aligned} \tag{4.15}$$

The two terms involving V_o have canceled, showing that the Schrödinger equation of (4.11) is regained, up to the physically irrelevant phase factor. The probability density is unchanged by the phase transformation, so that the classical property of energy represented by (1.10) is maintained in quantum mechanics.

4.2.2 Hermitian Hamiltonians

Because the Hamiltonian plays such a central role in the Schrödinger equation it is important to determine the conditions for which the Hamiltonian operator $\hat{H}$ appearing in (4.11) is Hermitian. This follows by noting that the Hermitian nature of $\hat{\boldsymbol{p}}$ and $\hat{\boldsymbol{x}}$ combined with result (3.159) for operator products gives

$$(\hat{p}^2)^\dagger = (\hat{\boldsymbol{p}} \cdot \hat{\boldsymbol{p}})^\dagger = \hat{\boldsymbol{p}}^\dagger \cdot \hat{\boldsymbol{p}}^\dagger = \hat{\boldsymbol{p}} \cdot \hat{\boldsymbol{p}} = \hat{p}^2, \tag{4.16}$$

with a similar result for products of $\hat{x}$. The potential energy $V(\hat{x}, t)$ is Hermitian as long as it is real valued, since that gives

$$V^{\dagger}(\hat{x}, t) = V^{*}(\hat{x}^{\dagger}, t) = V(\hat{x}, t). \tag{4.17}$$

Using these results and the general rule for linear sums (3.162) shows that the resulting Hamiltonian is Hermitian,

$$\begin{aligned} \hat{H}^{\dagger}(\hat{p}, \hat{x}, t) &= \left(\frac{\hat{p}^2}{2m} + V(\hat{x}, t)\right)^{\dagger} = \left(\frac{\hat{p}^2}{2m}\right)^{\dagger} + V^{\dagger}(\hat{x}, t) \\ &= \frac{\hat{p}^2}{2m} + V(\hat{x}, t) = \hat{H}(\hat{p}, \hat{x}, t). \end{aligned} \tag{4.18}$$

In the general case where $V(\hat{x}, t)$ is time-dependent it is not possible to assume that $\hat{H}$ has a set of constant eigenvalues. Demonstrating this property begins by assuming that there exists a time-dependent wave function $\Psi(E, x, t)$ that is also an eigenfunction of the time-dependent Hamiltonian $\hat{H}$, so that

$$\hat{H}(\hat{p}, \hat{x}, t)\Psi(E, x, t) = E\Psi(E, x, t). \tag{4.19}$$

Applying the time derivative to (4.19) gives

$$\begin{aligned} \frac{\partial \hat{H}(\hat{p}, \hat{x}, t)}{\partial t}\Psi(E, x, t) + \hat{H}(\hat{p}, \hat{x}, t)\frac{\partial}{\partial t}\Psi(E, x, t) &= \frac{\partial E}{\partial t}\Psi(E, x, t) \\ &\quad + E\frac{\partial}{\partial t}\Psi(E, x, t). \end{aligned} \tag{4.20}$$

Using the Schrödinger equation (4.11) gives

$$\begin{aligned} \hat{H}(\hat{p}, \hat{x}, t)\frac{\partial}{\partial t}\Psi(E, x, t) &= -\frac{i}{\hbar}\hat{H}^2(\hat{p}, \hat{x}, t)\Psi(E, x, t) = -\frac{i}{\hbar}E^2\Psi(E, x, t), \\ E\frac{\partial}{\partial t}\Psi(E, x, t) &= -\frac{i}{\hbar}E\hat{H}(\hat{p}, \hat{x}, t)\Psi(E, x, t) = -\frac{i}{\hbar}E^2\Psi(E, x, t), \end{aligned} \tag{4.21}$$

which shows that these two terms cancel in (4.20), leaving

$$\frac{\partial \hat{H}}{\partial t}\Psi(E, x, t) = \frac{\partial E}{\partial t}\Psi(E, x, t). \tag{4.22}$$

Only for the case that the left-hand side of (4.22) vanishes will the eigenvalues of $\hat{H}$ satisfy $\partial E/\partial t = 0$.

For a time-independent potential, results (3.167), (3.168), and (4.22) show that the eigenvalues of the Hermitian Hamiltonian $\hat{H}(\hat{p}, \hat{x})$ *will be real and constant* and the eigenfunctions *will be orthonormal*, up to the need to apply the Gram–Schmidt procedure for degenerate eigenfunctions demonstrated in (3.170). In the previous chapter it was shown that a complete set of commuting observables (CSCO) represents all possible simultaneously measurable properties of the system. For such a case the Hamiltonian can be added to the list of Hermitian observables for the particle. Any CSCO that includes the Hamiltonian must consist solely of observables that commute with the Hamiltonian. In the presence of the static potential $V(\boldsymbol{x})$ the momentum operator $\hat{\boldsymbol{p}}$ will not commute with $\hat{H}$ while the position operator $\hat{\boldsymbol{x}}$ will not commute with $\hat{p}^2/2m$. As a result, neither $\hat{\boldsymbol{p}}$ nor $\hat{\boldsymbol{x}}$ can be members of a CSCO that includes such a Hamiltonian. An example of a CSCO including the Hamiltonian will be presented when the hydrogen atom is analysed in the next chapter.

4.2.3 Conservation of probability and continuity conditions

Investigating the general properties of solutions to the Schrödinger equation (4.11) begins by noting that the complex conjugate of (4.11) for a Hermitian Hamiltonian yields a second equation, given by

$$\hat{H}^{\dagger}(\hat{\boldsymbol{p}}, \boldsymbol{x}, t)\, \Psi^*(\boldsymbol{x}, t) = \hat{H}(\hat{\boldsymbol{p}}, \boldsymbol{x}, t)\, \Psi^*(\boldsymbol{x}, t) = -i\hbar\frac{\partial}{\partial t}\, \Psi^*(\boldsymbol{x}, t), \tag{4.23}$$

where the process of complex conjugation for $\hat{H}$ uses $\hat{\boldsymbol{p}}^* = i\hbar\nabla$. Multiplying (4.11) by $\Psi^*(\boldsymbol{x}, t)$ and (4.23) by $\Psi(\boldsymbol{x}, t)$ and subtracting the latter from the former yields

$$\begin{aligned}\Psi^*(\boldsymbol{x}, t)\left(\hat{H}\Psi(\boldsymbol{x}, t)\right) - \left(\hat{H}^*\Psi^*(\boldsymbol{x}, t)\right)\Psi(\boldsymbol{x}, t) &= \Psi^*(\boldsymbol{x}, t)\left(i\hbar\frac{\partial}{\partial t}\, \Psi(\boldsymbol{x}, t)\right) \\ &\quad - \left(-i\hbar\frac{\partial}{\partial t}\, \Psi^*(\boldsymbol{x}, t)\right)\Psi(\boldsymbol{x}, t) \\ &= i\hbar\frac{\partial}{\partial t}(\Psi^*(\boldsymbol{x}, t)\Psi(\boldsymbol{x}, t)) = i\hbar\frac{\partial}{\partial t}\mathcal{P}_{\hat{x}}(\boldsymbol{x}, t).\end{aligned} \tag{4.24}$$

If the potential energy in the Hamiltonian of (4.11) has the property $V^*(\boldsymbol{x}, t) = V(\boldsymbol{x}, t)$, then the potential energy terms on the left-hand side of (4.24) cancel. The remaining Laplacian terms in (4.24) can be rewritten using the identity

$$\begin{aligned}&-\frac{\hbar^2}{2m}(\Psi^*(\boldsymbol{x}, t)\nabla^2\Psi(\boldsymbol{x}, t) - (\nabla^2\Psi^*(\boldsymbol{x}, t))\Psi(\boldsymbol{x}, t)) \\ &= \frac{\hbar^2}{2m}\nabla \cdot (\nabla\Psi^*(\boldsymbol{x}, t)\Psi(\boldsymbol{x}, t) - \Psi^*(\boldsymbol{x}, t)\nabla\Psi(\boldsymbol{x}, t)),\end{aligned} \tag{4.25}$$

so that, for the case $V^*(\boldsymbol{x}, t) = V(\boldsymbol{x}, t)$, dividing relation (4.24) by $i\hbar$ gives

$$\begin{aligned}-\nabla \cdot \frac{i\hbar}{2m}(\nabla\Psi^*(\boldsymbol{x}, t)\, \Psi(\boldsymbol{x}, t) - \Psi^*(\boldsymbol{x}, t)\, \nabla\Psi(\boldsymbol{x}, t)) &\equiv -\nabla \cdot \mathbf{J}(\boldsymbol{x}, t) \\ &= \frac{\partial}{\partial t}\mathcal{P}_{\hat{x}}(\boldsymbol{x}, t),\end{aligned} \tag{4.26}$$

where

$$\mathbf{J}(\boldsymbol{x}, t) = \frac{i\hbar}{2m}(\nabla\Psi^*(\boldsymbol{x}, t)\, \Psi(\boldsymbol{x}, t) - \Psi^*(\boldsymbol{x}, t)\, \nabla\Psi(\boldsymbol{x}, t)) \tag{4.27}$$

is referred to as the *spatial probability current density*. Result (4.26), when written $\nabla \cdot \mathbf{J} + \partial \mathcal{P}_{\hat{x}}/\partial t = 0$, is identical in form to the conservation law for electric charge given by (1.57). Since $\mathcal{P}_{\hat{x}}(\boldsymbol{x}, t) = \Psi^*(\boldsymbol{x}, t)\Psi(\boldsymbol{x}, t)$ is the probability density and is playing the role of the electric charge density, result (4.26) represents the *conservation of probability*. In three spatial dimensions the vector $\mathbf{J}(\boldsymbol{x}, t)$ of (4.27) has the cgs units of $1/(\text{cm}^2 \cdot \text{s})$ and can therefore be identified as the *current density for probability*. Result (4.26) satisfies the same relation as (1.58), so that any probability lost or gained in a volume over time is matched by the flow of probability out of or into that volume through its surfaces. This is stated using the divergence theorem,

$$\frac{\partial}{\partial t}\int \mathrm{d}^n x\, \mathcal{P}_{\hat{x}}(\boldsymbol{x}, t) = -\int \mathrm{d}^n x\, \nabla \cdot \mathbf{J}(\boldsymbol{x}, t) = -\oint_{S(V)} \mathrm{d}\boldsymbol{S} \cdot \mathbf{J}(\boldsymbol{x}, t). \tag{4.28}$$

The conservation of probability is essential to interpreting the solutions of the Schrödinger equation. An example for barrier penetration is given in the next chapter.

As an example of probability current, the momentum eigenfunction (3.127) for a particle in a volume V yields $\mathbf{J} = \boldsymbol{p}/(mV)$ and satisfies $\nabla \cdot \mathbf{J} = 0$. This is consistent with the right-hand side of (4.26), which vanishes since $\mathcal{P} = 1/V$ is a constant. The flow of probability through a surface associated with the momentum eigenfunction is therefore in the direction of the classical velocity $\boldsymbol{v} = \boldsymbol{p}/m$ and is equally into and out of *any volume*, including the total volume V of the system. The total volume can be chosen as a cube of side L, so that $V = L^3$. The flux of probability through a surface $\mathbf{S}$ of the cube with area L^2 is then given by

$$\mathbf{J} \cdot \mathbf{S} = \frac{\boldsymbol{p} \cdot \hat{\boldsymbol{n}}}{mL}, \tag{4.29}$$

where $\hat{\boldsymbol{n}}$ is the normal to the surface. This should be zero if the particle is to remain in the volume, and this can be true only if $L \to \infty$. As a result, (3.127) correctly describes the situation of a particle of momentum $\boldsymbol{p}$ contained in a volume V only if the volume becomes arbitrarily large. This is discussed further in the section on the Heisenberg uncertainty principle.

The physical interpretation of the wave function and the form of the spatial probability current (4.27) place a continuity requirement on Ψ and $\nabla\Psi$. These are summarized in Born's four conditions for the wave function. The first is that the wave function is single valued at all positions and times. This is necessary to obtain a unique probability density for the presence of the particle. The second is that the modulus of the wave function must belong to the space of square integrable functions L^2. This is necessary to enable normalization of the wave function. The third is that the wave function is continuous at all possible positions and times. This is necessary for the gradient $\nabla\Psi(\boldsymbol{x}, t)$ and the time derivative $\partial\Psi(\boldsymbol{x}, t)/\partial t$ to be defined at all points and times, and follows from the interpretation of the gradient as the momentum operator and the time derivative as the energy operator in the Schrödinger equation (4.11). If the wave function is discontinuous at a point or time, then $\nabla\Psi(\boldsymbol{x}, t)$ or $\partial\Psi(\boldsymbol{x}, t)/\partial t$ will be singular at that point or time, corresponding to a singular momentum or energy. This is considered unphysical. The fourth condition is that the gradient of the wave function is continuous at a point *if the potential is continuous at that point*. A continuous potential corresponds to a non-singular force and will therefore result in smooth changes in the momentum. A discontinuous potential corresponds to a singular force and therefore results in a discontinuous change in the momentum, which is considered unphysical.

Another physical argument for Born's conditions of continuity is found by examining the conservation of probability. If the spatial probability current $\mathbf{J}(\boldsymbol{x}, t)$ is discontinuous at a point, then that point will serve as a *source* or *sink* of probability. This is seen by considering an infinitesimal closed surface $\mathrm{d}\boldsymbol{S}$ around the point and noting that $\mathbf{J} \cdot \mathrm{d}\boldsymbol{S} \neq 0$ if $\mathbf{J}$ is discontinuous at that point. Therefore, in order to avoid creating or destroying probability, $\mathbf{J}$ must be *continuous*, and this *typically* requires that both Ψ and $\nabla\Psi$ must be continuous, justifying Born's fourth condition. However, if $\nabla\Psi(\boldsymbol{x}, t)$ is discontinuous at a point, which occurs for discontinuous potentials, then $\Psi(\boldsymbol{x}, t)$ must vanish at that point in such a way that $\mathbf{J}(\boldsymbol{x}, t)$ remains continuous.

4.2.4 Ehrenfest's theorem and classical mechanics

One of the most important properties of a quantum mechanical observable $\hat{\mathrm{O}}$ is its expectation value, defined in (3.202). Due to the fact that the wave function is time-dependent, the expection value will be time-dependent as well. This will be true even if the observable has no explicit time-dependence. The expectation value of a general time-dependent observable $\hat{\mathrm{O}}(t)$ is given by

$$\langle\hat{\mathrm{O}}\rangle(t) = \int \mathrm{d}^n x\ \Psi^*(\boldsymbol{x}, t)(\hat{\mathrm{O}}(t)\Psi(\boldsymbol{x}, t)), \tag{4.30}$$

where it is understood that $\hat{\mathrm{O}}$ acts to the right on $\Psi(\boldsymbol{x}, t)$. Using the Schrödinger equation (4.11) and its complex conjugate (4.23), along with the assumed Hermitian nature of the Hamiltonian, it follows that the time derivative is given by

$$\begin{aligned}
\frac{\mathrm{d}}{\mathrm{d}t}\langle\hat{\mathrm{O}}\rangle(t) &= \int \mathrm{d}^n x \left(\frac{\partial\Psi^*(\boldsymbol{x}, t)}{\partial t}\hat{\mathrm{O}}(t)\Psi(\boldsymbol{x}, t) + \Psi^*(\boldsymbol{x}, t)\frac{\partial\hat{\mathrm{O}}(t)}{\partial t}\Psi(\boldsymbol{x}, t) + \Psi^*(\boldsymbol{x}, t)\hat{\mathrm{O}}(t)\frac{\partial\Psi(\boldsymbol{x}, t)}{\partial t}\right) \\
&= \int \mathrm{d}^n x \left(\frac{i}{\hbar}\hat{H}^\dagger\Psi^*(\boldsymbol{x}, t)\hat{\mathrm{O}}(t)\Psi(\boldsymbol{x}, t) + \Psi^*(\boldsymbol{x}, t)\frac{\partial\hat{\mathrm{O}}(t)}{\partial t}\Psi(\boldsymbol{x}, t) - \frac{i}{\hbar}\Psi^*(\boldsymbol{x}, t)\hat{\mathrm{O}}(t)\hat{H}\Psi(\boldsymbol{x}, t)\right) \\
&= \int \mathrm{d}^n x \left(\Psi^*(\boldsymbol{x}, t)\frac{\partial\hat{\mathrm{O}}(t)}{\partial t}\Psi(\boldsymbol{x}, t) + \frac{i}{\hbar}\Psi^*(\boldsymbol{x}, t)(\hat{H}\hat{\mathrm{O}}(t) - \hat{\mathrm{O}}(t)\hat{H})\Psi(\boldsymbol{x}, t)\right) \\
&= \left\langle\frac{\partial\hat{\mathrm{O}}(t)}{\partial t}\right\rangle + \frac{i}{\hbar}\langle[\hat{H}, \hat{\mathrm{O}}(t)]\rangle.
\end{aligned} \tag{4.31}$$

This is identical in form to (3.195), which was anticipated using the correspondence to the Poisson bracket relation (1.26). It shows that the expectation values of these operators are related by the same classical expression for them. It also shows that if the observable $\hat{\mathrm{O}}$ has no explicit time-dependence, so that $\partial\hat{\mathrm{O}}/\partial t = 0$, and $\hat{\mathrm{O}}$ commutes with the Hamiltonian $\hat{H}$, so that $[\hat{H}, \hat{\mathrm{O}}] = 0$, then its expectation value will be time-independent. In the Copenhagen interpretation, this means that the statistical average of the values measured for $\hat{\mathrm{O}}$ will be independent of the time of measurement subsequent to preparation of the wave function. As another example, since $[\hat{H}, \hat{H}] = 0$, it follows that

$$\frac{\mathrm{d}}{\mathrm{d}t}\langle \hat{H} \rangle = \left\langle \frac{\partial \hat{H}}{\partial t} \right\rangle. \tag{4.32}$$

This shows that, if the Hamiltonian is not explicitly time-dependent, the expectation value of the energy is constant in time.

A simple example of (4.31) is the case of a free particle, so that the Hamiltonian is $\hat{H} = \hat{p}^2/2m$, and the observable position $\hat{\mathrm{O}} = \hat{\boldsymbol{x}}$. Result (3.199) gives

$$\frac{\mathrm{d}}{\mathrm{d}t}\langle \hat{\boldsymbol{x}} \rangle(t) = \frac{\langle \hat{\boldsymbol{p}} \rangle}{m} = \langle \hat{\boldsymbol{v}} \rangle, \tag{4.33}$$

which gives the usual relationship between velocity and momentum. A more specific version of (4.31) follows from the assumption that the Hamiltonian includes a potential $V(\boldsymbol{x}, t)$ and the observable is chosen to be the momentum, so that $\hat{\mathrm{O}} = \hat{\boldsymbol{p}} = -i\hbar\nabla$. For such a choice, it follows that the commutator in (4.31) is given by

$$\frac{i}{\hbar}[\hat{H}, \hat{\mathrm{O}}]\Psi(\boldsymbol{x}, t) = \hat{H}\nabla\Psi(\boldsymbol{x}, t) - \nabla(\hat{H}\Psi(\boldsymbol{x}, t)) = -(\nabla V(\boldsymbol{x}, t))\Psi(\boldsymbol{x}, t), \tag{4.34}$$

where $[\hat{\boldsymbol{p}}, \hat{p}^2/2\,m] = 0$ was used. Since the momentum has no explicit time-dependence, (4.31) yields

$$\frac{\mathrm{d}}{\mathrm{d}t}\langle \hat{\boldsymbol{p}} \rangle(t) = -\langle \nabla V(\hat{\boldsymbol{x}}, t) \rangle, \tag{4.35}$$

which is the quantum mechanical version of Newton's second law. Results (4.31) and (4.35) are known collectively as *Ehrenfest's theorem*.

4.2.5 The persistence of eigenvalues

It is now possible to answer the question raised in (3.169), which is to determine the circumstances under which the eigenvalues of an observable persist over time. This begins by assuming the observable $\hat{\mathrm{O}}(t)$ has the eigenvalue λ and the eigenfunction $\Psi(\lambda, \boldsymbol{x}, t)$ at the specific time t, so that

$$\hat{\mathrm{O}}(t)\Psi(\lambda, \boldsymbol{x}, t) = \lambda\Psi(\lambda, \boldsymbol{x}, t), \tag{4.36}$$

where λ is a constant. After an infinitesimal time ϵ passes, it is possible to expand both the operator and the wave function using a Taylor series,

$$\begin{aligned} \hat{\mathrm{O}}(t + \epsilon) &= \hat{\mathrm{O}}(t) + \epsilon\,\frac{\partial \hat{\mathrm{O}}(t)}{\partial t}, \\ \Psi(\boldsymbol{x}, t + \epsilon) &= \Psi(\boldsymbol{x}, t) + \epsilon\,\frac{\partial}{\partial t}\Psi(\boldsymbol{x}, t). \end{aligned} \tag{4.37}$$

Assuming that $\Psi(\boldsymbol{x}, t)$ obeys the Schrödinger equation and dropping terms of $O(\epsilon^2)$, this gives

$$\begin{aligned}
\hat{O}(t+\epsilon)\Psi(\lambda, \boldsymbol{x}, t+\epsilon) &= \hat{O}(t)\Psi(\lambda, \boldsymbol{x}, t) + \epsilon\frac{\partial \hat{O}(t)}{\partial t}\Psi(\lambda, \boldsymbol{x}, t) + \epsilon\, \hat{O}(t)\frac{\partial}{\partial t}\Psi(\boldsymbol{x}, t) \\
&= \lambda\Psi(\lambda, \boldsymbol{x}, t) + \epsilon\frac{\partial \hat{O}(t)}{\partial t}\Psi(\lambda, \boldsymbol{x}, t) - \epsilon\,\frac{i}{\hbar}\hat{O}(t)\hat{H}\Psi(\boldsymbol{x}, t) \\
&= \lambda\Psi(\lambda, \boldsymbol{x}, t) + \epsilon\frac{\partial \hat{O}(t)}{\partial t}\Psi(\lambda, \boldsymbol{x}, t) \\
&\quad + \epsilon\,\frac{i}{\hbar}[\hat{H}, \hat{O}(t)]\Psi(\boldsymbol{x}, t) - \epsilon\,\frac{i}{\hbar}\hat{H}\hat{O}(t)\Psi(\lambda, \boldsymbol{x}, t) \\
&= \lambda\Psi(\lambda, \boldsymbol{x}, t) - \epsilon\,\lambda\frac{i}{\hbar}\hat{H}\Psi(\lambda, \boldsymbol{x}, t) \\
&\quad + \epsilon\frac{\partial \hat{O}(t)}{\partial t}\Psi(\lambda, \boldsymbol{x}, t) + \epsilon\,\frac{i}{\hbar}[\hat{H}, \hat{O}(t)]\Psi(\lambda, \boldsymbol{x}, t) \\
&= \lambda\Psi(\lambda, \boldsymbol{x}, t) + \epsilon\,\lambda\frac{\partial}{\partial t}\Psi(\lambda, \boldsymbol{x}, t) \\
&\quad + \epsilon\frac{\partial \hat{O}(t)}{\partial t}\Psi(\lambda, \boldsymbol{x}, t) + \epsilon\,\frac{i}{\hbar}[\hat{H}, \hat{O}(t)]\Psi(\lambda, \boldsymbol{x}, t) \\
&= \lambda\Psi(\lambda, \boldsymbol{x}, t+\epsilon) + \epsilon\left(\frac{\partial \hat{O}(t)}{\partial t} + \frac{i}{\hbar}[\hat{H}, \hat{O}(t)]\right)\Psi(\lambda, \boldsymbol{x}, t).
\end{aligned} \tag{4.38}$$

This shows that $\Psi(\lambda, \boldsymbol{x}, t+\epsilon)$ persists as an eigenfunction of $\hat{O}$ if

$$\frac{\partial \hat{O}(t)}{\partial t} + \frac{i}{\hbar}[\hat{H}, \hat{O}(t)] = 0, \tag{4.39}$$

which requires conditions $\partial\hat{O}/\partial t = 0$ and $[\hat{H}, \hat{O}(t)] = 0$ to be satisfied. The eigenvalues asssociated with such an observable are referred to as *good quantum numbers* since their value does not change over time.

If the chosen CSCO includes the Hamiltonian, then, by definition, all observables in the CSCO commute with the Hamiltonian. As a result, the eigenfunctions of the Hamiltonian will also be eigenfunctions of the other observables in the CSCO *at all later times*. This is demonstrated in the next chapter where the hydrogen atom is analysed.

4.3 Stationary solutions to the Schrödinger equation

For the case that the potential energy function has no explicit time-dependence, the Schrödinger equation is given by

$$\left(-\frac{\hbar^2}{2m}\nabla^2 + V(\boldsymbol{x})\right)\Psi(\boldsymbol{x}, t) = i\hbar\frac{\partial}{\partial t}\Psi(\boldsymbol{x}, t). \tag{4.40}$$

4.3.1 Separation of variables

Equation (4.40) can be simplified by using the technique known as *separation of variables*. This begins by writing the wave function as the product

$$\Psi(\boldsymbol{x}, t) = T(t)\psi(\boldsymbol{x}). \tag{4.41}$$

Using this form in (4.40) and dividing by $T(t)\psi(\boldsymbol{x})$ gives

$$\frac{1}{\psi(\boldsymbol{x})}\left(-\frac{\hbar^2}{2m}\nabla^2 + V(\boldsymbol{x})\right)\psi(\boldsymbol{x}) = \frac{i\hbar}{T(t)}\frac{\partial T(t)}{\partial t}. \tag{4.42}$$

Because the left-hand side is a function of $\boldsymbol{x}$ and the right-hand side is a function of t, (4.42) can hold only if both sides are equal to a *constant*, denoted E, that is independent of $\boldsymbol{x}$ and t. The right-hand side gives

$$i\hbar\frac{\partial T(t)}{\partial t} = ET(t), \tag{4.43}$$

which has the solution

$$T(t) = Ce^{-iEt/\hbar}, \tag{4.44}$$

where C is another constant. The left-hand side of (4.42) gives the *eigenvalue equation*

$$\left(-\frac{\hbar^2}{2m}\nabla^2 + V(\boldsymbol{x})\right)\psi_E(\boldsymbol{x}) = \hat{H}(\boldsymbol{p}, \boldsymbol{x})\psi_E(\boldsymbol{x}) = E\psi_E(\boldsymbol{x}), \tag{4.45}$$

where $\psi_E(\boldsymbol{x})$ denotes the solution to (4.45) for a specific value of E. Absorbing the constant C into the function $\psi_E(\boldsymbol{x})$ shows that the solution to (4.40) can be written

$$\Psi(\boldsymbol{x}, t) = e^{-iEt/\hbar}\psi_E(\boldsymbol{x}), \tag{4.46}$$

where E is a constant with the units of energy. Both E and $\psi_E(\boldsymbol{x})$ are found by solving the eigenvalue equation (4.45). For consistency with the interpretation of the momentum operator and its eigenfunctions, the set of eigenvalues E and eigenfunctions $\psi_E(\boldsymbol{x})$ of the Hamiltonian are interpreted, respectively, as the *measurable energies* of the particle and the *spatial wave function* of a particle with that energy. It is important to remember that these results hold for the case that the Hamiltonian has no explicit time-dependence.

The probability density associated with the solution (4.46) is given by

$$\mathcal{P}_{\hat{x}}(E, \boldsymbol{x}, t) = |\Psi(\boldsymbol{x}, t)|^2 = |\psi_E(\boldsymbol{x})|^2, \tag{4.47}$$

which is independent of time but depends upon the eigenvalue E. The eigenfunction $\psi_E(\boldsymbol{x})$ is therefore referred to as a *stationary state* since its associated probability density has no time-dependence. This has the immediate implication that the conservation of probability equation (4.26) reduces to

$$\frac{\partial}{\partial t}\mathcal{P}_{\hat{x}}(\boldsymbol{x}, t) + \nabla \cdot \mathbf{J}(\boldsymbol{x}, t) = \nabla \cdot \mathbf{J}(\boldsymbol{x}, t) = 0. \tag{4.48}$$

As a result, for an energy eigenfunction there is no flux of probability through any surface.

4.3.2 Continuity conditions for stationary states

The continuity condition for the gradient of $\Psi(\boldsymbol{x}, t)$ can be examined for stationary states by performing an infinitesimal x integration of width ϵ of both sides of (4.45) around a point of interest $\boldsymbol{x} = (x_o, y, z)$. For $\epsilon \to 0$ using the mean value theorem for the integral of the left-hand side of (4.45) results in

$$\begin{aligned}
&-\frac{\hbar^2}{2m}\int_{x_o}^{x_o+\epsilon} \mathrm{d}x\, \frac{\partial^2}{\partial x^2}\psi_E(x, y, z) - \frac{\hbar^2}{2m}\left(\frac{\partial^2}{\partial y^2} + \frac{\partial^2}{\partial z^2}\right)\int_{x_o}^{x_o+\epsilon} \mathrm{d}x\, \psi_E(x, y, z) \\
&= -\frac{\hbar^2}{2m}\left(\frac{\partial\psi_E(x, y, z)}{\partial x}\bigg|_{x=x_o+\epsilon} - \frac{\partial\psi_E(x, y, z)}{\partial x}\bigg|_{x=x_o}\right) \\
&\quad -\frac{\hbar^2}{2m}\left(\frac{\partial^2}{\partial y^2} + \frac{\partial^2}{\partial z^2}\right)\epsilon\psi_E\left(x_o + \frac{1}{2}\epsilon, y, z\right) \\
&\to -\frac{\hbar^2}{2m}\left(\frac{\partial\psi_E(x, y, z)}{\partial x}\bigg|_{x=x_o+\epsilon} - \frac{\partial\psi_E(x, y, z)}{\partial x}\bigg|_{x=x_o}\right),
\end{aligned} \tag{4.49}$$

where it was assumed that $\psi_E(\boldsymbol{x})$ is continuous at the point of interest, so that $\epsilon\psi_E(x_o + \frac{1}{2}\epsilon, y, z)$ vanishes in the limit $\epsilon \to 0$.

Applying the mean value theorem for $\epsilon \to 0$ again, the integral of the right-hand side of (4.45) becomes

$$\lim_{\epsilon\to 0}\int_{x_o}^{x_o+\epsilon} \mathrm{d}x\, (E - V(x, y, z))\psi_E(x, y, z) = \epsilon\left(E - V\left(x_o + \frac{1}{2}\epsilon, y, z\right)\right)\psi_E\left(x_o + \frac{1}{2}\epsilon, y, z\right). \tag{4.50}$$

For the case that both $V(\boldsymbol{x})$ and the wave function $\psi_E(\boldsymbol{x})$ are continuous this term vanishes in the limit $\epsilon \to 0$. As a result, for a continuous potential and wave function, it follows that

$$\lim_{\epsilon\to 0}\left(\frac{\partial\psi_E(\boldsymbol{x})}{\partial x}\bigg|_{x=x_o+\epsilon} - \frac{\partial\psi_E(\boldsymbol{x})}{\partial x}\bigg|_{x=x_o}\right) = 0, \tag{4.51}$$

which shows that $\nabla\psi_E(\boldsymbol{x})$ is a *continuous function* for these conditions. This matches Born's fourth continuity condition.

However, for the case that the potential is singular in the interval $(x_o, x_o + \epsilon)$, the right side will not vanish. This is demonstrated using the properties of the Dirac delta function, which show that a singular potential of the form $V(\boldsymbol{x}) = V_o\, \delta^3(\boldsymbol{x} - \boldsymbol{x}_o)$, where V_o is a constant, gives

$$\int_{x_o}^{x_o+\epsilon} \mathrm{d}x\ V(\boldsymbol{x})\,\psi_E(\boldsymbol{x}) = \int_{x_o}^{x_o+\epsilon} \mathrm{d}x\ V_o\,\delta(x - x_o)\,\delta(y - y_o)\,\delta(z - z_o)\,\psi_E(x, y, z) \\ = \frac{1}{2} V_o\,\delta(y - y_o)\,\delta(z - z_o)\,\psi_E(x_o, y, z), \tag{4.52}$$

which does not vanish in the limit that $\epsilon \to 0$. The factor of one-half follows from by using the representation (3.89) for the Dirac delta, which gives

$$\lim_{\epsilon\to 0}\int_0^{\epsilon} \mathrm{d}x\ \delta(x - x_o) = \lim_{\epsilon\to 0}\lim_{\alpha\to 0}\int_0^{\epsilon} \mathrm{d}x\ \sqrt{\frac{\alpha}{\pi}}\ e^{-\alpha x^2} = \frac{1}{2}. \tag{4.53}$$

Such a singular potential gives rise to discontinuities in the derivatives of the wave function, corresponding to a discontinuity in the momentum of the particle.

In the remainder of this text it will be assumed that the stationary solutions are normalized, so that

$$\int \mathrm{d}^n x\ |\psi_E(\boldsymbol{x})|^2 = 1. \tag{4.54}$$

This ignores possible problems brought about by singular potentials.

4.3.3 Unitary time evolution

The Schrödinger equation (4.11) also has the *extremely important* property that it is linear in the time derivative. This means that the *time evolution* of the wave function is *determined* once its initial form is given or, in the Copenhagen interpretation, *prepared.* In other words, the time evolution of the wave function itself is *deterministic* and does not need additional initial or final information to be specified, unlike the electromagnetic wave equation (1.70). Assuming the wave function is not singular at the time t *and* that the Hamiltonian is not explicitly time-dependent, so that $\partial\hat{H}/\partial t = 0$, the Schrödinger equation allows the Taylor series expansion of the wave function around the time t to be written

$$\Psi(\boldsymbol{x}, t + T) = \sum_{n=0}^{\infty} \frac{T^n}{n!}\frac{\partial^n}{\partial t^n}\Psi(\boldsymbol{x}, t) \\ = \sum_{n=0}^{\infty} \frac{T^n}{n!}\left(-\frac{i}{\hbar}\hat{H}(\hat{\boldsymbol{p}}, \boldsymbol{x})\right)^n \Psi(\boldsymbol{x}, t) = e^{-i\hat{H}(\hat{\boldsymbol{p}},\, \hat{\boldsymbol{x}})\,T/\hbar}\ \Psi(\boldsymbol{x}, t), \tag{4.55}$$

where the time-independence of $\hat{H}$ allows the second step,

$$\hat{H}(\hat{\boldsymbol{p}}, \hat{\boldsymbol{x}})\,\frac{\partial^n}{\partial t^n}\Psi(\boldsymbol{x}, t) = \frac{\partial^n}{\partial t^n}(\hat{H}(\hat{\boldsymbol{p}}, \hat{\boldsymbol{x}})\,\Psi(\boldsymbol{x}, t)) = i\hbar\,\frac{\partial^{n+1}}{\partial t^{n+1}}\Psi(\boldsymbol{x}, t). \tag{4.56}$$

The operator appearing in the last step of (4.55),

$$\hat{\mathrm{U}}(T) = e^{-i\hat{H}(\hat{\boldsymbol{p}},\, \hat{\boldsymbol{x}})T/\hbar}, \tag{4.57}$$

is referred to as *the evolution operator* since it is responsible for the time evolution of the wave function,

$$\Psi(\boldsymbol{x},\, t) = \hat{\text{U}}(T)\Psi(\boldsymbol{x},\, 0). \tag{4.58}$$

This is particularly evident if $\Psi_E(\boldsymbol{x}, t = 0)$ is an eigenfunction of $\hat{H}$ with the eigenvalue E, since (4.45) gives

$$\hat{\text{U}}(t)\Psi_E(\boldsymbol{x},\, 0) = e^{-i\hat{H}(\hat{p},\, x)\, t/\hbar}\, \Psi_E(\boldsymbol{x},\, 0) = e^{-iEt/\hbar}\Psi_E(\boldsymbol{x},\, 0) = \Psi_E(\boldsymbol{x},\, t). \tag{4.59}$$

This result generalizes to mixed states of stationary solutions,

$$\Psi(\boldsymbol{x},\, 0) = \sum_E a_E \Psi_E(\boldsymbol{x}), \tag{4.60}$$

where the sum is over the energy eigenvalues found by solving the Schrödinger equation and a_E are the coefficients in the expansion. Using (4.59) the time evolution of this mixed state is given by

$$\Psi(\boldsymbol{x},\, t) = \hat{\text{U}}(t)\Psi(\boldsymbol{x},\, 0) = \hat{\text{U}}(t)\sum_E a_E \Psi_E(\boldsymbol{x}) = \sum_E a_E \Psi_E(\boldsymbol{x}) e^{-iEt/\hbar}. \tag{4.61}$$

This shows that the constituent wave functions making up the full wave function evolve in time individually. The Schrödinger equation therefore embodies the classical concept that the Hamiltonian is responsible for the time evolution of the physical system, an idea first encountered with the classical Poisson bracket formulation (1.25) involving the Hamiltonian.

For the case of a Hermitian Hamiltonian, the adjoint of the evolution operator, given by

$$\hat{\text{U}}^\dagger(T) = e^{i\hat{H}^\dagger(\hat{p},\, \hat{x})T/\hbar} = e^{i\hat{H}(\hat{p},\, \hat{x})T/\hbar}, \tag{4.62}$$

has the property that

$$\hat{\text{U}}^\dagger(T)\, \hat{\text{U}}(T) = 1, \tag{4.63}$$

so that the adjoint $\hat{\text{U}}^\dagger$ is the inverse of $\hat{\text{U}}(T)$,

$$\hat{\text{U}}^\dagger(T) = \hat{\text{U}}^{-1}(T). \tag{4.64}$$

An operator that obeys $\hat{\text{U}}^\dagger = \hat{\text{U}}^{-1}$ is referred to as a *unitary operator*.

The action of the evolution operator $\hat{\text{U}}(t)$ preserves the normalization of the wave function as long as the Hamiltonian is Hermitian,

$$(\Psi,\, \Psi) = (\Psi,\, \hat{\text{U}}^\dagger(t)\hat{\text{U}}(t)\Psi) = (\hat{\text{U}}^{\dagger\dagger}(t)\Psi,\, \hat{\text{U}}(t)\Psi) = (\hat{\text{U}}(t)\Psi,\, \hat{\text{U}}(t)\Psi). \tag{4.65}$$

This shows that the wave function $\Psi(\boldsymbol{x}, t) = \hat{\text{U}}(t)\Psi(\boldsymbol{x}, t = 0)$ remains normalized at all later times. This is referred to as *unitary time evolution.* It is worth noting that wave function collapse *does not necessarily* correspond to unitary time evolution since the wave function is discontinuous in its time development at the moment of measurement. Analysing the more complicated case where the potential energy is time-dependent is presented in chapter 8.

4.3.4 Mixed states and stationary solutions

It is important to note that a mixed state of two stationary solutions is *not* another stationary solution of the Schrödinger equation. This is easily seen by examining the linear combination

$$\Psi(x) = a\psi_{E_1}(x) + b\psi_{E_2}(x), \tag{4.66}$$

where a and b are constant. This immediately gives

$$\hat{H}\Psi(x) = a\hat{H}\psi_{E_1}(x) + b\hat{H}\psi_{E_2}(x) = aE_1\psi_{E_1}(x) + bE_2\psi_{E_2}(x) \neq E_3\Psi(x). \tag{4.67}$$

There is no solution for E_3 if $E_1 \neq E_2$.

However, it is straightforward to show that the mixed state

$$\Psi(x, t) = a\, e^{-iE_1t/\hbar}\psi_{E_1}(x) + b\, e^{-iE_2t/\hbar}\psi_{E_2}(x), \tag{4.68}$$

is a solution of the Schrödinger equation. The probability density for (4.68) exhibits the quantum interference terms first discussed in (3.245),

$$\begin{aligned} \mathcal{P}_{\hat{H}}(x, t) = &|a|^2|\psi_{E_1}(x)|^2 + |b|^2|\psi_{E_2}(x)|^2 \\ &+ a^*b\, e^{i(E_1-E_2)t/\hbar}\psi^*_{E_1}(x)\psi_{E_2}(x) + ab^*e^{-i(E_1-E_2)t/\hbar}\psi^*_{E_2}(x)\psi_{E_1}(x). \end{aligned} \tag{4.69}$$

This is a generalized version of the specific case (3.264) found for a mixed state of momentum eigenfunctions. The normalization of (4.68) is identical to that of the orthogonal momentum eigenfunctions of (3.263) since the two energy eigenfunctions are also orthonormal.

4.4 A simple example: the one-dimensional well

Solving for the stationary states of a particle can be demonstrated by picking a particularly simple physical system. In the case to be considered, a particle of mass m is placed in a one-dimensional well characterized by a constant potential $V(x)$ such that the $V(x) = 0$ for $x < 0$ and $x > L$ while $V(x) = -V_o$ for $0 < x < L$. For this simple case V_o is a constant. This is depicted in figure 4.1 and is referred to as the quantum well. While extraordinarily simple this potential will give insights into more complicated physical systems. For example, the neutrons and protons in an atomic nucleus can be viewed as moving in a well created by their interactions.

4.4.1 The classical well

Before analysing this system quantum mechanically it will be useful to examine what classical mechanics predicts for the motion of the point particle. For a particle with momentum p' outside the well, $x < 0$ or $x > L$, the total mechanical energy is given by

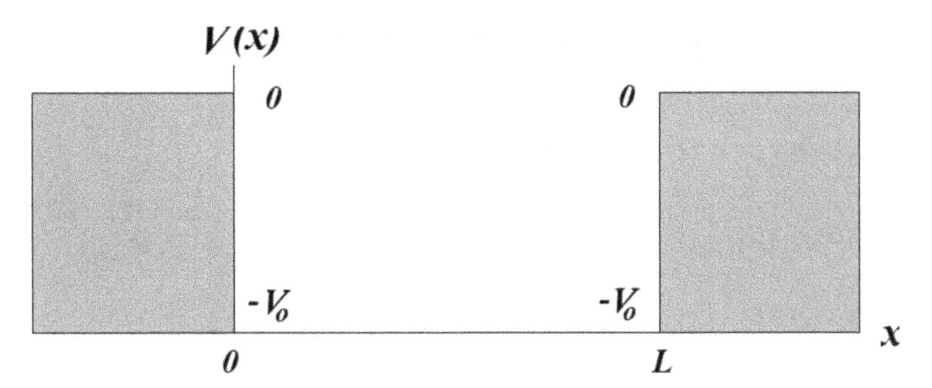

Figure 4.1. The one-dimensional well with depth $-V_0$.

$$E' = \frac{p'^2}{2m}. \tag{4.70}$$

For a particle with momentum p inside the well, $0 < x < L$, the total mechanical energy is given by

$$E = \frac{p^2}{2m} - V_o. \tag{4.71}$$

For a particle approaching the quantum well and passing over it, the conservation of total mechanical energy gives

$$E' = \frac{p'^2}{2m} = E = \frac{p^2}{2m} - V_o \implies p = \sqrt{p'^2 + 2mV_o} > p', \tag{4.72}$$

which corresponds to the particle picking up speed as it passes over the well.

However, there is a second case, which is that p is such that $E < 0$. This occurs if

$$-\sqrt{2mV_o} \leqslant p \leqslant +\sqrt{2mV_o}. \tag{4.73}$$

It is not possible to have $E = E'$, since this would correspond to an *imaginary value for* p'. As a result, for p satisfying (4.73), the regions $x < 0$ and $x > L$ are *classically forbidden* and the particle is bound in the well. The classical motion for $E > -V_o$ is given by the particle bouncing off the walls so that its momentum is reversed while its kinetic energy is conserved. This discontinuity in momentum occurs since the potential is discontinuous at $x = 0$ and $x = L$, giving rise to a singular force at the walls. However, the value of p is arbitrary as long as it is real and satisfies the inequality (4.73).

4.4.2 Stationary states for the well

Solving the stationary version of the Schrödinger equation for this system will focus on the bound state case, so that the momentum of the particle inside the well satisfies $|p| < \sqrt{2mV_o}$. The case $|p| > \sqrt{2mV_o}$ is treated in the next chapter in the section on transmission and tunneling. Since there is no force present on the particle in the interior of the well, the solution for $0 \leqslant x \leqslant L$, denoted $\Psi_{\rm w}(x, t)$, can be written as a superposition of momentum eigenfunctions,

$$\Psi_{\rm w}(x, t) = (Ae^{ipx/\hbar} + Be^{-ipx/\hbar})e^{-iE_{\rm p}t/\hbar} \equiv \psi_{\rm w}(p, x)e^{-iE_{\rm p}t/\hbar}. \tag{4.74}$$

where A and B are constants independent of x and t. In order for this to be a solution to the Schrödinger equation in the well, it follows that

$$\hat{H}\Psi_{\rm w}(x, t) = \left(-\frac{\hbar^2}{2m}\frac{{\rm d}^2}{{\rm d}x^2} - V_o\right)\Psi_{\rm w}(x, t) = i\hbar\frac{\partial}{\partial t}\Psi_{\rm w}(x, t) \implies E_{\rm p} = \frac{p^2}{2m} - V_o. \tag{4.75}$$

If the particle is energetically confined in the well, it is the case that $E_{\rm p} < 0$.

The total wave function of the particle must be continuous and have continuous derivatives at the boundaries of the well. If the wave function is treated as zero outside the well these continuity conditions cannot be satisfied. However, including a wave function component outside the well requires that it have the same energy found in (4.75). This is only possible if the wave function is exponentially damped. For example, the wave function for $x > L$ must take the form

$$\Psi(x, t) = \psi_{\rm L}(x)\, e^{-iE_{\rm p}t} = D\, e^{-\kappa(x-L)/\hbar}e^{-iE_{\rm p}t/\hbar}, \tag{4.76}$$

where D is also a constant and the presence of L in the exponential is for convenience in matching the continuity conditions. Schrödinger's equation for this solution gives

$$-\frac{\hbar^2}{2m}\nabla^2\Psi(x, t) = -\frac{\kappa^2}{2m}\psi_{\rm L}(x)\, e^{-iE_{\rm p}t/\hbar} = i\hbar\frac{\partial}{\partial t}\Psi(x, t) = \left(\frac{p^2}{2m} - V_o\right)\psi_{\rm L}(x)e^{-iE_{\rm p}t/\hbar}, \tag{4.77}$$

which holds if κ is given by the *real* value

$$\kappa = \sqrt{2\, mV_o - p^2}\,. \tag{4.78}$$

Similarly, in the region $x < 0$ the wave function takes the form

$$\Psi(x, t) = \psi_0(x)\, e^{-iE_pt} = Ce^{\kappa x/\hbar}e^{-iE_pt/\hbar}, \tag{4.79}$$

where once again $\kappa = \sqrt{2\, mV_o - p^2}$ satisfies the Schrödinger equation. As a result, the energy eigenvalue is the same for all three wave functions and they constitute the stationary solution in the three regions. The particular forms for $\psi_0(x)$ and $\psi_L(x)$ are dictated by the requirement of normalizability. Unlike the momentum eigenfunctions that make up the interior solution, the exterior solutions must both be exponentially damped to be square integrable over the respective infinite regions.

The next step is to match the wave functions in the three regions at the two boundaries $x = 0$ and $x = L$. The three solutions give

$$\begin{aligned} C &= A + B, \\ D &= Ae^{ipL/\hbar} + Be^{-ipL/\hbar}. \end{aligned} \tag{4.80}$$

Similarly, matching the derivatives at the boundaries gives

$$\begin{aligned} \kappa C &= ip(A - B), \\ -\kappa D &= ip(Ae^{ipL/\hbar} - Be^{-ipL/\hbar}). \end{aligned} \tag{4.81}$$

Solving this system of four equations is tedious but straightforward. Using the two equations for C gives B in terms of A,

$$B = -\left(\frac{\kappa - ip}{\kappa + ip}\right)A, \tag{4.82}$$

while repeating this using the equations for D gives a second expression for B in terms of A,

$$B = \left(\frac{ip + \kappa}{ip - \kappa}\right)e^{2ipL/\hbar}A. \tag{4.83}$$

Equating (4.82) and (4.83) and using Euler's relation (3.19) yields a transcendental equation for p,

$$\tan\left(\frac{pL}{\hbar}\right) = \frac{2\kappa p}{\kappa^2 - p^2} = \frac{p\sqrt{2\,mV_o - p^2}}{mV_o - p^2}. \tag{4.84}$$

Either of the expressions for B given by (4.82) and (4.83) can be used to find C and D in terms of A, with the result

$$\begin{aligned} C &= \left(\frac{2ip}{\kappa + ip}\right)A, \\ D &= -\left(\frac{2ip}{\kappa - ip}\right)e^{ipL/\hbar}A. \end{aligned} \tag{4.85}$$

As a result, all four constants are expressed in terms of the one constant A, which is then determined by normalizing the solution. This will not be presented for the general case.

It is instructive to contrast this description of the particle's behavior with that predicted by classical mechanics. For a finite value of κ the wave function is *not* zero in the classically forbidden regions. Instead, it is exponentially damped, corresponding to a rapidly decreasing probability for observing the particle further into the forbidden regions. In this regard, this phenomenon can be viewed as the presence of an *imaginary momentum* for the de Broglie wavelength. For such a case,

$$e^{ipx/\hbar} \rightarrow e^{-\kappa x/\hbar}, \tag{4.86}$$

and this is a valid solution to the Schrödinger equation for the case that the energy is negative. While imaginary momenta are unphysical in Newtonian mechanics, they are quantum mechanically allowed as long as the wave function remains normalizable for the entire spatial region.

These results simplify greatly for the case that $V_o \rightarrow \infty$, so that the particle is in an infinitely deep well. This limit corresponds to $\kappa \rightarrow \infty$, so that results (4.85) give

$C, D \to 0$. In this limit there is zero wave function in the classically forbidden regions. Equation (4.84) simplifies to

$$\tan\left(\frac{pL}{\hbar}\right) = 0, \tag{4.87}$$

which has the solutions

$$\frac{pL}{\hbar} = n\pi \implies p_n = \frac{n\pi\hbar}{L}, \tag{4.88}$$

where n is an arbitrary nonzero integer. The case $n = 0$ is excluded since (4.82) shows that the limit $\kappa \to \infty$ gives $B = -A$. For $p = 0$ this gives $\psi_w(x) = 0$, and is therefore excluded as unphysical. As a result, the eigenvalues of the Hamiltonian are given by

$$E_n = \frac{p_n^2}{2m} = \frac{n^2\pi^2\hbar^2}{2\,mL^2}, \quad n \neq 0. \tag{4.89}$$

Using $B = -A$ and (4.88) shows that the spatial wave functions for the infinitely deep well are given by

$$\psi_w(x) = Ae^{ipx/\hbar} - Ae^{-ipx/\hbar} = A\sin\left(\frac{n\pi x}{L}\right). \tag{4.90}$$

Normalizing the wave function $\psi_w(x)$ identifies $A = \sqrt{2/L}$. The normalized versions of the eigenfunctions $\psi_w(x)$ given by (4.90) are written

$$\psi_n(x) = \sqrt{\frac{2}{L}}\sin\left(\frac{n\pi x}{L}\right), \tag{4.91}$$

and are indexed by the integer n of the associated energy eigenvalue E_n given by (4.89). The integer n is therefore a *quantum number*, characterizing the momentum and the energy level.

The wave functions of (4.91) for the case of an infinitely deep well are eigenfunctions of the Hermitian Hamiltonian $\hat{H} = \hat{p}^2/2m$ associated with the energy eigenvalues E_n of (4.89). They are identical to the complete set of orthonormal functions appearing in the Fourier sine series (3.63) for a square integrable function periodic over the interval $x \in (0, L)$ and vanishing at the boundaries of the interval. As a result, the Hermitian Hamiltonian has both real eigenvalues and orthonormal eigenfunctions. It is important to note that it is the requirement of continuity that led to the energy eigenvalues being discrete. For this case, the energy eigenvalues are *quantized*, which is generally taken to mean they are characterized by an integer or set of integers. This is a validation of Planck's hypothesis (2.37) that the energy levels of atoms are quantized.

Contact is made with the de Broglie wavelength postulate (2.54) by comparing (4.94) to the spatial part of a *standing wave*, $\sin(2\pi x/\lambda)$, introduced in (1.76). Therefore, for each integer n, the basis function $\Psi_n(x, t = 0)$ corresponds to the wavelength $\lambda_n = 2L/n$. The de Broglie wavelength formula (2.54) then associates each of the wave functions (4.91) with the magnitude of momentum p_n given by

$$p_n = \frac{h}{\lambda_n} = \frac{nh}{2L} = \frac{n\pi\hbar}{L},\; n = 1, 2, 3..., \tag{4.92}$$

which is identical to (4.88). Further insight into the basis wave functions is obtained from the identity

$$\sin\left(\frac{n\pi x}{L}\right) = \sin\left(\frac{p_n x}{\hbar}\right) = \frac{1}{2i}(e^{ip_n x/\hbar} - e^{-ip_n x/\hbar}). \tag{4.93}$$

The basis wave functions (4.94) are themselves linear superpositions of the momentum eigenfunctions (3.173) found earlier, with the momenta restricted to the discrete or quantized values $p_n = n\pi\hbar/L$. This is consistent with the idea of a *standing wave* formed by left and right moving waves reflected from the walls of the well. It follows that the Schrödinger equation is consistent with the de Broglie wavelength hypothesis.

4.4.3 Wave function preparation for the particle in a well

The act of placing the particle inside the infinitely deep well involves *preparing the initial wave function.* This means specifying the initial wave function, $\Psi(x, t = 0) \equiv \psi(x)$, to represent the initial spatial probability distribution, $\mathcal{P}(x, t = 0) = |\psi(x)|^2$, that is consistent with the experimental situation. For the case of the infinitely deep well $\psi(x)$ must vanish outside the well and at the walls of the well. In addition, it must also satisfy the normalization condition (3.2) so that it is *square integrable.* However, in all other ways the function $\psi(x)$ is *arbitrary*, since the particle can be placed in the well or prepared in an arbitrary manner. Since the wave function must be continuous and square integrable, it can be represented by a Fourier sine series or, *equivalently*, a superposition of the Hamiltonian eigenfunctions. Since $\psi(0) = \psi(L) = 0$, the Fourier series representation (3.63) of $\psi(x)$ in the interior of the well is given by a superposition of the eigenfunctions,

$$\psi(x) = \sqrt{\frac{2}{L}}\sum_{n=1}^{\infty} a_n \sin\left(\frac{n\pi x}{L}\right), \tag{4.94}$$

where the coefficients a_n are *dimensionless* complex constants given by

$$a_n = \sqrt{\frac{2}{L}}\int_0^L \mathrm{d}x\, \psi(x) \sin\left(\frac{n\pi x}{L}\right). \tag{4.95}$$

The next step is to implement the normalization condition (3.2), which gives

$$\begin{aligned}\int_0^L \mathrm{d}x\, \psi^*(x)\, \psi(x) &= \frac{2}{L}\sum_{n,j=1}^{\infty} a_j^*\, a_n \int_0^L \mathrm{d}x \sin\left(\frac{j\pi x}{L}\right)\sin\left(\frac{n\pi x}{L}\right)\\ &= \sum_{n,j=1}^{\infty} a_j^*\, a_n\, \delta_{jn} = \sum_{n=1}^{\infty} |a_n|^2 = 1.\end{aligned} \tag{4.96}$$

Condition (4.96) can be satisfied only if each of the Fourier coefficients has a modulus less than or equal to one. Apart from condition (4.96), there are no restrictions on the coefficients, and so the initial wave function $\psi(x)$ can be prepared in *an arbitrary way*. Recalling Born's rule (3.243), which showed that $|a_n|^2$ gives the probability of measuring the particle in the *n*th *state*, this result shows that an arbitrary spatial preparation of the particle's wave function results in a mixed state of energy eigenfunctions.

For the case that the particle in the well is prepared in an initial state where $a_n = 1$ and all others zero, then the probability density at any later time t is given by

$$\mathcal{P}_{\hat{x}}(x, t) = |\Psi(x, t)|^2 = |\psi_n(x)|^2 = \frac{2}{L}\sin^2\left(\frac{n\pi x}{L}\right). \tag{4.97}$$

The function (4.97) gives the probability density that the particle is at the position x at t, and so it can be used to find the expectation value for observables using the recipe (3.202). For the choice of preparation given by (4.97), the expectation value for the position of the particle is given by

$$\langle x\rangle = \int x\, \mathrm{d}P(x, t) = \int_0^L \mathrm{d}x\; x\; \mathcal{P}(x, t) = \frac{2}{L}\int_0^L \mathrm{d}x\; x\, \sin^2\left(\frac{n\pi x}{L}\right) = \frac{1}{2}L. \tag{4.98}$$

Result (4.98) is independent of n and corresponds to the center of the well. It is important to understand that this does *not* mean the particle is predicted to be sitting stationary at the center of the well. In the Copenhagen interpretation, the particle cannot be considered to be at a specific location *until a measurement of its position occurs*. Instead, (4.98) is understood to mean that many observations of the particle's position results in the *average value* $\langle x\rangle = L/2$, which is the center of the well. Although classical intuition can be misleading, it is possible to visualize the particle described by (4.97) as being *in motion back and forth in the well.* This is consistent with the nature of a standing wave such as (1.76), and so the particle has a nonzero probability density (4.97) to be at *any* location inside the well. The statistical average of many measurements of the position will result in the center of the well.

To further investigate the nature of expectation values for the quantum well, it is possible to prepare the particle in the mixed state at $t = 0$ given by

$$\Psi(x, t = 0) = a\psi_1(x) + b\psi_2(x), \tag{4.99}$$

where a and b are real. Because $\psi_1(x)$ and $\psi_2(x)$ are orthonormal, the normalization condition (4.96) reduces to the requirement that $a^2 + b^2 = 1$. This is consistent with Born's rule, so that the values of a^2 and b^2 give the *relative probabilities* of observing the particle in the energy eigenstate $\psi_1(x)$ and $\psi_2(x)$ respectively. However, now the wave function at a later time t is given by

$$\Psi(x, t) = a\psi_1(x)e^{-iE_1t/\hbar} + b\psi_2(x)e^{-iE_2t/\hbar}, \tag{4.100}$$

which satisfies the Schrödinger equation and boundary conditions for the quantum well at all later times. The expectation value $\langle x\rangle$ at t for this case of preparation is found using

$$\mathcal{P}(x, t) = |\Psi(x, t)|^2 = a^2\psi_1^2(x) + b^2\psi_2^2(x) + 2ab\cos\left(\frac{(E_1 - E_2)t}{\hbar}\right)\psi_1(x)\psi_2(x). \tag{4.101}$$

Using the forms for $\psi_1(x)$ and $\psi_2(x)$ gives the expectation value $\langle x\rangle(t)$,

$$\begin{aligned}\langle x\rangle(t) &= \frac{2}{L}\int_0^L \mathrm{d}x\, x\left(a^2\sin^2\left(\frac{\pi x}{L}\right) + b^2\sin^2\left(\frac{2\pi x}{L}\right) + 2ab\cos\left(\frac{(E_1 - E_2)t}{\hbar}\right)\sin\left(\frac{\pi x}{L}\right)\sin\left(\frac{2\pi x}{L}\right)\right)\\ &= (a^2 + b^2)\left(\frac{L}{2}\right) - 2ab\left(\frac{16L}{9\pi^2}\right)\cos\left(\frac{3\pi^2\hbar t}{2mL^2}\right) = \frac{L}{2} - 2ab\left(\frac{16L}{9\pi^2}\right)\cos\left(\frac{3\pi^2\hbar t}{2mL^2}\right),\end{aligned} \tag{4.102}$$

where the normalization condition of the mixed state, $a^2 + b^2 = 1$, has been used. The expectation value is no longer the center of the well and depends on the choice of the mixed state coefficients as well as the time of observation. The quantum interference terms discussed earlier, manifested by the term proportional to ab, affect the outcome of measurements in an oscillatory manner. For example, if $a = b = 1/\sqrt{2}$, then

$$\langle x\rangle(t = 0) = (9\pi^2 - 32)L/18\pi^2. \tag{4.103}$$

This initial wave function therefore corresponds to preparing the particle initially to be located closer to $x = 0$ *on average*. Similarly, the choice $a = -b = 1/\sqrt{2}$ gives

$$\langle x\rangle(t = 0) = (9\pi^2 + 32)L/18\pi^2, \tag{4.104}$$

corresponding to initially placing the particle closer *on average* to the location $x = L$. In the Copenhagen interpretation it is the initial preparation of the particle and its wave function that determines the mixed state coefficients.

It is possible to find the average energy for such a mixed state. Using the Copenhagen interpretation that a^2 and b^2 are the probabilities of observing the two possible energy states, the expectation value for the energy is given by

$$\langle E\rangle = a^2E_1 + b^2E_2. \tag{4.105}$$

For the specific choice $a = b = 1/\sqrt{2}$, this gives $\langle E\rangle = (E_1 + E_2)/2$, the arithmetic average. The other choice, $a = -b = 1/\sqrt{2}$, gives the same result. It is important to note that preparing the particle's location more toward one edge of the well has required a tradeoff, which is that the particle is no longer in a *single* energy state. Increasing the certainty in the particle's position has come at the expense of decreasing the certainty in the particle's momentum.

4.4.4 Bohr's correspondence principle and the well

Similar to Bohr's atom, there are only discrete energy levels available to the particle in an infinite well and the energy levels are *quantized*. In this case the allowed energies take the general form $E_n = n^2E_1$, so that they are integer multiples of the *ground state energy*,

$$E_1 = \frac{\pi^2\hbar^2}{2mL^2}. \tag{4.106}$$

Requiring the wave function associated with the particle to satisfy Schrödinger's equation and satisfying the boundary conditions has prevented a continuum of energies, which was the case for a Newtonian particle confined in a well. As with the Bohr atom, the change from one energy level to another is associated with a discrete amount of energy or quantum. If the particle is charged, it is possible to find the wavelengths of light associated with such *quantum jumps* by associating the energy difference with that of the photon emitted or absorbed. This is identical in approach to the method (2.67) used to find the spectrum of the hydrogen atom. Assuming that the initial energy level, E_{n_i}, is higher than the final energy level, E_{n_f}, then the frequency ω of the emitted photon is determined by the formula

$$\omega = \frac{E_{n_i} - E_{n_f}}{\hbar} = \frac{(n_i^2 - n_f^2)\pi^2\hbar}{2\,mL^2}. \tag{4.107}$$

Like the result (2.68) for atoms, the emission and absorption spectra of a particle trapped in a well are discrete and characterized by integers.

It is important to examine both the microscopic and macroscopic cases for these formulas to determine both the quantitative and qualitative nature of the energy levels. For the case that these results are used to model the rough features of a proton confined to a nucleus, the mass is given by 1.7×10^{-27} kg and the size of the well is $\approx 10^{-14}$ m. The ground state energy of the proton would be $E_1 \approx 3 \times 10^{-13}$ J or ≈ 2 MeV. A photon emitted during the transition from the *first excited state* E_2 to the ground state E_1 would therefore have the energy $(2^2 - 1^2)E_1 \approx 6$ MeV. Photons with energies in the MeV range correspond to *gamma rays*, which are common in nuclear processes. Similarly, it is also possible to model the hydrogen atom as a well containing the electron. In this case, the mass of the electron is $\approx 9 \times 10^{-31}$ kg and the size of the well is $\approx 10^{-10}$ m. The ground state energy of the electron would be $E_1 \approx 5 \times 10^{-18}$ J or ≈ 30 eV and energy transitions would correspond to frequencies such as those of visible light. These are typical of atomic spectra. Of course, the nucleus and the atom are far more complicated than simply a well trapping nucleons or electrons. However, this simple quantum mechanical model exhibits the correct range of observed photon energies associated with both systems.

At the other extreme, it is instructive to apply these same formulas to a kilogram ball placed in a box with width one meter. For such a case, the ground state energy is given by $E_1 \approx 4 \times 10^{-68}$ J, which is 55 orders of magnitude smaller than the ground state energy of a proton in a nucleus and far too small for any device to measure. For the case that the ball is treated as a quantum object, it is possible to associate a quantum number with an everyday energy level. If the ball is moving in the box with the speed of 6 m s^{-1}, then its state would correspond to the quantum number

$$n = \frac{p_n L}{\pi\hbar} = \frac{mvL}{\pi\hbar} \approx 1.8 \times 10^{34}. \tag{4.108}$$

Using this quantum number in (4.89) yields approximately the same energy as the Newtonian value of $E = \frac{1}{2}mv^2 = 18$ J. The energy level immediately below differs by the amount of energy

$$\Delta E = E_n - E_{n-1} = (n^2 - (n-1)^2)E_1 = (2n-1)E_1 \approx 1.4 \times 10^{-33} \text{ J}. \qquad (4.109)$$

This means that the energy levels available to the macroscopic ball are roughly 20 orders of magnitude *closer together* than those that resulted from similarly modeling the proton in the nucleus. The neighboring velocities available to the ball differ by the amount

$$v_n - v_{n-1} = \frac{\pi\hbar}{mL} \approx 3 \times 10^{-34} \text{ m s}^{-1}, \qquad (4.110)$$

a difference too small to be observable. In effect, the macroscopic ball has energies and velocities that are so close together that they appear to be virtually continuous. The frequency of a photon associated with such small energy level differences is the extremely low frequency (ELF) $f \approx 3$ Hz, which corresponds to the extremely long wavelength ($\lambda \approx 10^8$ m) sector of the radio wave spectrum. Once again, applying quantum mechanics to an everyday object has yielded results virtually indistinguishable from the continuum of values available to the same object in Newtonian mechanics.

Results similar to these led Bohr to postulate the correspondence principle first discussed in chapter 2 in the context of the de Broglie wavelength. In the setting here the correspondence principle argues that classical behavior emerges from quantum mechanics in the limit of large quantum numbers. In the example of the quantum well, the extremely large quantum numbers correspond to energy levels and velocities so close together for the macroscopic ball that their difference cannot be measured. It is in this manner that Bohr's correspondence principle allows classical behavior to emerge from quantum mechanics.

4.5 The Heisenberg uncertainty principle

The Heisenberg uncertainty principle is one of the more dramatic outcomes of quantum mechanics. It places limitations on the simultaneous measurability of pairs of observables.

4.5.1 Motivating the Heisenberg uncertainty principle

In the case of a particle in an infinitely deep well, using (4.88) shows that the product $p_1 L$ gives

$$p_1 L = \pi\hbar = \frac{1}{2}h, \qquad (4.111)$$

a result that is *independent* of the well size. Heisenberg pointed out that L can be viewed as the *uncertainty in the particle's position*, i.e. it quantifies the amount Δx by which the particle has been localized. The particle has been put into the well, but could be anywhere within the well depending on its preparation. In the process of localizing the particle into the well, wave–particle duality results in a nonzero momentum that is greater than or equal to p_1. In effect, p_1 is the minimum uncertainty Δp in the particle's momentum. In this specific case, the product gives

$$\Delta x \, \Delta p = \frac{1}{2}h. \tag{4.112}$$

Similarly, a measurement of the particle's energy requires the detector used for this to *encounter the particle.* The particle may need to bounce off the wall in order to reach the detector, traveling the distance $2L$. In its lowest momentum state the particle can take the approximate time,

$$T \approx \frac{2L}{v_1} = \frac{2Lm}{p_1} = \frac{2L^2m}{\pi\hbar}, \tag{4.113}$$

before it arrives at the detector. This corresponds to the *uncertainty in the time Δt required to measure the particle's energy*. It then follows that that the uncertainty in energy ΔE is at least the ground state energy $E_1 = \pi^2\hbar^2/2\,mL^2$, which gives

$$\Delta E \, \Delta t = \frac{\pi^2\hbar^2}{2\,mL^2}\frac{2L^2m}{\pi\hbar} = \pi\hbar = \frac{1}{2}h, \tag{4.114}$$

another product that is independent of the width of the well.

At the core of the Heisenberg uncertainty principle is the failure of two observables to share a mutual set of eigenfunctions *if they do not commute.* In the case of (4.112) it is the observables $\hat{p}$ and $\hat{x}$ that do not commute. The earlier result (3.207) showed that the expectation value of their product yields complex values for such a case and is thus unmeasurable. In the second case (4.114) it is the time-derivative action of the Hamiltonian $\hat{H}$ that does not commute with the time t, since

$$\begin{aligned}\hat{H}(t\Psi(x, t)) = i\hbar\frac{\partial}{\partial t}(t\Psi(x, t)) &= i\hbar\Psi(x, t) \\ + i\hbar t\frac{\partial}{\partial t}\Psi(x, t) &= i\hbar\Psi(x, t) + t\hat{H}\Psi(x, t),\end{aligned} \tag{4.115}$$

so that

$$[\hat{H}, t] = i\hbar. \tag{4.116}$$

This result will lead to an uncertainty in energy that depends on the time spent in observation.

However, even if two observables do not commute it is still possible to devise an experiment that will attempt to measure both of them simultaneously. An example is depicted in figure 4.2. Passing electrons through a slit of width Δx determines their position up to the slit width Δx. This can be accompanied by a measurement of the range of their momentum *parallel* to the slit, designated Δp. A simple way to find this range of momentum is to measure the angle θ at which the electrons strike the screen

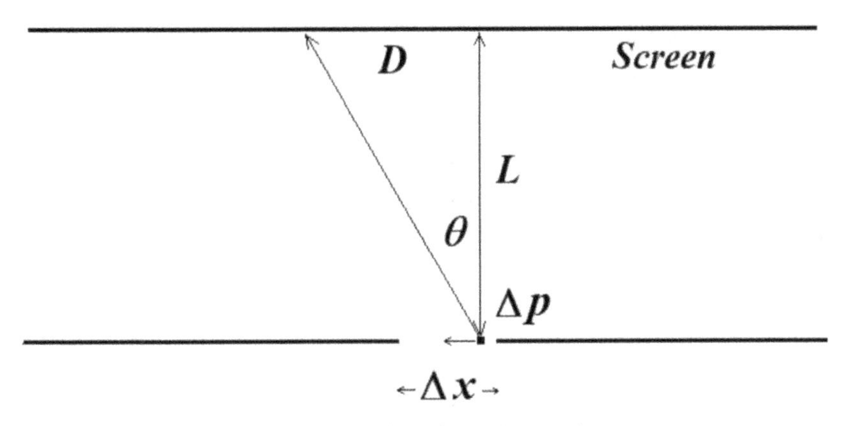

Figure 4.2. Slit diffraction and uncertainty.

which is located a distance L behind the slit, as depicted in figure 4.2. If the velocity perpendicular to the slit is v, assumed to be much greater than $\Delta p/m = \Delta v$, then the time to arrive at the screen is $T = L/v$, so that $L/T = v$. The distance D traveled parallel to the screen before striking the screen is $D \approx L \sin\theta$, and so the momentum *parallel* to the slit must have been

$$\Delta p = m\,\Delta v \approx m\left(\frac{D}{T}\right) = m\left(\frac{L}{T}\right)\sin\theta = mv\sin\theta. \tag{4.117}$$

However, the wave aspect of the electron undergoes diffraction when passing through the slit, spreading the beam so that their impact on the screen occurs within the first diffraction miminum of (3.80), which gives

$$\sin\theta = \frac{\lambda}{\Delta x} = \frac{h}{mv\,\Delta x}, \tag{4.118}$$

where the de Broglie wavelength formula (2.54) was used. Using this result for $\sin\theta$ in the expression for Δp gives the result

$$\Delta x\,\Delta p \approx h, \tag{4.119}$$

which is nearly identical to (4.112).

Heisenberg interpreted the slit width as the *uncertainty in the position* of the electron as it passes through the slit, while Δp represents the *uncertainty in the momentum* parallel to the slit *subsequent* to passing through the slit. If $\Delta x \to 0$, reducing the uncertainty in position to zero, it results in $\Delta p \to \infty$, and the uncertainty in the subsequent momentum of the electron parallel to the slit becomes arbitrarily large. The uncertainty in momentum can be reduced to zero, $\Delta p \to 0$, only if $\Delta x \to \infty$, which requires the slit width or uncertainty in position to be arbitrarily large. It is important to note that this occurs even if the source of the electrons is far from the slit, which would classically preclude electrons with any initial momentum parallel to the slit from passing through. As a result, it is the *simultaneous observation of both momentum and position* that is involved in this effect. The results of this section highlight the failure of Newtonian determinism in quantum mechanics and will be elevated to the *Heisenberg uncertainty principle*.

4.5.2 Defining the uncertainty in measurement

It will now be shown that the two observables, momentum and position, cannot be measured simultaneously to arbitrary accuracy because they do not commute. In order to demonstrate this, it is necessary to quantify mathematically the uncertainty, denoted $\Delta\hat{O}$, in the measurement of an observable $\hat{O}$, and then to derive a relationship between the uncertainties for observables that involves their commutator. The definition of $\Delta\hat{O}$ is borrowed from statistics, where deviation from the average value is an important aspect of a statistical correlation. This begins by noting that the expectation value of *any* expectation value is once again the expectation value, which follows from

$$\langle\langle\hat{O}\rangle\rangle = (\Psi, \langle\hat{O}\rangle\Psi) = \langle\hat{O}\rangle(\Psi, \Psi) = \langle\hat{O}\rangle. \tag{4.120}$$

Quantifying the deviation from the average or expectation value begins by introducing the related operator

$$\hat{o} = \hat{O} - \langle\hat{O}\rangle. \tag{4.121}$$

For the case that $\hat{O}$ is Hermitian, it follows that $\hat{o}$ is also Hermitian,

$$\hat{o}^{\dagger} = \hat{O}^{\dagger} - \langle\hat{O}^{\dagger}\rangle = \hat{O} - \langle\hat{O}\rangle = \hat{o}. \tag{4.122}$$

This operator has an expectation value of zero, which follows from

$$\langle\hat{o}\rangle = \langle\hat{O} - \langle\hat{O}\rangle\rangle = \langle\hat{O}\rangle - \langle\langle\hat{O}\rangle\rangle = \langle\hat{O}\rangle - \langle\hat{O}\rangle = 0. \tag{4.123}$$

However, an individual measurement of $\hat{o}$ will not necessarily vanish, and the uncertainty $\Delta\hat{O}$ in measuring $\hat{O}$ is found by averaging the *absolute value* of $\hat{o}$ over many measurements, thereby preventing positive and negative results from canceling. This is achieved by finding the expectation value of its square and then taking the square root. Formally quantifying the average deviation from the expectation value or *uncertainty* for the observable $\hat{O}$, denoted $\Delta\hat{O}$, therefore uses the statistical definition

$$\Delta\hat{O} = \sqrt{\langle\hat{o}^2\rangle}, \tag{4.124}$$

where it is the positive root that is used. This gives the result

$$\Delta\hat{O} = \sqrt{\langle(\hat{O} - \langle\hat{O}\rangle)^2\rangle} = \sqrt{\langle\hat{O}^2 - 2\langle\hat{O}\rangle\hat{O} + \langle\hat{O}\rangle^2\rangle} = \sqrt{\langle\hat{O}^2\rangle - \langle\hat{O}\rangle^2}. \tag{4.125}$$

The uncertainty of an observable depends upon the observable chosen *and* the wave function or *state* of the system.

It is instructive to consider several simple cases. For the case that the wave function for the system is prepared in a normalized eigenfunction of $\hat{O}$, denoted $f_\lambda(\boldsymbol{x}, t)$ and obeying $\hat{O}f_\lambda(\boldsymbol{x}, t) = \lambda f_\lambda(\boldsymbol{x}, t)$, the associated uncertainty is given by

$$\begin{aligned}\Delta\hat{\mathrm{O}} &= \sqrt{\langle\hat{\mathrm{O}}^2\rangle - \langle\hat{\mathrm{O}}\rangle^2} = \sqrt{(f_\lambda, \hat{\mathrm{O}}^2 f_\lambda) - (f_\lambda, \hat{\mathrm{O}} f_\lambda)^2} \\ &= \sqrt{\lambda^2 (f_\lambda, f_\lambda) - \lambda^2 (f_\lambda, f_\lambda)^2} = 0.\end{aligned} \tag{4.126}$$

For such a case, the uncertainty in the measured value of $\hat{\mathrm{O}}$ is zero. In the second example, a particle in an infinite well is prepared in its ground state eigenfunction $\psi_1(x)$ given by (4.91). The uncertainty in the momentum $\hat{p}$ is found using $\hat{p} = -i\hbar\partial/\partial x$, which gives

$$\begin{aligned}\langle\hat{p}^2\rangle &= -\frac{2\hbar^2}{L}\int_0^L \mathrm{d}x \sin\left(\frac{\pi x}{L}\right)\frac{\partial^2}{\partial x^2}\sin\left(\frac{\pi x}{L}\right) \\ &= \frac{2\pi^2\hbar^2}{L^3}\int_0^L \mathrm{d}x \sin^2\left(\frac{\pi x}{L}\right) = \frac{\pi^2\hbar^2}{L^2}, \\ \langle\hat{p}\rangle &= -i\frac{2\hbar}{L}\int_0^L \mathrm{d}x \sin\left(\frac{\pi x}{L}\right)\frac{\partial}{\partial x}\sin\left(\frac{\pi x}{L}\right) \\ &= -i\frac{2\pi\hbar}{L^2}\int_0^L \mathrm{d}x \sin\left(\frac{\pi x}{L}\right)\cos\left(\frac{\pi x}{L}\right) = 0.\end{aligned} \tag{4.127}$$

The uncertainty in the momentum for the ground state of the infinite well is therefore given by

$$\Delta\hat{p} = \sqrt{\frac{\pi^2\hbar^2}{L^2}} = \frac{\pi\hbar}{L}. \tag{4.128}$$

It is worth noting that $\Delta\hat{p}$ would vanish in the limit $\hbar \to 0$.

4.5.3 The formal Heisenberg uncertainty principle

Having defined the uncertainty in measurement for an observable a relationship can now be established between uncertainties for two Hermitian observables $\hat{\mathrm{A}}$ and $\hat{\mathrm{B}}$. The first step is to define two operators $\hat{\alpha}$ and $\hat{\beta}$ as

$$\begin{aligned}\hat{\alpha} &= \hat{\mathrm{A}} - \langle\hat{\mathrm{A}}\rangle, \\ \hat{\beta} &= \hat{\mathrm{B}} - \langle\hat{\mathrm{B}}\rangle,\end{aligned} \tag{4.129}$$

which are the operators introduced in (4.121) and used in the definition (4.124) of the uncertainty. Both $\hat{\alpha}$ and $\hat{\beta}$ are Hermitian since both $\hat{\mathrm{A}}$ and $\hat{\mathrm{B}}$ are Hermitian. Using the wave function $\Psi(\boldsymbol{x}, t)$ two functions are defined using these operators,

$$\begin{aligned}f(\boldsymbol{x}, t) &= \hat{\alpha}\Psi(\boldsymbol{x}, t), \\ g(\boldsymbol{x}, t) &= \hat{\beta}\Psi(\boldsymbol{x}, t).\end{aligned} \tag{4.130}$$

The Cauchy–Schwarz inequality for functions (3.51) then gives

$$(f, f)(g, g) \geqslant |(f, g)|^2 \implies (\hat{\alpha}\Psi, \hat{\alpha}\Psi)(\hat{\beta}\Psi, \hat{\beta}\Psi) \geqslant |(\hat{\alpha}\Psi, \hat{\beta}\Psi)|^2. \tag{4.131}$$

By virtue of hermiticity, the inner products satisfy

$$\begin{aligned}(\hat{\alpha}\Psi, \hat{\alpha}\Psi) &= (\Psi, \hat{\alpha}^2\Psi) = \langle \hat{\alpha}^2 \rangle = (\Delta\hat{\mathrm{A}})^2, \\ (\hat{\beta}\Psi, \hat{\beta}\Psi) &= (\Psi, \hat{\beta}^2\Psi) = \langle \hat{\beta}^2 \rangle = (\Delta\hat{\mathrm{B}})^2,\end{aligned} \tag{4.132}$$

where the definition of the uncertainty made in (4.124) was used.

The previous result (3.209) for the product of operators allows the inner product on the right-hand side of (4.131) to be written

$$(\hat{\alpha}\Psi, \hat{\beta}\Psi) = (\Psi, \hat{\alpha}\hat{\beta}\Psi) = X + iY, \tag{4.133}$$

where X and Y are *real numbers* given by

$$\begin{aligned}X &= \frac{1}{2}\langle \{\hat{\alpha}, \hat{\beta}\} \rangle, \\ Y &= -\frac{1}{2} i \langle [\hat{\alpha}, \hat{\beta}] \rangle.\end{aligned} \tag{4.134}$$

Using the definitions of (4.129) gives

$$\begin{aligned}X &= \frac{1}{2}\langle \{\hat{\mathrm{A}} - \langle\hat{\mathrm{A}}\rangle, \hat{\mathrm{B}} - \langle\hat{\mathrm{B}}\rangle\} \rangle = \frac{1}{2}\langle \{\hat{\mathrm{A}}, \hat{\mathrm{B}}\} \rangle - \langle\hat{\mathrm{A}}\rangle\langle\hat{\mathrm{B}}\rangle, \\ Y &= -\frac{1}{2} i \langle [\hat{\mathrm{A}} - \langle\hat{\mathrm{A}}\rangle, \hat{\mathrm{B}} - \langle\hat{\mathrm{B}}\rangle] \rangle = -\frac{1}{2} i \langle [\hat{\mathrm{A}}, \hat{\mathrm{B}}] \rangle.\end{aligned} \tag{4.135}$$

The right-hand side of (4.131) then becomes

$$|(\hat{\alpha}\Psi, \hat{\beta}\Psi)|^2 = X^2 + Y^2 = \left(\frac{1}{2}\langle \{\hat{\mathrm{A}}, \hat{\mathrm{B}}\} \rangle - \langle\hat{\mathrm{A}}\rangle\langle\hat{\mathrm{B}}\rangle\right)^2 + \frac{1}{4}|\langle [\hat{\mathrm{A}}, \hat{\mathrm{B}}] \rangle|^2. \tag{4.136}$$

Combining the results (4.132) and (4.136) in (4.131) and taking the square root yields the final form of the Heisenberg uncertainty principle,

$$\Delta\hat{\mathrm{A}}\, \Delta\hat{\mathrm{B}} \geqslant \sqrt{\left(\frac{1}{2}\langle \{\hat{\mathrm{A}}, \hat{\mathrm{B}}\} \rangle - \langle\hat{\mathrm{A}}\rangle\langle\hat{\mathrm{B}}\rangle\right)^2 + \frac{1}{4}|\langle [\hat{\mathrm{A}}, \hat{\mathrm{B}}] \rangle|^2}. \tag{4.137}$$

It is common in the literature to ignore the first term in the square root involving the anticommutator and abbreviate result (4.137) to

$$\Delta\hat{\mathrm{A}}\, \Delta\hat{\mathrm{B}} \geqslant \frac{1}{2}|\langle [\hat{\mathrm{A}}, \hat{\mathrm{B}}] \rangle|. \tag{4.138}$$

This is consistent with (4.137), but overlooks the contribution of the first term to the measurement limitation.

For the choice of position and momentum, which have the commutation relation (3.184), (4.138) gives the uncertainty relation

$$\Delta\hat{x}_j\,\Delta\hat{p}_k \geqslant \frac{1}{2}|(\Psi,\,[\hat{x}_j,\,\hat{p}_k]\Psi)| = \frac{1}{2}\hbar\,\delta_{jk}(\Psi,\,\Psi) = \frac{1}{2}\hbar\,\delta_{jk}. \tag{4.139}$$

The uncertainty principle (4.139) for momentum and position explains why the volume V in the momentum eigenfunction (3.127) must be arbitrarily large in order to be consistent with a vanishing probability flux into and out of the volume V found earlier in (4.29). Since the particle is contained in the cubic volume $V = L^3$, the length L determines the uncertainty in the components of position, so that $L \approx \Delta\hat{x}_k$. However, in the case of a momentum eigenfunction it follows that $\Delta\hat{p}_k = 0$ since $\langle\hat{p}_k^2\rangle = \langle\hat{p}_k\rangle^2$. As a result, $\Delta\hat{x}_k \to \infty$ for the case that $\Delta\hat{p}_k = 0$, and this requires an arbitrarily large volume.

It is instructive to evaluate $\Delta\hat{x}$ and combine it with the $\Delta\hat{p}$ found earlier (4.128) for the simple case that the wave function corresponds to the ground state of a particle in an infinite one-dimensional well of width L, given by (4.91) with $n = 1$. The uncertainty squared in the particle's position is then

$$\begin{aligned}(\Delta\hat{x})^2 = \langle\hat{x}^2\rangle - \langle\hat{x}\rangle^2 &= \frac{2}{L}\int_0^L \mathrm{d}x\; x^2 \sin^2\left(\frac{\pi x}{L}\right) - \left(\frac{2}{L}\int_0^L \mathrm{d}x\; x\sin^2\left(\frac{\pi x}{L}\right)\right)^2 \\ &= \frac{2}{L}\left(\frac{L^3}{6} - \frac{L^3}{4\pi^2}\right) - \left(\frac{2}{L}\frac{L^2}{4}\right)^2 = L^2\left(\frac{1}{12} - \frac{1}{2\pi^2}\right),\end{aligned} \tag{4.140}$$

so that the uncertainty in position in the ground state is

$$\Delta\hat{x} = \left(\frac{L}{2\pi}\right)\sqrt{\frac{\pi^2}{3} - 2}\,. \tag{4.141}$$

Using the uncertainty in momentum $\Delta\hat{p} = \pi\hbar/L$ found in (4.128) gives

$$\Delta\hat{x}\,\Delta\hat{p} = \frac{1}{2}\hbar\sqrt{\frac{\pi^2}{3} - 2}\,, \tag{4.142}$$

which satisfies the inequality (4.138) since the square root is ≈ 1.14, slightly larger than one. A more careful evaluation of the exact statement (4.137) is done in the section on mixed state uncertainty.

4.5.4 Energy uncertainty and duration of measurement

Quantum mechanics was developed to be consistent with Newtonian physics, where the time t is a *parameter* rather than an observable. As a result, the time t is not converted into a non-relativistic quantum mechanical operator. However, result (4.116) indicates that simultaneously measuring energy and time is similar to simultaneously measuring momentum and position. In order to clarify this an uncertainty relation will be derived involving time for the case that an observable $\hat{\mathrm{O}}$ has no explicit time-dependence, $\partial\hat{\mathrm{O}}/\partial t = 0$, but does *not* commute with the Hamiltonian $\hat{H}$. For such a case, the Ehrenfest theorem of (4.31) gives

$$\langle [\hat{H}, \hat{O}] \rangle = i\hbar \frac{d\langle \hat{O} \rangle(t)}{dt}. \tag{4.143}$$

Using this relationship shows that the abbreviated Heisenberg uncertainty relation (4.138) for the energy $\hat{H}$ and the observable $\hat{O}$ is given by

$$\Delta E \, \Delta\hat{O} \geqslant \frac{\hbar}{2} \left| \frac{d}{dt} \langle \hat{O} \rangle(t) \right|, \tag{4.144}$$

where $\Delta\hat{H} = \Delta E$ is the uncertainty in the energy of the system.

Result (4.144) can be restated by introducing a time interval Δt corresponding to the time spent measuring the observable $\hat{O}$. During the time interval Δt the expectation value of the observable $\hat{O}$ will change by the approximate amount

$$\Delta\langle \hat{O} \rangle = \left| \frac{d}{dt} \langle \hat{O} \rangle \right| \Delta t. \tag{4.145}$$

The time interval Δt will now be chosen to correspond to the time interval required so that the change in the expectation value, $\Delta\langle \hat{O} \rangle$, matches the uncertainty in the observable $\Delta\hat{O}$. This is the time interval required for the time variation in the expectation value to be *experimentally distinguishable* from the uncertainty in the observable. If the uncertainty in the observable is large then the time spent in observation Δt will need to be large in order to distinguish the time-dependence of the expectation value from the uncertainty. Of course, the details of this depend upon the system and the observable chosen to measure. In this case, there is no requirement that the observable $\hat{O}$ and $\hat{H}$ are both members of the CSCO. If Δt is chosen to be this specific time interval, then

$$\Delta\hat{O} = \left| \frac{d}{dt} \langle \hat{O} \rangle \right| \Delta t, \tag{4.146}$$

and the abbreviated uncertainty relation (4.144) becomes the commonly used inequality

$$\Delta E \, \Delta t \geqslant \frac{1}{2}\hbar. \tag{4.147}$$

It is tempting to view (4.146) as the outgrowth of applying a Taylor series expansion and keeping the first term, but it is not. Instead it is a definition of the time interval Δt. As a result, it must be understood that the time interval Δt has an interpretation that depends on the choice for the observable $\hat{O}$ and the state of the system with which it is associated.

The interpretation of (4.147) is clarified by the following simple example. Choosing the observable to be the position $\hat{x}$ of a *free particle* moving in one dimension, Ehrenfest's theorem (4.33) allows $d\langle \hat{x} \rangle / dt$ to be interpreted as the average velocity of the particle, denoted $v = \langle \hat{p} \rangle / m$. For this case (4.146) gives

$$\Delta t = \frac{\Delta\hat{x}}{v} = m\frac{\Delta\hat{x}}{\langle\hat{p}\rangle}, \tag{4.148}$$

which is interpreted as the time interval a particle detector must interact with the particle in order to compensate for the uncertainty $\Delta\hat{x}$ in the particle's position. The uncertainty in the energy ΔE can be expressed as a function of the uncertainty $\Delta\hat{p}$,

$$\Delta E = \frac{(\langle\hat{p}\rangle + \Delta\hat{p})^2}{2m} - \frac{\langle\hat{p}\rangle^2}{2m} \approx \frac{\langle\hat{p}\rangle}{m}\Delta\hat{p}, \tag{4.149}$$

where the term $O(\Delta\hat{p}^2)$ was dropped as small. Combining the two expressions gives

$$\Delta E\ \Delta t = \left(\frac{\langle\hat{p}\rangle}{m}\ \Delta\hat{p}\right)\left(m\frac{\Delta\hat{x}}{\langle\hat{p}\rangle}\right) = \Delta\hat{p}\ \Delta\hat{x} \geqslant \frac{1}{2}\hbar, \tag{4.150}$$

which reflects the previous uncertainty relation (4.139). In order for Δt to be small, there must be a small uncertainty in the particle's position, and this requires a large uncertainty in the particle's momentum, which in turn creates a large uncertainty in the particle's energy. This interpretation is therefore consistent with the dynamics of a free particle, but should not be extrapolated to particles undergoing a force.

The Heisenberg uncertainty principle is further evidence that it is the CSCO that represents all that can be measured simultaneously to arbitrary accuracy, since if the two observables do not commute then the right-hand side of (4.137) will not vanish regardless of the contribution of the anticommutator term. This is solely an outgrowth of wave–particle duality and the probabilistic interpretation of quantum mechanics. It does not incorporate any limitations on accuracy for the measuring devices, nor does it reflect any effect from the chosen experimental set-up that a measurement of one observable may have on the properties of another observable. In that sense, it is a fundamental difference between quantum mechanics and Newtonian mechanics, where all observables are considered simultaneously measurable to arbitrary accuracy.

4.5.5 Classical objects and quantum uncertainty

As discussed in chapter 1, classical mechanics assumes that the measurements of observables are ideally unlimited in accuracy and that all aspects of an object's trajectory are simultaneously measurable. During the development of classical mechanics there was no experimentally supported reason to believe otherwise.

Using Bohr's correspondence principle this can be viewed as an outgrowth of the macroscopic limit of quantum mechanics, where the Heisenberg uncertainty principle becomes irrelevant through unobservable effects. A simple thought experiment indicates why this limitation on simultaneous measurements manifests itself in microscopic systems and not in macroscopic systems. In order to find the location of the particle in a well an observer might shine light into the well and determine the location from which a reflected photon returns. A visible photon of wavelength 700 nm has the momentum

$$p = \frac{h}{\lambda} \approx 10^{-22} \text{ gm cm s}^{-1}. \tag{4.151}$$

For a macroscopic object such as a one kilogram ball, the act of reflecting the photon changes the momentum of the object by $2p$, which results in the change of velocity given by

$$m\Delta v = 2p \implies \Delta v = \frac{2p}{m} \approx 2 \times 10^{-25} \text{ cm s}^{-1}. \tag{4.152}$$

Being struck by 10^{15} such photons, corresponding to 3×10^4 ergs of energy, still gives the unmeasurably small change in velocity $\Delta v \approx 2 \times 10^{-10}$ cm s^{-1}. However, a single photon would alter the velocity of a free electron by $\approx 2 \times 10^5$ cm s^{-1}, creating a significant effect on the electron's position. Treating the momentum of the photon as the uncertainty in the momentum of the object being probed, the uncertainty relation gives

$$\Delta \hat{x} \geqslant \frac{\hbar}{2\Delta \hat{p}} \approx 5 \times 10^{-6} \text{ cm}. \tag{4.153}$$

For the case of a macroscopic one kilogram ball made of lead with a diameter of ≈ 5 cm this is roughly one million times smaller than the ball. However, for an atom it corresponds to roughly five hundred times the size of the atom, so that it is a far more significant uncertainty compared to the size of the atom. Therefore, it should be no surprise that quantum mechanics manifests the result that measuring both the position and momentum of a quantum mechanical object to arbitrary accuracy is impossible, while classical level energies and objects manifest little if any observable restrictions.

4.5.6 Uncertainty for eigenstates and mixed states

For the case that the wave function $\Psi(\boldsymbol{x}, t)$ of the system is a single eigenfunction, *not* a mixed state of eigenfunctions, then all observables in the CSCO will have a well-defined expectation value corresponding to the eigenvalue associated with that eigenfunction. It has the property

$$\begin{aligned} \hat{\mathrm{A}}\Psi(\boldsymbol{x}, t) = e_a\Psi(\boldsymbol{x}, t) &\implies \langle \hat{\mathrm{A}} \rangle = (\Psi, \hat{\mathrm{A}}\Psi) = e_a(\Psi, \Psi) = e_a, \\ \hat{\mathrm{B}}\Psi(\boldsymbol{x}, t) = e_b\Psi(\boldsymbol{x}, t) &\implies \langle \hat{\mathrm{B}} \rangle = (\Psi, \hat{\mathrm{B}}\Psi) = e_b(\Psi, \Psi) = e_b, \end{aligned} \tag{4.154}$$

where e_a and e_b are the eigenvalues of $\hat{\mathrm{A}}$ and $\hat{\mathrm{B}}$ and $[\hat{A}, \hat{B}] = 0$ since they belong to the CSCO. It follows that the first term in the square root of (4.137) vanishes since

$$\left\langle \left\{ \hat{\mathrm{A}}, \hat{\mathrm{B}} \right\} \right\rangle = 2e_a e_b. \tag{4.155}$$

Since the commutator also vanishes, the right-hand side of (4.137) vanishes. This is consistent with result (4.126), which shows the uncertainty vanishes for each observable.

It is useful to recall when the equality condition of the Cauchy–Schwarz inequality is met. This occurs when the two functions appearing in (4.131) are proportional, so that $f(\boldsymbol{x}, t) = \gamma g(\boldsymbol{x}, t)$, where γ is an arbitrary complex constant. For the case of the uncertainty principle, this means that

$$\hat{\alpha}\Psi(\boldsymbol{x}, t) = \gamma\hat{\beta}\Psi(\boldsymbol{x}, t). \tag{4.156}$$

If the wave function $\Psi(\boldsymbol{x}, t)$ is an eigenfunction of *both* observables $\hat{\mathrm{A}}$ and $\hat{\mathrm{B}}$, it follows that

$$\hat{\alpha}\Psi(\boldsymbol{x}, t) = \hat{\beta}\Psi(\boldsymbol{x}, t) = 0, \tag{4.157}$$

and (4.156) is satisfied trivially. However, for other wave functions (4.156) will not necessarily hold.

To examine this further, it is instructive to examine the uncertainty associated with measurements of a simple mixed state of eigenfunctions for two commuting observables $\hat{\mathrm{A}}$ and $\hat{\mathrm{B}}$. Since the two observables commute, $[\hat{\mathrm{A}}, \hat{\mathrm{B}}] = 0$, result (3.194) shows that they share a set of mutual eigenfunctions. These observables do not necessarily commute with the Hamiltonian, and so consideration in the following is limited to the initial time $t = 0$ at which the wave function of the system is prepared as a mixed state of two mutual normalized eigenfunctions $\psi_1(\boldsymbol{x}, t = 0)$ and $\psi_2(\boldsymbol{x}, t = 0)$,

$$\Psi(\boldsymbol{x}, t = 0) = c_1\psi_1(\boldsymbol{x}, t = 0) + c_2\psi_2(\boldsymbol{x}, t = 0). \tag{4.158}$$

The normalization of the wave function requires that

$$|c_1|^2 + |c_2|^2 = 1. \tag{4.159}$$

The respective eigenvalues of $\hat{\mathrm{A}}$ are denoted e_{a1} and e_{a2}, while the eigenvalues of $\hat{\mathrm{B}}$ are e_{b1} and e_{b2}, so that

$$\begin{aligned}
\hat{\mathrm{A}}\Psi(\boldsymbol{x}, 0) &= c_1 e_{a1}\psi_1(\boldsymbol{x}, 0) + c_2 e_{a2}\psi_2(\boldsymbol{x}, 0),\\
\hat{\mathrm{B}}\Psi(\boldsymbol{x}, 0) &= c_1 e_{b1}\psi_1(\boldsymbol{x}, 0) + c_2 e_{b2}\psi_2(\boldsymbol{x}, 0),\\
\hat{\mathrm{A}}^2\Psi(\boldsymbol{x}, 0) &= c_1 e_{a1}^2\psi_1(\boldsymbol{x}, 0) + c_2 e_{a2}^2\psi_2(\boldsymbol{x}, 0),\\
\hat{\mathrm{B}}^2\Psi(\boldsymbol{x}, 0) &= c_1 e_{b1}^2\psi_1(\boldsymbol{x}, 0) + c_2 e_{b2}^2\psi_2(\boldsymbol{x}, 0),\\
\hat{\mathrm{A}}\hat{\mathrm{B}}\Psi(\boldsymbol{x}, 0) &= c_1 e_{a1}e_{b1}\psi_1(\boldsymbol{x}, 0) + c_2 e_{a2}e_{b2}\psi_2(\boldsymbol{x}, 0),
\end{aligned} \tag{4.160}$$

Using the orthonormality of the eigenfunctions allows the expectation values of the operators and their products to be calculated from (4.160) in a straightforward way. The results are

$$\begin{aligned}
\langle\hat{\mathrm{A}}\rangle &= |c_1|^2 e_{a1} + |c_2|^2 e_{a2},\\
\langle\hat{\mathrm{B}}\rangle &= |c_1|^2 e_{b1} + |c_2|^2 e_{b2},\\
\langle\hat{\mathrm{A}}^2\rangle &= |c_1|^2 e_{a1}^2 + |c_2|^2 e_{a2}^2,\\
\langle\hat{\mathrm{B}}^2\rangle &= |c_1|^2 e_{b1}^2 + |c_2|^2 e_{b2}^2,\\
\langle\hat{\mathrm{A}}\hat{\mathrm{B}}\rangle &= |c_1|^2 e_{a1}e_{b1} + |c_2|^2 e_{a2}e_{b2}.
\end{aligned} \tag{4.161}$$

Using (4.159) and (4.161), the initial uncertainties in $\hat{A}$ and $\hat{B}$ are found to be

$$\begin{aligned} \Delta\hat{A} &= \sqrt{\langle\hat{A}^2\rangle - \langle\hat{A}\rangle^2} = |c_1||c_2||e_{a1} - e_{a2}|, \\ \Delta\hat{B} &= \sqrt{\langle\hat{B}^2\rangle - \langle\hat{B}\rangle^2} = |c_1||c_2||e_{b1} - e_{b2}|. \end{aligned} \tag{4.162}$$

The mixed state has resulted in nonzero uncertainties for both observables despite the fact that the two observables commute. Only if $|c_1|$ or $|c_2|$ vanishes will the uncertainties of the two operators become zero. For the case that $|c_1| = |c_2| = 1/\sqrt{2}$ the uncertainties are one-half of the difference in the eigenvalues.

Using (4.162) the left-hand side of (4.137) is given by

$$\Delta\hat{A}\,\Delta\hat{B} = |c_1|^2|c_2|^2|(e_{a1} - e_{a2})(e_{b1} - e_{b2})|. \tag{4.163}$$

Before proceeding to calculate the right-hand side of the Heisenberg uncertainty principle (4.137), it will be useful to note that, for the product of Hermitian operators, the following identity holds,

$$\langle\hat{A}\hat{B}\rangle^* = (\Psi, \hat{A}\hat{B}\Psi)^* = (\hat{A}\hat{B}\Psi, \Psi) = (\Psi, \hat{B}\hat{A}\Psi) = \langle\hat{B}\hat{A}\rangle. \tag{4.164}$$

Using (4.159), (4.161), and (4.164), it is another straightforward exercise to show that

$$\begin{aligned} \frac{1}{2}\langle\{\hat{A}, \hat{B}\}\rangle - \langle\hat{A}\rangle\langle\hat{B}\rangle &= \frac{1}{2}\langle\hat{A}\hat{B}\rangle + \frac{1}{2}\langle\hat{A}\hat{B}\rangle^* - \langle\hat{A}\rangle\langle\hat{B}\rangle \\ &= |c_1|^2|c_2|^2(e_{a1} - e_{a2})(e_{b1} - e_{b2}). \end{aligned} \tag{4.165}$$

It follows from $[\hat{A}, \hat{B}] = 0$ that the Heisenberg uncertainty principle (4.137) is an exact equality for this mixed state,

$$\Delta\hat{A}\,\Delta\hat{B} = \frac{1}{2}\langle\{\hat{A}, \hat{B}\}\rangle - \langle\hat{A}\rangle\langle\hat{B}\rangle = |c_1|^2|c_2|^2|(e_{a1} - e_{a2})(e_{b1} - e_{b2})|. \tag{4.166}$$

However, in the case of a mixed state there is a nonzero product even if the two operators commute.

It is useful to note that, for this specific example, the equality holds in the uncertainty principle because condition (4.156) is satisfied. This follows from (4.159), (4.160), and (4.161), which give

$$\begin{aligned} \hat{\alpha}\Psi(\boldsymbol{x}, 0) &= (\hat{A} - \langle\hat{A}\rangle)\Psi(\boldsymbol{x}, 0) = (e_{a1} - e_{a2})(c_1|c_2|^2\psi_1(\boldsymbol{x}, 0) - |c_1|^2c_2\psi_2(\boldsymbol{x}, 0)), \\ \hat{\beta}\Psi(\boldsymbol{x}, 0) &= (\hat{B} - \langle\hat{B}\rangle)\Psi(\boldsymbol{x}, 0) = (e_{b1} - e_{b2})(c_1|c_2|^2\psi_1(\boldsymbol{x}, 0) - |c_1|^2c_2\psi_2(\boldsymbol{x}, 0)). \end{aligned} \tag{4.167}$$

It immediately follows that

$$\hat{\alpha}\Psi(\boldsymbol{x}, 0) = \left(\frac{e_{a1} - e_{a2}}{e_{b1} - e_{b2}}\right)\hat{\beta}\Psi(\boldsymbol{x}, 0) = \gamma\hat{\beta}\Psi(\boldsymbol{x}, 0), \tag{4.168}$$

so that condition (4.156) for the equality in the Heisenberg uncertainty principle (4.137) is satisfied. It is stressed that result (4.168) followed from the choice of the initial wave function as a mixed state of eigenfunctions.

4.6 Minimum uncertainty wave functions

In the Copenhagen interpretation of quantum mechanics the wave function is prepared at some initial time and then undergoes unitary time evolution according to (4.55) until an observation forces the wave function to collapse onto a set of possible values for a complete set of commuting observables. This section discusses a particular choice in wave function preparation. The goal is to understand what sort of wave function minimizes the product of uncertainties appearing the Heisenberg uncertainty principle. To be specific, the goal is to determine the wave function for which the Heisenberg uncertainty principle (4.137) becomes exact,

$$\Delta\hat{\mathrm{A}}\,\Delta\hat{\mathrm{B}} = \frac{1}{2}|\langle[\hat{\mathrm{A}}, \hat{\mathrm{B}}]\rangle|. \tag{4.169}$$

In the remainder of what follows it will be assumed that this equality is to be achieved during preparation of the wave function at an initial time, chosen to be $t = 0$ for simplicity.

4.6.1 Deriving the minimum uncertainty wave function

The first step in deriving the minimum uncertainty wave function is to attain the equality in the Cauchy–Schwarz inequality, and this occurs if condition (4.156) is satisfied. Upon choosing $\hat{\mathrm{A}}$ and $\hat{\mathrm{B}}$, this requires the wave function to satisfy

$$\hat{\alpha}\Psi(\boldsymbol{x}, 0) = (\hat{\mathrm{A}} - \langle\hat{\mathrm{A}}\rangle)\Psi(\boldsymbol{x}, 0) = \gamma(\hat{\mathrm{B}} - \langle\hat{\mathrm{B}}\rangle)\Psi(\boldsymbol{x}, 0) = \gamma\hat{\beta}\Psi(\boldsymbol{x}, 0), \tag{4.170}$$

where the equality in (4.137) will hold if γ is any complex constant. Condition (4.170) typically generates a differential equation for the wave function. Subsequent to solving this differential equation the wave function is then normalized.

However, the analysis of the mixed state wave function in a previous section showed that even the simultaneous measurements of commuting observables can be accompanied by uncertainties if the term involving the anticommutator appearing in (4.137) does not vanish. Since the goal is to minimize the product of uncertainties, the next step is to determine the conditions under which the anticommutator term on the right-hand side of (4.137) vanishes. This can be combined with (4.170), so that only the commutator remains on the right-hand side of (4.169). Using the notation of (4.129), this requires

$$\langle\{\hat{\alpha}, \hat{\beta}\}\rangle = \langle\{\hat{\mathrm{A}}, \hat{\mathrm{B}}\}\rangle - 2\langle\hat{\mathrm{A}}\rangle\langle\hat{\mathrm{B}}\rangle = 0. \tag{4.171}$$

Using result (4.164) allows (4.171) to be written

$$\langle\{\hat{\alpha}, \hat{\beta}\}\rangle = \langle\hat{\alpha}\hat{\beta}\rangle + \langle\hat{\beta}\hat{\alpha}\rangle = \langle\hat{\alpha}\hat{\beta}\rangle + \langle\hat{\alpha}\hat{\beta}\rangle^* = 0. \tag{4.172}$$

Combining this with the requirement $\hat{\alpha} = \gamma\hat{\beta}$ and the hermiticity of $\hat{\beta}$ simplifies this to

$$\langle\{\hat{\alpha}, \hat{\beta}\}\rangle = \langle\gamma\hat{\beta}^2\rangle + \langle\gamma\hat{\beta}^2\rangle^* = \gamma\langle\hat{\beta}^2\rangle + \gamma^*\langle\hat{\beta}^2\rangle^* = (\gamma + \gamma^*)\langle\hat{\beta}^2\rangle = 0. \tag{4.173}$$

As a result, the anticommutator term (4.171) vanishes if

$$\gamma + \gamma^* = 0. \tag{4.174}$$

The Heisenberg uncertainty principle therefore reduces to (4.169) if γ is pure imaginary and the wave function is chosen to satisfy (4.170).

4.6.2 Position and momentum minimum uncertainty wave function

This process is clarified with a specific example. For simplicity, the system is reduced to one spatial dimension and $\hat{\mathrm{A}} = \hat{x}$ and $\hat{\mathrm{B}} = \hat{p}$. For convenience, the constant γ is written as the pure imaginary quantity $\gamma = -i\Gamma^2/\hbar$, where Γ is a real constant with the units of length. Denoting $\langle\hat{p}\rangle = p_o$ and $\langle\hat{x}\rangle = x_o$, requirement (4.170) becomes the first order differential equation

$$(x - x_o)\Psi(x, 0) = -\frac{i}{\hbar}\Gamma^2(\hat{p} - p_o)\Psi(x, 0) = -\frac{i}{\hbar}\Gamma^2\left(-i\hbar\frac{\mathrm{d}}{\mathrm{d}x} - p_o\right)\Psi(x, 0). \tag{4.175}$$

This is readily solved by noting that

$$\mathrm{d}\Psi(x, 0) = \frac{\mathrm{d}\Psi(x, 0)}{\mathrm{d}x}\mathrm{d}x, \tag{4.176}$$

and this allows (4.175) to be written

$$\frac{\mathrm{d}\Psi(x, 0)}{\Psi(x, 0)} = \left(-\frac{(x - x_o)}{\Gamma^2} + \frac{ip_o}{\hbar}\right)\mathrm{d}x. \tag{4.177}$$

Integrating both sides from x_o to x results in

$$\Psi(x, 0) = \Psi_0\, e^{ip_o(x-x_o)/\hbar}e^{-\frac{1}{2}(x-x_o)^2/\Gamma^2}, \tag{4.178}$$

where $\Psi_o = \Psi(x_o, 0)$ is a constant determined by normalization. This gives

$$\int_{-\infty}^{\infty}\mathrm{d}x\,|\Psi(x, 0)|^2 = 1 \implies \Psi_o = \frac{1}{\sqrt{\Gamma\sqrt{\pi}}} \tag{4.179}$$

so that the initial wave function is given by

$$\Psi(x, 0) = \frac{1}{\sqrt{\Gamma\sqrt{\pi}}}\, e^{ip_o(x-x_o)/\hbar}\, e^{-\frac{1}{2}(x-x_o)^2/\Gamma^2}. \tag{4.180}$$

Using the wave function (4.180) shows that the initial expectation values for momentum and position are given by

$$
\begin{aligned}
\langle\hat{p}\rangle &= -i\hbar\left(\Psi, \frac{\mathrm{d}}{\mathrm{d}x}\Psi\right) = \frac{1}{\Gamma\sqrt{\pi}}\int_{-\infty}^{\infty}\mathrm{d}x\left(p_o + \frac{i\hbar}{\Gamma^2}(x - x_o)\right)e^{-(x-x_o)^2/\Gamma^2} = p_o, \\
\langle\hat{x}\rangle &= (\Psi, x\Psi) = \frac{1}{\Gamma\sqrt{\pi}}\int_{-\infty}^{\infty}\mathrm{d}x\; x\; e^{-(x-x_o)^2/\Gamma^2} = x_o,
\end{aligned}
\tag{4.181}
$$

which are consistent with the earlier designations. These also show that requirement (4.170) is satisfied since both sides reduce to zero, which allows Γ to be arbitrary. Additional expectation values are given by

$$
\begin{aligned}
\langle\hat{p}^2\rangle &= \frac{1}{\Gamma\sqrt{\pi}}\int_{-\infty}^{\infty}\mathrm{d}x\left(p_o^2 + \frac{i\hbar p_o}{\Gamma^2}(x - x_o) + \frac{\hbar^2}{\Gamma^2} - \frac{\hbar^2}{\Gamma^4}(x - x_o)^2\right)e^{-(x-x_o)^2/\Gamma^2} \\
&= p_o^2 + \frac{\hbar^2}{2\Gamma^2}, \\
\langle x^2\rangle &= \frac{1}{\Gamma\sqrt{\pi}}\int_{-\infty}^{\infty}\mathrm{d}x\; x^2\; e^{-(x-x_o)^2/\Gamma^2} = \frac{1}{2}\Gamma^2 + x_o^2.
\end{aligned}
\tag{4.182}
$$

Combining results (4.181) and (4.182) gives

$$
\begin{aligned}
\Delta\hat{p} &= \sqrt{\langle\hat{p}^2\rangle - \langle\hat{p}\rangle^2} = \frac{\hbar}{\sqrt{2}\Gamma}, \\
\Delta\hat{x} &= \sqrt{\langle\hat{x}^2\rangle - \langle\hat{x}\rangle^2} = \frac{\Gamma}{\sqrt{2}},
\end{aligned}
\tag{4.183}
$$

which correspond to the desired minimum uncertainty relation,

$$
\Delta\hat{x}\;\Delta\hat{p} = \frac{1}{2}|\langle[\hat{x}, \hat{p}]\rangle| = \frac{1}{2}|i\hbar| = \frac{1}{2}\hbar. \tag{4.184}
$$

The initial probability density is given by

$$
\mathcal{P}_{\hat{x}}(x, 0) = |\Psi(x, 0)|^2 = \frac{1}{\Gamma\sqrt{\pi}}e^{-(x-x_o)^2/\Gamma^2}, \tag{4.185}
$$

which is peaked around the position $x = x_o$, with the parameter Γ determining how sharply it is peaked. Comparing (4.185) to the representation of the Dirac delta (3.89) gives

$$
\lim_{\Gamma\to 0}\mathcal{P}_{\hat{x}}(x, 0) = \delta(x - x_o), \tag{4.186}
$$

showing that the minimum uncertainty wave function describes a point-like object in the limit $\Gamma \to 0$, corresponding to $\Delta\hat{x} \to 0$ and $\Delta\hat{p} \to \infty$. At the other extreme, setting $\Gamma = \sqrt{\pi}\,V$, where V is the one-dimensional volume, shows that the $V \to \infty$ limit of $\Psi(x, 0)$ becomes

$$
\lim_{V\to\infty}\Psi(x, 0) = \frac{1}{\sqrt{V}}e^{ip_o(x-x_o)/\hbar}. \tag{4.187}
$$

In this limit the initial wave function becomes the momentum eigenfunction (3.127), corresponding to the results $\Delta\hat{p} \rightarrow 0$ and $\Delta\hat{x} \rightarrow \infty$.

4.6.3 Wave packets

The initial wave function can be understood as a mixed state of the one-dimensional version of the free particle momentum eigenfunctions introduced in (3.173). Because it is a mixture of momentum eigenfunctions it is often referred to as a *wave packet*. For simplicity x_o is chosen as the origin and the initial wave function is written as a Fourier transform,

$$\Psi(x, 0) = \int_{-\infty}^{\infty} \frac{\mathrm{d}p}{2\pi\hbar}\, \tilde{\Psi}(p)\, e^{ipx\hbar}. \tag{4.188}$$

This expression is inverted using the orthonormality of the momentum eigenfunctions,

$$\begin{aligned}\int_{-\infty}^{\infty} \mathrm{d}x\, \Psi(x, 0)\, e^{-ipx/\hbar} &= \int_{-\infty}^{\infty} \mathrm{d}x \int_{-\infty}^{\infty} \frac{\mathrm{d}p'}{2\pi\hbar}\, \tilde{\Psi}(p')\, e^{i(p-p')x/\hbar} \\ &= \int_{-\infty}^{\infty} \mathrm{d}p'\, \tilde{\Psi}(p')\, \delta(p - p') = \tilde{\Psi}(p).\end{aligned} \tag{4.189}$$

Using the wave function (4.180) for $x_o = 0$ gives

$$\tilde{\Psi}(p) = \frac{1}{\sqrt{\Gamma\sqrt{\pi}}} \int_{-\infty}^{\infty} \mathrm{d}x\, e^{-\frac{1}{2}x^2/\Gamma^2} e^{-i(p-p_o)x/\hbar} = \sqrt{2\Gamma\sqrt{\pi}}\; e^{-\frac{1}{2}\Gamma^2(p-p_o)^2/\hbar^2}. \tag{4.190}$$

The earlier result (3.228) showed that the initial probability density in momentum space is given by $\mathcal{P}(p, 0) = |\tilde{\Psi}(p)|^2$, so that the momentum distribution is peaked around $p = p_o$.

4.6.4 The time evolution of wave packets

Having prepared the initial wave function in this manner it is instructive to examine how it will evolve in time. The general time evolution of the wave function is given by expression (4.61). For the case that the particle is free, the Hamiltonian is given by $\hat{H}(\hat{p}) = \hat{p}^2/2\,m$. Because $\hat{H}(\hat{p})$ has no explicit time-dependence the solution to the Schrödinger equation (4.11) that coincides with the preparation (4.180) at $t = 0$ is obtained by writing

$$\Psi(x, t) = \hat{\mathrm{U}}(t)\Psi(x, 0) = e^{-i\hat{H}(\hat{p})t/\hbar}\Psi(x, 0) = \int_{-\infty}^{\infty} \frac{\mathrm{d}p}{2\pi\hbar}\, \tilde{\Psi}(p)\, e^{i(px - E_{\mathrm{p}}t)/\hbar}, \tag{4.191}$$

where $\tilde{\Psi}(p)$ is given by (4.190) and $E_{\mathrm{p}} = p^2/2m$. In effect, the wave function is a linear superposition of free particle momentum eigenfunctions, but is not itself an eigenfunction of momentum. Inserting (4.190) into (4.191) gives

$$\Psi(x, t) = \sqrt{2\Gamma\sqrt{\pi}} \int_{-\infty}^{\infty} \frac{\mathrm{d}p}{2\pi\hbar}\, e^{-\frac{1}{2}\Gamma^2(p-p_o)^2/\hbar^2}\, e^{i(px - E_{\mathrm{p}}t)/\hbar}. \tag{4.192}$$

The integral (4.192) is readily evaluated to obtain

$$\Psi(x, t) = \sqrt{\frac{\Gamma}{\sqrt{\pi}(\Gamma^2 + \frac{i\hbar t}{m})}} \exp\left(\frac{i}{\hbar}\left(p_o x - \frac{p_o^2 t}{2m}\right)\right) \exp\left(-\frac{1}{2}\frac{(x - p_o t/m)^2}{\Gamma^2 + \frac{i\hbar t}{m}}\right). \quad (4.193)$$

From (4.193) the time-dependent spatial probability density is given by

$$\mathcal{P}_{\hat{x}}(x, t) = |\Psi(x, t)|^2 = \frac{1}{\sqrt{\pi}}\left(\Gamma^2 + \frac{\hbar^2 t^2}{m^2\Gamma^2}\right)^{-1/2} \exp\left(-\frac{(x - p_o t/m)^2}{\left(\Gamma^2 + \frac{\hbar^2 t^2}{m^2\Gamma^2}\right)}\right). \quad (4.194)$$

Result (4.194) shows that the probability density remains normalized, but the Γ^2 in the initial wave function (4.180) is replaced by the time-dependent shape factor $\Gamma^2(t)$, where

$$\Gamma^2(t) = \Gamma^2 + \frac{\hbar^2 t^2}{m^2\Gamma^2}. \quad (4.195)$$

Even if $\Gamma \approx 0$ initially, so that the initial uncertainty in location is $\Delta x \approx 0$, as time passes the probability density (4.194) becomes less sharply peaked. The width of the probability distribution grows, signaling a growing uncertainty in the location determined by the increasing value of $\Gamma(t)$ in (4.195). After sufficient time has passed, the argument of the exponential tends to zero and the probability density becomes the same everywhere. This is a direct outgrowth of the uncertainty principle. If the particle's initial uncertainty in position is very small, i.e. $\Gamma \approx 0$, it will possess a large initial uncertainty in momentum, $\Delta p(0) = \hbar/(\sqrt{2}\,\Gamma)$, and the uncertainty in its position will therefore grow rapidly since Newtonian dynamics gives $\Delta x \approx \Delta p(0)\, t/m$. This is consistent with (4.195) since at later times $\Delta x(t) = \Gamma(t) \approx \hbar t/m\Gamma \approx \Delta p(0)\, t/m$ for $\Gamma \approx 0$. The spreading of the wave packet is also a reflection that the individual momentum eigenfunctions that make up the wave packet have phases that travel at different speeds since $px - E_p t = p(x - (p/2m)t)$. The constituent momentum eigenfunctions therefore separate, resulting in a dispersing wave packet.

It is an interesting historical note that Schrödinger was the first to consider localized wave packets as a way to model localized particles in quantum mechanics. Upon discovering that such a wave packet disperses and no longer represents a localized particle, Schrödinger began considering alternatives to the equation that bears his name. Current research indicates that the interaction of the particle with its environment can maintain the localization of a wave packet. Environmental decoherence is discussed in chapter 7 and the reader should consult the references there.

References and recommended further reading

Schrödinger published the equation that bears his name in

- Schrödinger E 1926 Quantisierung als Eigenwertproblem *Ann. Phys.* **384** 361 (https://doi.org/10.1002/andp.19263840404)

Differential equations of the Sturm–Liouville type are analysed in

- Haberman R 2012 *Applied Partial Differential Equations with Fourier Series and Boundary Value Problems* 5th edn (London: Pearson)
- Arfken G, Weber H–J and Harris F 2013 *Mathematical Methods for Physicists* 7th edn (Cambridge, MA: Elsevier)

Many excellent textbooks on quantum mechanics begin with Schrödinger's equation and construct quantum mechanics from it. These include

- Schiff L 1968 *Quantum Mechanics* 3rd edn (New York: McGraw–Hill)
- Schwabl F 2007 *Quantum Mechanics* 4th edn (New York: Springer)
- Mahan D 2009 *Quantum Mechanics in a Nutshell* (Princeton, NJ: Princeton University Press)
- Griffiths D and Schroeter D 2018 *Introduction to Quantum Mechanics* 3rd edn (New York: Cambridge University Press)

Born's conditions for the wave function were presented in

- Born M 1927 Physical aspects of quantum mechanics *Nature* **119** 354 (https://doi.org/10.1038/119354a0)

Ehrenfest's theorem was published in

- Ehrenfest P 1927 Bemerkung über die angenäherte Gültigkeit der klassischen Mechanik innerhalb der Quantenmechanik *Z. Phys.* **45** 455 (https://doi.org/10.1007/BF01329203)

Heisenberg's uncertainty principle was published in

- Heisenberg W 1927 Über den anschaulichen Inhalt der quantentheoretischen Kinematik und Mechanik *Z. Phys.* **43** 172 (https://doi.org/10.1007/BF01397280)

IOP Publishing

A Concise Introduction to Quantum Mechanics (Second Edition)

Mark S Swanson

Chapter 5

Applications of wave mechanics

The range of quantum mechanical systems that have been analysed and compared to experiment is vast. In this chapter a handful of very basic applications of wave mechanics will bc presented. The choice of topics has been made to provide completely solvable and important applications of quantum mechanics, as well as to demonstrate techniques useful for solving quantum mechanical problems. Wave mechanics often reduces to solving the Schrödinger energy eigenvalue equation (4.45) for a given system, and the so-called *special functions* of applied mathematics play a large role in expressing the solutions.

5.1 Barrier reflection and tunneling

One of the more unusual aspects of quantum mechanics is that it allows classically forbidden processes to occur. In the case of the finite well it was seen in (4.86) that the wave function is nonzero in the classically forbidden zones, albeit exponentially damped. This section will examine barrier penetration by particles that are not sufficiently energetic for this to occur classically. In the simplest one-dimensional case the barrier is described by a potential energy of constant height V_o and width a, as depicted in figure 5.1,

$$U(x) = \begin{cases} V_o & \text{for } x \in (0, a) \\ 0 & \text{for } x < 0, \ \ x > a \end{cases}, \tag{5.1}$$

where V_o is assumed to be positive. This potential can be written in terms of the Heaviside step function θ defined in (3.70), so that

$$U(x) = V_o(\theta(x) - \theta(x - a)). \tag{5.2}$$

Classically, the motion of the particle is determined by the conservation of total energy $E = p^2/2\,m + U(x)$. In the regions $x < 0$ and $x > a$ where $U(x) = 0$ the energy and momentum have the classical relationship $p^2/2\,m = E$ and all positive

doi:10.1088/978-0-7503-5663-3ch5

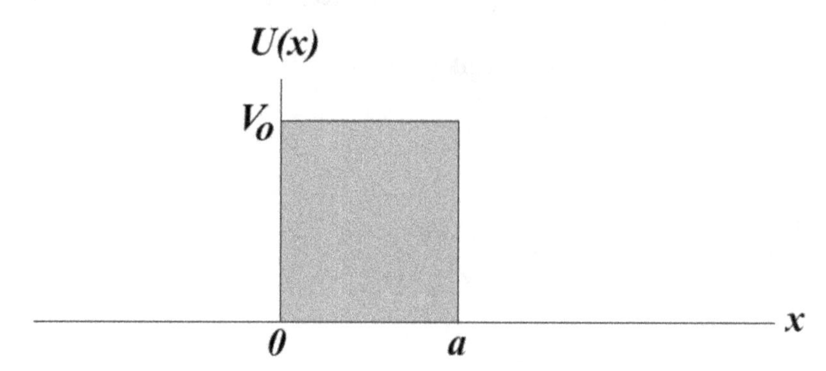

Figure 5.1. The one-dimensional barrier.

energies are allowed. However, in the region $0 < x < a$, a total energy such that $E < V_o$ is *classically forbidden* since it gives $p^2/2\,m = E - V_o < 0$, corresponding to an imaginary momentum.

It is important to understand the behavior of the wave function for the simple potential of (5.1) since it can be used to approximate a number of important physical processes that occur in nuclear fusion and semiconductor devices. Because the potential (5.1) is time-independent, the wave function can be written using the separation of variables technique of (4.46), so that $\Psi(x, t) = \psi(x)e^{-iEt/\hbar}$. The spatial part of the wave function therefore satisfies the eigenvalue equation

$$\left(-\frac{\hbar^2}{2m}\frac{d^2}{dx^2} + U(x)\right)\psi(x) = E\,\psi(x). \tag{5.3}$$

5.1.1 The infinite width barrier

For the moment, the case $a \to \infty$ will be considered. Solving for the wave function in the stationary case follows a procedure similar to that of the well treated in the previous chapter. In the region $x < 0$, where $U(x) = 0$, the most general solution to (5.3) for a fixed value of $E = p^2/2m$ is given by the linear combination

$$\psi(x) = Ae^{ipx/\hbar} + Be^{-ipx/\hbar}. \tag{5.4}$$

Nonzero values for both A and B create a mixed state of negative and positive momentum eigenfunctions, and these are understood to represent left moving and right moving particle momentum states, respectively. This allows the wave function to model the experimental result that the incoming particle with momentum p may be *reflected* from the barrier and leave with the momentum $-p$. In the region $x > 0$, where $U(x) = V_o$, the spatial part of the wave function that satisfies (5.3) is given by

$$\psi(x) = Ce^{ikx/\hbar} + De^{-ikx/\hbar}, \tag{5.5}$$

where $k = \sqrt{2\,m(E - V_o)}$. For the case that $E \geqslant V_o$ the constant k will be real and (5.5) is again a mixed state of momentum eigenfunctions. However, in the classically forbidden case, where $E < V_o$, the constant k becomes imaginary. For that case it is useful to denote $k = i\kappa$, where

$$\kappa = \sqrt{2\,mV_o - p^2}\,. \tag{5.6}$$

The constant κ is both real and positive for the case under consideration. This results in the stationary state (5.5) becoming

$$\psi(x) = Ce^{-\kappa x/\hbar} + De^{\kappa x/\hbar}. \tag{5.7}$$

Since (5.7) occurs for $x > 0$ and the case $a \to \infty$ is under consideration, the term proportional to D is divergent as $x \to \infty$, and its presence would therefore prevent normalizing the wave function. It therefore must be suppressed by setting $D = 0$.

For the classically forbidden case of an infinitely wide barrier, the wave function must be continuous at $x = 0$, and this requires (5.4) and (5.7) to match at $x = 0$, giving the requirement

$$C = A + B. \tag{5.8}$$

Similarly, the derivatives of the wave function must be continuous at $x = 0$, which gives

$$ip(A - B) = -\kappa C. \tag{5.9}$$

This pair of equations immediately gives

$$\begin{aligned} C &= \left(\frac{2ip}{ip - \kappa}\right)A, \\ B &= \left(\frac{ip + \kappa}{ip - \kappa}\right)A, \end{aligned} \tag{5.10}$$

which give

$$\begin{aligned} |B|^2 &= |A|^2, \\ |C|^2 &= \frac{4p^2}{p^2 + \kappa^2}|A|^2 = \frac{2p^2}{mV_o}|A|^2. \end{aligned} \tag{5.11}$$

In order to understand the physical meaning of these solutions, it is useful to evaluate the spatial probability current given by (4.27). In the region $x < 0$ the solution (5.4) gives the spatial component of the probability current density in the x-direction as

$$J_x = \frac{i\hbar}{2m}\left(\frac{\mathrm{d}}{\mathrm{d}x}\psi^*\psi - \psi^*\frac{\mathrm{d}}{\mathrm{d}x}\psi\right) = \frac{p}{m}(|A|^2 - |B|^2) = v|A|^2 - v|B|^2 = 0. \tag{5.12}$$

Result (5.12) shows that the probability current for $x < 0$ is composed of an *incident* right moving flux $J_\mathrm{I} = v|A|^2$ and a *reflected* left moving flux $J_\mathrm{R} = -v|B|^2$, and that these cancel since $|A|^2 = |B|^2$.

Recalling Born's rule, the ratio of the two fluxes, given by

$$\frac{J_R}{J_I} = \frac{v|B|^2}{v|A|^2} = \frac{|B|^2}{|A|^2}, \tag{5.13}$$

gives the relative probability that the particle is reflected from the barrier since it is the ratio of the reflected probability flux to the incident probability flux. It is therefore referred to as the *reflection coefficient*. For the case of an infinitely wide barrier the reflection coefficient is given by $|B|^2/|A|^2 = 1$, showing that quantum mechanically the particle cannot remain in the infinitely wide classically forbidden region and is reflected back into the incident region. Using (5.7) in (4.27) shows that $J = 0$ for $x > 0$ even though $\psi(x) \neq 0$ in that region, and so the probability current is continuous at $x = 0$. This requirement was discussed in the previous chapter where the probability current was derived and discussed in the context of Born's continuity conditions. It is essential in order to have physically meaningful solutions.

However, there is a nonzero probability density of observing the particle in the classically forbidden region, even in the case that the barrier has an infinite width. This is given explicitly by

$$\mathcal{P}_{\hat{x}}(x, t) = |\psi(x)|^2 = |C|^2 e^{-2\kappa x/\hbar} = \frac{4E}{V_o}|A|^2 e^{-2\kappa x/\hbar}. \tag{5.14}$$

The wave function is nonzero in the classically forbidden region where it takes the form of a momentum eigenfunction with an *imaginary momentum*. Such a component is not normalizable over *all space*, but for the case of either a finite volume or for the case that it is *always* exponentially damped, it can be part of the wave function. This was seen earlier in the case of the quantum well, where the classically forbidden region wave function (4.76) was nonzero. This is put in perspective by normalizing the wave function. For $x < 0$ the probability density is given by

$$\begin{aligned}\mathcal{P}_{\hat{x}}(x, t) &= |Ae^{ipx/\hbar} + Be^{-ipx/\hbar}|^2 \\ &= 2|A|^2\left(1 + \frac{\kappa^2 - p^2}{\kappa^2 + p^2}\cos\left(\frac{2px}{\hbar}\right) + \frac{2p\kappa}{\kappa^2 + p^2}\sin\left(\frac{2px}{\hbar}\right)\right),\end{aligned} \tag{5.15}$$

where the expression for B given by (5.10) was used. It follows that

$$\lim_{L\to\infty}\int_{-L}^{0} \mathrm{d}x\, \mathcal{P}_{\hat{x}}(x, t) \to 2L|A|^2, \tag{5.16}$$

where the Riemann–Lebesgue lemma (3.222) shows the oscillatory terms contribute zero in the large L limit as long as $p \neq 0$. The contribution from the forbidden region is found by using (5.14), which gives

$$\int_0^\infty \mathrm{d}x\, \frac{4E}{V_o}|A|^2 e^{-2\kappa x/\hbar} = \frac{2E\hbar}{V_o\kappa}|A|^2. \tag{5.17}$$

Normalization therefore requires

$$|A|^2 = \left(2L + \frac{2E\hbar}{V_o\kappa}\right)^{-1}, \tag{5.18}$$

where the large L limit is understood. This gives the fractional probability P_f of observing the particle in the forbidden region $x > 0$ as

$$P_f = \frac{2E\hbar}{V_o\kappa}|A|^2 \approx \frac{E\hbar}{V_o\kappa L}. \tag{5.19}$$

For an infinitely high barrier κ diverges as $V_o \to \infty$. Even if $E \to V_o$ the probability (5.19) vanishes and classical behavior emerges. For such a case there are no energies for which the term in (5.7) is present since $e^{-\kappa x} \to 0$ for $x > 0$.

The case of an infinitely high and wide barrier therefore effectively corresponds to $C = 0$. This requires $A = -B$, finally yielding for the case that $V_o \to \infty$,

$$\begin{aligned} \psi(x) &= A(e^{ipx/\hbar} - e^{-ipx/\hbar}) = 2iA\sin\left(\frac{px}{\hbar}\right) \text{ for } x < 0, \\ \psi(x) &= 0 \text{ for } x > 0. \end{aligned} \tag{5.20}$$

Such a stationary state corresponds to a mixed state of equal amounts of left and right momentum states and describes the quantum mechanical *reflection* of the particle from the barrier. The normalization of (5.20) requires the region of integration to the left of the barrier is large but finite, so that

$$\begin{aligned} \int_{-L}^{0} \mathrm{d}x\, 4|A|^2 \sin^2\left(\frac{px}{\hbar}\right) &= 4|A|^2\left(\frac{L}{2} - \frac{\hbar}{p}\sin\left(\frac{2pL}{\hbar}\right)\right) \\ &= 2|A|^2 L = 1 \implies |A|^2 = \frac{1}{2L}, \end{aligned} \tag{5.21}$$

where the Riemann–Lebesgue lemma (3.222) allowed the second term to be discarded for large L and $p \neq 0$. Since the wave function vanishes for $x > 0$ there is no penetration into the region $x > 0$ for the case of an infinitely high and wide barrier and it does not contribute to normalization. The normalized form of the wave function (5.20) for $x < 0$ is therefore given by

$$\psi(x) = \sqrt{\frac{2}{L}}\sin\left(\frac{px}{\hbar}\right), \tag{5.22}$$

which is identical to the mixed state of two momentum eigenfunctions in an infinite volume. It is worth noting that the derivative of (5.4) for $x < 0$ is

$$\frac{\mathrm{d}\psi(x)}{\mathrm{d}x} = \frac{2i}{\hbar}pA\cos\left(\frac{px}{\hbar}\right), \tag{5.23}$$

which yields $2iAp/\hbar$ at $x = 0$. The derivative of the wave function (5.7) vanishes for $x > 0$ in the case $V_o \to \infty$, and so the derivative is therefore not continuous at $x = 0$. This possibility was discussed earlier in (4.53). However, since the sine function vanishes at $x = 0$, the probability current J given by (4.27) also vanishes at $x = 0$ for both (5.4) and (5.5), and this allows Born's continuity conditions to be met.

5.1.2 The finite width barrier

The second case for analysis is a finite value for both V_o and the barrier width a. Since classically forbidden barrier penetration is of interest, the case that $E = p^2/2m < V_o$ is once again considered. The wave function in the three regions is written

$$\begin{aligned}\psi(x) &= Ae^{ipx/\hbar} + Be^{-ipx/\hbar} \quad \text{for } x < 0,\\ \psi(x) &= Ce^{-\kappa x/\hbar} + De^{\kappa x/\hbar} \quad \text{for } 0 < x < a,\\ \psi(x) &= Fe^{ipx/\hbar} \quad \text{for } x > a.\end{aligned} \tag{5.24}$$

The choice of the third form in (5.24) is justified by using (4.27) to find the probability flux of $J_T = (p/m)|F|^2 = v|F|^2$ for $x > a$, which represents only a *transmitted* probability flux *past the barrier*. In addition, the finite barrier width a allows the term proportional to D to be present in (5.24). The continuity of the wave function at both $x = 0$ and $x = a$ requires

$$A + B = C + D, \tag{5.25}$$

$$Ce^{-\kappa a/\hbar} + De^{\kappa a/\hbar} = F\, e^{ipa/\hbar}, \tag{5.26}$$

while continuity of the derivatives of the wave function at both $x = 0$ and $x = a$ requires

$$ip(A - B) = -\kappa(C - D) \tag{5.27}$$

$$-\kappa(Ce^{-\kappa a/\hbar} - De^{\kappa a/\hbar}) = ip\, F\, e^{ipa/\hbar}. \tag{5.28}$$

Since there are five coefficients and four equations of continuity, it is possible to express four of the coefficients in terms of the fifth. The three coefficients of interest are A, B, and F, since these represent the amplitudes of the incident, reflected, and transmitted wave function, respectively. The most convenient choice is to express A, B, and F in terms of D. After tedious but straightforward algebra it can be shown that

$$\begin{aligned}A &= \left(\left(\frac{\kappa + ip}{2ip}\right)e^{2\kappa a/\hbar} - \left(\frac{(\kappa - ip)^2}{2ip(\kappa + ip)}\right)\right)D,\\ B &= \left(\frac{\kappa - ip}{2ip}\right)(e^{\kappa a/\hbar} - 1)D,\\ F &= \left(\frac{2\kappa}{\kappa + ip}\right)e^{(\kappa - ip)a/\hbar}D.\end{aligned} \tag{5.29}$$

Using these allows the following ratios to be calculated,

$$R \equiv \frac{|B|^2}{|A|^2} = \left(1 + \frac{4p^2\kappa^2}{(\kappa^2 + p^2)^2 \sinh^2(\kappa a/\hbar)}\right)^{-1} = \left(1 + \frac{4E(V_o - E)}{V_o^2 \sinh^2(\kappa a/\hbar)}\right)^{-1}, \quad (5.30)$$

$$T \equiv \frac{|F|^2}{|A|^2} = \left(1 + \frac{(\kappa^2 + p^2)^2 \sinh^2(\kappa a/\hbar)}{4p^2\kappa^2}\right)^{-1} = \left(1 + \frac{V_o^2 \sinh^2(\kappa a/\hbar)}{4E(V_o - E)}\right)^{-1}, \quad (5.31)$$

where the hyperbolic sine is defined as $\sinh x = \frac{1}{2}(e^x - e^{-x})$ and $E = p^2/2m$.

Born's rule states that R represents the relative probability that the particle will be reflected from the barrier and is therefore referred to as the *reflection coefficient*. The quantity T is the *transmission coefficient*, and represents the probability that the particle will *tunnel through the barrier* despite the fact that it is classically forbidden due to insufficient energy. For the case that $\hbar \to 0$ the hyperbolic sine diverges, $\sinh(\kappa a/\hbar) \to \infty$, giving $R = 1$ and $T = 0$. Therefore, as expected, *barrier tunneling is a quantum mechanical phenomenon*. For the case that $a \to \infty$ the hyperbolic sine again diverges, and the previous results for an infinitely wide barrier, $R = 1$ and $T = 0$, are once again obtained. It is important to note that $R + T = 1$ for all values of E, so that the total probability of scattering and not scattering is unity, as it must be. This conservation of total probability is referred to as *unitarity*, not to be confused with a unitary operator. In this case unitarity is brought about by normalization of the wave function.

This can be related to the Heisenberg uncertainty principle, which states that a particle with an exact momentum is *delocalized*, i.e. its spatial uncertainty diverges. In such a case, its wave function can be nonzero past the barrier. It is also possible to consider the behavior of a localized wave packet as it encounters the barrier. Examining the momentum distribution of the minimum uncertainty wave packet (4.190) shows that it possesses momentum components that are energetically greater than the barrier height, $p > \sqrt{2mV_o}$, and that transmission of part of the wave packet will occur.

This simple system can be viewed as a one-dimensional version of *scattering*. It shows that the interaction of the incoming probability current with the potential representing the barrier results in some of the incoming wave reflected or scattered back in the direction it came from, while some of the incoming wave simply passes through the barrier as if it was unaffected. This will be adapted in chapter 9 to the three-dimensional case, resulting in the scattered wave function spreading over a spherical shell around the potential.

5.1.3 The one-dimensional well

The formulas of (5.30) and (5.31) can be easily adapted to the case of a potential well of depth $-V_o$ and width a by simply replacing V_o by $-V_o$ in the two formulas. For such a replacement $\kappa \to iq$, where $q = \sqrt{2m(E + V_o)}$, and $\sinh(\kappa a/\hbar) \to i\sin(qa/\hbar)$. The final results are the reflection and transmission coefficients for a particle of momentum p and kinetic energy $E = p^2/2m$ encountering a potential well of depth $-V_o$ and width a,

$$R = \left(1 + \frac{4E(V_o + E)}{V_o^2 \sin^2(qa/\hbar)}\right)^{-1}, \tag{5.32}$$

$$T = \left(1 + \frac{V_o^2 \sin^2(qa/\hbar)}{4E(V_o + E)}\right)^{-1}. \tag{5.33}$$

In classical mechanics the momentum of the particle would simply increase as it passed over the well. However, result (5.32) shows that a quantum mechanical particle can be reflected by a well. An interesting property of (5.33) is that it becomes unity, corresponding to the absence of reflection, for values of q such that $\sin(qa/\hbar) = 0$, and these occur if $qa/\hbar = n\pi$, where n is an integer. Solving this for the associated incident energy E gives

$$E = \frac{n^2\pi^2\hbar^2}{2\,ma^2} - V_o, \tag{5.34}$$

and all of these incident energies will give rise to total transmission past the well. The result (4.89) for the energies available to a particle in an infinitely deep one-dimensional well of width a has appeared. This phenomenon is referred to as *resonance*, and it also occurs in the presence of a potential barrier for the case that the particle has an energy E such that $E > V_0$. Resonance will occur when the node to node distance of the de Broglie wavelength matches the width of the potential well and the conditions required for a standing wave are met.

As mentioned earlier, both results (5.30) and (5.32) are the one-dimensional version of a *quantum mechanical scattering process*, where the particle encounters a localized potential well and its motion is altered. In the one-dimensional case, the only possible alteration in motion is reflection. Obviously, in three dimensions the particle can undergo a change in direction characterized by a scattering angle. Methods for determining the probability of scattering into a particular angle, as well as the associated cross-sectional area of the target, are important quantum mechanical tools for analysing scattering experiments and identifying the forces present during the collision process. This is the topic of chapter 9.

5.2 The one-dimensional harmonic oscillator

Quantum mechanics began with Planck's analysis of blackbody radiation. In order to explain the nature of the blackbody spectrum Planck postulated that the radiation is generated by the atoms of the cavity wall changing energy levels that are separated by the discrete quantum of energy $\hbar\omega$. The Schrödinger equation will justify his assumption.

A simple mechanical model for the atoms comprising a solid, such as the blackbody cavity walls, is that of a chain of point masses coupled to their nearest neighbors by an attractive spring-like force. Restricting attention to one dimension, each atom is considered bound to its site in the chain by a harmonic oscillator potential, $U(x) = \frac{1}{2}kx^2$. The classical Hamiltonian for the atom undergoing a small displacement x away from equilibrium is given by the approximation

$$H = \frac{p^2}{2m} + \frac{1}{2}kx^2 = \frac{p^2}{2m} + \frac{1}{2}m\omega^2x^2, \tag{5.35}$$

where m is the atomic mass and $\omega = \sqrt{k/m}$ is the natural frequency of the classical oscillator. Because the potential energy $U = \frac{1}{2}m\omega^2x^2$ has no explicit time-dependence, the Schrödinger equation can undergo the separation of variables (4.42),

$$\Psi(x, t) = \psi(x)\, e^{-iEt/\hbar}, \tag{5.36}$$

and therefore becomes an eigenvalue equation for the allowed energies E,

$$\hat{H}\,\psi(x) = \left(\frac{\hat{p}^2}{2m} + U(x)\right)\psi(x) = \left(-\frac{\hbar^2}{2m}\frac{\mathrm{d}^2}{\mathrm{d}x^2} + \frac{1}{2}m\omega^2x^2\right)\psi(x) = E\,\psi(x). \tag{5.37}$$

A useful step in analysing (5.37) is to introduce the dimensionless coordinate ξ, defined by

$$\xi = \left(\frac{mk}{\hbar^2}\right)^{\frac{1}{4}} x = \left(\frac{m^2\omega^2}{\hbar^2}\right)^{\frac{1}{4}} x = \sqrt{\frac{m\omega}{\hbar}}\, x \equiv \alpha x, \tag{5.38}$$

Simultaneously, the energy E is scaled the dimensionless quantity λ according to

$$E = \frac{1}{2}\hbar\sqrt{\frac{k}{m}}\,\lambda = \frac{1}{2}\hbar\omega\lambda. \tag{5.39}$$

This generates a common factor of $\frac{1}{2}\hbar\omega$ in all three terms of (5.37) and, after canceling it, the eigenvalue equation (5.37) becomes

$$\frac{\mathrm{d}^2}{\mathrm{d}\xi^2}\psi(\xi) + (\lambda - \xi^2)\psi(\xi) = 0. \tag{5.40}$$

5.2.1 Solving the oscillator eigenvalue equation

An important aspect of solving (5.40) is imposing a consistent set of boundary conditions. Clearly, the solution to (5.37) must be normalizable, i.e. a member of L^2. The wave function should therefore vanish at great distances from the equilibrium position at $\xi = 0$ faster than any power of ξ, giving $\lim_{\xi\to\pm\infty} \xi^n\psi(\xi) = 0$. As a result, the wave function is given the general form

$$\psi(\xi) = \mathrm{H}(\xi)\, e^{-\frac{1}{2}\xi^2}, \tag{5.41}$$

where $H(\xi)$ is a polynomial in ξ to be determined from (5.40). Substituting (5.41) into (5.40) gives

$$\frac{d^2}{d\xi^2}H(\xi) - 2\xi \frac{d}{d\xi}H(\xi) + (\lambda - 1) H(\xi) = 0. \tag{5.42}$$

Equation (5.42) is Hermite's differential equation and its solutions are the *Hermite polynomials*, denoted $H_n(\xi)$ and indexed by the integer n, where $n \geqslant 0$. Each Hermite polynomial $H_n(\xi)$ is paired with the eigenvalue $\lambda_n = 2n + 1$, where n is the same integer that indexes $H_n(\xi)$.

In order to define the Hermite polynomials and show that they are the solution to (5.42) for the eigenvalues $\lambda_n = 2n + 1$, it is useful to define them using the *Rodriguez formula*, which states that

$$H_n(\xi) = (-1)^n e^{\xi^2} \frac{d^n}{d\xi^n} e^{-\xi^2}. \tag{5.43}$$

It follows from (5.43) that $H_0(\xi) = 1$, $H_1(\xi) = 2\xi$, and $H_2(\xi) = 4\xi^2 - 2$. These three polynomials are easily demonstrated to satisfy (5.42) for the respective eigenvalues for λ of 1, 3, and 5.

In order to show that the Rodriguez formula generates the full sequence of solutions to (5.42) it is first useful to derive several *recurrence relations* among the Hermite polynomials. These are derived from the mathematical identity

$$\frac{d^{n+1}}{d\xi^{n+1}} e^{-\xi^2} = \frac{d^n}{d\xi^n}\left(\frac{d}{d\xi} e^{-\xi^2}\right) = -2\frac{d^n}{d\xi^n}\left(\xi e^{-\xi^2}\right) = -2n\frac{d^{n-1}}{d\xi^{n-1}} e^{-\xi^2} - 2\xi\frac{d^n}{d\xi^n} e^{-\xi^2}. \tag{5.44}$$

Result (5.44) clearly holds for $n = 1$, where $H_1(\xi) = 2\xi$, and the general proof can be established through induction by differentiating both sides of (5.44),

$$\begin{aligned} \frac{d^{n+2}}{d\xi^{n+2}} e^{-\xi^2} &= \frac{d}{d\xi}\left(\frac{d^{n+1}}{d\xi^{n+1}} e^{-\xi^2}\right) = \frac{d}{d\xi}\left(-2n\frac{d^{n-1}}{d\xi^{n-1}} e^{-\xi^2} - 2\xi\frac{d^n}{d\xi^n} e^{-\xi^2}\right) \\ &= -2(n+1)\frac{d^n}{d\xi^n} e^{-\xi^2} - 2\xi\frac{d^{n+1}}{d\xi^{n+1}} e^{-\xi^2}, \end{aligned} \tag{5.45}$$

which is the $n + 1$ version of (5.44), completing the inductive proof. Multiplying the identity (5.44) by $(-1)^{n+1} e^{\xi^2}$ gives

$$(-1)^n e^{\xi^2} \frac{d^{n+1}}{d\xi^{n+1}} e^{-\xi^2} = -2n(-1)^{n-1} e^{\xi^2} \frac{d^{n-1}}{d\xi^{n-1}} e^{-\xi^2} + 2\xi(-1)^n e^{\xi^2} \frac{d^n}{d\xi^n} e^{-\xi^2}. \tag{5.46}$$

The Rodriguez formula shows that (5.46) can be written as the first recurrence relation,

$$H_{n+1}(\xi) - 2\xi H_n(\xi) + 2n H_{n-1}(\xi) = 0. \tag{5.47}$$

Differentiating the Rodriguez formula (5.43) gives the identity

$$\frac{d}{d\xi} H_n(\xi) = 2\xi(-1)^n e^{\xi^2} \frac{d^n}{d\xi^n} e^{-\xi^2} + (-1)^n e^{\xi^2} \frac{d^{n+1}}{d\xi^{n+1}} e^{-\xi^2} = 2\xi H_n(\xi) - H_{n+1}(\xi). \tag{5.48}$$

Combining (5.48) with the first recurrence relation (5.47) gives the second recurrence relation,

$$\frac{\mathrm{d}}{\mathrm{d}\xi}\mathrm{H}_n(\xi) = 2n\,\mathrm{H}_{n-1}(\xi). \tag{5.49}$$

Substituting (5.49) into (5.47) gives

$$\frac{\mathrm{d}}{\mathrm{d}x}H_n(\xi) - 2\xi H_n(\xi) + H_{n+1}(\xi) = 0. \tag{5.50}$$

Differentiating this identity gives

$$\frac{\mathrm{d}^2}{\mathrm{d}x^2}H_n(\xi) - 2\xi\frac{\mathrm{d}}{\mathrm{d}x}H_n(\xi) - 2H_n(\xi) + \frac{\mathrm{d}}{\mathrm{d}x}H_{n+1}(\xi) = 0. \tag{5.51}$$

The second recurrence relation (5.49) gives $\mathrm{d}H_{n+1}(\xi)/\mathrm{d}x = (2n+2)H_n(\xi)$, which reduces (5.51) to

$$\frac{\mathrm{d}^2}{\mathrm{d}x^2}H_n(\xi) - 2\xi\frac{\mathrm{d}}{\mathrm{d}x}H_n(\xi) - 2nH_n(\xi) = 0. \tag{5.52}$$

This is Hermite's differential equation (5.42) with the identification $\lambda_n - 1 = 2n$.

Changing back to the variable x by using $\xi = \alpha x$, the full solution to (5.37) is therefore given by

$$\psi_n(x) = N_n\,\mathrm{H}_n(\alpha x)\,e^{-\frac{1}{2}\alpha^2x^2}, \tag{5.53}$$

where N_n is the normalization constant. A general expression for N_n will be derived in the next section. Using (5.40) shows that each stationary state $\psi_n(x)$ is paired with the energy eigenvalue

$$E_n = \frac{1}{2}\hbar\omega\lambda = \left(n + \frac{1}{2}\right)\hbar\omega, \quad n = 0, 1, 2, \ldots. \tag{5.54}$$

Result (5.54) is identical to Planck's original assumption that the atomic oscillators in the walls of the blackbody cavity have energy levels separated by an integer multiple of $\hbar\omega$. Because the energy eigenvalues are not degenerate, result (3.168) immediately demonstrates that the energy eigenfunctions (5.53) are orthonormal. Although the proof will not be presented, it can be shown that they are also a complete basis set of functions for the space of functions L^2, allowing the wave function to be prepared in any form. Unlike the continuum of energies (1.111) available to the classical harmonic oscillator, the quantum mechanical harmonic oscillator has a discrete set of energy levels available to it. This is similar to the infinite well, which also creates a discrete set of energy levels given by (4.89).

5.2.2 Uncertainty and the harmonic oscillator

It is instructive to evaluate the uncertainty in position for the ground state of the harmonic oscillator. This requires normalizing the ground state wave function,

$$\int_{-\infty}^{\infty} \mathrm{d}x \; |\psi_0(x)|^2 = |N_0|^2 \int_{-\infty}^{\infty} \mathrm{d}x \; e^{-\alpha^2 x^2} = |N_0|^2 \frac{\sqrt{\pi}}{\alpha} = 1 \implies N_0 = \sqrt{\frac{\alpha}{\sqrt{\pi}}}. \tag{5.55}$$

The ground state expectation value for $\hat{x}$ is given by

$$\langle \hat{x} \rangle = \int_{-\infty}^{\infty} \mathrm{d}x \; x|\psi_0(x)|^2 = |N_0|^2 \int_{-\infty}^{\infty} \mathrm{d}x \; x \; e^{-\alpha^2 x^2} = 0, \tag{5.56}$$

which is consistent with oscillation placing the particle on either side of the origin with equal probability. The ground state expectation value for $\hat{x}^2$ is given by

$$\langle \hat{x}^2 \rangle = \int_{-\infty}^{\infty} \mathrm{d}x \; x^2|\psi_0(x)|^2 = |N_0|^2 \int_{-\infty}^{\infty} \mathrm{d}x \; x^2 \; e^{-\alpha^2 x^2} = \frac{1}{2\alpha^2}. \tag{5.57}$$

The uncertainty in $\hat{x}$ is therefore given by

$$\Delta\hat{x} = \sqrt{\langle \hat{x}^2 \rangle - \langle \hat{x} \rangle^2} = \frac{\sqrt{2}}{2\alpha} = \sqrt{\frac{\hbar}{2 \, m\omega}}. \tag{5.58}$$

The ground state expectation value for the momentum $\hat{p}$ is given by

$$\langle \hat{p} \rangle = -i\hbar \int_{-\infty}^{\infty} \mathrm{d}x \; \psi_0(x) \frac{\mathrm{d}}{\mathrm{d}x} \psi_0(x) = i\hbar\alpha |N_0|^2 \int_{-\infty}^{\infty} \mathrm{d}x \; x \; e^{-\alpha^2 x^2} = 0. \tag{5.59}$$

Similarly, the ground state expectation value for the momentum squared is given by

$$\begin{aligned} \langle \hat{p}^2 \rangle &= -\hbar^2 \int_{-\infty}^{\infty} \mathrm{d}x \; \psi_0(x) \frac{\mathrm{d}^2}{\mathrm{d}x^2} \psi_0(x) \\ &= \hbar^2 \alpha^2 |N_0|^2 \int_{-\infty}^{\infty} \mathrm{d}x \; (1 - \alpha^2 x^2) \; e^{-\alpha^2 x^2} = \frac{1}{2} \hbar^2 \alpha^2. \end{aligned} \tag{5.60}$$

Combining these results gives

$$\Delta\hat{p} = \sqrt{\langle \hat{p}^2 \rangle - \langle \hat{p} \rangle^2} = \frac{\sqrt{2}\,\hbar\alpha}{2} = \sqrt{\frac{1}{2} m\hbar\omega}. \tag{5.61}$$

The minimum uncertainty principle is therefore met, since (5.58) and (5.61) give

$$\Delta\hat{p} \; \Delta\hat{x} = \frac{1}{2}\hbar. \tag{5.62}$$

The quantum mechanical harmonic oscillator also has a nonzero ground state energy, given by $E_0 = \frac{1}{2}\hbar\omega$. This result is consistent with the minimum uncertainty version of the Heisenberg uncertainty principle (4.139). An expression for the energy is obtained by expanding the Hamiltonian around the classical equilibrium position at $p = 0$ and $x = 0$ using the uncertainties in p and x, so that the energy is given by

$$E = \frac{(\Delta\hat{p})^2}{2m} + \frac{1}{2} k \, (\Delta\hat{x})^2 = \frac{\hbar^2}{8 \, m \, (\Delta\hat{x})^2} + \frac{1}{2} k \, (\Delta\hat{x})^2. \tag{5.63}$$

Elementary calculus shows that expression (5.63) is minimized by the value

$$\Delta\hat{x} = \sqrt{\frac{\hbar}{2\sqrt{km}}} = \sqrt{\frac{\hbar}{2\,m\omega}}, \tag{5.64}$$

which is identical to the formal result (5.58). Using this in (5.63) gives the minimum energy consistent with the uncertainties in $\hat{p}$ and $\hat{x}$,

$$E = \frac{1}{2}\hbar\omega, \tag{5.65}$$

which matches the ground state energy. Result (5.65) should not be mistaken for the uncertainty in energy in the ground state, which is zero since the ground state is an eigenfunction of the Hamiltonian.

In order to apply the time–energy uncertainty principle (4.147), it is necessary to consider a mixed state of the ground state and the first excited state,

$$\Psi(x,\,t) = c_0\psi_0(x)e^{-iE_0t/\hbar} + c_1\psi_1(x)e^{-iE_1t/\hbar}, \tag{5.66}$$

where normalization requires $|c_0|^2 + |c_1|^2 = 1$. Result (4.162) shows that the uncertainty in the energy at $t = 0$ is given by

$$\Delta E = |c_0||c_1||E_1 - E_0| = |c_0||c_1|\hbar\omega. \tag{5.67}$$

Result (4.162) vanishes if either c_0 or c_1 is zero, which is consistent with a pure energy state. Using (5.65) in the time–energy uncertainty principle (4.147) gives

$$\Delta t \geqslant \frac{\hbar}{2\,\Delta E} = \frac{1}{2|c_0||c_1|\omega}. \tag{5.68}$$

For the case that $c_0 = c_1 = 1/\sqrt{2}$, which corresponds to an equal mixture of the ground state and the first excited state, the uncertainty Δt becomes

$$\Delta t = \frac{1}{\omega}. \tag{5.69}$$

This is the time interval required for detection of the particle in the mixed state since the particle is oscillating with the angular frequency ω. This follows by considering

$$\sin(\omega\Delta t) = \sin(1) \approx 0.8, \tag{5.70}$$

so that Δt corresponds to the time for the particle to transit from the origin to roughly its full amplitude.

5.2.3 An algebraic approach to the harmonic oscillator

A very useful alternative approach to analysing the harmonic oscillator starts by defining the differential operator $\hat{a}$ and its Hermitian adjoint $\hat{a}^\dagger$ by

$$\hat{a} = \frac{1}{\sqrt{2}}\left(\alpha\hat{x} + \frac{i\hat{p}}{\hbar\alpha}\right) \implies \hat{a}^\dagger = \frac{1}{\sqrt{2}}\left(\alpha\hat{x}^\dagger - \frac{i\hat{p}^\dagger}{\hbar\alpha}\right) = \frac{1}{\sqrt{2}}\left(\alpha\hat{x} - \frac{i\hat{p}}{\hbar\alpha}\right), \tag{5.71}$$

where $\alpha = \sqrt{m\omega/\hbar}$. Using $[\hat{x},\,\hat{p}] = i\hbar$ shows that $\hat{a}$ and $\hat{a}^\dagger$ satisfy the *algebra* given by

$$[\hat{a}, \hat{a}^\dagger] = \frac{1}{2}\alpha^2[\hat{x}, \hat{x}] - \frac{i}{2\hbar}[\hat{x}, \hat{p}] + \frac{i}{2\hbar}[\hat{p}, \hat{x}] + \frac{1}{2\hbar^2\alpha^2}[\hat{p}, \hat{p}] = 1. \tag{5.72}$$

It also follows that

$$\begin{aligned} \hat{a}^\dagger\hat{a} &= \frac{1}{2\hbar^2\alpha^2}\hat{p}^2 + \frac{1}{2}\alpha^2\hat{x}^2 + \frac{i}{2\hbar}\hat{x}\hat{p} - \frac{i}{2\hbar}\hat{p}\hat{x} \\ &= \frac{1}{\hbar\omega}\left(\frac{\hat{p}^2}{2m} + \frac{1}{2}m\omega^2\hat{x}^2\right) + \frac{i}{2\hbar}[\hat{x}, \hat{p}] = \frac{1}{\hbar\omega}\hat{H} - \frac{1}{2}, \end{aligned} \tag{5.73}$$

so that the Hamiltonian of (5.37) can be written

$$\hat{H} = \hbar\omega\left(\hat{a}^\dagger\hat{a} + \frac{1}{2}\right). \tag{5.74}$$

Understanding the details of how the Hamiltonian (5.74) solves the eigenvalue equation (5.37) and reproduces the results of the previous analysis begins by defining the function $f_0(x)$,

$$f_0(x) = e^{-\frac{1}{2}\alpha^2x^2}, \tag{5.75}$$

which is the unnormalized version of the ground state eigenfunction $\psi_0(x)$. It follows that the function $f_0(x)$ satisfies

$$\hat{a}\, f_0(x) = \frac{1}{\sqrt{2}}\left(\alpha\hat{x} + \frac{i\hat{p}}{\hbar\alpha}\right)f_0(x) = \frac{1}{\sqrt{2}}\left(\alpha x + \frac{1}{\alpha}\frac{\partial}{\partial x}\right)\left(e^{-\frac{1}{2}\alpha^2x^2}\right) = 0. \tag{5.76}$$

For that reason $\hat{a}$ is known as an *annihilation operator*, since it annihilates the harmonic oscillator ground state $f_0(x)$.

The next step is to use the algebra defined by (5.72) to show that

$$\hat{a}(\hat{a}^\dagger)^n f_0(x) = n(\hat{a}^\dagger)^{n-1}f_0(x). \tag{5.77}$$

The proof is inductive, and begins by noting that the (5.77) holds for $n = 1$ since (5.72) and (5.76) give

$$\hat{a}\hat{a}^\dagger f_0(x) = (\hat{a}^\dagger\hat{a} + 1)f_0(x) = \hat{a}^\dagger\hat{a}f_0(x) + f_0(x) = f_0(x). \tag{5.78}$$

Applying $\hat{a}^\dagger$ to the both sides of (5.77) gives

$$n(\hat{a}^\dagger)^{n+1}f_0(x) = \hat{a}^\dagger\hat{a}(\hat{a}^\dagger)^n f_0(x) = (\hat{a}\hat{a}^\dagger - 1)(\hat{a}^\dagger)^n f_0(x) = \hat{a}(\hat{a}^\dagger)^{n+1}f_0(x) - (\hat{a}^\dagger)^n f_0(x), \tag{5.79}$$

which immediately gives

$$\hat{a}(\hat{a}^\dagger)^{n+1}f_0(x) = (n + 1)(\hat{a}^\dagger)^n f_0(x), \tag{5.80}$$

completing the inductive proof.

The next step is to introduce the Hermitian operator $\hat{N}$, defined by

$$\hat{N} = \hat{a}^\dagger\hat{a}. \tag{5.81}$$

Using (5.77) shows that $\hat{N}$ has the property

$$\hat{N}(\hat{a}^{\dagger})^n f_0(x) = \hat{a}^{\dagger}\hat{a}(\hat{a}^{\dagger})^n f_0(x) = \hat{a}^{\dagger}\big(n(\hat{a}^{\dagger})^{n-1} f_0(x)\big) = n(\hat{a}^{\dagger})^n f_0(x). \tag{5.82}$$

The operator $\hat{N}$ is known as the *number operator* since its eigenvalues are the number of factors of $\hat{a}^{\dagger}$ acting on $f_0(x)$. Combining (5.82) with the expression (5.74) for the Hamiltonian gives

$$\hat{H}\big[(\hat{a}^{\dagger})^n f_0(x)\big] = \hbar\omega\left(\hat{N} + \frac{1}{2}\right)\big[(\hat{a}^{\dagger})^n f_0(x)\big] = \left(n + \frac{1}{2}\right)\hbar\omega\big[(\hat{a}^{\dagger})^n f_0(x)\big]. \tag{5.83}$$

This shows that $(\hat{a}^{\dagger})^n$ acting on the function $f_0(x)$ creates an eigenfunction of the Hamiltonian that is proportional to the excited state wave function $\psi_n(x)$, and so $\hat{a}^{\dagger}$ is referred to as a *creation operator.*

The next step is to normalize the wave function appearing in (5.83), so that the normalized wave function $\psi_n(x)$ is written

$$\psi_n(x) = \lambda(\hat{a}^{\dagger})^n f_0(x), \tag{5.84}$$

where λ is the normalization constant. Since $\psi_n(x)$ is normalized it follows that

$$(\psi_n, \psi_n) = |\lambda|^2((\hat{a}^{\dagger})^n f_0, (\hat{a}^{\dagger})^n f_0) = 1. \tag{5.85}$$

Using the property of the Hermitian adjoint, $((\hat{a}^{\dagger})^n)^{\dagger} = (\hat{a})^n$, this gives

$$|\lambda|^2(f_0, (\hat{a})^n(\hat{a}^{\dagger})^n f_0) = 1. \tag{5.86}$$

Applying result (5.77) n times gives

$$\begin{aligned} |\lambda|^2(f_0, (\hat{a})^n(\hat{a}^{\dagger})^n f_0) &= |\lambda|^2 n(f_0, (\hat{a})^{n-1}(\hat{a}^{\dagger})^{n-1} f_0) = |\lambda|^2 n(n-1)(f_0, (\hat{a})^{n-2}(\hat{a}^{\dagger})^{n-2} f_0) = \cdots \\ &= |\lambda|^2\, n!\,(f_0, f_0) = |\lambda|^2\, n! \int_{-\infty}^{\infty} \mathrm{d}x\, e^{-\alpha^2 x^2} = |\lambda|^2\, n!\, \frac{\sqrt{\pi}}{\alpha} = 1. \end{aligned} \tag{5.87}$$

This gives the normalized energy eigenfunctions,

$$\psi_n(x) = \sqrt{\frac{\alpha}{n!\sqrt{\pi}}}\,(\hat{a}^{\dagger})^n e^{-\frac{1}{2}\alpha^2 x^2}. \tag{5.88}$$

The final step is to note that the Hermite polynomials are given by

$$H_n(x)e^{-\frac{1}{2}\alpha^2 x^2} = (\sqrt{2})^n(\hat{a}^{\dagger})^n e^{-\frac{1}{2}\alpha^2 x^2}. \tag{5.89}$$

This can be demonstrated by applying $\hat{a}^{\dagger}$ to the left-hand side of (5.89) expressed in terms of $\xi = \alpha x$, which gives

$$\hat{a}^{\dagger} H_n(\xi)e^{-\frac{1}{2}\xi^2} = \frac{1}{\sqrt{2}}\left(\xi - \frac{\mathrm{d}}{\mathrm{d}\xi}\right)\left(H_n(\xi)e^{-\frac{1}{2}\xi^2}\right) = \frac{1}{\sqrt{2}}\left(2\xi H_n(\xi) - \frac{\mathrm{d}H_n(\xi)}{\mathrm{d}\xi}\right)e^{-\frac{1}{2}\xi^2}. \tag{5.90}$$

Using the Hermite polynomial identity (5.48) this immediately gives

$$\hat{a}^{\dagger} H_n(\xi) e^{-\frac{1}{2}\xi^2} = \frac{1}{\sqrt{2}} H_{n+1}(\xi)\, e^{-\frac{1}{2}\xi^2}, \tag{5.91}$$

which verifies (5.89). As a result, the normalized wave functions for the harmonic oscillator can be written in terms of the Hermite polynomials as

$$\psi_n(x) = \sqrt{\frac{\alpha}{n!\, 2^n \sqrt{\pi}}}\; H_n(\alpha x) e^{-\frac{1}{2}\alpha^2 x^2}, \tag{5.92}$$

where $\alpha = \sqrt{m\omega/\hbar}$.

Because of its power the algebraic approach is used to solve numerous quantum mechanical problems, and will be applied again in the section on the quantum mechanical behavior of a charge in a magnetic field. It is also used in quantum field theory, where the annihilation and creation operators represent the particle content of a field.

5.3 The hydrogen atom

The Bohr formula (2.65) for the energy levels of the electron in a hydrogen atom can be derived using the Schrödinger equation. In addition, the associated wave functions can be found, revealing a wealth of information regarding atomic structure. The starting point is the classical Lagrangian for a two particle system consisting of a positive charge q with mass m_1 and position $\boldsymbol{r}_1$ and negative charge $-q$ with mass m_2 and position $\boldsymbol{r}_2$ interacting through a potential given by $U(|\boldsymbol{r}_1 - \boldsymbol{r}_2|)$,

$$\mathcal{L} = \frac{1}{2} m_1 \dot{\boldsymbol{r}}_1^{\,2} + \frac{1}{2} m_2 \dot{\boldsymbol{r}}_2^{2} - U(|\boldsymbol{r}_1 - \boldsymbol{r}_2|). \tag{5.93}$$

5.3.1 Multiple particle wave functions

The classical Lagrangian (5.93) describes the motion of a two particle system and its quantum mechanical description therefore requires a two particle wave function. The wave function for a two particle system is written

$$\Psi = \Psi(\boldsymbol{r}_1, \boldsymbol{r}_2, t), \tag{5.94}$$

and its interpretation is that it gives the probability density for finding the first particle at $\boldsymbol{r}_1$ and the second particle at $\boldsymbol{r}_2$ at the time t,

$$\mathcal{P}(\boldsymbol{r}_1, \boldsymbol{r}_2, t) = |\Psi(\boldsymbol{r}_1, \boldsymbol{r}_2, t)|^2. \tag{5.95}$$

There is an important issue with (5.95) that stems from the nature of *elementary particles* such as the electron. All electrons are identical in their properties and their behavior. In a system where there are two electrons there is no experimental way to distinguish the two electrons from one another in the sense that there is no experimental way to tag one electron as the *first electron* and the other as the *second electron*. This means that the probability of observing one electron at $\boldsymbol{r}_1$ and the other at $\boldsymbol{r}_2$ must be identical to the probability of observing one electron at $\boldsymbol{r}_2$ and the other at $\boldsymbol{r}_1$. This assures that the two particles are truly experimentally indistinguishable. This means that (5.95) must obey

$$\mathcal{P}(\boldsymbol{r}_1, \boldsymbol{r}_2, t) = \mathcal{P}(\boldsymbol{r}_2, \boldsymbol{r}_1, t). \tag{5.96}$$

This has an immediate implication for the nature of the wave function for two identical particles. Because the wave function is squared to find the probability density, there are two ways to insure that (5.96) holds. The first is to *symmetrize* the wave function,

$$\Psi_S(\boldsymbol{r}_1, \boldsymbol{r}_2, t) = \frac{1}{2}\Psi(\boldsymbol{r}_1, \boldsymbol{r}_2, t) + \frac{1}{2}\Psi(\boldsymbol{r}_2, \boldsymbol{r}_1, t), \tag{5.97}$$

where $\Psi(\boldsymbol{r}_1, \boldsymbol{r}_2, t)$ is found from the Schrödinger equation for the system. The wave function (5.97) has the property that

$$\Psi_S(\boldsymbol{r}_1, \boldsymbol{r}_2, t) = \Psi_S(\boldsymbol{r}_2, \boldsymbol{r}_1, t), \tag{5.98}$$

which satisfies (5.95) since

$$\mathcal{P}(\boldsymbol{r}_1, \boldsymbol{r}_2, t) = |\Psi_S(\boldsymbol{r}_1, \boldsymbol{r}_2, t)|^2 = |\Psi_S(\boldsymbol{r}_2, \boldsymbol{r}_1, t)|^2 = \mathcal{P}(\boldsymbol{r}_2, \boldsymbol{r}_1, t). \tag{5.99}$$

The second option is to *antisymmetrize* the wave function,

$$\Psi_A(\boldsymbol{r}_1, \boldsymbol{r}_2, t) = \frac{1}{2}\Psi(\boldsymbol{r}_1, \boldsymbol{r}_2, t) - \frac{1}{2}\Psi(\boldsymbol{r}_2, \boldsymbol{r}_1, t). \tag{5.100}$$

The wave function (5.100) has the property that

$$\Psi_A(\boldsymbol{r}_1, \boldsymbol{r}_2, t) = -\Psi_A(\boldsymbol{r}_2, \boldsymbol{r}_1, t), \tag{5.101}$$

which also satisfies (5.95) since

$$\mathcal{P}(\boldsymbol{r}_1, \boldsymbol{r}_2, t) = |\Psi_A(\boldsymbol{r}_1, \boldsymbol{r}_2, t)|^2 = |\Psi_A(\boldsymbol{r}_2, \boldsymbol{r}_1, t)|^2 = \mathcal{P}(\boldsymbol{r}_2, \boldsymbol{r}_1, t). \tag{5.102}$$

However, the two wave functions have a very different outcome for the case that $\boldsymbol{r}_1 = \boldsymbol{r}_2$. The symmetric wave function gives

$$\mathcal{P}(\boldsymbol{r}_1, \boldsymbol{r}_1, t) = |\Psi(\boldsymbol{r}_1, \boldsymbol{r}_1, t)|^2, \tag{5.103}$$

which is not necessarily zero, while the antisymmetric wave function gives

$$\mathcal{P}(\boldsymbol{r}_1, \boldsymbol{r}_1, t) = 0. \tag{5.104}$$

The symmetric wave function allows the two particles to have a nonzero probability of sharing the same location, while the antisymmetric wave function prevents the two particles from ever sharing the same location. This is a distinctive property of all elementary particles. The symmetric wave function corresponds to a type of elementary particle generically known as a *boson*, while the antisymmetric function corresponds to a type of elementary particle generically known as a *fermion*. In the case of fermions it leads to the *Pauli exclusion principle*, since it states in this simple case that two identical fermions will not share the same property of location. This is discussed in much more detail when intrinsic angular momentum is introduced in chapter 7 and the nature of elementary particle wave functions becomes the final assumption of nonrelativistic quantum mechanics.

As it turns out both the electron and the proton are fermions. For the hydrogen atom wave function this is not an issue, since the proton and the electron are distinguishable by both electric charge and mass, and so it is not necessary to antisymmetrize the wave function. However, in the case of multielectron atoms it is necessary to antisymmetrize the wave function governing the electrons.

5.3.2 The Schrödinger equation for the hydrogen atom

Because the potential $U(|\boldsymbol{r}_1 - \boldsymbol{r}_2|)$ depends only on the magnitude of the vector $\boldsymbol{r}_1 - \boldsymbol{r}_2$ joining the two particles, the associated force will act along the line joining the two particles and, for that reason, it is often referred to as a *central force*. In the case of the neutral hydrogen atom, the central force is created between the two charges through the attractive *Coulomb potential* given by (1.35),

$$U(|\boldsymbol{r}_1 - \boldsymbol{r}_2|) = -\frac{q^2}{|\boldsymbol{r}_1 - \boldsymbol{r}_2|}. \tag{5.105}$$

The process of defining the Hamiltonian for the hydrogen atom from the Lagrangian (5.93) is simplified considerably by rewriting it in terms of the more convenient *center of mass variables*. First, the vector from the electron to the proton is denoted as

$$\boldsymbol{r} = \boldsymbol{r}_1 - \boldsymbol{r}_2, \tag{5.106}$$

while the *center of mass position* $\boldsymbol{R}$ is defined as

$$\boldsymbol{R} = \frac{m_1\boldsymbol{r}_1 + m_2\boldsymbol{r}_2}{m_1 + m_2}, \tag{5.107}$$

where $M = m_1 + m_2$ is the total mass of the two particle system. The *reduced mass* is defined as

$$\mu = \frac{m_1 m_2}{m_1 + m_2} = \frac{m_1 m_2}{M}, \tag{5.108}$$

while r is defined as

$$r = |\boldsymbol{r}| = |\boldsymbol{r}_1 - \boldsymbol{r}_2|. \tag{5.109}$$

Using the identity

$$\frac{1}{2}m_1\dot{\boldsymbol{r}}_1^2 + \frac{1}{2}m_2\dot{\boldsymbol{r}}_2 = \frac{1}{2}\frac{(m_1\dot{\boldsymbol{r}}_1 + m_1\dot{\boldsymbol{r}}_2)^2}{(m_1 + m_2)} + \frac{1}{2}\frac{m_1 m_2(\dot{\boldsymbol{r}}_1 - \dot{\boldsymbol{r}}_2)^2}{(m_1 + m_2)} \tag{5.110}$$

allows the Lagrangian (5.93) for the hydrogen atom to be rewritten as

$$\mathcal{L} = \frac{1}{2}\frac{(m_1\dot{\boldsymbol{r}}_1 + m_1\dot{\boldsymbol{r}}_2)^2}{(m_1 + m_2)} + \frac{1}{2}\frac{m_1 m_2(\dot{\boldsymbol{r}}_1 - \dot{\boldsymbol{r}}_2)^2}{(m_1 + m_2)} - U(r) = \frac{1}{2}M\dot{\boldsymbol{R}}^2 + \frac{1}{2}\mu\dot{\boldsymbol{r}}^2 - U(r). \quad (5.111)$$

The classical Lagrangian (5.111) describes the center of mass $\boldsymbol{R}$ moving as a *free particle*, since its equation of motion is

$$\frac{\partial \mathcal{L}}{\partial R_j} - \frac{\mathrm{d}}{\mathrm{d}t}\frac{\partial \mathcal{L}}{\partial \dot{R}_j} = -\frac{\mathrm{d}}{\mathrm{d}t}M\dot{R}_j = -M\ddot{R}_j = 0, \quad (5.112)$$

while the reduced mass μ moves in a static Coulomb potential (1.35) located at $r = 0$. This is easy to visualize if $m_2/m_1 \to 0$, since for that case $M \approx m_1$, $\mu \approx m_2$, and $\boldsymbol{R} \approx \boldsymbol{r}_1$. This holds for the hydrogen atom since the proton mass is roughly 1836 times that of the electron mass. As a result, the center of mass for the hydrogen atom is roughly the position of the proton, while the reduced mass is roughly the electron mass. In this case, the electron can be viewed as moving in the Coulomb potential created by the proton as the latter moves at a constant velocity. While this mass disparity is true of the hydrogen atom, the Lagrangian (5.111) is correct for all mass ratios and central potentials.

Combining the definition of the canonical momentum (1.15) with the definition of the Hamiltonian (1.16) and the Coulomb potential (1.35) gives the classical Hamiltonian for the hydrogen atom,

$$H(\boldsymbol{P}, \boldsymbol{R}, \boldsymbol{p}, \boldsymbol{r}) = \frac{\boldsymbol{P}^2}{2M} + \frac{\boldsymbol{p}^2}{2\mu} - \frac{q^2}{r}, \quad (5.113)$$

where $\boldsymbol{P}$ is the center of mass momentum that is canonically conjugate to $\boldsymbol{R}$, $\boldsymbol{p}$ is the reduced mass momentum canonically conjugate to $\boldsymbol{r}$, and the charges are measured in esu. The Hamiltonian operator for the Schrödinger equation is found by combining the Hamiltonian (5.113) with the substitutions

$$\begin{aligned} \boldsymbol{P} &\to \hat{\boldsymbol{P}} = -i\hbar\,\nabla_R, \\ \boldsymbol{p} &\to \hat{\boldsymbol{P}} = -i\hbar\nabla_r. \end{aligned} \quad (5.114)$$

It is important to note that the two three-dimensional gradients act on the two different spatial coordinates, ∇_R on $\boldsymbol{R}$ and ∇_r on $\boldsymbol{r}$. Since the wave function must be a function of both $\boldsymbol{R}$ and $\boldsymbol{r}$, the resulting Schrödinger equation is given by

$$\begin{aligned} \hat{H}(\hat{\boldsymbol{P}}, \hat{\boldsymbol{p}}, \boldsymbol{R}, \boldsymbol{r})\Psi(\boldsymbol{R}, \boldsymbol{r}, t) &= \left(-\frac{\hbar^2}{2M}\nabla_R^2 - \frac{\hbar^2}{2\mu}\nabla_r^2 - \frac{q^2}{r}\right)\Psi(\boldsymbol{R}, \boldsymbol{r}, t) \\ &= i\hbar\frac{\partial}{\partial t}\Psi(\boldsymbol{R}, \boldsymbol{r}, t). \end{aligned} \quad (5.115)$$

Since the potential is time-independent and the two spatial variables $\boldsymbol{R}$ and $\boldsymbol{r}$ are independent, the wave function $\Psi(\boldsymbol{R}, \boldsymbol{r}, t)$ can be written using separation of variables as

$$\Psi(\boldsymbol{R}, \boldsymbol{r}, t) = \Psi_R(\boldsymbol{R}, t)\,\Psi_r(\boldsymbol{r}, t) = \psi_R(\boldsymbol{R})\psi_r(\boldsymbol{r})e^{-iE_T t/\hbar}, \quad (5.116)$$

where E_T is the total energy of the two particles. After dividing by $\psi_R(\boldsymbol{R})\psi_r(\boldsymbol{r})$, the Schrödinger equation (5.115) becomes the sum,

$$-\frac{1}{\psi_R(\boldsymbol{R})}\frac{\hbar^2}{2M}\nabla_R^2\,\psi_R(\boldsymbol{R}) + \frac{1}{\psi_r(\boldsymbol{r})}\left(-\frac{\hbar^2}{2\mu}\nabla_r^2 - \frac{q^2}{r}\right)\psi_r(\boldsymbol{r}) = E_T. \tag{5.117}$$

Since $\boldsymbol{R}$ and $\boldsymbol{r}$ are independent of each other, (5.117) yields two eigenvalue equations,

$$-\frac{\hbar^2}{2M}\,\nabla_R^2\psi_R(\boldsymbol{R}) = E_c\,\psi_R(\boldsymbol{R}), \tag{5.118}$$

$$\left(-\frac{\hbar^2}{2\mu}\nabla_r^2 - \frac{q^2}{r}\right)\psi_r(\boldsymbol{r}) = E\,\psi_r(\boldsymbol{r}), \tag{5.119}$$

where the individual energies must sum to the total energy, $E_T = E + E_c$. Equation (5.118) is the quantum mechanical equation for a free particle. As a result, the solutions for the center of mass wave function ψ_R are given by the free particle momentum eigenfunctions,

$$\psi_R(\boldsymbol{R}) = \frac{1}{\sqrt{V}}e^{i\boldsymbol{P}\cdot\boldsymbol{R}/\hbar}, \tag{5.120}$$

where $P^2/2M = E_c$ is the center of mass energy. The free wave function (5.120) has been discussed in previous sections, and so analysis will proceed to the *Coulomb wave function* equation (5.119).

Center of mass variables reduces the two-body central force problem to the problem of finding the wave function of the reduced mass μ in a static Coulomb field. Because the potential is a function of r, the analysis of (5.119) requires using spherical coordinates (r, θ, ϕ), for which the Laplacian is given by (1.52),

$$\nabla_r^2 = \frac{\partial^2}{\partial r^2} + \frac{2}{r}\frac{\partial}{\partial r} + \frac{1}{r^2}\frac{\partial^2}{\partial\theta^2} + \frac{1}{r^2}\cot\theta\,\frac{\partial}{\partial\theta} + \frac{1}{r^2\sin^2\theta}\frac{\partial^2}{\partial\phi^2}. \tag{5.121}$$

The Coulomb potential eigenvalue equation (5.119) is solved using separation of variables, so that the wave function is written

$$\psi_r(\boldsymbol{r}) = \psi_r(r, \theta, \phi) = R(r)Z(\theta, \phi). \tag{5.122}$$

Substituting this into equation (5.119), dividing by $R(r)Z(\theta, \phi)$, and multiplying by r^2 gives

$$\begin{aligned}&\frac{r^2}{R(r)}\left(-\frac{\hbar^2}{2\mu}\left(\frac{\partial^2}{\partial r^2} + \frac{2}{r}\frac{\partial}{\partial r}\right) - \frac{Kq^2}{r} - E\right)R(r)\\&-\frac{\hbar^2}{2\mu Z(\theta, \phi)}\left(\frac{\partial^2}{\partial\theta^2} + \cot\theta\,\frac{\partial}{\partial\theta} + \frac{1}{\sin^2\theta}\frac{\partial^2}{\partial\phi^2}\right)Z(\theta, \phi) = 0.\end{aligned} \tag{5.123}$$

The first term of (5.123) is a function solely of r, while the second term is a function of θ and ϕ. Since these variables are independent of each other, the two terms must be equal and opposite *constants*. Equating the first term to $-\hbar^2\lambda/2\,m$ and the second term to $\hbar^2\lambda/2m$, equation (5.123) yields both a radial and an angular eigenvalue equation,

$$\left(\frac{\partial^2}{\partial r^2} + \frac{2}{r}\frac{\partial}{\partial r} + \frac{2\mu Kq^2}{\hbar^2 r} + \frac{2\mu E}{\hbar^2}\right)R(r) = \frac{\lambda}{r^2}R(r), \tag{5.124}$$

$$\left(\frac{\partial^2}{\partial \theta^2} + \cot\theta\,\frac{\partial}{\partial \theta} + \frac{1}{\sin^2\theta}\frac{\partial^2}{\partial \phi^2}\right)Z(\theta,\,\phi) = -\lambda\,Z(\theta,\,\phi). \tag{5.125}$$

Since (5.125) involves only the single constant λ it will be solved first.

5.3.3 The angular wave function

Understanding the angular eigenvalue equation (5.125) is greatly facilitated by examining its relationship to the orbital angular momentum operator of the reduced mass,

$$\hat{\boldsymbol{L}} = \boldsymbol{r} \times \hat{\boldsymbol{p}} = -i\hbar\,\boldsymbol{r} \times \nabla_r. \tag{5.126}$$

The three Cartesian components of the angular momentum operator do not commute among themselves, instead inheriting the Poisson bracket structure (1.31). For example,

$$\begin{aligned}[\hat{L}_x,\,\hat{L}_y] &= [\hat{y}\hat{p}_z - \hat{z}\hat{p}_y,\,\hat{z}\hat{p}_x - \hat{x}\hat{p}_z] \\ &= \hat{y}\hat{p}_x[\hat{p}_z,\,\hat{z}] + \hat{x}\hat{p}_y[\hat{z},\,\hat{p}_z] = i\hbar(\hat{x}\hat{p}_y - \hat{y}\hat{p}_x) = i\hbar\hat{L}_z,\end{aligned} \tag{5.127}$$

with similar results for the other commutators,

$$[\hat{L}_y,\,\hat{L}_z] = i\hbar\hat{L}_x, \quad [\hat{L}_z,\,\hat{L}_x] = i\hbar\hat{L}_y. \tag{5.128}$$

It will be shown in chapter 7 that this algebra is characteristic of rotations. Because the components of angular momentum do not commute among themselves, *at best* only one component can be included in the complete set of commuting observables (CSCO) for the hydrogen atom. This is unlike the previous one-dimensional examples of the harmonic oscillator and barrier penetration, where there was no angular momentum and the only observable in the CSCO was the energy.

Using the commutation relations of (5.128) and the commutator identity (3.197) it follows that

$$\begin{aligned}\left[\hat{L}_z,\,\hat{L}_x^2\right] &= \left[\hat{L}_z,\,\hat{L}_x\right]\hat{L}_x + \hat{L}_x\left[\hat{L}_z,\,\hat{L}_x\right] = i\hbar\left(\hat{L}_y\hat{L}_x + \hat{L}_x\hat{L}_y\right) = i\hbar\left\{\hat{L}_x,\,\hat{L}_y\right\}, \\ \left[\hat{L}_z,\,\hat{L}_y^2\right] &= \left[\hat{L}_z,\,\hat{L}_y\right]\hat{L}_y + \hat{L}_y\left[\hat{L}_z,\,\hat{L}_y\right] = -i\hbar\left(\hat{L}_x\hat{L}_y + \hat{L}_y\hat{L}_x\right) = -i\hbar\left\{\hat{L}_x,\,\hat{L}_y\right\}, \\ \left[\hat{L}_z,\,\hat{L}_z^2\right] &= 0.\end{aligned} \tag{5.129}$$

Adding the results of (5.129) shows that

$$\left[\hat{L}_z, \hat{L}_x^2 + \hat{L}_y^2 + \hat{L}_z^2\right] = \left[\hat{L}_z, \hat{L}^2\right] = 0. \tag{5.130}$$

As a result, $\hat{L}_z$ commutes with the total angular momentum squared $\hat{L}^2$. It should be noted that the other two components, $\hat{L}_x$ and $\hat{L}_y$, also commute with $\hat{L}^2$, but not with $\hat{L}_z$.

Further analysis is facilitated by expressing the angular momentum in the spherical coordinates introduced in chapter 1, although complications still occur. The angular momentum operator is obtained from the gradient in spherical coordinates (1.51), so that

$$\begin{aligned}\hat{\boldsymbol{L}} = \boldsymbol{r} \times \boldsymbol{p} = -i\hbar r\, \boldsymbol{e}_r \times \nabla_r = -i\hbar r\, \boldsymbol{e}_r \times \left(\boldsymbol{e}_r \frac{\partial}{\partial r} + \boldsymbol{e}_\theta \frac{1}{r}\frac{\partial}{\partial \theta} + \boldsymbol{e}_\phi \frac{1}{r \sin\theta}\frac{\partial}{\partial \phi}\right) \\ = -i\hbar\left(\boldsymbol{e}_\phi \frac{\partial}{\partial \theta} - \boldsymbol{e}_\theta \frac{1}{\sin\theta}\frac{\partial}{\partial \phi}\right),\end{aligned} \tag{5.131}$$

where the vector products listed in (1.49) were used. The Cartesian components of $\hat{\boldsymbol{L}}$ in spherical coordinates can now be found from (5.131) by using the expression (1.48) for the spherical unit vectors in terms of the Cartesian unit vectors. This gives

$$\begin{aligned}\hat{L}_x &= \boldsymbol{e}_x \cdot \hat{\boldsymbol{L}} \\ &= -i\hbar\left(\boldsymbol{e}_x \cdot \boldsymbol{e}_\phi \frac{\partial}{\partial \theta} - \boldsymbol{e}_x \cdot \boldsymbol{e}_\theta \frac{1}{\sin\theta}\frac{\partial}{\partial \phi}\right) = i\hbar\left(\sin\phi \frac{\partial}{\partial \theta} + \cot\theta \cos\phi \frac{\partial}{\partial \phi}\right),\end{aligned} \tag{5.132}$$

$$\begin{aligned}\hat{L}_y &= \boldsymbol{e}_y \cdot \hat{\boldsymbol{L}} \\ &= -i\hbar\left(\boldsymbol{e}_y \cdot \boldsymbol{e}_\phi \frac{\partial}{\partial \theta} - \boldsymbol{e}_y \cdot \boldsymbol{e}_\theta \frac{1}{\sin\theta}\frac{\partial}{\partial \phi}\right) = -i\hbar\left(\cos\phi \frac{\partial}{\partial \theta} - \cot\theta \sin\phi \frac{\partial}{\partial \phi}\right),\end{aligned} \tag{5.133}$$

$$\hat{L}_z = \boldsymbol{e}_z \cdot \hat{\boldsymbol{L}} = -i\hbar\left(\boldsymbol{e}_z \cdot \boldsymbol{e}_\phi \frac{\partial}{\partial \theta} - \boldsymbol{e}_z \cdot \boldsymbol{e}_\theta \frac{1}{\sin\theta}\frac{\partial}{\partial \phi}\right) = -i\hbar \frac{\partial}{\partial \phi}. \tag{5.134}$$

Since the Hamiltonian (5.113) is independent of θ and ϕ, it follows that each of the individual components of $\hat{\boldsymbol{L}}$ commute with the Hamiltonian and therefore any one of them could be included with $\hat{H}$ in the CSCO. Because of its relatively simple expression in spherical coordinates, it is universal practice to choose $\hat{L}_z$ for inclusion in the CSCO.

Squaring (5.131) and then using the expressions for the spherical unit vectors given by (1.48) shows that $\hat{L}^2$ is given by

$$\begin{aligned}\hat{L}^2 &= -\hbar^2\left(e_\phi\frac{\partial}{\partial\theta} - e_\theta\frac{1}{\sin\theta}\frac{\partial}{\partial\phi}\right)\cdot\left(e_\phi\frac{\partial}{\partial\theta} - e_\theta\frac{1}{\sin\theta}\frac{\partial}{\partial\phi}\right)\\ &= -\hbar^2\left(\frac{\partial^2}{\partial\theta^2} + \left(e_\phi\cdot\frac{\partial e_\phi}{\partial\theta} - \frac{e_\theta}{\sin\theta}\cdot\frac{\partial e_\phi}{\partial\phi}\right)\frac{\partial}{\partial\theta} + \left(\frac{e_\theta}{\sin^2\theta}\cdot\frac{\partial e_\theta}{\partial\phi} - \frac{e_\phi}{\sin\theta}\cdot\frac{\partial e_\theta}{\partial\theta}\right)\frac{\partial}{\partial\phi} + \frac{1}{\sin^2\theta}\frac{\partial^2}{\partial\phi^2}\right)\\ &= -\hbar^2\left(\frac{\partial^2}{\partial\theta^2} + \cot\theta\frac{\partial}{\partial\theta} + \frac{1}{\sin^2\theta}\frac{\partial^2}{\partial\phi^2}\right).\end{aligned} \tag{5.135}$$

Apart from the factor of $-\hbar^2$, result (5.135) is precisely the differential operator appearing in (5.125). As a result, the angular part of the wave function, $Z(\theta, \phi)$, is an eigenfunction of $\hat{L}^2$ since equation (5.125) can be written

$$\hat{L}^2 Z(\theta, \phi) = \lambda\hbar^2 Z(\theta, \phi). \tag{5.136}$$

Because the Hamiltonian (5.113) has no angular dependence, it follows that $[\hat{L}^2, \hat{H}] = 0$. Since (5.135) contains no function of ϕ, it verifies that $[\hat{L}^2, \hat{L}_z] = 0$. Therefore, the set $\{\hat{H}, \hat{L}^2, \hat{L}_z\}$ forms a CSCO for the hydrogen atom. Recalling result (3.194) for commuting observables, this means that the wave function for the hydrogen atom can be prepared as a *simultaneous eigenfunction* of all three of these observables.

Like the momentum operator, the eigenfunctions of $\hat{L}_z$ are not difficult to find since it is a first order derivative. The eigenfunctions of the CSCO chosen for the hydrogen atom must therefore be proportional to $\Phi_m(\phi) = e^{im\phi}$, since it gives

$$\hat{L}_z\,\Phi_m(\phi) = -i\hbar\frac{\partial}{\partial\phi}\Phi_m(\phi) = m\hbar\,\Phi_m(\phi). \tag{5.137}$$

Born's continuity conditions require the function $\Phi_m(\phi)$ to be single valued, so that $\Phi_m(\phi + 2\pi) = \Phi_m(\phi)$. As a result, m must be *an integer* since only then is $e^{i2\pi m} = 1$. These eigenfunctions can be made orthonormal by writing

$$\Phi_m(\phi) = \frac{1}{\sqrt{2\pi}}e^{im\phi}. \tag{5.138}$$

It follows then that

$$\int_0^{2\pi} \mathrm{d}\phi\,\Phi^*_{m'}(\phi)\Phi_m(\phi) = \frac{1}{2\pi}\int_0^{2\pi}\mathrm{d}\phi\, e^{i(m-m')\phi} = \delta_{m'm}, \tag{5.139}$$

showing that the functions $\Phi_m(\phi)$ are orthonormal. It can be shown that they are also complete, allowing a function $f(\phi)$ to be expanded in terms of a Fourier series in the $\Phi_m(\phi)$.

The angular part of the wave function can now be written as a product,

$$Z(\theta, \phi) = \Theta(\theta)\Phi_m(\phi), \tag{5.140}$$

so that the eigenvalue equation (5.125) becomes

$$\left(\frac{\mathrm{d}^2}{\mathrm{d}\theta^2} + \cot\theta\frac{\mathrm{d}}{\mathrm{d}\theta} - \frac{m^2}{\sin^2\theta}\right)\Theta(\theta) = -\lambda\,\Theta(\theta). \tag{5.141}$$

Equation (5.141) can be put into a more commonly analysed form by changing variables from θ to $\omega = \cos\theta$. Using

$$\begin{aligned}
\frac{\mathrm{d}}{\mathrm{d}\theta} &= \frac{\mathrm{d}\omega}{\mathrm{d}\theta}\frac{\mathrm{d}}{\mathrm{d}\omega} = -\sin\theta\,\frac{\mathrm{d}}{\mathrm{d}\omega},\\
\cot\theta\,\frac{\mathrm{d}}{\mathrm{d}\theta} &= -\cos\theta\,\frac{\mathrm{d}}{\mathrm{d}\omega} = -\omega\,\frac{\mathrm{d}}{\mathrm{d}\omega},\\
\frac{\mathrm{d}^2}{\mathrm{d}\theta^2} &= \frac{\mathrm{d}}{\mathrm{d}\theta}\left(-\sin\theta\,\frac{\mathrm{d}}{\mathrm{d}\omega}\right) = -\cos\theta\,\frac{\mathrm{d}}{\mathrm{d}\omega} + \sin^2\theta\,\frac{\mathrm{d}^2}{\mathrm{d}\omega^2} = -\omega\frac{\mathrm{d}}{\mathrm{d}\omega} + (1-\omega^2)\frac{\mathrm{d}^2}{\mathrm{d}\omega^2},
\end{aligned} \tag{5.142}$$

equation (5.141) becomes

$$\left((1-\omega^2)\frac{\mathrm{d}^2}{\mathrm{d}\omega^2} - 2\omega\frac{\mathrm{d}}{\mathrm{d}\omega} - \frac{m^2}{1-\omega^2}\right)\Theta(\omega) = -\lambda\,\Theta(\omega), \tag{5.143}$$

which is known as the *associated Legendre equation.*

5.3.4 Legendre polynomials

The strategy for solving (5.143) consists of first finding the solutions for the case that $m = 0$ and then generalizing them to the case that $m \neq 0$. For the case that $m = 0$ equation (5.143) becomes the *Legendre equation.* The solutions are the *Legendre polynomials*, $P_\ell(\omega)$, where ℓ can be any non-negative integer ($\ell = 0, 1, 2, \ldots$). Like the Hermite polynomials of (5.43), the Legendre polynomials are given by the *Rodriguez formula*,

$$P_\ell(\omega) = \frac{1}{2^\ell \ell!}\frac{\mathrm{d}^\ell}{\mathrm{d}\omega^\ell}(\omega^2-1)^\ell, \tag{5.144}$$

so that $P_0 = 1$, $P_1 = \omega$, and $P_2 = \frac{1}{2}(3\omega^2 - 1)$. These three polynomials are easily shown to satisfy the $m = 0$ version of (5.143) for the respective eigenvalues $\lambda = 0, 2,$ and 6.

Showing that the polynomial defined by (5.144) for a general ℓ satisfies the $m = 0$ version of (5.143) begins by introducing an identity similar to (5.44) used for the Hermite polynomials,

$$\frac{\mathrm{d}^\ell}{\mathrm{d}\omega^\ell}(\omega f(\omega)) = \ell\frac{\mathrm{d}^{\ell-1}}{\mathrm{d}\omega^{\ell-1}}f(\omega) + \omega\frac{\mathrm{d}^\ell}{\mathrm{d}\omega^\ell}f(\omega). \tag{5.145}$$

This identity is proved using induction by first noting that it holds for $\ell = 1$. Differentiating it gives

$$\begin{aligned}
\frac{\mathrm{d}^{\ell+1}}{\mathrm{d}\omega^{\ell+1}}(\omega f(\omega)) &= \ell\frac{\mathrm{d}^\ell}{\mathrm{d}\omega^\ell}f(\omega) + \frac{\mathrm{d}^\ell}{\mathrm{d}\omega^\ell}f(\omega) + \omega\frac{\mathrm{d}^{\ell+1}}{\mathrm{d}\omega^{\ell+1}}f(\omega)\\
&= (\ell+1)\frac{\mathrm{d}^\ell}{\mathrm{d}\omega^\ell}f(\omega) + \omega\frac{\mathrm{d}^{\ell+1}}{\mathrm{d}\omega^{\ell+1}}f(\omega),
\end{aligned} \tag{5.146}$$

which is the identity (5.145) for $\ell + 1$, completing the proof by induction.

Taking the derivative of the Rodriguez formula (5.144) gives

$$\frac{\mathrm{d}}{\mathrm{d}\omega}P_\ell(\omega) = \frac{1}{2^\ell\,\ell!}\frac{\mathrm{d}^{\ell+1}}{\mathrm{d}\omega^{\ell+1}}(\omega^2-1)^\ell = \frac{1}{2^{\ell-1}\,(\ell-1)!}\frac{\mathrm{d}^\ell}{\mathrm{d}\omega^\ell}(\omega(\omega^2-1)^{\ell-1}). \quad (5.147)$$

Using the identity (5.145) gives

$$\begin{aligned}\frac{1}{2^{\ell-1}\,(\ell-1)!}\frac{\mathrm{d}^\ell}{\mathrm{d}\omega^\ell}(\omega(\omega^2-1)^{\ell-1}) &= \frac{\ell}{2^{\ell-1}\,(\ell-1)!}\frac{\mathrm{d}^{\ell-1}}{\mathrm{d}\omega^{\ell-1}}(\omega^2-1)^{\ell-1}\\ &\quad+\frac{\omega}{2^{\ell-1}\,(\ell-1)!}\frac{\mathrm{d}^\ell}{\mathrm{d}\omega^\ell}(\omega^2-1)^{\ell-1}\\ &= \ell P_{\ell-1}(\omega) + \omega\frac{\mathrm{d}}{\mathrm{d}\omega}P_{\ell-1}(\omega).\end{aligned} \quad (5.148)$$

Combining this with (5.147) gives the first Legendre polynomial recurrence relation,

$$\frac{\mathrm{d}P_\ell(\omega)}{\mathrm{d}\omega} - \ell P_{\ell-1}(\omega) - \omega\frac{\mathrm{d}P_{\ell-1}(\omega)}{\mathrm{d}\omega} = 0. \quad (5.149)$$

The second Legendre recurrence relation follows by first proving the identity

$$\omega P_\ell(\omega) - P_{\ell-1}(\omega) = \frac{(\ell+1)}{2^\ell\,\ell!}\frac{\mathrm{d}^{\ell-1}}{\mathrm{d}\omega^{\ell-1}}(\omega^2-1)^\ell. \quad (5.150)$$

The first step in proving (5.150) is obtained by combining the Rodriguez formula (5.144) with (5.145),

$$\begin{aligned}\omega P_\ell(\omega) &= \frac{\omega}{2^\ell\,\ell!}\frac{\mathrm{d}^\ell}{\mathrm{d}\omega^\ell}(\omega^2-1)^\ell\\ &= \frac{1}{2^\ell\,\ell!}\frac{\mathrm{d}^\ell}{\mathrm{d}\omega^\ell}(\omega(\omega^2-1)^\ell) - \frac{\ell}{2^\ell\,\ell!}\frac{\mathrm{d}^{\ell-1}}{\mathrm{d}\omega^{\ell-1}}(\omega^2-1)^\ell\\ &= \frac{1}{2^\ell\,\ell!}\frac{\mathrm{d}^{\ell-1}}{\mathrm{d}\omega^{\ell-1}}((\omega^2-1)^\ell + 2\omega^2\ell(\omega^2-1)^{\ell-1}) - \frac{\ell}{2^\ell\,\ell!}\frac{\mathrm{d}^{\ell-1}}{\mathrm{d}\omega^{\ell-1}}(\omega^2-1)^\ell\\ &= \frac{1}{2^\ell\,\ell!}\frac{\mathrm{d}^{\ell-1}}{\mathrm{d}\omega^{\ell-1}}((1-\ell)(\omega^2-1)^\ell + 2\omega^2\ell(\omega^2-1)^{\ell-1}).\end{aligned} \quad (5.151)$$

Combining (5.151) with the Rodriguez formula for $P_{\ell-1}(\omega)$ gives the desired identity (5.150),

$$\begin{aligned}\omega P_\ell(\omega) - P_{\ell-1}(\omega) &= \frac{1}{2^\ell\,\ell!}\frac{\mathrm{d}^{\ell-1}}{\mathrm{d}\omega^{\ell-1}}((1-\ell)(\omega^2-1)^\ell + 2\omega^2\ell(\omega^2-1)^{\ell-1} - 2\ell(\omega^2-1)^{\ell-1})\\ &= \frac{(\ell+1)}{2^\ell\,\ell!}\frac{\mathrm{d}^{\ell-1}}{\mathrm{d}\omega^{\ell-1}}(\omega^2-1)^\ell.\end{aligned} \quad (5.152)$$

Differentiating both sides of (5.152) and using the Rodriguez formula gives

$$
\begin{aligned}
\frac{\mathrm{d}}{\mathrm{d}\omega}(\omega P_\ell(\omega) - P_{\ell-1}(\omega)) &= P_\ell(\omega) + \omega\frac{\mathrm{d}P_\ell(\omega)}{\mathrm{d}\omega} - \frac{\mathrm{d}P_{\ell-1}(\omega)}{\mathrm{d}\omega} \\
&= \frac{(\ell+1)}{2^\ell\,\ell!}\frac{\mathrm{d}^\ell}{\mathrm{d}\omega^\ell}(\omega^2-1)^\ell = (\ell+1)P_\ell(\omega),
\end{aligned}
\tag{5.153}
$$

which gives the second Legendre recurrence relation,

$$
\omega\frac{\mathrm{d}}{\mathrm{d}\omega}P_\ell(\omega) - \frac{\mathrm{d}}{\mathrm{d}\omega}P_{\ell-1}(\omega) - \ell\, P_\ell(\omega) = 0. \tag{5.154}
$$

The two recurrence relations can now be used to demonstrate that the Rodriguez formula solves the Legendre equation.

Multiplying (5.154) by ω and subtracting it from (5.149) gives

$$
(1-\omega^2)\frac{\mathrm{d}}{\mathrm{d}\omega}P_\ell(\omega) = \ell P_{\ell-1}(\omega) - \ell\omega P_\ell(\omega). \tag{5.155}
$$

Differentiating both sides of (5.155) gives

$$
(1-\omega^2)\frac{\mathrm{d}^2}{\mathrm{d}\omega^2}P_\ell(\omega) - 2\omega\frac{\mathrm{d}}{\mathrm{d}\omega}P_\ell(\omega) = \ell\left(\frac{\mathrm{d}}{\mathrm{d}\omega}P_{\ell-1}(\omega) - \omega\frac{\mathrm{d}}{\mathrm{d}\omega}P_\ell(\omega) - P_\ell(\omega)\right). \tag{5.156}
$$

Adding $\lambda\, P_\ell(\omega) = \ell(\ell+1)\, P_\ell(\omega)$ to both sides of (5.156) gives

$$
(1-\omega^2)\frac{\mathrm{d}^2}{\mathrm{d}\omega^2}P_\ell(\omega) - 2\omega\frac{\mathrm{d}}{\mathrm{d}\omega}P_\ell(\omega) + \ell(\ell+1)P_\ell(\omega) = \ell\left(\frac{\mathrm{d}}{\mathrm{d}\omega}P_{\ell-1}(\omega) - \omega\frac{\mathrm{d}}{\mathrm{d}\omega}P_\ell(\omega) + \ell\, P_\ell(\omega)\right). \tag{5.157}
$$

The right-hand side vanishes since it is the second Legendre recurrence relation (5.154). As a result, the $P_\ell(\omega)$ generated by (5.144) are solutions of the equation

$$
(1-\omega^2)\frac{\mathrm{d}^2}{\mathrm{d}\omega^2}P_\ell(\omega) - 2\omega\frac{\mathrm{d}}{\mathrm{d}\omega}P_\ell(\omega) + \ell(\ell+1)P_\ell(\omega) = 0, \tag{5.158}
$$

which is the Legendre equation (5.143) for the case $m = 0$ and the eigenvalue

$$
\lambda = \ell(\ell+1). \tag{5.159}
$$

In order for the derivatives appearing in the Rodriguez formula (5.144) to be defined the value of ℓ must be an integer such that $\ell \geqslant 0$.

The Legendre polynomials are orthogonal over the interval $(-1, 1)$. This follows from using the definition (5.144) and, assuming $\ell' \geqslant \ell$, integrating by parts ℓ' times to find

$$
\begin{aligned}
\int_{-1}^{1}\mathrm{d}\omega\, P_{\ell'}(\omega)\, P_\ell(\omega) &= \frac{(-1)^{\ell'}}{2^{\ell+\ell'}\ell!\ell'!}\int_{-1}^{1}\mathrm{d}\omega\,(\omega^2-1)^{\ell'}\frac{\mathrm{d}^{\ell+\ell'}}{\mathrm{d}\omega^{\ell+\ell'}}(\omega^2-1)^\ell \\
&= \delta_{\ell'\ell}\frac{(2\ell)!}{2^{2\ell}\ell!^2}\int_0^\pi \mathrm{d}\theta\,\sin^{2\ell+1}(\theta) = \frac{2}{2\ell+1}\delta_{\ell'\ell}.
\end{aligned}
\tag{5.160}
$$

The Kronecker delta appears because $\mathrm{d}^{\ell+\ell'}\omega^{2\ell}/\mathrm{d}\omega^{\ell+\ell'} = 0$ if $\ell' > \ell$. For the case $\ell > \ell'$ the integrations by parts can be reversed, yielding zero again. Only the case $\ell' = \ell$ is nonzero, and that gives the factor $\mathrm{d}^{2\ell}(\omega^2 - 1)^\ell/\mathrm{d}\omega^{2\ell} = (2\ell)!$. Using (5.160) shows that the Legendre polynomials are complete,

$$\sum_{\ell=0}^{\infty} \frac{2\ell + 1}{2} P_\ell(x) P_\ell(y) = \delta(x - y). \tag{5.161}$$

5.3.5 Associated Legendre polynomials

Using the eigenvalue (5.159) in the associated Legendre equation (5.143) shows that for $m \neq 0$ it is given by

$$\left((1 - \omega^2)\frac{\mathrm{d}^2}{\mathrm{d}\omega^2} - 2\omega\frac{\mathrm{d}}{\mathrm{d}\omega} - \frac{m^2}{1 - \omega^2} + \ell(\ell + 1)\right)\Theta(\omega) = 0. \tag{5.162}$$

Finding the solution to (5.162) begins by defining the quantity

$$\pi_\ell^m(\omega) = \frac{\mathrm{d}^{m+\ell}}{\mathrm{d}\omega^{m+\ell}}(\omega^2 - 1)^\ell = \frac{\mathrm{d}^m}{\mathrm{d}\omega^m}\left(\frac{\mathrm{d}^\ell}{\mathrm{d}\omega^\ell}(\omega^2 - 1)^\ell\right) = \frac{\mathrm{d}^m}{\mathrm{d}\omega^m}\pi_\ell^0(\omega). \tag{5.163}$$

The Rodriguez formula (5.144) shows that $\pi_\ell^0(\omega)$ is proportional to the Legendre polynomial $P_\ell(\omega)$, and so it is a solution to the $m = 0$ Legendre equation,

$$(1 - \omega^2)\frac{\mathrm{d}^2}{\mathrm{d}\omega^2}\pi_\ell^0(\omega) - 2\omega\frac{\mathrm{d}}{\mathrm{d}\omega}\pi_\ell^0(\omega) + \ell(\ell + 1)\pi_\ell^0(\omega) = 0. \tag{5.164}$$

Differentiating (5.164) once and using

$$\frac{\mathrm{d}^3}{\mathrm{d}\omega^3}\pi_\ell^0 = \frac{\mathrm{d}^2}{\mathrm{d}\omega^2}\pi_\ell^1, \quad \frac{\mathrm{d}^2}{\mathrm{d}\omega^2}\pi_\ell^0 = \frac{\mathrm{d}}{\mathrm{d}\omega}\pi_\ell^1, \quad \frac{\mathrm{d}}{\mathrm{d}\omega}\pi_\ell^0 = \pi_\ell^1, \tag{5.165}$$

results in (5.164) becoming

$$(1 - \omega^2)\frac{\mathrm{d}^2}{\mathrm{d}\omega^2}\pi_\ell^1(\omega) - 4\omega\frac{\mathrm{d}}{\mathrm{d}\omega}\pi_\ell^1(\omega) + (\ell(\ell + 1) - 2)\pi_\ell^1(\omega) = 0. \tag{5.166}$$

After m differentiations it will now be shown that (5.164) becomes

$$\begin{aligned}&(1 - \omega^2)\frac{\mathrm{d}^2}{\mathrm{d}\omega^2}\pi_\ell^m(\omega) - 2(m + 1)\omega\frac{\mathrm{d}}{\mathrm{d}\omega}\pi_\ell^m(\omega)\\ &+ (\ell(\ell + 1) - m(m + 1))\pi_\ell^m(\omega) = 0.\end{aligned} \tag{5.167}$$

Comparing (5.167) to (5.166) shows that it holds for $m = 1$. Differentiating the general formula (5.167) once more and collecting terms yields

$$\begin{aligned}&(1 - \omega^2)\frac{\mathrm{d}^3}{\mathrm{d}\omega^3}\pi_\ell^m(\omega) - 2(m + 2)\omega\frac{\mathrm{d}^2}{\mathrm{d}\omega^2}\pi_\ell^m(\omega)\\ &+ (\ell(\ell + 1) - (m + 1)(m + 2))\frac{\mathrm{d}}{\mathrm{d}\omega}\pi_\ell^m(\omega) = 0,\end{aligned} \tag{5.168}$$

which immediately yields the $m + 1$ version of (5.167),

$$(1 - \omega^2)\frac{d^2}{d\omega^2}\pi_\ell^{m+1}(\omega) - 2(m + 2)\omega\frac{d}{d\omega}\pi_\ell^{m+1}(\omega) + (\ell(\ell + 1) - (m + 1)(m + 2))\pi_\ell^{m+1}(\omega) = 0. \tag{5.169}$$

The general formula (5.167) therefore holds by induction.

The next step is to define the *associated Legendre polynomials* $P_\ell^m(\omega)$ by

$$\pi_\ell^m(\omega) = (-1)^m 2^\ell \ell!\,(1 - \omega^2)^{-\frac{1}{2}m} P_\ell^m(\omega). \tag{5.170}$$

It then follows that the first derivative of $\pi_\ell^m(\omega)$ satisfies

$$(-1)^m\frac{1}{2^\ell \ell!}\frac{d}{d\omega}\pi_\ell^m(\omega) = m\omega(1 - \omega^2)^{-\frac{1}{2}m-1}P_\ell^m(\omega) + (1 - \omega^2)^{-\frac{1}{2}m}\frac{d}{d\omega}P_\ell^m(\omega), \tag{5.171}$$

and from (5.171) the second derivative is given by

$$(-1)^m\frac{1}{2^\ell \ell!}\frac{d^2}{d\omega^2}\pi_\ell^m(\omega) = \left(\omega^2\, m(m + 2)(1 - \omega^2)^{-\frac{1}{2}m-2} + m(1 - \omega^2)^{-\frac{1}{2}m-1}\right)P_\ell^m(\omega) + 2\,m\omega(1 - \omega^2)^{-\frac{1}{2}m-1}\frac{d}{d\omega}P_\ell^m(\omega) + (1 - \omega^2)^{-\frac{1}{2}m}\frac{d^2}{d\omega^2}P_\ell^m(\omega). \tag{5.172}$$

Inserting (5.170), (5.171), and (5.172) into (5.167) and dropping the common factor of $(-1)^m 2^\ell \ell!(1 - \omega^2)^{-\frac{1}{2}m}$ gives

$$\left((1 - \omega^2)\frac{d^2}{d\omega^2} - 2\omega\frac{d}{d\omega} - \frac{m^2}{1 - \omega^2} + \ell(\ell + 1)\right)P_\ell^m(\omega) = 0, \tag{5.173}$$

showing that the associated Legendre polynomials of (5.170) are solutions to the associated Legendre equation (5.162).

Combining (5.170) with (5.144) shows that the associated Legendre polynomials are derived from the Legendre polynomials,

$$\begin{aligned} P_\ell^m(\omega) &= (-1)^m\frac{1}{2^\ell \ell!}(1 - \omega^2)^{\frac{1}{2}m}\pi_\ell^m(\omega) \\ &= (-1)^m(1 - \omega^2)^{\frac{1}{2}m}\frac{d^m}{d\omega^m}\left(\frac{1}{2^\ell \ell!}\frac{d^\ell}{d\omega^\ell}(1 - \omega^2)^\ell\right) \\ &= (-1)^m(1 - \omega^2)^{\frac{1}{2}m}\frac{d^m}{d\omega^m}P_\ell(\omega). \end{aligned} \tag{5.174}$$

While the integer m is arbitrary in the equation (5.162), it follows that (5.174) gives $P_\ell^m(\omega) = 0$ if $m > \ell$. This occurs since the largest power of ω appearing in (5.174) is $\omega^{2\ell}$, and

$$m > \ell \implies \frac{d^{m+\ell}}{d\omega^{\ell+m}}\,\omega^{2\ell} = 0. \tag{5.175}$$

The integer m can also be negative in the related angular wave function $\Phi_m(\phi)$ since only m^2 enters equation (5.143). However, for the case that m is a negative integer, expression (5.174) breaks down if $\ell + m < 0$. As a result, in order to obtain a nonzero function the integer m must be restricted to a value in the sequence of $2\ell + 1$ integers ranging from $-\ell$ to ℓ, so that m is any integer such that $|m| \leqslant \ell$.

The associated Legendre polynomials are also orthogonal, with the proof following from (5.174) in a manner similar to (5.160). Rather than determining the normalization constant solely for the associated Legendre polynomials, it is more convenient to combine them with the function $\Phi_m(\phi) = e^{im\phi}$.

5.3.6 Spherical harmonics

Combining the associated Legendre polynomial $P_\ell^m(\cos\theta)$ with the eigenfunction $\Phi_m(\phi)$ defined in (5.138) gives the full angular part of the hydrogen wave function, $Z(\theta, \phi)$, which solves (5.136). The resulting angular wave functions are known as the *spherical harmonics*, and are denoted

$$Z(\theta, \phi) = Y_{\ell,m}(\theta, \phi) = C_{\ell m} P_\ell^m(\cos\theta) e^{im\phi}, \tag{5.176}$$

where $C_{\ell m}$ is a normalization constant. By construction, the spherical harmonics are the eigenfunctions of $\hat{L}^2$ and $\hat{L}_z$, and they satisfy

$$\hat{L}^2 Y_{\ell,m}(\theta, \phi) = \ell(\ell+1)\hbar^2\, Y_{\ell,m}(\theta, \phi), \tag{5.177}$$

$$\hat{L}_z Y_{\ell,m}(\theta, \phi) = m\hbar\, Y_{\ell,m}(\theta, \phi). \tag{5.178}$$

The spherical harmonics inherit the orthogonality of $\Phi_m(\phi)$ and $P_\ell(\theta)$ given by (5.139) and (5.160). For the choice

$$C_{\ell m} = (-1)^m \sqrt{\frac{(2\ell+1)}{4\pi}\frac{(\ell-m)!}{(\ell+m)!}}, \tag{5.179}$$

the $Y_{\ell m}(\theta, \phi)$ are orthonormal over the angular sphere, $\mathrm{d}\Omega = \sin\theta\, \mathrm{d}\theta\, \mathrm{d}\phi$, satisfying

$$\begin{aligned}\oint \mathrm{d}\Omega\, Y^*_{\ell,m}(\theta, \phi)\, Y_{\ell',m'}(\theta, \phi) &= \int_0^{2\pi} \mathrm{d}\phi \int_0^{\pi} \mathrm{d}\theta \sin\theta\, Y^*_{\ell,m}(\theta, \phi)\, Y_{\ell',m'}(\theta, \phi) \\ &= \delta_{\ell'\ell}\delta_{m'm}.\end{aligned} \tag{5.180}$$

For example, using (5.174) and (5.179) yields the first three spherical harmonics,

$$Y_{0,0} = \frac{1}{\sqrt{4\pi}}, \quad Y_{1,0} = \sqrt{\frac{3}{4\pi}}\cos\theta, \quad Y_{1,\pm1} = \mp\sqrt{\frac{3}{8\pi}}\sin\theta\, e^{\pm i\phi}. \tag{5.181}$$

These are easily shown to satisfy (5.180).

The two integers ℓ and m are referred to as *angular momentum quantum numbers*, with the restrictions that $\ell \geqslant 0$ and m is chosen so that $|m| \leqslant \ell$. This restriction shows that the maximum value of the z-component of angular momentum, $|m|\hbar = \ell\hbar$, is less than or equal to the magnitude of the total orbital angular momentum

$\sqrt{\ell(\ell+1)}\hbar$. For the choice $\ell = 0$, which also requires $m = 0$, (5.177) and (5.178) show that the hydrogen atom has a quantum mechanical state of zero angular momentum available to it. This result has *no classical counterpart*, since it would correspond to the electron somehow oscillating through the nucleus rather than orbiting it, which is classically impossible for the Coulomb potential.

5.3.7 The radial equation

The eigenvalue $\lambda = \ell(\ell + 1)$ obtained from the angular analysis can now be returned to (5.124), so that the radial eigenvalue equation becomes

$$-\frac{\hbar^2}{2\mu r^2}\frac{\mathrm{d}}{\mathrm{d}r}\left(r^2\frac{\mathrm{d}R(r)}{\mathrm{d}r}\right) + \left(\frac{\ell(\ell+1)\hbar^2}{2\mu r^2} - \frac{q^2}{r}\right)R(r) = E\ R(r). \tag{5.182}$$

In addition to the Coulomb potential, the radial equation (5.182) now has an effective potential,

$$V_\ell(r) = \frac{\ell(\ell+1)\hbar^2}{2\mu r^2}, \tag{5.183}$$

which is known the *centrifugal barrier*. The centrifugal barrier is also present in the classical analysis of radial motion for Kepler orbits.

5.3.8 Bound state solutions

In the classical case, the Kepler equations of motion have solutions for both positive and negative values of E, with the former corresponding to *a particle scattering from an attractive potential* and the latter corresponding to *bound state motion*. As it will turn out, the same is true of (5.182). Since the behavior of the neutral hydrogen *atom* is of interest, the negative energy or bound state solutions to (5.182) will be investigated first, so that the energy eigenvalue is written $E = -|E|$. Defining the dimensionless variable $\rho = 2\alpha r$, where $\alpha = \sqrt{2\mu|E|}/\hbar$, allows (5.182) to be written

$$\frac{1}{\rho^2}\frac{\mathrm{d}}{\mathrm{d}\rho}\left(\rho^2\frac{\mathrm{d}R(\rho)}{\mathrm{d}\rho}\right) + \left(\frac{n}{\rho} - \frac{\ell(\ell+1)}{\rho^2} - \frac{1}{4}\right)R(\rho) = 0, \tag{5.184}$$

where the dimensionless quantity n is given by

$$n = \frac{q^2}{\hbar}\sqrt{\frac{\mu}{2|E|}}. \tag{5.185}$$

In the limit $\rho \to \infty$, where all terms with inverse powers of ρ are suppressed, the radial equation (5.184) simplifies to

$$\left(\frac{\mathrm{d}^2}{\mathrm{d}\rho^2} - \frac{1}{4}\right)R(\rho) \approx 0. \tag{5.186}$$

Since $R(\rho) = e^{-\frac{1}{2}\rho}$ is the solution of (5.186) that is bounded for large ρ, the full solution $R(\rho)$ will be written

$$R(\rho) = \rho^s L(\rho)\, e^{-\frac{1}{2}\rho}. \tag{5.187}$$

Substituting this form into (5.184) gives

$$\left(\rho^2 \frac{d^2}{d\rho^2} + \rho(2(s+1) - \rho)\frac{d}{d\rho} + (n - s - 1)\rho + s(s+1) - \ell(\ell+1)\right)L(\rho) = 0. \tag{5.188}$$

Assuming that $L(\rho)$, $dL(\rho)/d\rho$, and $d^2L(\rho)/d\rho^2$ are finite as $\rho \to 0$, equation (5.188) is satisfied at $\rho = 0$ only if

$$s(s+1) - \ell(\ell+1) = 0. \tag{5.189}$$

This quadratic equation has two solutions, $s = \ell$ and $s = -(\ell + 1)$. Because $\ell + 1$ is a positive integer, choosing $s = -(\ell + 1)$ results in ρ^s not being bounded as $\rho \to 0$. As a result, ρ^s is bounded as $\rho \to 0$ only if the solution $s = \ell$ is chosen. Equation (5.188) then further simplifies to

$$\left(\rho \frac{d^2}{d\rho^2} + (2(\ell+1) - \rho)\frac{d}{d\rho} + (n - \ell - 1)\right)L(\rho) = 0, \tag{5.190}$$

which is referred to as the *associated Laguerre equation.*

The *normalizable* solutions to (5.190) are the *generalized Laguerre polynomials*, $L_k^\ell(\rho)$, which are given by the *Rodriguez formula*

$$L_k^\ell(\rho) = \frac{1}{k!}\rho^{-(2\ell+1)}e^{\rho}\frac{d^k}{d\rho^k}(e^{-\rho}\rho^{k+2\ell+1}). \tag{5.191}$$

In order for the derivative appearing in (5.191) to be defined k must be an integer such that $k \geqslant 0$. In order to understand how these polynomials solve (5.190), it is useful to examine the first three polynomials generated by (5.191), which are given by

$$\begin{aligned} L_0^\ell(\rho) &= 1, \\ L_1^\ell(\rho) &= 2\ell + 2 - \rho, \\ L_2^\ell(\rho) &= \frac{1}{2}\rho^2 - (2\ell+3)\rho + \frac{1}{2}(2\ell+2)(2\ell+3). \end{aligned} \tag{5.192}$$

Each of these functions is easily shown to solve (5.190) if *n is the integer* given by $\ell + 1$, $\ell + 2$, and $\ell + 3$ respectively. For the first three polynomials generated by (5.191) this relationship can be written $n - \ell - 1 = k$.

Demonstrating that (5.191) gives an eigenfunction solution to (5.190) for all integer values of k begins by first using (5.191) to derive the recurrence relation

$$L_{k+1}^\ell(\rho) = \left(\frac{\rho}{k+1}\right)\frac{d}{d\rho}L_k^\ell(\rho) + \left(\frac{2\ell + 2 + k - \rho}{k+1}\right)L_k^\ell(\rho). \tag{5.193}$$

Relation (5.193) is derived by applying $\rho\, \mathrm{d}/\mathrm{d}\rho$ to (5.191) and using the identity (5.145), rewritten as

$$\rho\,\frac{\mathrm{d}^k f(\rho)}{\mathrm{d}\rho^k} = \frac{\mathrm{d}^{k+1}(\rho f(\rho))}{\mathrm{d}\rho^{k+1}} - (k+1)\,\frac{\mathrm{d}^k f(\rho)}{\mathrm{d}\rho^k}. \tag{5.194}$$

The identity (5.194) shows that L_{k+1}^{ℓ} and L_k^{ℓ} appear in (5.193) as a result of the terms generated by applying the derivative to (5.191). An inductive proof showing that (5.191) solves (5.190) starts by assuming that $L_k^{\ell}(\rho)$ satisfies (5.190) for $n - \ell - 1 = k$, which (5.192) shows is true for $k = 0$, 1, and 2. Differentiating (5.193) gives

$$\begin{aligned}\frac{\mathrm{d}L_{k+1}^{\ell}(\rho)}{\mathrm{d}\rho} &= \left(\frac{\rho}{k+1}\right)\frac{\mathrm{d}^2 L_k^{\ell}(\rho)}{\mathrm{d}\rho^2} + \left(\frac{2\ell + 3 + k - \rho}{k+1}\right)\frac{\mathrm{d}L_k^{\ell}(\rho)}{\mathrm{d}\rho} - \frac{1}{k+1}L_k^{\ell}(\rho)\\ &= \frac{\mathrm{d}L_k^{\ell}(\rho)}{\mathrm{d}\rho} + \frac{1}{k+1}\left[\rho\frac{\mathrm{d}^2 L_k^{\ell}(\rho)}{\mathrm{d}\rho^2} + (2\ell + 2 - \rho)\frac{\mathrm{d}L_k^{\ell}(\rho)}{\mathrm{d}\rho} - L_k^{\ell}(\rho)\right]\\ &= \frac{\mathrm{d}L_k^{\ell}(\rho)}{\mathrm{d}\rho} + \left(\frac{\ell - n}{k+1}\right)L_k^{\ell}(\rho) = \frac{\mathrm{d}L_k^{\ell}(\rho)}{\mathrm{d}\rho} - L_k^{\ell}(\rho).\end{aligned} \tag{5.195}$$

Differentiating (5.195) gives

$$\rho\frac{\mathrm{d}^2 L_{k+1}^{\ell}(\rho)}{\mathrm{d}\rho^2} = \rho\frac{\mathrm{d}^2 L_k^{\ell}(\rho)}{\mathrm{d}\rho^2} - \rho\frac{\mathrm{d}L_k^{\ell}(\rho)}{\mathrm{d}\rho}, \tag{5.196}$$

and using (5.195) again gives

$$\begin{aligned}&\rho\frac{\mathrm{d}^2 L_{k+1}^{\ell}(\rho)}{\mathrm{d}\rho^2} + (2\ell + 2 - \rho)\frac{\mathrm{d}L_{k+1}^{\ell}(\rho)}{\mathrm{d}\rho}\\ &= \rho\frac{\mathrm{d}^2 L_k^{\ell}(\rho)}{\mathrm{d}\rho^2} + (2\ell + 2 - \rho)\frac{\mathrm{d}L_k^{\ell}(\rho)}{\mathrm{d}\rho} + kL_k^{\ell}\\ &- \rho\frac{\mathrm{d}L_k^{\ell}(\rho)}{\mathrm{d}\rho} - (2\ell + 2 - k - \rho)L_k^{\ell}\\ &= -\rho\frac{\mathrm{d}L_k^{\ell}(\rho)}{\mathrm{d}\rho} - (2\ell + 2 - k - \rho)L_k^{\ell} = -(k+1)L_{k+1}^{\ell},\end{aligned} \tag{5.197}$$

where the final step in (5.197) used the recurrence relation (5.193). As a result, the L_{k+1}^{ℓ} obtained from (5.191) satisfies (5.190) for $n - \ell - 1 = k + 1$ *as long as* L_k^{ℓ} *satisfies* (5.190) for

$$n = k + \ell + 1, \tag{5.198}$$

completing the inductive proof.

Since $L_0^{\ell} = 1$ satisfies (5.190), it can be used to generate the entire sequence of generalized Laguerre polynomials using (5.193), as is easily verified for L_1^{ℓ} and L_2^{ℓ}.

Since the differential operator appearing in (5.182) is self-adjoint, result (3.168) shows that the radial functions that solve (5.182) will be orthogonal for different values of energy. As an aside, if n is not the integer $\ell + 1 + k$, a solution to (5.190) still exists, but it is an infinite order polynomial. For such a case the radial part of the wave function is unbounded as $\rho \to \infty$ and the wave function is not normalizable.

The minimum values of ℓ and k are both zero, so that the integer $n = k + \ell + 1$ must satisfy $n \geqslant 1$, with its minimum value, $n = 1$, occurring for $\ell = 0$ and $k = 0$. Recalling the definition (5.185) of the variable n, the negative energy levels of the hydrogen atom are given by

$$E_n = -\frac{\mu q^4}{2\hbar^2 n^2}, \qquad n = 1, 2, 3, \dots, \tag{5.199}$$

a result identical to the Bohr energy levels (2.65) discussed earlier, with the mild exception that it is the reduced mass μ appearing in the formula. The wave functions associated with these energy levels are therefore often referred to as *Rydberg states.* Recalling that the dimensionless radial variable ρ is defined as $\rho = 2\alpha r$ shows that α is an inverse length. Using $\alpha = \sqrt{2\mu|E|}/\hbar$, it follows that there are a sequence of possible values for α, given by

$$\alpha_n = \sqrt{\frac{2\mu|E_n|}{\hbar^2}} = \frac{\mu q^2}{n\hbar^2} = \frac{1}{na_0}, \tag{5.200}$$

where $a_0 = \hbar^2/\mu q^2$ is the Bohr radius found in (2.63). It is $\alpha_n = 1/na_0$ that appears in the radial wave function $R_n^\ell(\rho) = R_m^\ell(2\alpha_n r)$.

The full solution to the radial equation (5.182) is therefore

$$R_{n\ell}(r) = C_{n\ell}\,\rho^\ell L_{n-\ell-1}^\ell(\rho)\,e^{-\frac{1}{2}\rho} = C_{n\ell}\,(2\alpha_n r)^\ell L_{n-\ell-1}^\ell(2\alpha_n r)\,e^{-\alpha_n r}, \tag{5.201}$$

where $C_{n\ell}$ is a normalization factor. A valid choice of the *quantum numbers,* comsisting of the *integers* n, ℓ, and m, requires $\ell \geqslant 0$, $|m| \leqslant \ell$, and $n \geqslant \ell + 1$. The hydrogen atom bound states are therefore characterized by the set of quantum numbers $\{n, \ell, m\}$, which characterize the eigenvalues of the CSCO. The full solution to (5.119) associated with one of the possible negative energies of (5.199) is given by the product of the radial and the angular functions,

$$\Psi_{n\ell m}(r, \theta, \phi) = C_{n\ell}\,e^{-\alpha_n r}\,(2\alpha_n r)^\ell\,L_{n-\ell-1}^\ell(2\alpha_n r)\,Y_{\ell m}(\theta, \phi) = R_{n\ell}(r)\,Y_{\ell m}(\theta, \phi). \tag{5.202}$$

It can be shown that the normalization constant $C_{n\ell}$ is given by

$$C_{n\ell} = \left(\frac{2\alpha_n}{(n+\ell)!}\right)^{3/2} \sqrt{\frac{(n-\ell-1)!}{2n}}, \tag{5.203}$$

where a general proof is given in the references. The choice of (5.203) gives the orthonormality relation

$$\int_0^\infty \mathrm{d}r\, r^2 \int_0^\pi \mathrm{d}\theta\,\sin\theta \int_0^{2\pi} \mathrm{d}\phi\,\Psi_{n'\ell'm'}^*(r, \theta, \phi)\,\Psi_{n\ell m}(r, \theta, \phi) = \delta_{n'n}\delta_{\ell'\ell}\delta_{m'm}, \tag{5.204}$$

which is easily verified using the simple cases given by (5.181) and (5.192). Since the negative energies of (5.199) are associated with the bound states of hydrogen, any possible state of the neutral hydrogen atom can be written as a linear superposition of the basis wave functions (5.202).

It is important to note that the bound state energy levels of the hydrogen atom are degenerate since multiple values of ℓ and m are possible for a given value of n, with the exception of $n = 1$. For $n = 2$, it is possible to choose $\ell = 0$ or $\ell = 1$. The choice of $\ell = 1$ allows three possible values for m, given by -1, 0, and 1. The $n = 2$ energy level therefore has a fourfold degeneracy. In general, for a given n the quantum number ℓ has n possible values from 0 to $n - 1$, and each choice of ℓ has $2\ell + 1$ possible choices for m. The total degeneracy of the nth energy level is therefore

$$\sum_{\ell=0}^{n-1}(2\ell + 1) = n^2. \tag{5.205}$$

It will be shown in chapter 7 that the spin angular momentum of the electron will double this degeneracy.

5.3.9 A first look at quantum mechanical scattering

The set of bound state wave functions given by (5.202) are not the only possible solutions to the radial equation (5.182). As mentioned earlier, it is possible to find solutions corresponding to positive values of E. Such a solution is interpreted as a scattering wave function since a classical particle with positive energy can escape to infinity, but with a deflection induced by the Coulomb potential. Since the energy eigenvalue does not appear in the angular eigenvalue equation (5.125), the spherical harmonics $Y_{\ell m}(\theta, \phi)$ remain the angular eigenfunctions. Assuming that $E \geqslant 0$ in (5.182) and defining $\alpha = \sqrt{2\mu E}/\hbar$, a repetition of the steps that were taken to obtain (5.184) yields the radial equation for positive energies,

$$\frac{1}{\rho^2}\frac{\mathrm{d}}{\mathrm{d}\rho}\left(\rho^2\frac{\mathrm{d}R(\rho)}{\mathrm{d}\rho}\right) + \left(\frac{n}{\rho} - \frac{\ell(\ell + 1)}{\rho^2} + \frac{1}{4}\right)R(\rho) = 0, \tag{5.206}$$

where once again $\rho = 2\alpha r$ and $n = (q^2/\hbar)\sqrt{\mu/2E}$. A scattering solution to (5.206) should represent the positive energy particle escaping to arbitrarily large distances. Assuming that $R(\rho)$ and its derivative are bounded for $\rho \to \infty$, the eigenvalue equation (5.206) reduces in that limit to

$$\left(\frac{\mathrm{d}^2}{\mathrm{d}\rho^2} + \frac{1}{4}\right)R(\rho) \approx 0, \tag{5.207}$$

and this has oscillatory solutions of the form

$$\lim_{\rho\to\infty} R(\rho) = Ce^{\frac{1}{2}i\rho} = Ce^{i\alpha r}. \tag{5.208}$$

At large distances the Coulomb potential tends to zero and it is *arguable* that the behavior of the electron should be like that of a free particle. The energy eigenvalue E is therefore parameterized in terms of the continuous free particle energy expression, $E = p^2/2\mu$. Using this parameterization gives

$$\alpha = \frac{p}{\hbar}, \quad n = \frac{\mu q^2}{p\hbar}, \tag{5.209}$$

both of which have a continuum of positive and negative values determined by the value of the continuous momentum variable p. For such a choice of parameterization, the solution of (5.207) becomes the familiar form

$$R(r) = Ce^{ipr/\hbar}. \tag{5.210}$$

Noting that the momentum in spherical coordinates is found from the spherical coordinate gradient (1.51),

$$\hat{\boldsymbol{p}} = -i\hbar\nabla = -i\hbar\left(\boldsymbol{e}_r\frac{\partial}{\partial r} + \boldsymbol{e}_\theta\frac{1}{r}\frac{\partial}{\partial\theta} + \boldsymbol{e}_\phi\frac{1}{r\sin\theta}\frac{\partial}{\partial\phi}\right), \tag{5.211}$$

so that the radial component of the momentum is given by

$$\hat{p}_r = -i\hbar\frac{\partial}{\partial r}. \tag{5.212}$$

The function (5.210) is therefore an eigenfunction of $\hat{p}_r$,

$$\hat{p}_r R(r) = -i\hbar\frac{\partial}{\partial r}\, Ce^{ipr/\hbar} = p\, R(r). \tag{5.213}$$

However, while (5.210) is an eigenfunction of the radial momentum, it does not correspond to a solution of the Schrödinger equation for the case that $V(r) = 0$ and $\ell = 0$. This follows from

$$-\frac{\hbar^2}{2\mu}\nabla^2 R(r) = -\frac{\hbar^2}{2\mu}\left(\frac{\mathrm{d}^2}{\mathrm{d}r^2} + \frac{2}{r}\frac{\mathrm{d}}{\mathrm{d}r}\right)Ce^{ipr/\hbar} = \left(\frac{p^2}{2\mu} - \frac{i\hbar p}{\mu r}\right)Ce^{ipr/\hbar}. \tag{5.214}$$

Instead, a useful solution of the $\ell = 0$ version of the Schrödinger equation in spherical coordinates is given by

$$R(r) = \frac{C}{r}e^{ipr/\hbar}, \tag{5.215}$$

which satisfies

$$-\frac{\hbar^2}{2\mu}\nabla^2 R(r) = -\frac{\hbar^2}{2\mu}\left(\frac{\mathrm{d}^2}{\mathrm{d}r^2} + \frac{2}{r}\frac{\mathrm{d}}{\mathrm{d}r}\right)\left(\frac{C}{r}e^{ipr/\hbar}\right) = \frac{p^2}{2\mu}R(r). \tag{5.216}$$

It therefore corresponds to a particle with the usual Newtonian energy $p^2/2\mu$ and a spherically symmetric wave function. The solution (5.215) is known as a *spherical wave*. It will be modified in chapter 9 to include an angular deflection, yielding the critical component to characterize scattering processes. Normalizing (5.215) is similar to the original free particle energy eigenfunction (3.127). This requires normalizing the wave function in the sphere of radius a, so that

$$\int_V \mathrm{d}V\,|R(r)|^2 = 4\pi|C|^2\int_0^a \mathrm{d}r = 4\pi|C|^2 a = 1 \implies C = \frac{1}{\sqrt{4\pi a}}, \tag{5.217}$$

with the understanding that the limit $a \to \infty$ is to be taken.

It is possible to use the spherical wave solution (5.215) as a solution to the radial equation (5.206) for the case that the momentum p becomes large, so that $n = \mu K q^2/p\hbar \to 0$. Choosing $\ell = 0$ in (5.223), corresponding to no centrifugal barrier, and suppressing the n/ρ term for large p, a solution to (5.206) is given by the normalized spherical wave

$$R(r) = \frac{1}{\sqrt{4\pi a}}\frac{1}{r}e^{ipr/\hbar}, \tag{5.218}$$

with the understanding that $a \to \infty$. It is important to note that (5.218) is *not* an eigenfunction of the radial component of momentum, $\hat{p}_r = -i\hbar\,\partial/\partial r$, since

$$\hat{p}_r\, R(r) = \left(p + \frac{i\hbar}{r}\right)R(r). \tag{5.219}$$

The appearance of complex numbers in (5.219) is a direct result of the fact that $\hat{p}_r$ is *not* a Hermitian operator. This is obvious in spherical coordinates, where the volume element $\mathrm{d}V$ contains the factor $r^2\,\mathrm{d}r$, so that integrating by parts results in

$$\begin{aligned}\int_V \mathrm{d}V\,\Psi^*(\boldsymbol{r})\left(\hat{p}_r\Psi(\boldsymbol{r})\right) &= \int_V \mathrm{d}V\left[\left(\hat{p}_r^* + \frac{2i\hbar}{r}\right)\Psi^*(\boldsymbol{r})\right]\Psi(\boldsymbol{r}) \\ &\neq \int_V \mathrm{d}V\,\left(\hat{p}_r\Psi(\boldsymbol{r})\right)^*\Psi(\boldsymbol{r}),\end{aligned} \tag{5.220}$$

where it was assumed that

$$\lim_{r\to\infty} r^2\,|\Psi(\boldsymbol{r})|^2 - \lim_{r\to 0} r^2|\Psi(\boldsymbol{r})|^2 = 0, \tag{5.221}$$

as it does for the case of (5.218). However, as $r \to \infty$ the spherical wave (5.218) tends to an eigenfunction of $\hat{p}_r$ since $i\hbar/r \to 0$ in both (5.214) and (5.220). These results are consistent with the interpretation that $\ell = 0$ and large p corresponds to a fast moving free electron passing directly through the position of the positive charge, becoming asymptotically free as $r \to \infty$ with the energy $E = p^2/2\mu$. The spherical wave therefore corresponds to the outgoing wave function of the electron, describing a particle scattered uniformly into all possible angles. This will form the basis of scattered wave functions in chapter 9.

The general solutions to (5.206) for arbitrary n and ℓ provide a basis for analysing the quantum mechanical scattering of a free electron from a positive charge. This can be briefly sketched by borrowing the $1/r$ dependence of the spherical wave, so that the radial function appearing in (5.206) is rewritten as

$$R(\rho) = \frac{1}{\rho}F(\rho). \tag{5.222}$$

Inserting this form into (5.206) gives

$$\frac{d^2}{d\rho^2}F(\rho) + \left(\frac{n}{\rho} - \frac{\ell(\ell+1)}{\rho^2} + \frac{1}{4}\right)F(\rho) = 0. \tag{5.223}$$

Changing variables by setting $\rho = 2x$ and defining $n = -\eta$ places (5.223) into the standard form,

$$\frac{d^2}{dx^2}F(x) + \left(1 - \frac{2\eta}{x} - \frac{\ell(\ell+1)}{x^2}\right)F(x) = 0, \tag{5.224}$$

which is known as the *Coulomb wave equation.* Denoting $C_{\eta\ell}$ as the normalization constant for given values of η and ℓ, the regular solution of (5.224) is the power series expression

$$F_{\eta\ell}(x) = C_{\eta\ell}\sum_{k=\ell+1}^{\infty} A_k^{\ell}(\eta)\, x^k \equiv C_{\eta\ell}\, x^{\ell+1}e^{-ix}M(\ell+1-i\eta, 2\ell+2, 2ix), \tag{5.225}$$

where $M(\ell + 1 - i\eta, 2\ell + 2, 2ix)$ is known as a *confluent hypergeometric function.*

The A_k^{ℓ} appearing in the power series (5.225) are determined by substituting it into (5.224) and setting the coefficients of each power of x equal to zero. The coefficient of $x^{\ell-1}$ vanishes automatically by virtue of the lowest power of x having been chosen to be $x^{\ell+1}$. This allows $A_{\ell+1}^{\ell}(\eta)$ to be set to unity. The coefficient of x^{ℓ} then gives

$$A_{\ell+2}^{\ell}(\eta) = \frac{\eta}{\ell+1}A_{\ell+1}^{\ell}(\eta) = \frac{\eta}{\ell+1}. \tag{5.226}$$

Setting the coefficient of higher powers of x to zero gives a relation between three of the $A_{\ell+k}^{\ell}(\eta)$. This follows from examining the case where the integer $k > 0$, so that the coefficient of the power of $x^{\ell+k}$ is the sum of the four terms generated by the differential equation (5.225), given sequentially by

$$\begin{aligned}(\ell + k + 1)(\ell + k + 2)A_{\ell+k+2}^{\ell}(\eta) + A_{\ell+k}^{\ell}(\eta)\\ - 2\eta A_{\ell+k+1}^{\ell}(\eta) - \ell(\ell+1)A_{\ell+k+2}^{\ell}(\eta) = 0,\end{aligned} \tag{5.227}$$

These terms combine into a recurrence relation,

$$A_{\ell+k}^{\ell}(\eta) + (k+1)(2\ell+2+k)A_{\ell+k+2}^{\ell}(\eta) - 2\eta A_{\ell+1+k}^{\ell} = 0. \tag{5.228}$$

Since $A_{\ell+1}^{\ell}(\eta)$ and $A_{\ell+2}^{\ell}(\eta)$ are known, the recurrence relation (5.228) uniquely generates all higher order coefficients in the power series (5.225). A much more complete presentation of the numerous properties and behavior of the confluent hypergeometric function generated by this power series is available in the references.

The solution of the Coulomb potential stands as one of the great early triumphs of formal quantum mechanics, and its extension to arbitrary atoms created the subdiscipline of atomic physics. Combining the bound state results (5.199) and (5.202) with the concept of electron spin, to be discussed in chapter 7, provides an understanding of how the atom's quantum properties determine the periodic table of the elements.

5.4 A charged particle in a uniform and constant magnetic field

The interaction of charged particles with an externally applied electromagnetic field was an early determining factor in the formulation of quantum mechanics. It remains a source of insight into quantum behavior.

5.4.1 The Hamiltonian and gauge invariance

The classical Hamiltonian (1.108) for a particle with the charge q and mass m in the presence of an electromagnetic field was derived from the Lagrangian (1.102) and given by

$$H = \frac{\boldsymbol{p}^2}{2m} - \frac{q}{mc}\boldsymbol{p} \cdot \mathbf{A}(\boldsymbol{x}, t) + \frac{q^2}{2\,mc^2}\mathbf{A}(\boldsymbol{x}, t) \cdot \mathbf{A}(\boldsymbol{x}, t) + q\,\varphi(\boldsymbol{x}, t), \qquad (5.229)$$

where $\mathbf{A}$ and φ are the electromagnetic vector and scalar potentials. Making the replacement $\boldsymbol{p} \to -i\hbar\nabla$ gives the Schrödinger equation that governs the system,

$$\hat{H}\Psi(\boldsymbol{x}, t) = \left[\frac{1}{2m}\left(-i\hbar\nabla - \frac{q}{c}\mathbf{A}(\boldsymbol{x}, t)\right)^2 + q\,\varphi(\boldsymbol{x}, t)\right]\Psi(\boldsymbol{x}, t) = i\hbar\frac{\partial}{\partial t}\Psi(\boldsymbol{x}, t). \qquad (5.230)$$

Solving the Schrödinger equation (5.230) provides the wave function of an electric charge in the presence of an electromagnetic field. The neutral hydrogen atom analysed in the previous section corresponds to choosing φ to be the Coulomb potential and setting $\mathbf{A}$ to zero.

Formulating (5.230) has ignored any effect that the charged particle may have on the electromagnetic field. Instead, the electromagnetic fields are treated as *external potentials* that remain unchanged by the state of the particle. It also ignores the photon content of the electromagnetic field unless there is a frequency ω present in the potentials that can be associated with photon energies through the Einstein relation $\omega = E/\hbar$. Such an approach, referred to as the *semiclassical treatment of radiation*, therefore has drawbacks, but it is expected to give the general features of the quantum mechanical behavior of the particle in the presence of an electromagnetic field. If $\mathbf{A}$ or φ are time dependent, the wave function appearing in (5.230) will not be separable between $\boldsymbol{x}$ and t as in (4.46). This is consistent with the fact that a time-dependent electromagnetic field corresponds to an explicitly time-dependent Hamiltonian for the particle. For such a case, the right-hand side of Ehrenfest's energy theorem (4.32) no longer vanishes, and the expectation value for the

Hamiltonian is no longer constant since the energy of the particle is changing as it interacts with the external field.

It is important to note that equation (5.230) should be invariant under the gauge transformations (1.66) for the potentials. However, in order for (5.230) to be gauge invariant, it is necessary to also transform the wave function according to

$$\Psi(\boldsymbol{x}, t) \to \Psi'(\boldsymbol{x}, t) = e^{iq\Lambda(x,\, t)/\hbar c}\, \Psi(\boldsymbol{x}, t). \tag{5.231}$$

By virtue of

$$\begin{aligned} -i\hbar\nabla\Psi'(\boldsymbol{x}, t) &= e^{iq\Lambda(x,\, t)/\hbar c}\left(-i\hbar\nabla + \frac{q}{c}\nabla\Lambda(\boldsymbol{x}, t)\right)\Psi(\boldsymbol{x}, t), \\ i\hbar\frac{\partial}{\partial t}\Psi'(\boldsymbol{x}, t) &= e^{iq\Lambda(x,\, t)/\hbar c}\left(i\hbar\frac{\partial}{\partial t} - \frac{q}{c}\frac{\partial}{\partial t}\Lambda(\boldsymbol{x}, t)\right)\Psi(\boldsymbol{x}, t), \end{aligned} \tag{5.232}$$

it follows that

$$\begin{aligned} &\left[\frac{1}{2m}\left(-i\hbar\nabla - \frac{q}{c}\mathbf{A}'(\boldsymbol{x}, t)\right)^2 + q\,\varphi'(\boldsymbol{x}, t) - i\hbar\frac{\partial}{\partial t}\right]\Psi'(\boldsymbol{x}, t) = \\ &e^{iq\Lambda(x,\, t)/\hbar c}\left[\frac{1}{2m}\left(-i\hbar\nabla - \frac{q}{c}\mathbf{A}(\boldsymbol{x}, t)\right)^2 + q\,\varphi(\boldsymbol{x}, t) - i\hbar\frac{\partial}{\partial t}\right]\Psi(\boldsymbol{x}, t) = 0. \end{aligned} \tag{5.233}$$

The wave function $\Psi(\boldsymbol{x}, t)$ is therefore not invariant under a gauge transformation, just as it is not invariant under the Galilean transformation of (3.114). However, the probability density $\mathcal{P}(\boldsymbol{x}, t) = \Psi^*(\boldsymbol{x}, t)\Psi(\boldsymbol{x}, t)$ is invariant under a gauge transformation, just as it is in (3.115) for a Galilean transformation. This is further indication that the wave function is not directly observable, while the probability density is.

5.4.2 The Schrödinger equation for a constant magnetic field

A very useful and fully solvable case for (5.230) is that of a uniform and constant magnetic field, assumed to be oriented in the z-direction, given by

$$\mathbf{B} = \mathrm{B}_o\mathbf{e}_z, \tag{5.234}$$

where the $\boldsymbol{e}_i$ denote the Cartesian unit vectors. Such a magnetic field corresponds to a scalar potential given by $\varphi(\boldsymbol{x}, t) = 0$, while the static vector potential is given by

$$\mathbf{A}(\boldsymbol{x}) = \mathrm{A}_x(\boldsymbol{x})\,\mathbf{e}_x + \mathrm{A}_y(\boldsymbol{x})\,\mathbf{e}_y = -\frac{1}{2}y\,\mathrm{B}_o\,\mathbf{e}_x + \frac{1}{2}x\,\mathrm{B}_o\,\mathbf{e}_y. \tag{5.235}$$

Using (1.65) shows that the vector potential of (5.235) gives

$$\mathrm{B}_z = (\nabla \times \mathbf{A})_z = \frac{\partial \mathrm{A}_y}{\partial x} - \frac{\partial \mathrm{A}_x}{\partial y} = \mathrm{B}_o, \tag{5.236}$$

while the other two components vanish. In addition, it also satisfies the required Coulomb condition

$$\nabla \cdot \mathbf{A}(\boldsymbol{x}) = 0. \tag{5.237}$$

Using this form for $\mathbf{A}(\boldsymbol{x})$ gives

$$\mathbf{A}(\boldsymbol{x}) \cdot \mathbf{A}(\boldsymbol{x}) = \frac{1}{4}\mathrm{B}_o^2(x^2 + y^2), \tag{5.238}$$

as well as

$$\mathbf{A}(\boldsymbol{x}) \cdot \hat{\boldsymbol{p}} = \frac{1}{2}\mathrm{B}_o\,(x\hat{p}_y - y\hat{p}_x) = \frac{1}{2}\mathrm{B}_o\,\hat{L}_z, \tag{5.239}$$

where $\hat{L}_z$ is the z-component of the angular momentum defined in (1.30) and found in (5.127). Using these results and the definition of ω allows the Schrödinger equation (5.230) to be written

$$\begin{aligned}\hat{H}\Psi(\boldsymbol{x}, t) &= \left(-\frac{\hbar^2}{2m}\nabla^2 + \frac{1}{8}m\left(\frac{qB_o}{mc}\right)^2(x^2 + y^2) - \frac{1}{2}\frac{qB_o}{mc}\hat{L}_z\right)\Psi(\boldsymbol{x}, t)\\ &= i\hbar\frac{\partial}{\partial t}\Psi(\boldsymbol{x}, t).\end{aligned} \tag{5.240}$$

It is convenient to define the *cyclotron frequency*

$$\omega = \frac{qB_o}{mc}. \tag{5.241}$$

The cyclotron frequency is the same as the angular velocity $\omega = v/R$ of the charge obtained from elementary physics by equating the Lorentz force $qv\mathrm{B}_o/c$ with the centripetal force mv^2/R required for circular motion of radius R. If q is positive then ω is positive and the circular motion is *clockwise or left-handed*, while if q is negative then ω is negative and the circular motion is *counterclockwise or right-handed.* The left-hand side of (5.240) is time-independent, so the separation of variables (4.46) can be performed. This results in

$$\Psi(\boldsymbol{x}, t) = \Psi(\boldsymbol{x})e^{-iEt/\hbar}, \tag{5.242}$$

and (5.240) becomes the energy eigenvalue equation $\hat{H}\Psi(\boldsymbol{x}) = E\Psi(\boldsymbol{x})$. Since $\hat{H}$ is Hermitian, the eigenvalues E will be real.

5.4.3 Classical behavior

Before solving (5.240) for the quantum mechanical behavior of the charged particle, it is useful to review the properties of the classical motion. Using (5.235) and $\varphi(\boldsymbol{x}, t) = 0$ in the classical Lagrangian (1.102) gives

$$\mathcal{L} = \frac{1}{2}m\dot{\boldsymbol{x}}^2 + \frac{q}{c}\mathbf{A}(\boldsymbol{x}, t) \cdot \dot{\boldsymbol{x}} = \frac{1}{2}m(\dot{x}^2 + \dot{y}^2 + \dot{z}^2) + \frac{1}{2}m\omega(x\dot{y} - y\dot{x}). \tag{5.243}$$

The Euler–Lagrange equations (1.14) give

$$\frac{\partial \mathcal{L}}{\partial x_i} - \frac{\mathrm{d}}{\mathrm{d}t}\frac{\partial \mathcal{L}}{\partial \dot{x}_i} = 0 \implies \ddot{z} = 0, \quad \ddot{x} - \omega\dot{y} = 0, \quad \ddot{y} + \omega\dot{x} = 0, \tag{5.244}$$

where ω is given by (5.241) and the notation of (1.1) is in use. The motion in the z-direction is that of a free particle, so that

$$z = v_z t + z_o. \tag{5.245}$$

A particularly simple solution for x and y occurs for the initial conditions $x(0) = R$ and $y(0) = 0$, so that the last two equations of (5.244) are solved by

$$x = R\cos(\omega t), \quad y = -R\sin(\omega t). \tag{5.246}$$

It will be useful in the quantum mechanical analysis to note that each component resembles classical harmonic oscillation with the angular frequency ω. These two solutions combine to give $x^2 + y^2 = R^2$, and therefore describe clockwise circular motion in the x–y plane with the angular velocity ω and the tangential velocity $v = \omega R$. This is consistent with a *positive charge* undergoing a centripetal acceleration due to the Lorentz force. It is important to note that counterclockwise rotation for a positive charge is not a solution to (5.243). The situation is reversed if the charge is negative, since the change of signs in q results in $\omega \to -\omega$, and the solutions of (5.246) transform into counterclockwise rotation. Combining this result with the uniform motion in the z-direction results in a helical or spiral motion in the z-direction for either sign of the charge.

The canonical momenta are given by

$$\begin{aligned}
p_x &= \frac{\partial \mathcal{L}}{\partial \dot{x}} = m\dot{x} - \frac{1}{2}m\omega y \implies \dot{x} = \frac{p_x}{m} + \frac{1}{2}\omega y, \\
p_y &= \frac{\partial \mathcal{L}}{\partial \dot{y}} = m\dot{y} + \frac{1}{2}m\omega x \implies \dot{y} = \frac{p_y}{m} - \frac{1}{2}\omega x, \\
p_z &= \frac{\partial \mathcal{L}}{\partial \dot{z}} = m\dot{z} \implies \dot{z} = \frac{p_z}{m}.
\end{aligned} \tag{5.247}$$

Using the results of (5.247) in the definition of the Hamiltonian gives

$$\begin{aligned}
H &= p_x\dot{x} + p_y\dot{y} + p_z\dot{z} - \mathcal{L}(\boldsymbol{x}, \dot{\boldsymbol{x}}) \\
&= \frac{p_x^2}{2m} + \frac{p_y^2}{2m} + \frac{p_z^2}{2m} + \frac{1}{8}m\omega^2(x^2 + y^2) - \frac{1}{2}\omega L_z,
\end{aligned} \tag{5.248}$$

where $L_z = xp_y - yp_x$. It is important to note that the canonical momentum of the particle given by (5.247) is not the elementary result $\boldsymbol{p} = m\boldsymbol{v}$, but also includes a contribution from its interaction with the vector potential of the magnetic field. It is important to remember that, in the process of quantizing the system, it is the canonical momentum $\boldsymbol{p}$ that is replaced by $-i\hbar\nabla$, rather than $m\boldsymbol{v}$.

Evaluating the classical energy begins by using the two solutions (5.246) in (5.247) to find

$$
\begin{aligned}
p_x &= \frac{\partial \mathcal{L}}{\partial \dot{x}} = m\dot{x} - \frac{qB_o}{2c}y = -\frac{1}{2}m\omega R\sin(\omega t), \\
p_y &= \frac{\partial \mathcal{L}}{\partial \dot{y}} = m\dot{y} + \frac{qB_o}{2c}x = -\frac{1}{2}m\omega R\cos(\omega t).
\end{aligned}
\tag{5.249}
$$

For clockwise circular motion it follows that the angular momentum of the system is given by

$$
L_z = xp_y - yp_x = -\frac{1}{2}m\omega R^2\cos^2(\omega t) - \frac{1}{2}m\omega R^2\sin^2(\omega t) = -\frac{1}{2}m\omega R^2. \tag{5.250}
$$

Combining this with (5.246) and (5.249) to evaluate the Hamiltonian gives the total classical energy

$$
H = \frac{1}{2m}(p_x^2 + p_y^2 + p_z^2) + \frac{1}{8}m\omega^2(x^2 + y^2) - \frac{1}{2}\omega L_z = \frac{p_z^2}{2m} + \frac{1}{2}mR^2\omega^2. \tag{5.251}
$$

The second term is simply the kinetic energy of the charge as it moves in a circle of radius R with the velocity $v = \omega R$.

Using (5.241) and (5.250) enables the orbital angular momentum energy term in (5.251) to be written

$$
U_d = -\frac{1}{2}\omega L_z = -\frac{qB_o}{2\,mc}L_z = -\frac{q}{2mc}\boldsymbol{B}_o \cdot \boldsymbol{L} \equiv -\boldsymbol{M} \cdot \boldsymbol{B}_o, \tag{5.252}
$$

where the quantity $\boldsymbol{M}$ is known as the *magnetic dipole moment* of the circular motion,

$$
\boldsymbol{M} = \frac{q}{2mc}\boldsymbol{L}. \tag{5.253}
$$

Since the orbital angular momentum is quantized in the case of an atom, it is expected that a quantum mechanical treatment will yield $L_z = n\hbar$, where n is an integer such that $n \leqslant 0$ to reflect the clockwise motion. The presence of simple harmonic motion in two directions will also contribute factors of $n\hbar\omega$. For such a case, the energy eigenvalues are expected to take the form $E \approx |n|\hbar\omega$, up to a ground state energy.

Apart from the free motion in the z-direction, the *classical Hamiltonian* of (5.251) is the energy due to the magnetic dipole moment $\boldsymbol{M}$ created by the circular motion. This formula has its origin in the behavior of current loops in a magnetic field, where the force creating a torque on the current loop results from the dipole energy (5.252). For the case of a clockwise current I circulating in a loop of area A, the Lorentz force law can be used to show that M_z is defined as

$$
M_z = -\frac{IA}{c}. \tag{5.254}
$$

For the case of a charge q moving in a circle of radius R, the area of the current loop is $A = \pi R^2$ and the electric current I of the rotating charge is

$$I = \frac{\Delta q}{\Delta t} = \frac{q}{(2\pi R/v)} = \frac{q}{(2\pi R/\omega R)} = \frac{\omega q}{2\pi}. \tag{5.255}$$

Using (5.250) shows that the associated magnetic dipole moment is given by

$$M_z = -\frac{IA}{c} = -\frac{\omega q}{2\pi}\frac{\pi R^2}{c} = -\frac{\omega q R^2}{2c} = \frac{q}{2mc}L_z, \tag{5.256}$$

which is precisely result (5.253).

5.4.4 Quantum behavior

In order to determine the quantum mechanical behavior, it is important to first identify the CSCO associated with the physical situation. Doing so will determine what eigenvalues and quantum numbers are associated with the spatial part of the wave function $\Psi(\boldsymbol{x})$. Clearly, the CSCO must contain the Hamiltonian $\hat{H}$ as a result of the separation of variables. Because L_z was critical to the classical solution, it is noted that

$$\begin{aligned}
[\hat{H}, \hat{L}_z] &= \frac{1}{2m}[(\hat{p}_x^2 + \hat{p}_y^2), (\hat{x}\hat{p}_y - \hat{y}\hat{p}_x)] + \frac{1}{8}m\omega^2[(\hat{x}^2 + \hat{y}^2), (\hat{x}\hat{p}_y - \hat{y}\hat{p}_x)] - \frac{1}{2}\omega[\hat{L}_z, \hat{L}_z] \\
&= -\frac{i\hbar}{m}(\hat{p}_x\hat{p}_y - \hat{p}_y\hat{p}_x) + \frac{1}{4}i\hbar m\omega^2(\hat{y}\hat{x} - \hat{x}\hat{y}) = 0, \\
[\hat{H}, \hat{p}_z] &= \frac{1}{2m}\left[\hat{p}^2, \hat{p}_z\right] + \frac{1}{8}m\omega^2[(\hat{x}^2 + \hat{y}^2), \hat{p}_z] - \frac{1}{2}\omega[\hat{L}_z, \hat{p}_z] = 0.
\end{aligned} \tag{5.257}$$

However, due to the presence of the factor $(\hat{x}^2 + \hat{y}^2)$ in $\hat{H}$ neither $\hat{L}_x^2$ nor $\hat{L}_y^2$ commutes with $\hat{H}$. As a result, $[\hat{L}^2, \hat{H}] \neq 0$. Unlike the hydrogen atom, $\hat{L}^2$ cannot be part of the CSCO that includes the Hamiltonian $\hat{H}$. The CSCO is therefore the set $\{\hat{H}, \hat{p}_z, \hat{L}_z\}$, and the spatial part of the wave function $\Psi(\boldsymbol{x})$ can be a simultaneous eigenfunction of all three of these observables.

The next step is to perform a second separation of variables, $\Psi(\boldsymbol{x}) = \psi(x, y)Z(z)$. Since $\hat{L}_z$ acts on the x and y coordinates, dividing (5.240) by $\Psi(\boldsymbol{x})$ separates it into

$$\frac{1}{\psi(x, y)}\left(-\frac{\hbar^2}{2m}\left(\frac{\partial^2}{\partial x^2} + \frac{\partial^2}{\partial y^2}\right) + \frac{1}{8}m\omega^2(x^2 + y^2) - \frac{1}{2}\omega\hat{L}_z\right)\psi(x, y) - \frac{1}{Z(z)}\frac{\hbar^2}{2m}\frac{\partial^2}{\partial z^2}Z(z) = E. \tag{5.258}$$

Because x and y are independent of z, the two terms in (5.258) must be constants. Since $\hat{p}_z$ is a member of the CSCO, the function $Z(z)$ can be chosen to be an eigenfunction of $\hat{p}_z$. This yields

$$Z(z) = Ce^{ip_z z/\hbar}, \tag{5.259}$$

and the second term in (5.258) becomes the contribution $p_z^2/2m$ to the total energy E. In effect, the motion in the direction of the magnetic field is that of a free particle, just as it is in the classical case.

The other term in (5.258) gives the eigenvalue equation

$$\hat{H}_{xy}\psi(x, y) = \left(\frac{1}{2m}\left(\hat{p}_x^2 + \hat{p}_y^2\right) + \frac{1}{8}m\omega^2(x^2 + y^2) - \frac{1}{2}\omega\hat{L}_z\right)\psi(x, y) = E'\psi(x, y), \quad (5.260)$$

where E' is the contribution from the quantum mechanical motion in the x–y plane to the total energy eigenvalue E. Because $\hat{L}_z$ is also a member of the CSCO, the function $\psi(x, y)$ can be chosen to be an eigenfunction of $\hat{L}_z$. In that regard, it is useful to note that the Cartesian form for (5.260) reduces to two copies of the simple harmonic oscillator Hamiltonian of (5.37) for the case that $\hat{L}_z\psi(x, y) = 0$, with both harmonic oscillators associated with the angular frequency $\omega_c = q\mathrm{B}_o/2mc = \frac{1}{2}\omega$. This reflects the classical circular motion at a constant angular velocity, which can be viewed as the combination of two harmonic oscillations with the relative phase of $\pi/2$.

5.4.5 An algebraic solution

Although (5.260) can be solved by converting it into a differential equation in cylindrical coordinates, the physical meaning of the results obtained are not readily apparent. Because of the similarity between rotational motion and two-dimensional harmonic oscillations, a simpler and more insightful approach is to recast (5.260) in terms of creation and annihilation operators like those of (5.71) for the harmonic oscillator. Because the problem contains a harmonic oscillator potential in both x and y, there are two annihilation operators,

$$\hat{a}_x = \frac{1}{\sqrt{2}}\left(\alpha\hat{x} + \frac{i\hat{p}_x}{\hbar\alpha}\right), \quad \hat{a}_y = \frac{1}{\sqrt{2}}\left(\alpha\hat{y} + \frac{i\hat{p}_y}{\hbar\alpha}\right), \quad (5.261)$$

where

$$\alpha = \sqrt{\frac{m\omega}{2\hbar}} = \sqrt{\frac{qB_o}{2\hbar c}}. \quad (5.262)$$

The operators of (5.261) give two *independent* copies of the one-dimensional harmonic oscillator algebra (5.72),

$$\begin{aligned} \left[\hat{a}_x, \hat{a}_y^\dagger\right] &= 0, \\ \left[\hat{a}_x, \hat{a}_x^\dagger\right] &= 1, \\ \left[\hat{a}_y, \hat{a}_y^\dagger\right] &= 1. \end{aligned} \quad (5.263)$$

In order to obtain a simple representation of $\hat{L}_z$, the operators of (5.261) are used to define

$$\hat{a}_\mathrm{L} = \frac{1}{\sqrt{2}}(\hat{a}_x + i\hat{a}_y), \quad \hat{a}_\mathrm{R} = \frac{1}{\sqrt{2}}(\hat{a}_x - i\hat{a}_y). \quad (5.264)$$

Noting that $\hat{a}_\mathrm{L}^\dagger = (\hat{a}_x^\dagger - i\hat{a}_y^\dagger)/\sqrt{2}$ and $\hat{a}_\mathrm{R}^\dagger = (\hat{a}_x^\dagger + i\hat{a}_y^\dagger)/\sqrt{2}$, the algebra of $\hat{a}_x$ and $\hat{a}_y$ gives

$$\begin{aligned}
[\hat{a}_R, \hat{a}_R^\dagger] &= \frac{1}{2}[\hat{a}_x - i\hat{a}_y, \hat{a}_x^\dagger + i\hat{a}_y^\dagger] = \frac{1}{2}[\hat{a}_x, \hat{a}_x^\dagger] + \frac{1}{2}[\hat{a}_y, \hat{a}_y^\dagger] = 1,\\
[\hat{a}_L, \hat{a}_L^\dagger] &= \frac{1}{2}[\hat{a}_x + i\hat{a}_y, \hat{a}_x^\dagger - i\hat{a}_y^\dagger] = \frac{1}{2}[\hat{a}_x, \hat{a}_x^\dagger] + \frac{1}{2}[\hat{a}_y, \hat{a}_y^\dagger] = 1,\\
[\hat{a}_L, \hat{a}_R^\dagger] &= \frac{1}{2}[\hat{a}_x + i\hat{a}_y, \hat{a}_x^\dagger + i\hat{a}_y^\dagger] = \frac{1}{2}[\hat{a}_x, \hat{a}_x^\dagger] - \frac{1}{2}[\hat{a}_y, \hat{a}_y^\dagger] = 0.
\end{aligned} \tag{5.265}$$

The $\hat{a}_L$ and $\hat{a}_R$ operators can be combined to give two number operators, $\hat{N}_R$ and $\hat{N}_L$, similar to the one-dimensional definition (5.81),

$$\begin{aligned}
\hat{N}_R &= \hat{a}_R^\dagger\hat{a}_R = \frac{1}{2}\left(\hat{a}_x^\dagger + i\hat{a}_y^\dagger\right)\left(\hat{a}_x - i\hat{a}_y\right) = \frac{1}{2}\left(\hat{a}_x^\dagger\hat{a}_x + i\hat{a}_y^\dagger\hat{a}_x - i\hat{a}_x^\dagger\hat{a}_y + \hat{a}_y^\dagger\hat{a}_y\right),\\
\hat{N}_L &= \hat{a}_L^\dagger\hat{a}_L = \frac{1}{2}\left(\hat{a}_x^\dagger - i\hat{a}_y^\dagger\right)\left(\hat{a}_x + i\hat{a}_y\right) = \frac{1}{2}\left(\hat{a}_x^\dagger\hat{a}_x - i\hat{a}_y^\dagger\hat{a}_x + i\hat{a}_x^\dagger\hat{a}_y + \hat{a}_y^\dagger\hat{a}_y\right).
\end{aligned} \tag{5.266}$$

These two definitions show that

$$\begin{aligned}
\hat{N}_R + \hat{N}_L &= \hat{a}_x^\dagger\hat{a}_x + \hat{a}_y^\dagger\hat{a}_y,\\
\hat{N}_R - \hat{N}_L &= i(\hat{a}_y^\dagger\hat{a}_x - \hat{a}_x^\dagger\hat{a}_y).
\end{aligned} \tag{5.267}$$

Using (5.267), the definitions (5.261), and $[\hat{x}, \hat{p}_x] = [\hat{y}, \hat{p}_y] = i\hbar$ shows that

$$\begin{aligned}
\frac{1}{2m}(\hat{p}_x^2 + \hat{p}_y^2) + \frac{1}{8}m\omega^2(\hat{x}^2 + y^2) &= \frac{1}{2}\hbar\omega(\hat{a}_x^\dagger\hat{a}_x + \hat{a}_y^\dagger\hat{a}_y + 1)\\
&= \frac{1}{2}\hbar\omega(\hat{N}_R + \hat{N}_L + 1),
\end{aligned} \tag{5.268}$$

while

$$\begin{aligned}
\hat{L}_z = \hat{x}\hat{p}_y - \hat{y}\hat{p}_x &= -\frac{i\hbar}{2}(\hat{a}_x + \hat{a}_x^\dagger)(\hat{a}_y - \hat{a}_y^\dagger) + \frac{i\hbar}{2}(\hat{a}_y + \hat{a}_y^\dagger)(\hat{a}_x - \hat{a}_x^\dagger) = i\hbar(\hat{a}_y^\dagger\hat{a}_x - \hat{a}_x^\dagger\hat{a}_y)\\
&= \hbar(\hat{N}_R - \hat{N}_L).
\end{aligned} \tag{5.269}$$

The two quantum number operators, $\hat{N}_R$ and $\hat{N}_L$, therefore correspond to the amount of angular momentum measured in units of $\hbar$. This follows from the fact that the two number operators will have positive eigenvalues, so that $\hat{N}_R$ corresponds to positive L_z or right-handed rotation, while $\hat{N}_L$ corresponds to negative L_z or left-handed rotation.

Inserting (5.268) and (5.269) into the Hamiltonian present in (5.260) gives

$$\hat{H}_{xy} = \frac{1}{2m}\left(\hat{p}_x^2 + \hat{p}_y^2\right) + \frac{1}{8}m\omega^2(x^2 + y^2) - \frac{1}{2}\omega\hat{L}_z = \hbar\omega\left(\hat{N}_L + \frac{1}{2}\right). \tag{5.270}$$

The number operator $\hat{N}_R$ has disappeared from the energy eigenvalue problem, but remains in the definition of the observable $\hat{L}_z$. This is consistent with the fact that a positive charge will undergo only clockwise or left-handed rotation in a magnetic field directed along the z-axis. As a result, only the left-handed number operator $\hat{N}_{\mathrm{L}}$ will contribute to the energy.

Solving (5.260) *and* finding the simultaneous eigenfunctions of $\hat{L}_z$ has been reduced to the much simpler problem of constructing the eigenfunctions of $\hat{N}_{\mathrm{R}}$ and $\hat{N}_{\mathrm{L}}$. This is done identically to the one-dimensional simple harmonic oscillator case (5.76) by first finding the function $\psi_0(x, y)$ that is simultaneously annihilated by $\hat{a}_x$ and $\hat{a}_y$, so that

$$\hat{a}_x\psi_0(x, y) = \hat{a}_y\psi_0(x, y) = 0. \tag{5.271}$$

The function $\psi_0(x, y)$ will then also satisfy

$$\begin{aligned}\hat{a}_{\mathrm{L}}\psi_0(x, y) &= \frac{1}{\sqrt{2}}(\hat{a}_x + i\hat{a}_y)\psi_0(x, y) = 0,\\ \hat{a}_{\mathrm{R}}\psi_0(x, y) &= \frac{1}{\sqrt{2}}(\hat{a}_x - i\hat{a}_y)\psi_0(x, y) = 0.\end{aligned} \tag{5.272}$$

This function will also satisfy

$$\hat{L}_z\psi_0(x, y) = \hbar(\hat{N}_{\mathrm{R}} - \hat{N}_{\mathrm{L}})\psi_0(x, y) = \hbar(\hat{a}_{\mathrm{R}}^\dagger\hat{a}_{\mathrm{R}} - \hat{a}_{\mathrm{L}}^\dagger\hat{a}_{\mathrm{L}})\psi_0(x, y) = 0, \tag{5.273}$$

and therefore corresponds to zero angular momentum, while

$$\hat{H}_{xy}\psi_0(x, y) = \hbar\omega\left(\hat{N}_{\mathrm{L}} + \frac{1}{2}\right)\psi_0(x, y) = \hbar\omega\left(\hat{a}_{\mathrm{L}}^\dagger\hat{a}_{\mathrm{L}} + \frac{1}{2}\right)\psi_0(x, y) = \frac{1}{2}\hbar\omega\ \psi_0(x, y), \tag{5.274}$$

revealing that the lowest possible energy is $E_0 = \frac{1}{2}\hbar\omega$. The function $\psi_0(x, y)$ is therefore the ground state wave function for $\hat{H}_{xy}$.

In the classical solution to the motion all radii R of motion are possible for a given value of ω. Larger radii correspond to higher kinetic energies since the tangential velocity is given by $v = \omega R$ so that $K = \frac{1}{2}mv^2 = \frac{1}{2}m\omega^2R^2$. Classically, the $R \to 0$ limit corresponds to zero energy. However, quantum mechanically the uncertainty relation prevents a zero energy. Demonstrating this begins by noting that the product form of the wave function (5.279) results in the uncertainties in the x and y position and momentum being identical, so that $\Delta\hat{x} = \Delta\hat{y}$ and $\Delta\hat{p}_x = \Delta\hat{p}_y$. This corresponds to zero uncertainty in the classical angular momentum, since

$$\Delta L_z \approx \Delta\hat{x}\ \Delta\hat{p}_y - \Delta\hat{y}\ \Delta\hat{p}_x = 0, \tag{5.275}$$

which is consistent with the wave functions being eigenfunctions of the angular momentum. Assuming $\Delta\hat{p}_z = 0$, the classical Hamiltonian (5.251) has the value

$$\begin{aligned} H &= \frac{(\Delta\hat{p}_x)^2}{2m} + \frac{(\Delta\hat{p}_y)^2}{2m} + \frac{1}{8}m\omega^2((\Delta\hat{x})^2 + (\Delta\hat{y})^2) - \frac{1}{2}\omega\Delta L_z\\ &= \frac{(\Delta\hat{p}_x)^2}{m} + \frac{1}{4}m\omega^2(\Delta\hat{x})^2.\end{aligned} \tag{5.276}$$

Assuming the minimal uncertainty relation $\Delta\hat{p}_x = \hbar/2\Delta\hat{x}$ holds gives the energy due to the uncertainties in $\hat{p}$ and $\hat{x}$,

$$H = \frac{\hbar^2}{4\,m(\Delta\hat{x})^2} + \frac{1}{4}m\omega^2(\Delta\hat{x})^2. \tag{5.277}$$

This is minimized by $(\Delta\hat{x})^2 = \hbar/m\omega$. This is the uncertainty in position found for the harmonic oscillator in (5.58) when the angular velocity ω is identified as $2\omega_o$, where ω_o is the harmonic oscillator frequency. Using this uncertainty gives

$$H = \frac{1}{2}\hbar\omega, \tag{5.278}$$

which is identical to the quantum mechanical ground state energy found in (5.270).

5.4.6 Cylindrical coordinate eigenfunctions

The function $\psi_0(x, y)$ is found by noting that separation of variables, given by

$$\psi_0(x, y) = X_0(x)Y_0(y), \tag{5.279}$$

will satisfy (5.271) if both

$$\begin{aligned} \hat{a}_x X_0(x) &= \frac{1}{\sqrt{2}}\left(\alpha\hat{x} + \frac{i}{\hbar\alpha}\hat{p}_x\right)X_0(x) = \frac{1}{\sqrt{2}}\left(\alpha x + \frac{1}{\alpha}\frac{\partial}{\partial x}\right)X_0(x) = 0, \\ \hat{a}_y Y_0(y) &= \frac{1}{\sqrt{2}}\left(\alpha\hat{y} + \frac{i}{\hbar\alpha}\hat{p}_y\right)Y_0(y) = \frac{1}{\sqrt{2}}\left(\alpha y + \frac{1}{\alpha}\frac{\partial}{\partial y}\right)Y_0(y) = 0. \end{aligned} \tag{5.280}$$

This yields two copies of equation (5.76), and the normalized solutions are

$$X_0(x) = Ce^{-\frac{1}{2}\alpha^2x^2}, \quad Y_0(y) = Ce^{-\frac{1}{2}\alpha^2y^2}, \tag{5.281}$$

where $C = \sqrt{\alpha/\sqrt{\pi}}$. The normalized ground state wave function of $\hat{H}_{xy}$ is therefore given by

$$\psi_0(x, y) = C^2e^{-\frac{1}{2}\alpha^2(x^2+y^2)} = \frac{\alpha}{\sqrt{\pi}}e^{-\frac{1}{2}\alpha^2\rho^2}, \tag{5.282}$$

where $\rho^2 = x^2 + y^2$ is the radial coordinate in cylindrical coordinates. Since $\hat{L}_z = -i\hbar\,\partial/\partial\phi$ in both cylindrical and spherical coordinates, (5.282) obviously satisfies the claim $\hat{L}_z\psi_0(x, y) = 0$ made earlier.

Like the harmonic oscillator eigenfunctions found in (5.83), the eigenfunctions of $\hat{H}_{xy}$ and $\hat{L}_z$ are obtained by applying the creation operators $\hat{a}_\mathrm{R}^\dagger$ and $\hat{a}_\mathrm{L}^\dagger$ to the ground state wave function (5.282). The result is

$$\psi_{n_\mathrm{R}n_\mathrm{L}}(x, y) = C_{n_\mathrm{R}n_\mathrm{L}}(\hat{a}_\mathrm{R}^\dagger)^{n_\mathrm{R}}(\hat{a}_\mathrm{L}^\dagger)^{n_\mathrm{L}}\psi_0(x, y), \tag{5.283}$$

where $C_{n_\mathrm{R}n_\mathrm{L}}$ is the required normalization factor. By virtue of the two copies of the harmonic oscillator algebra given by (5.263), (5.283) is a simultaneous eigenfunction of both number operators $\hat{N}_{\rm R}$ and $\hat{N}_{\rm L}$ with the eigenvalues or quantum numbers $n_{\rm R}$ and $n_{\rm L}$, respectively. The demonstration of this is identical to the one-dimensional harmonic oscillator case given by (5.82). Therefore, the eigenvalues of the Hamiltonian (5.270) are

$$E_{n_{\rm L}} = \left(n_{\rm L} + \frac{1}{2}\right)\hbar\omega, \tag{5.284}$$

and these are referred to as *Landau levels*. The eigenvalues of $\hat{L}_z$ are

$$\ell_z = (n_{\rm R} - n_{\rm L})\hbar. \tag{5.285}$$

It is evident that $\hat{a}_{\rm L}^\dagger$ creates an eigenfunction with a negative z-component of angular momentum, while $\hat{a}_{\rm R}^\dagger$ creates one with a positive z-component. The result for the energy is consistent with the argument made earlier regarding the results expected from the classical energy for a quantized angular momentum. It is also easy to see that the harmonic oscillator algebra generates orthogonal eigenfunctions. For example,

$$\int {\rm d}^2x\, \psi_{10}^*\psi_{01} = \int {\rm d}^2x\, (\hat{a}_{\rm R}^\dagger\psi_0)^*(\hat{a}_{\rm L}^\dagger\psi_0) = \int {\rm d}^2x\, (\psi_0)^*(\hat{a}_{\rm R}\hat{a}_{\rm L}^\dagger\psi_0) = 0, \tag{5.286}$$

since $\hat{a}_{\rm R}$ commutes with $\hat{a}_{\rm L}^\dagger$ and annihilates the ground state.

The creation operators can be expressed in terms of cylindrical coordinates, where $\rho = \sqrt{x^2 + y^2}$ and $\phi = \tan^{-1} y/x$, so that

$$\begin{aligned}\frac{\partial}{\partial x} &= \frac{\partial \rho}{\partial x}\frac{\partial}{\partial \rho} + \frac{\partial \phi}{\partial x}\frac{\partial}{\partial \phi} = \cos\phi\,\frac{\partial}{\partial \rho} - \sin\phi\,\frac{1}{\rho}\frac{\partial}{\partial \phi},\\ \frac{\partial}{\partial y} &= \frac{\partial \rho}{\partial y}\frac{\partial}{\partial \rho} + \frac{\partial \phi}{\partial y}\frac{\partial}{\partial \phi} = \sin\phi\,\frac{\partial}{\partial \rho} + \cos\phi\,\frac{1}{\rho}\frac{\partial}{\partial \phi}.\end{aligned} \tag{5.287}$$

Using the two results

$$\begin{aligned} x - iy &= \rho\cos\phi - i\rho\sin\phi = \rho\, e^{-i\phi},\\ \frac{\partial}{\partial x} - i\frac{\partial}{\partial y} &= (\cos\phi - i\sin\phi)\,\frac{\partial}{\partial \rho} - i(\cos\phi - i\sin\phi)\,\frac{1}{\rho}\frac{\partial}{\partial \phi}\\ &= e^{-i\phi}\left(\frac{\partial}{\partial \rho} - \frac{i}{\rho}\frac{\partial}{\partial \phi}\right),\end{aligned} \tag{5.288}$$

shows that $\hat{a}_{\rm L}^\dagger$ is given by

$$\hat{a}_{\rm L}^\dagger = \frac{1}{2}\alpha\left(x - iy - \frac{1}{\alpha^2}\left(\frac{\partial}{\partial x} - i\frac{\partial}{\partial y}\right)\right) = \frac{1}{2}\alpha e^{-i\phi}\left(\rho - \frac{1}{\alpha^2}\left(\frac{\partial}{\partial \rho} - \frac{i}{\rho}\frac{\partial}{\partial \phi}\right)\right), \tag{5.289}$$

Applying (5.289) to the ground state wave function (5.282) written in terms of cylindrical coordinates yields the explicit form

$$\psi_{01}(\rho, \phi) = \hat{a}_{\rm L}^{\dagger}\psi_0(\rho) = \frac{\alpha^2}{\sqrt{\pi}}\rho\, e^{-\frac{1}{2}\alpha^2\rho^2}e^{-i\phi}. \tag{5.290}$$

It is obvious that (5.290) is an eigenfunction of $\hat{L}_z = -i\hbar\, \partial/\partial\phi$ with the eigenvalue $-\hbar$. Rewriting the Hamiltonian $\hat{H}_{xy}$ appearing in (5.260) in cylindrical coordinates,

$$\hat{H}_{xy} = -\frac{\hbar^2}{2m}\left(\frac{\partial^2}{\partial\rho^2} + \frac{1}{\rho}\frac{\partial}{\partial\rho} + \frac{1}{\rho^2}\frac{\partial^2}{\partial\phi^2}\right) + \frac{1}{8}m\omega^2\rho^2 + \frac{1}{2}i\hbar\omega\frac{\partial}{\partial\phi}, \tag{5.291}$$

shows that (5.290) satisfies

$$\hat{H}_{xy}\psi_{01}(\rho, \phi) = \frac{3}{2}\hbar\omega\, \psi_{01}(\rho, \phi), \tag{5.292}$$

which is the eigenvalue predicted by (5.284). In general, the total energy E of a charge in a uniform magnetic field is consistent with helical motion, since it is the sum of $(n_{\rm L} + \frac{1}{2})\hbar\omega$, the energy of rotation in the magnetic field, and $p_z^2/2\,m$, the energy of free motion in the z-direction.

The classical Lorentz force on a positive charge supplies a centripetal force that *always* creates a left-handed rotation, so that a right-handed circular rotation is *not* classically possible. The quantum mechanical Hamiltonian (5.270) correctly reflects the absence of right-handed rotations in determining the energy spectrum. However, the Hilbert space contains both left and right-handed excitations, and so each energy eigenvalue is *infinitely degenerate*, and there is an infinite-dimensional *subspace* of radial functions for each allowed energy eigenvalue. The basis states of the subspace are found by applying an arbitrary power of $\hat{a}_{\rm R}^{\dagger}$ to the $\psi_{0n_{\rm L}}$ of (5.283). At the classical level the kinetic energy of right-handed rotation, $\frac{1}{2}mR^2\omega^2$, is exactly canceled by the classical magnetic dipole energy (5.252) of right-handed rotation since the orbital angular momentum $mR^2\omega$ is parallel to the magnetic field, so that

$$U_d = -\frac{1}{2}\omega L_z = -\frac{1}{2}mR^2\omega^2, \tag{5.293}$$

accounting for the degeneracy of the quantum levels.

References and recommended further reading

There are many excellent introductory textbooks on quantum mechanics that present numerous additional applications of quantum mechanics. These include

- Schiff L 1968 *Quantum Mechanics* 3rd edn (New York: McGraw–Hill)
- Greiner W 2001 *Quantum Mechanics: An Introduction* 4th edn (New York: Springer)
- Gasiorowicz S 2003 *Quantum Physics* 3rd edn (New York: Wiley)
- Schwabl F 2007 *Quantum Mechanics* 4th edn (New York: Springer)
- Mahan D 2009 *Quantum Mechanics in a Nutshell* (Princeton: Princeton University Press)
- Messiah A 2014 *Quantum Mechanics* (New York: Dover)

- Griffiths D and Schroeter D 2018 *Introduction to Quantum Mechanics* 3rd edn (New York: Cambridge University Press)
- Cohen–Tannoudji C, Diu B and Laloë F 2019 *Quantum Mechanics* vol 1, 2nd edn (New York: Wiley)
- Sakurai J and Napolitano J 2020 *Modern Quantum Mechanics* (New York: Cambridge University Press)

The mathematical aspects of special functions relevant to quantum mechanics, such as completeness and normalization, are detailed in

- Courant R and Hilbert D 1989 *Methods of Mathematical Physics* vols I and II (New York: Wiley)
- Byron F and Fuller R 1992 *Mathematics of Classical and Quantum Physics* (New York: Dover)
- Arfken G, Weber H–J and Harris F 2013 *Mathematical Methods for Physicists* 7th edn (Massachusetts: Elsevier)

In addition, there are books that consist entirely of problems and their solutions demonstrating how quantum mechanics is applied. These examples are of great value in understanding the many aspects of quantum mechanics. These include

- Squires G 1995 *Problems in Quantum Mechanics with Solutions* (Cambridge: Cambridge University Press)
- Galitski V, Karnakov B, Kogan V and Galitski V Jr 2013 *Exploring Quantum Mechanics* (Oxford: Oxford University Press)
- Angelini L 2019 *Solved Problems in Quantum Mechanics* (New York: Springer)

IOP Publishing

A Concise Introduction to Quantum Mechanics (Second Edition)

Mark S Swanson

Chapter 6

Dirac notation and the matrix formulation

The wave mechanical approach developed and applied in the last two chapters is a powerful method to determine the quantum mechanical behavior of a system. It is characterized by the quantization axiom that associates differential operators with classical observables via the replacement $\boldsymbol{p} \to \hat{\boldsymbol{p}} = -i\hbar\nabla$, and then solves the Schrödinger equation (4.11) to find the wave function of the system as well as the eigenvalues of commuting observables.

However, there is a more general formulation that shows wave mechanics is a *specific representation of quantum mechanics*. The basis of this more general formulation is the idea that the *state* of a system is a member of an *abstract Hilbert space*. In this approach both the classical momentum $\boldsymbol{p}$ *and* position $\boldsymbol{x}$ are elevated to *operators*, denoted $\hat{\boldsymbol{P}}$ and $\hat{\boldsymbol{X}}$, that act on the Hilbert space. Combining these steps with *Dirac notation* results in giving quantum mechanics its most general formulation, one that can be used to derive an operator and a matrix representation for quantum mechanics that is entirely equivalent to wave mechanics. The chapter begins with a brief overview of Hilbert space theory.

6.1 Hilbert space

Hilbert space theory is an important component of the mathematical area known as *functional analysis*, and many of its results are relevant to quantum mechanics. This section gives a brief overview of the properties of a Hilbert space and numerous important details and subtleties will be omitted. However, many of the properties of Hilbert space have already been used in the previous chapter, so that these properties will be familiar.

6.1.1 Generalized inner products and the dual space

A *Hilbert space* can be viewed as a generalization of the familiar Euclidean vector space. Many of the properties of Hilbert space were first introduced in the section on

doi:10.1088/978-0-7503-5663-3ch6

complex variables in chapter 3, and will now be reviewed using a more abstract notation. Forsaking mathematical rigor, a Hilbert space can be envisioned as a *linear space* characterized by a *complete basis set* of members and an *inner product.* Linearity means that the sum of two members is also a member of the space. As a simple example of a Hilbert space, the Euclidean space of three-dimensional position vectors emanating from the origin forms a linear space under addition. It has a *complete set of orthonormal basis vectors* $\boldsymbol{e}_i$ at each point. For simplicity, attention will be restricted to Cartesian basis vectors for the moment. For that case *all vectors* belonging to the Hilbert space of vectors can be written as a *linear superposition,*

$$\boldsymbol{x} = x_1\boldsymbol{e}_1 + x_2\boldsymbol{e}_2 + x_3\boldsymbol{e}_3 = \sum_{i=1}^{3} x_i\boldsymbol{e}_i, \tag{6.1}$$

where the x_i are the *vector components* familiar from basic vector analysis. There is a variety of choices for the basis set, so that in curvilinear coordinates the basis vectors change orientation from point to point, while in Cartesian coordinates they do not.

A critical property of Euclidean space and the more general Hilbert space is that the basis vectors $\boldsymbol{e}_i$ are *orthonormal*, which means that the basis vectors are chosen so that their *scalar or inner product* at least locally satisfies

$$\boldsymbol{e}_j \cdot \boldsymbol{e}_k = (\boldsymbol{e}_j, \boldsymbol{e}_k) = \delta_{jk}. \tag{6.2}$$

The inner product notation $(\boldsymbol{e}_j, \boldsymbol{e}_k)$ was introduced in chapter 3 and generalized to complex functions. It serves as a better notation and forms the basis of the Dirac notation to be introduced. The scalar product of two arbitrary vectors is a *linear operation,* so that it is given by

$$(\boldsymbol{x}, \boldsymbol{y}) = \sum_{i,j=1}^{3} x_i y_j (\boldsymbol{e}_i, \boldsymbol{e}_j) = \sum_{i,j=1}^{3} x_i y_j \delta_{ij} = \sum_{i=1}^{3} x_i y_i, \tag{6.3}$$

which is the familiar form of the vector scalar product. The vector *norm* or magnitude $|\boldsymbol{x}|$ is obtained by defining

$$|\boldsymbol{x}|^2 = (\boldsymbol{x}, \boldsymbol{x}) = \sum_{i=1}^{3} x_i^2. \tag{6.4}$$

It is important to note that the orthonormality of the basis vectors allows the components of a position vector to be written

$$x_j = (\boldsymbol{e}_j, \boldsymbol{x}) = \sum_{i=1}^{3} x_i (\boldsymbol{e}_j, \boldsymbol{e}_i) = \sum_{i=1}^{3} x_i \delta_{ij} = x_j. \tag{6.5}$$

If sufficient care is taken, many of the intuitions gained by working with finite-dimensional sets of spatial vectors carry over into the infinite-dimensional complex Hilbert spaces encountered in quantum mechanics and discussed in chapter 3. In that regard, a *complex function* in $\mathbb{L}^2$, the set of square integrable functions, can be viewed as a vector in a Hilbert space with its components obtained from the generalization of the inner product of two complex functions defined by (3.50) along with the identification of a complete set of basis functions. This is identical to the Fourier representation of a function discussed in chapter 3 and its generalization employed in chapters 4 and 5. Like Euclidean position space, there are numerous possible sets of basis functions for L^2, such as the harmonic oscillator and hydrogen wave functions discussed in chapter 5.

6.1.2 The Riesz representation theorem

An operation on Hilbert spaces relevant to quantum mechanics is the set of linear mappings of elements of the Hilbert space into the field of complex numbers. The set of all such linear mappings is referred to as the *dual space* of the Hilbert space under consideration. The *Riesz representation theorem* gives the useful result that the dual space can be represented *uniquely* by the inner product with members of the Hilbert space. As a result, the basis of the Hilbert space is also the basis of the dual space.

The formal proof of the Riesz representation theorem requires significant mathematical development that precludes its inclusion here, but it can be motivated from the properties of Euclidean position vectors, which form a Hilbert space. If f is a member of the dual space, then it satisfies the property of linearity,

$$f(a\boldsymbol{x} + b\boldsymbol{y}) = af(\boldsymbol{x}) + bf(\boldsymbol{y}), \tag{6.6}$$

where a and b are constants and $\boldsymbol{x}$ and $\boldsymbol{y}$ are two arbitrary members of the space of position vectors. The value $f(\boldsymbol{x})$ is a real number for the case of the Euclidean position vectors. The Riesz representation theorem states that, for any linear mapping function f, there exists a *unique* vector $\boldsymbol{F}$ in the space of position vectors such that it satisfies

$$(\boldsymbol{F}, \boldsymbol{x}) = f(\boldsymbol{x}), \tag{6.7}$$

for *all* possible vectors $\boldsymbol{x}$. The demonstration of (6.7) begins by noting that the linearity of f gives

$$f(\boldsymbol{x}) = f\left(\sum_{j=1}^{3} x_j \boldsymbol{e}_j\right) = \sum_{j=1}^{3} x_j f(\boldsymbol{e}_j). \tag{6.8}$$

The three values $f(\boldsymbol{e}_j)$ are assumed to be unique and are denoted

$$F_j = f(\boldsymbol{e}_j). \tag{6.9}$$

The vector $\boldsymbol{F}$ is then defined as

$$\boldsymbol{F} = \sum_{j=1} F_j \boldsymbol{e}_j. \tag{6.10}$$

It follows immediately that combining the linearity of the inner product with (6.8) gives

$$(\boldsymbol{F}, \boldsymbol{x}) = \sum_{i,j=1}^{3} F_j x_i(\boldsymbol{e}_j, \boldsymbol{e}_i) = \sum_{j=1}^{3} F_j x_i \delta_{ji} = \sum_{j=1}^{3} F_j x_j = \sum_{j=1}^{3} x_j f(\boldsymbol{e}_j) = f(\boldsymbol{x}), \tag{6.11}$$

which shows that (6.7) is satisfied.

While this demonstration took place in the context of a *real valued* vector space, the Riesz representation theorem carries over to the complex function spaces relevant to quantum mechanics, where the elements are written abstractly as f. The inner product is given by the abstract version (3.50), so that it is written (f, g). The basis functions for the space are assumed to be complete and orthonormal, satisfying $(f_j, f_k) = \delta_{jk}$. If the function space is complex, then the inner product is a complex number and has the property $(f, g)^*=(g, f)$. The dual space basis functions are found by taking the complex conjugate of the Hilbert space basis functions, f_j, so that $f_j \to f_j^*$. This mapping of f_j into the dual space is referred to as *the adjoint* of f_j.

6.1.3 Operators on Hilbert spaces

Of particular importance in quantum mechanics are the *operators* $\hat{\text{O}}$ on the relevant Hilbert spaces. An operator $\hat{\text{O}}$ on an abstract Hilbert space maps a member of the Hilbert space onto a member of a different or the same Hilbert space. In quantum mechanics the latter type of operator is of particular interest. If f is a member of the Hilbert space, then the action of such an operator can be written $\hat{\text{O}} \cdot f = g$, where g is another member of the same Hilbert space.

In generalizing quantum mechanics, it is important to define a *projection operator* $\hat{1}$ that projects the function space onto itself. Projection operators are easily visualized for the set of spatial vectors, where the complete set of basis vectors are denoted $\boldsymbol{e}_j$. An arbitrary spatial vector $\boldsymbol{x}$ can be written $\boldsymbol{x} = \sum_j x_j \boldsymbol{e}_j$, where the sum is over the complete set of basis vectors and $x_j = (\boldsymbol{e}_j, \boldsymbol{x})$. This result allows the definition of a *unit projection operator*,

$$\hat{1} = \sum_{j=1}^{3} \boldsymbol{e}_j\, (\boldsymbol{e}_j, \cdot), \tag{6.12}$$

where the dot refers to the action of the inner product of $\boldsymbol{e}_j$ with an *arbitrary* member of the Hilbert space. Using (6.5) shows that

$$\hat{1} \cdot \boldsymbol{x} = \sum_{j=1}^{3} \boldsymbol{e}_j\, (\boldsymbol{e}_j, \boldsymbol{x}) = \sum_{j=1}^{3} x_j\, \boldsymbol{e}_j = \boldsymbol{x}. \tag{6.13}$$

This holds for an arbitrary vector $\boldsymbol{x}$ in the vector space, justifying the identification of (6.12) as a *unit projection operator* for the vector space. As a further demonstration, it follows that

$$\hat{1} \cdot \hat{1} = \sum_{j=1}^{3} e_j \left(e_j, \sum_{k=1}^{3} e_k (e_k, \cdot) \right)$$
$$= \sum_{j,k=1}^{3} e_j (e_j, e_k)(e_k, \cdot) = \sum_{j,k=1}^{3} e_j \delta_{jk} (e_k, \cdot) = \sum_{j=1}^{3} e_j \, (e_j, \cdot) = \hat{1}. \tag{6.14}$$

An operator satisfying $\hat{O} \cdot \hat{O} = \hat{O}$ is said to be *idempotent*, so that $\hat{1}$ is idempotent. The cumbersome notation used in (6.12) will be rectified by using Dirac notation in the next section.

For a function space there are typically an infinite number of *basis functions*, denoted f_j. The earlier statement of completeness (3.84) shows that a similar unit projection operator can be constructed by writing

$$\hat{1} = \sum_j f_j \, (f_j, \cdot), \tag{6.15}$$

where the sum is over a complete infinite-dimensional basis set of orthonormal functions f_j and the inner product is defined by (3.50). These basis functions are assumed to be orthonormal, so that they satisfy

$$(f_j, f_k) = \delta_{jk}. \tag{6.16}$$

It is important to note that there is no explicit reference to the argument of the function g or the basis functions f_j, so that these functions have become *abstract quantities*. The claim that (6.15) is a unit projection operator is equivalent to the earlier statement (3.84) that the basis functions are *complete*, yielding an infinite series that *converges* for an arbitrary function. For example,

$$\hat{1} \cdot g = \sum_j (f_j, g) f_j = \sum_j a_j f_j = g, \tag{6.17}$$

where $a_j = (f_j, g)$ is the generalization of the coefficients (3.64) appearing in the Fourier series representation (3.63) of the function g. This is understood to be true for the function regardless of its argument. Result (3.63) is therefore the abstract version of a series representation for the function g in terms of a linear superposition of the basis functions f_j.

6.1.4 Projecting onto position

Prior to choosing an argument $\boldsymbol{x}$ for the function, the g and f_j appearing in (3.63) can be thought of as abstract members of a *linear* function space equipped with an inner product. The process of choosing a value for the argument $\boldsymbol{x}$ can be viewed as a mapping of the abstract function f, so that it expressed

$$f(\boldsymbol{x}) = \mathbf{X}(f). \tag{6.18}$$

If f and g are two functions in the Hilbert space, then the linearity of the Hilbert space has the property that $h = af + b\,g$, where a and b are constants, will be another function in the space. This means the mapping $\mathbf{X}$ must be linear, since

$$\mathbf{X}(h) = h(\boldsymbol{x}) = af(\boldsymbol{x}) + b\,g(\boldsymbol{x}) = a\,\mathbf{X}(f) + b\,\mathbf{X}(g) = \mathbf{X}(af + b\,g). \tag{6.19}$$

Since $\mathbf{X}$ is a linear mapping it can be implemented through the Riesz representation theorem by introducing the member of the space $\boldsymbol{x}$ such that its adjoint dual space member satisfies

$$(\boldsymbol{x}, f) = f(\boldsymbol{x}). \tag{6.20}$$

Using the unit projection operator (3.63), it follows that

$$g(\boldsymbol{x}) = (\boldsymbol{x}, g) = (\boldsymbol{x}, \hat{1} \cdot g) = \sum_j (\boldsymbol{x}, f_j)\,(f_j, g) = \sum_j (f_j, g)\,f_j(\boldsymbol{x}) = \sum_j a_j f_j(\boldsymbol{x}). \tag{6.21}$$

In a mild abuse of terminology, $\boldsymbol{x}$ is said to project the function f onto its *coordinate or configuration space representation*. For a complex function space, it follows that

$$(\boldsymbol{x}, f)^* = (f, \boldsymbol{x}) = f^*(\boldsymbol{x}), \tag{6.22}$$

which is consistent with identifying the mapping $f \rightarrow f^*$ as the adjoint of f. This allows the definition of a useful *unit projection operator*, given by

$$\hat{1} = \int \mathrm{d}^n x\; \boldsymbol{x}(\boldsymbol{x}, \cdot), \tag{6.23}$$

where n is the spatial dimension of the function space.

This is identical in nature to the projection operator in a Euclidean vector space given by (6.12). Showing that (6.23) is a unit operator follows by noting that it projects the abstract inner product into a coordinate representation,

$$(f, g) = (f, \hat{1} \cdot g) = \int \mathrm{d}^n x\,(f, \boldsymbol{x})(\boldsymbol{x}, g) = \int \mathrm{d}^n x\, f^*(\boldsymbol{x})\, g(\boldsymbol{x}). \tag{6.24}$$

Such projection operators will be of great use when quantum mechanics is placed into Dirac notation.

6.2 Dirac notation

The Hilbert space structures of the previous section are manifestly present in wave mechanics, including linearity, the inner product, and a complete set of basis functions. The functions that comprise the quantum mechanical wave function space L^2 form an *abstract Hilbert space* and there are many possible *complete basis sets of functions* $\{f_j\}$ that span L^2. In effect, the wave function $\Psi(\boldsymbol{x}, t)$ is replaced by a member of the abstract Hilbert space.

6.2.1 States and their adjoints

The key step in generalizing the formulation of quantum mechanics and preserving wave mechanics is to treat the state of the system as a member of an abstract Hilbert space. The notation used in statements such as (6.23) is cumbersome and is improved dramatically by using *Dirac notation*. In Dirac notation the mathematical object that is equivalent to the wave function is designated

$$\Psi(\boldsymbol{x}, t) \rightarrow |\, \Psi, t \,\rangle, \tag{6.25}$$

where $|\, \Psi, t \,\rangle$ is often referred to as a *ket* and represents the *state* of the system. It is important to note that the *state* $|\, \Psi, t \,\rangle$ has no reference to $\boldsymbol{x}$ or $\boldsymbol{p}$. It does have a reference to the time t, since the state will satisfy an extension of the Schrödinger equation. The inner product of the state $|\, \Psi, t \,\rangle$ and another state $|\, \Psi', t \,\rangle$ is written by extending the original definition (3.50) for complex functions into Dirac notation,

$$(\Psi', \Psi) \rightarrow \langle\, \Psi', t \,|\, \Psi, t \,\rangle, \tag{6.26}$$

which is often referred to as a *bracket*. Like (3.50), the inner product satisfies

$$\langle\, \Psi', t \,|\, \Psi, t \,\rangle^* = \langle\, \Psi, t \,|\, \Psi', t \,\rangle, \tag{6.27}$$

so that, in general, the inner product is a complex number. The required normalization of the wave function becomes

$$\int \mathrm{d}^n x \, \Psi^*(\boldsymbol{x}, t)\Psi(\boldsymbol{x}, t) = 1 \rightarrow \langle\, \Psi, t \,|\, \Psi, t \,\rangle = 1. \tag{6.28}$$

Further properties of the notation can now be developed.

It is important to stress that $|\, \Psi, t \,\rangle$ is *not the wave function* developed and analysed in the preceding chapters. However, the state $|\, \Psi, t \,\rangle$ contains the same information about the system, no more and no less, that the wave function does. Since the state is a member of a Hilbert space, it can be expressed as a linear superposition of a *complete set of basis states*, denoted $|\, \Psi_j, t \,\rangle$. Completeness means that an arbitrary member of the Hilbert space, $|\, \Psi, t \,\rangle$, can be written as a linear superposition of the basis states,

$$|\, \Psi, t \,\rangle = \sum_j a_j \,|\, \Psi_j, t \,\rangle, \tag{6.29}$$

where the a_j are *complex numbers* and the sum is over the *complete set of basis kets*. The summation appearing in (6.29) assumes that the basis states are indexed by a discrete set of *quantum numbers*, denoted collectively as j. This assumption reflects the typical results of wave mechanics, where the eigenfunctions of the Hamiltonian were often indexed by a set of discrete quantum numbers. Examples are the integer n that indexed the energy eigenfunctions of the harmonic oscillator or the set $\{n, \ell, m\}$ that indexed the energy eigenfunctions of the hydrogen atom. In some cases, such as the free particle momentum eigenfunctions (3.127) or the positive energy wave functions (5.128) associated with the Coulomb potential, the basis states are indexed by a continuous quantum number. For such a case the sum of (6.29) is replaced with an integral over the continuous quantum numbers, just as the Fourier series was replaced by the continuous Fourier transform in (3.98).

The next step is to use the Riesz representation theorem to express the members of the dual space of linear functions on the Hilbert space in terms of the *adjoint states* that are present in the inner product (6.26),

$$(a|\,\Psi, t\,\rangle)^{\dagger} = a^{*}\langle\,\Psi, t\,|, \tag{6.30}$$

where a is an arbitrary complex number. The state $\langle\,\Psi, t\,|$ is known as the *adjoint* of $|\,\Psi, t\,\rangle$ and is sometimes referred to as a *bra*. Consistent with the Riesz representation theorem the action of this dual space member is represented by the *inner product*, and in Dirac notation the inner product forms a *bra-ket* or bracket. The adjoint mapping requires that

$$(\langle\,\Psi, t\,|)^{\dagger} = ((|\,\Psi, t\,\rangle)^{\dagger})^{\dagger} = (|\,\Psi, t\,\rangle)^{\dagger\dagger} = |\,\Psi, t\,\rangle. \tag{6.31}$$

This allows the space dual to the dual space to be the original Hilbert space. The adjoint operation is then consistent with complex conjugation of the inner product, since it follows that

$$\langle\,\Psi', t\,|\,\Psi, t\,\rangle^{\dagger} = \langle\,\Psi, t\,|\,\Psi', t\,\rangle = \langle\,\Psi', t\,|\,\Psi, t\,\rangle^{*}. \tag{6.32}$$

Because the inner product is linear, the action of the dual space will also be linear in accordance with the Riesz representation theorem.

The state of the system $|\,\Psi, t\,\rangle$ is *not a unique object*. All of the operations on $|\,\Psi, t\,\rangle$ to extract observational information, such as its inner product, are invariant under a phase transformation,

$$|\,\Psi, t\,\rangle \to e^{i\theta}|\,\Psi, t\,\rangle. \tag{6.33}$$

This is identical to the phase transformation of the wave function (3.113), which enables the realization of Galilean relativity. As a result, the state of the system is referred to as a *ray* in the Hilbert space.

6.2.2 Operators

Dirac notation includes the presence of *linear operators* denoted by $\hat{O}$. In the case of quantum mechanics, operators map states into the same Hilbert space, so that the mapping associated with the operator $\hat{O}$, perhaps an observable, can be written

$$\hat{O}\,|\,\Psi, t\,\rangle = |\,\Psi', t\,\rangle. \tag{6.34}$$

Speaking loosely, (6.34) is the *effect* the operator $\hat{O}$ has on the state $|\,\Psi, t\,\rangle$ of the system. Similar to (6.26), the expression (3.154) defining the adjoint of the operator is given in Dirac notation by

$$(\Psi_1, \hat{O}\Psi_2) = (\hat{O}^{\dagger}\Psi_1, \Psi_2) \;\rightarrow\; \langle\,\Psi_1, t\,|(\hat{O}\,|\,\Psi_2, t\,\rangle) = \left(\langle\,\Psi_1, t\,|\hat{O}^{\dagger}\right)|\,\Psi_2, t\,\rangle, \tag{6.35}$$

from which it follows that

$$(\hat{O}\,|\,\Psi_1, t\,\rangle)^{\dagger} = \langle\,\Psi_1, t\,|\hat{O}^{\dagger}, \tag{6.36}$$

so that $\hat{O}^\dagger$ is acting on the adjoint state. For the case that the operator is Hermitian or self-adjoint, $\hat{O}^\dagger = \hat{O}$, it follows that

$$\langle \Psi_1, t \mid \hat{O} \mid \Psi_2, t \rangle = (\langle \Psi_1, t \mid \hat{O}) \mid \Psi_2, t \rangle = \langle \Psi_1, t \mid (\hat{O} \mid \Psi_2, t \rangle), \tag{6.37}$$

and so the first term can be understood as the operator acting upon either the state or its adjoint. This allows the expectation value of a Hermitian observable to be written

$$\langle \hat{O} \rangle = \langle \Psi, t \mid \hat{O} \mid \Psi, t \rangle, \tag{6.38}$$

where $\mid \Psi, t \rangle$ is the state of the system.

A time-independent Hermitian observable $\hat{O}$ will have a set of *eigenstates*, $\mid \lambda_j \rangle$, which belong to the Hilbert space and which satisfy

$$\hat{O} \mid \lambda_j \rangle = \lambda_j \mid \lambda_j \rangle. \tag{6.39}$$

The definition of the adjoint state (6.30) and the hermiticity of $\hat{O}$ give

$$\langle \lambda_j \mid \hat{O} = \langle \lambda_j \mid \hat{O}^\dagger = \left(\hat{O} \mid \lambda_j \rangle\right)^\dagger = \left(\lambda_j \mid \lambda_j \rangle\right)^\dagger = \lambda_j^* \langle \lambda_j \mid. \tag{6.40}$$

Combining this with (6.37) immediately gives the Dirac notation statement

$$\lambda_j^* \langle \lambda_j \mid \lambda_j \rangle = \left(\langle \lambda_j \mid \hat{O}\right) \mid \lambda_j \rangle = \langle \lambda_j \mid \left(\hat{O} \mid \lambda_j \rangle\right) = \lambda_j \langle \lambda_j \mid \lambda_j \rangle. \tag{6.41}$$

This shows that $\lambda_j^* = \lambda_j$, which is the previous result (3.167) for the reality of the eigenvalues of a Hermitian observable. Similarly, this shows that

$$\lambda_j \langle \lambda_j \mid \lambda_k \rangle = \left(\langle \lambda_j \mid \hat{O}\right) \mid \lambda_k \rangle = \langle \lambda_j \mid (\hat{O} \mid \lambda_k \rangle) = \lambda_k \langle \lambda_j \mid \lambda_k \rangle. \tag{6.42}$$

For $\lambda_j \neq \lambda_k$ the two states must be orthogonal. This immediately gives

$$\langle \lambda_j \mid \lambda_k \rangle = \delta_{jk}, \tag{6.43}$$

where the eigenstates are normalized and j stands for the set of quantum numbers that are required to index the state. Statement (6.43) also assumes that the eigenvalues and eigenstates are indexed by a set of discrete quantum numbers. For the case that there is a continuum of eigenvalues and eigenstates, (6.43) is stated using a Dirac delta. This is discussed in the next section.

The eigenstates of an observable $\hat{O}$ are typically complete, so that they can be used to represent any other member of the Hilbert space. The previous statement of completeness, (6.15), can now be stated in Dirac notation by the replacement

$$\hat{1} = \sum_j f_j\, (f_j, \cdot) \rightarrow \hat{1} = \sum_j \mid \lambda_j \rangle \langle \lambda_j \mid. \tag{6.44}$$

This unit projection operator can be used to place an arbitrary state $\mid \Psi, t \rangle$ into a superposition of the eigenstates $\mid \lambda_j \rangle$ in a form identical to the Fourier series (3.63),

$$\mid \Psi, t \rangle = \hat{1} \mid \Psi, t \rangle = \sum_j \mid \lambda_j \rangle \langle \lambda_j \mid \Psi, t \rangle = \sum_j \psi_j(t) \mid \lambda_j \rangle, \tag{6.45}$$

where the coefficients $\psi_j(t)$ are given by

$$\psi_j(t) = \langle \lambda_j \mid \Psi, t \rangle. \tag{6.46}$$

It is useful to note that the time-dependence of the state of the system $\mid \Psi, t \rangle$ has been subsumed into the time-dependence of the coefficients $\psi_j(t)$ in the expansion (6.45).

6.2.3 Position and momentum operators and eigenstates

In the wave mechanical version of quantum mechanics the position is treated as a trivial operator, satisfying

$$\hat{x}\, \Psi(\boldsymbol{x}, t) = \boldsymbol{x}\, \Psi(\boldsymbol{x}, t), \tag{6.47}$$

while the momentum operator takes the form of the gradient,

$$\hat{p}\, \Psi(\boldsymbol{x}, t) = -i\hbar \nabla \Psi(\boldsymbol{x}, t). \tag{6.48}$$

In the generalized version of quantum mechanics both momentum and the position are elevated to *Hermitian operators on the Hilbert space of states.* To distinguish them from the operators of wave mechanics they are denoted $\hat{x} \to \hat{\boldsymbol{Q}}$ and $\hat{p} \to \hat{\boldsymbol{P}}$. It is important to note that $\hat{\boldsymbol{P}}$ is no longer automatically represented by the gradient as it is in wave mechanics. However, the commutation relation between $\hat{\boldsymbol{P}}$ and $\hat{\boldsymbol{Q}}$ is assumed to be identical to the commutator (3.184) of wave mechanics, so that their components satisfy

$$\left[\hat{Q}_j, \hat{P}_k\right] = i\hbar\, \delta_{jk}. \tag{6.49}$$

Statement (6.49) is often referred to as a *quantization condition* and is simply an extension of the second postulate (3.201) made earlier.

The quantization procedure now states that the quantum mechanical observable Ô is obtained from the classical observable O using the prescription

$$\mathrm{O}(\boldsymbol{p}, \boldsymbol{x}, t) \to \hat{\mathrm{O}}(\hat{\boldsymbol{P}}, \hat{\boldsymbol{Q}}, t). \tag{6.50}$$

Coupling this recipe with the commutation relation (6.49) insures that the commutation relationships among *all* observables will coincide with the Poisson bracket results (1.25) obtained from classical physics for the same observables, up to the factor of $-i\hbar$ for each factor of $\boldsymbol{p}$ in the classical observable.

The Hermitian vector operators $\hat{\boldsymbol{Q}}$ and $\hat{\boldsymbol{P}}$ appearing in the recipe of (6.50) possess eigenstates $\mid \boldsymbol{x} \rangle$ and $\mid \boldsymbol{p} \rangle$, respectively, which satisfy

$$\begin{aligned} \hat{\boldsymbol{Q}} \mid \boldsymbol{x} \rangle = \boldsymbol{x} \mid \boldsymbol{x} \rangle \;&\Longrightarrow\; \hat{Q}_j \mid \boldsymbol{x} \rangle = x_j \mid \boldsymbol{x} \rangle, \\ \hat{\boldsymbol{P}} \mid \boldsymbol{p} \rangle = \boldsymbol{p} \mid \boldsymbol{p} \rangle \;&\Longrightarrow\; \hat{P}_j \mid \boldsymbol{p} \rangle = p_j \mid \boldsymbol{p} \rangle. \end{aligned} \tag{6.51}$$

The Hermitian property of $\hat{\boldsymbol{Q}}$ and $\hat{\boldsymbol{P}}$ ensures that the eigenvalues $\boldsymbol{x}$ and $\boldsymbol{p}$ are real valued, as they must be. It is stressed that these eigenstates are also members of the same Hilbert space to which $| \Psi, t \rangle$ belongs. These states correspond to possible states of the quantum mechanical particle, so that $| \boldsymbol{x} \rangle$ is the quantum mechanical state where the particle has the exact position $\boldsymbol{x}$. Similarly, the momentum eigenstate $| \boldsymbol{p} \rangle$ corresponds to a particle with the exact momentum $\boldsymbol{p}$. Like the eigenstates of the observable $\hat{\mathrm{O}}$ considered in the previous section, the position and momentum eigenstates have no intrinsic time-dependence, whereas the state of the system $| \Psi, t \rangle$ does. This is because the state of the system evolves in time, while the position and momentum eigenstates form a constant basis that spans the Hilbert space *at all times*. This is different from classical mechanics, where $\boldsymbol{x}(t)$ describes a trajectory. The quantum mechanical particle no longer has a unique trajectory and it is the observer of a quantum mechanical system who chooses a position to measure for the presence of the particle. This is discussed in more detail in chapter 8, where the path integral formulation of quantum mechanics is discussed.

Because they are members of the Hilbert space, these eigenstates also possess adjoint states $\langle \boldsymbol{x} |$ and $\langle \boldsymbol{p} |$. Because the operators $\hat{\boldsymbol{Q}}$ and $\hat{\boldsymbol{P}}$ are Hermitian, their eigenvalues are real, so that their adjoint states satisfy

$$\langle \boldsymbol{x} | \hat{\boldsymbol{Q}} = \langle \boldsymbol{x} | \boldsymbol{x}, \quad \langle \boldsymbol{p} | \hat{\boldsymbol{P}} = \langle \boldsymbol{p} | \boldsymbol{p}. \tag{6.52}$$

Concentrating on the position eigenstates gives

$$\left(\langle \boldsymbol{x} | \hat{\boldsymbol{Q}} \right) | \boldsymbol{y} \rangle = \boldsymbol{x} \langle \boldsymbol{x} | \boldsymbol{y} \rangle = \langle \boldsymbol{x} | \left(\hat{\boldsymbol{Q}} | \boldsymbol{y} \rangle \right) = \boldsymbol{y} \langle \boldsymbol{x} | \boldsymbol{y} \rangle \implies (\boldsymbol{x} - \boldsymbol{y}) \langle \boldsymbol{x} | \boldsymbol{y} \rangle = 0. \tag{6.53}$$

The continuous nature of $\boldsymbol{x}$ and $\boldsymbol{y}$ requires that the inner product of the position eigenstates be given by

$$\langle \boldsymbol{x} | \boldsymbol{y} \rangle = \delta^n(\boldsymbol{x} - \boldsymbol{y}), \tag{6.54}$$

where n is the spatial dimension of the system. A similar statement holds for the momentum eigenstates,

$$\langle \boldsymbol{p} | \boldsymbol{q} \rangle = \delta^n(\boldsymbol{p} - \boldsymbol{q}). \tag{6.55}$$

It is worth noting that the position and momentum states $| \boldsymbol{x} \rangle$ and $| \boldsymbol{p} \rangle$ have different units. The position eigenstates $| \boldsymbol{x} \rangle$ have the same units as $1/\sqrt{V}$, where V is the spatial volume of the system. The momentum eigenstates $| \boldsymbol{p} \rangle$ have the same units as $\sqrt{V/\hbar^n}$, where n is the spatial dimension.

Result (6.54) is consistent with the unit projection operator (6.23), which is written in Dirac notation as

$$\hat{1} = \int \mathrm{d}^n x \, | \boldsymbol{x} \rangle \langle \boldsymbol{x} |, \tag{6.56}$$

where the integral is over the volume V of the system. It follows that

$$\begin{aligned} \hat{1} \cdot \hat{1} &= \int \mathrm{d}^n x \int \mathrm{d}^n y \, | \boldsymbol{x} \rangle \langle \boldsymbol{x} | \boldsymbol{y} \rangle \langle \boldsymbol{y} | = \int \mathrm{d}^n x \int \mathrm{d}^n y \, | \boldsymbol{x} \rangle \, \delta^n(\boldsymbol{x} - \boldsymbol{y}) \, \langle \boldsymbol{y} | \\ &= \int \mathrm{d}^n x \, | \boldsymbol{x} \rangle \langle \boldsymbol{x} | = \hat{1}, \end{aligned} \tag{6.57}$$

showing that the projection operator (6.56) is idempotent. Similar results follow for the momentum states, so that

$$\hat{1} = \int \mathrm{d}^n p \; | \, \boldsymbol{p} \, \rangle\langle \, \boldsymbol{p} \, |, \tag{6.58}$$

where the integration is over the entirity of momentum space. The completeness of the states $| \, \boldsymbol{x} \, \rangle$ and $| \, \boldsymbol{p} \, \rangle$, as stated by (6.56) and (6.58), will be left as an assumption. The operator of (6.56) reflects the fact that the set of all position states is complete, so that they span the entire Hilbert space.

The system state $| \, \Psi, \, t \, \rangle$ is mapped onto the wave function $\Psi(\boldsymbol{x}, t)$ in the same manner as abstract functions were in (6.18), so that

$$\mathbf{X}(| \, \Psi, \, t \, \rangle) = \Psi(\boldsymbol{x}, \, t). \tag{6.59}$$

The Riesz representation theorem is satisfied by showing that this is accomplished using the inner product (6.20) with the position state $\langle \, \boldsymbol{x} \, |$, so that

$$\langle \, \boldsymbol{x} \, | \, \Psi, \, t \, \rangle = \Psi(\boldsymbol{x}, \, t). \tag{6.60}$$

The proof of (6.60) follows by showing that it is consistent with the definition of the expectation value for position. Combining the expectation value of $\hat{\boldsymbol{Q}}$ with the projection operator (6.56) gives

$$\begin{aligned} \langle \hat{\boldsymbol{Q}} \rangle &= \langle \, \Psi, \, t \, | \, \hat{\boldsymbol{Q}} \, | \, \Psi, \, t \, \rangle = \langle \, \Psi, \, t \, | \, \hat{\boldsymbol{Q}} \cdot \hat{1} \, | \, \Psi, \, t \, \rangle = \int \mathrm{d}^n x \; \langle \, \Psi, \, t \, | \, \hat{\boldsymbol{Q}} \, | \, \boldsymbol{x} \, \rangle\langle \, \boldsymbol{x} \, | \, \Psi, \, t \, \rangle \\ &= \int \mathrm{d}^n x \; \langle \, \Psi, \, t \, | \, \boldsymbol{x} \, \rangle \, \boldsymbol{x} \, \langle \, \boldsymbol{x} \, | \, \Psi, \, t \, \rangle. \end{aligned} \tag{6.61}$$

The basic wave mechanics definition (3.129) shows that the expectation value of position is given by

$$\langle \hat{\boldsymbol{Q}} \rangle = \int \mathrm{d}^n x \; \Psi^*(\boldsymbol{x}, \, t) \, \boldsymbol{x} \, \Psi(\boldsymbol{x}, \, t), \tag{6.62}$$

which immediately verifies the identification (6.60). The wave function $\Psi(\boldsymbol{x}, t)$ is therefore the projection of the Hilbert space state $| \, \Psi, \, t \, \rangle$ onto the position state $| \, \boldsymbol{x} \, \rangle$ obtained using the inner product.

Similarly, the completeness of the momentum eigenstates shows that the norm of the state $| \, \Psi, \, t \, \rangle$ can be written

$$\langle \, \Psi, \, t \, | \, \Psi, \, t \, \rangle = \langle \, \Psi, \, t \, | \hat{1} | \, \Psi, \, t \, \rangle = \int \mathrm{d}^n p \; \langle \, \Psi, \, t \, | \, \boldsymbol{p} \, \rangle\langle \, \boldsymbol{p} \, | \, \Psi, \, t \, \rangle = 1. \tag{6.63}$$

Comparing this result to the Fourier transform $\tilde{\Psi}(\boldsymbol{p}, t)$ of the wave function defined by (3.225) and appearing in the normalization statement (3.229) shows that

$$\langle \, \boldsymbol{p} \, | \, \Psi, \, t \, \rangle = \frac{1}{(2\pi\hbar)^{n/2}} \tilde{\Psi}(\boldsymbol{p}, \, t). \tag{6.64}$$

The momentum space wave function $\tilde{\Psi}(\boldsymbol{p}, t)$ is understood as the projection of the Hilbert space state $| \, \Psi, \, t \, \rangle$ onto the momentum state $| \, \boldsymbol{p} \, \rangle$.

The two results (6.60) and (6.64) can be related by using the projection operator (6.56) to give

$$\begin{aligned}\frac{1}{(2\pi\hbar)^{n/2}}\tilde{\Psi}(\boldsymbol{p},\, t) &= \langle\, \boldsymbol{p} \mid \Psi,\, t\,\rangle = \langle\, \boldsymbol{p} \mid \hat{1} \mid \Psi,\, t\,\rangle \\ &= \int \mathrm{d}^n x\, \langle\, \boldsymbol{p} \mid \boldsymbol{x}\,\rangle\langle\, \boldsymbol{x} \mid \Psi,\, t\,\rangle = \int \mathrm{d}^n x\, \langle\, \boldsymbol{p} \mid \boldsymbol{x}\,\rangle \Psi(\boldsymbol{x},\, t).\end{aligned} \tag{6.65}$$

Once again, comparing this expression to the Fourier transform (3.225) gives the identification

$$\langle\, \boldsymbol{p} \mid \boldsymbol{x}\,\rangle = \frac{1}{(2\pi\hbar)^{n/2}} e^{-i\boldsymbol{p}\cdot\boldsymbol{x}/\hbar}, \tag{6.66}$$

along with its complex conjugate,

$$\langle\, \boldsymbol{x} \mid \boldsymbol{p}\,\rangle = \langle\, \boldsymbol{p} \mid \boldsymbol{x}\,\rangle^{*} = \frac{1}{(2\pi\hbar)^{n/2}} e^{i\boldsymbol{p}\cdot\boldsymbol{x}/\hbar}. \tag{6.67}$$

These two results are consistent with the inner product (6.54) and the integral representation of the Dirac delta first made in (3.95) and extended to quantum mechanics in (3.176),

$$\delta^n(\boldsymbol{x} - \boldsymbol{y}) = \langle\, \boldsymbol{x} \mid \boldsymbol{y}\,\rangle = \int \mathrm{d}^n p\, \langle\, \boldsymbol{x} \mid \boldsymbol{p}\,\rangle\langle\, \boldsymbol{p} \mid \boldsymbol{y}\,\rangle = \int \frac{\mathrm{d}^n p}{(2\pi\hbar)^n} e^{i\boldsymbol{p}\cdot(\boldsymbol{x}-\boldsymbol{y})/\hbar}. \tag{6.68}$$

A similar statement follows for the momentum eigenstates,

$$\delta^n(\boldsymbol{p} - \boldsymbol{q}) = \langle\, \boldsymbol{p} \mid \boldsymbol{q}\,\rangle = \int \mathrm{d}^n x\, \langle\, \boldsymbol{p} \mid \boldsymbol{x}\,\rangle\langle\, \boldsymbol{x} \mid \boldsymbol{q}\,\rangle = \int \frac{\mathrm{d}^n x}{(2\pi\hbar)^n} e^{i\boldsymbol{x}\cdot(\boldsymbol{q}-\boldsymbol{p})/\hbar}, \tag{6.69}$$

which verifies the continuum normalization of the two states. Additional properties of these states are developed in the next sections.

6.2.4 The Schrödinger equation and quantum mechanical pictures

The Dirac notation version of the Schrödinger equation is obtained by applying the recipe (6.50) to the classical Hamiltonian, $H(\boldsymbol{p},\, \boldsymbol{x},\, t)$, and is identical in form to the wave mechanical version (4.11),

$$\hat{H}(\hat{\boldsymbol{P}},\, \hat{\boldsymbol{Q}},\, t) \mid \Psi,\, t\,\rangle = i\hbar\frac{\partial}{\partial t} \mid \Psi,\, t\,\rangle. \tag{6.70}$$

Assuming the Hamiltonian is Hermitian, the adjoint of the Schrödinger equation (6.70) gives

$$\langle\, \Psi,\, t \mid \hat{H} = \langle\, \Psi,\, t \mid \hat{H}^{\dagger} = \left(\hat{H} \mid \Psi,\, t\,\rangle\right)^{\dagger} = \left(i\hbar\frac{\partial}{\partial t} \mid \Psi,\, t\,\rangle\right)^{\dagger} = -i\hbar\frac{\partial}{\partial t}\langle\, \Psi,\, t \mid, \tag{6.71}$$

which is the counterpart of the wave mechanical result (4.23). A state that satisfies (6.70) is said to be in the *Schrödinger picture* of quantum mechanics, and the state is sometimes designated $| \Psi, t \rangle_S$ to indicate that its time development is governed by the Schrödinger equation.

Relation (6.70) means that the operator version of (4.59) holds for Schrödinger picture states, so that if $\partial \hat{H} / \partial t = 0$ then

$$| \Psi, t \rangle_S = e^{-i\hat{H}(\hat{P},\, \hat{Q})t/\hbar} | \Psi, 0 \rangle_S \equiv \hat{U}(t) | \Psi, 0 \rangle_S, \tag{6.72}$$

where

$$\hat{U}(t) = e^{-i\hat{H}(\hat{P},\, \hat{Q})t/\hbar}. \tag{6.73}$$

The operator $\hat{U}(t)$ is the Dirac notation version of the *evolution operator* defined in (4.59). Direct substitution shows that (6.72) satisfies (6.70) as long as $\partial \hat{H} / \partial t = 0$. In what follows a state with an explicit indication of time t will be assumed to be in the Schrödinger picture.

Result (4.64) showed that the operator $\hat{U}(t)$ is unitary, and this follows from

$$\hat{U}^{\dagger}(t) = e^{i\hat{H}(\hat{P},\, \hat{Q})t/\hbar} = \hat{U}^{-1}(t). \tag{6.74}$$

Using (6.72) and its adjoint expression

$$\langle \Psi, t | = | \Psi, t \rangle^{\dagger} = (\hat{U}(t) | \Psi, 0 \rangle)^{\dagger} = \langle \Psi, 0 | \hat{U}^{\dagger}(t), \tag{6.75}$$

shows that

$$\langle \Psi, t | \Psi, t \rangle = \langle \Psi, 0 | \hat{U}^{\dagger}(t) \hat{U}(t) | \Psi, 0 \rangle = \langle \Psi, 0 | \Psi, 0 \rangle. \tag{6.76}$$

Result (6.76) shows that the *unitary time development* given by (6.72) preserves the inner product as long as $\hat{H}$ is time-independent. If $\hat{H}$ is explicitly time-dependent the evolution operator can still be defined, but this will be deferred to chapter 8.

The action of the unitary operator $\hat{U}(t)$ can be inverted to give

$$| \Psi, 0 \rangle_S = \hat{U}^{\dagger}(t) \hat{U}(t) | \Psi, 0 \rangle_S = \hat{U}^{\dagger}(t) | \Psi, t \rangle_S. \tag{6.77}$$

Combining this result with the basic definition (6.34) of an operator allows the action of an operator or observable $\hat{O}$ to be written

$$\begin{aligned} | \Psi', 0 \rangle_S &= \hat{U}^{\dagger}(t) | \Psi', t \rangle_S = \hat{U}^{\dagger}(t) \hat{O} | \Psi, t \rangle_S \\ &= \hat{U}^{\dagger}(t)\, \hat{O}\, \hat{U}(t) | \Psi, 0 \rangle_S \equiv \hat{O}_H(t) | \Psi, 0 \rangle_S. \end{aligned} \tag{6.78}$$

The operator $\hat{O}_H(t)$, given by

$$\hat{O}_H(t) = \hat{U}^{\dagger}(t)\, \hat{O}\, \hat{U}(t), \tag{6.79}$$

is referred to as the *Heisenberg picture operator*. The original operator $\hat{O}$ obtained from the recipe (6.50) is then referred to as the *Schrödinger picture operator* and designated $\hat{O}_S$.

This definition shows that if $\hat{\text{O}}$ commutes with the Hamiltonian appearing in the definition of $\hat{U}(t)$, then $\hat{\text{O}}_\text{S}$ and $\hat{\text{O}}_\text{H}(t)$ will coincide. For the case that $\partial \hat{H}/\partial t = 0$, it follows from the definition of $\hat{U}$ that

$$\frac{\partial}{\partial t}\hat{U}(t) = -\frac{i}{\hbar}\hat{H}\hat{U}(t) = -\frac{i}{\hbar}\hat{U}(t)\hat{H} \implies \frac{\partial}{\partial t}\hat{U}^\dagger(t) = \frac{i}{\hbar}\hat{H}\hat{U}^\dagger(t). \tag{6.80}$$

Using (6.80) shows that the Heisenberg picture operator satisfies

$$\frac{\mathrm{d}}{\mathrm{d}t}\hat{\text{O}}_\text{H}(t) = \frac{\partial}{\partial t}\left(\hat{U}^\dagger(t)\hat{\text{O}}_\text{S}\hat{U}(t)\right) = \frac{i}{\hbar}[\hat{H}, \hat{\text{O}}_\text{H}(t)] + \hat{U}^\dagger(t)\frac{\partial\hat{\text{O}}_\text{S}}{\partial t}\hat{U}(t). \tag{6.81}$$

Result (6.81) is the quantum mechanical version of the classical result (1.24). The state $|\, \Psi, 0 \,\rangle_\text{S}$ is often designated as $|\, \Psi \,\rangle_\text{H}$, *the state in the Heisenberg picture.* In the Heisenberg picture the observables are explicitly time-dependent, while the states are static. In the Schrödinger picture, the states evolve in time while the observables are typically static. The two pictures are entirely equivalent, since the expectation value of an observable calculated by either method is identical,

$$\begin{aligned}\langle\hat{\text{O}}_\text{H}\rangle &= {}_\text{H}\langle\, \Psi \,|\, \hat{\text{O}}_\text{H}(t) \,|\, \Psi \,\rangle_\text{H} = {}_\text{S}\langle\, \Psi, 0 \,|\, \hat{U}^\dagger(t)\hat{\text{O}}_\text{S}\hat{U}(t) \,|\, \Psi, 0 \,\rangle_\text{S} \\ &= {}_\text{S}\langle\, \Psi, t \,|\, \hat{\text{O}}_\text{S} \,|\, \Psi, t \,\rangle_\text{S} = \langle\hat{\text{O}}_\text{S}\rangle.\end{aligned} \tag{6.82}$$

The choice of picture can therefore be made for convenience of calculation.

For the case that the Hamiltonian appearing in (6.70) is time-independent, it will typically have a complete set of Schrödinger picture energy eigenstates, $|\, E_n \,\rangle$, which satisfy

$$\hat{H}(\hat{\boldsymbol{P}}, \hat{\boldsymbol{Q}})|\, E_n \,\rangle = E_n|\, E_n \,\rangle. \tag{6.83}$$

Because the Hamiltonian is assumed to be Hermitian and time-independent, the energy eigenvalues E_n are real and constant and the eigenstates are orthonormal. Because $\hat{H}$ commutes with itself, its eigenstates remain eigenstates under the action of the time evolution operator defined in (6.72),

$$\begin{aligned}\hat{H}(\hat{\boldsymbol{P}}, \hat{\boldsymbol{Q}})|\, E_n, t \,\rangle &= \hat{H}(\hat{\boldsymbol{P}}, \hat{\boldsymbol{Q}})\hat{U}(t)|\, E_n \,\rangle \\ &= \hat{U}(t)\hat{H}(\hat{\boldsymbol{P}}, \hat{\boldsymbol{Q}})|\, E_n \,\rangle = E_n\hat{U}(t)|\, E_n \,\rangle = E_n|\, E_n, t \,\rangle.\end{aligned} \tag{6.84}$$

Since the energy eigenstates are assumed to be complete, it is possible to identify a unit projection operator, denoted at the time t as $\hat{1}(t)$, given explicitly by

$$\hat{1}(t) = \sum_j |\, E_j, t \,\rangle\langle\, E_j, t \,|. \tag{6.85}$$

This projection operator allows a general state $|\, \Psi, t \,\rangle$ to be expressed as a linear superposition of the energy eigenstates $|\, E_j, t \,\rangle$, as in (6.29),

$$| \Psi, t \rangle = \hat{1}(t) | \Psi, t \rangle = \sum_j | E_j, t \rangle\langle E_j, t | \Psi, t \rangle = \sum_j | E_j, t \rangle\langle E_j, 0 | \Psi, 0 \rangle = \sum_j a_j | E_j, t \rangle, \quad (6.86)$$

where the coefficients are given by the time-independent expression

$$a_j = \langle E_j, 0 | \Psi, 0 \rangle. \quad (6.87)$$

In this case the time evolution of the state appears in the eigenstates, which are given by

$$| E_j, t \rangle = e^{-i\hat{H}t/\hbar} | E_j, 0 \rangle = e^{-iE_j t/\hbar} | E_j, 0 \rangle. \quad (6.88)$$

This shows that the projection operator (6.85) satisfies

$$\begin{aligned} i\hbar\frac{\partial}{\partial t}\hat{1}(t) &= \sum_j \left(i\hbar\frac{\partial}{\partial t} | E_j, t \rangle\langle E_j, t | + | E_j, t \rangle i\hbar\frac{\partial}{\partial t}\langle E_j, t | \right) \\ &= \sum_j \left(E_j | E_j, t \rangle\langle E_j, t | - | E_j, t \rangle\langle E_j, t | E_j \right) = 0, \end{aligned} \quad (6.89)$$

which shows that the unit projection operator is time-independent as long as $\hat{H}$ is Hermitian and time-independent.

This projection operator is consistent with the inner product (6.54) and the definition of the energy eigenfunction obtained from (6.60). This follows from

$$\delta^n(\boldsymbol{x} - \boldsymbol{y}) = \langle \boldsymbol{x} | \boldsymbol{y} \rangle = \sum_j \langle \boldsymbol{x} | E_j, t \rangle\langle E_j, t | \boldsymbol{y} \rangle = \sum_j \Psi_j^*(\boldsymbol{y}, t)\Psi_j(\boldsymbol{x}, t), \quad (6.90)$$

which is identical to the earlier statement of completeness (3.84) found from Fourier series.

6.2.5 Representations of quantum mechanics

The operator version of the Schrödinger equation (6.70) offers a number of useful insights into the nature of quantum mechanics. It is occasionally possible to work directly with the form of the Hamiltonian and its operator content to solve (6.70). It is more common to place the Schrödinger equation into a particular *representation* by projecting it onto the eigenstates of a chosen observable, which is referred to as a *representation space*. In so doing, the position and momentum operators appearing in the Hamiltonian are given a *representation* derived from the choice of representation space.

This process is demonstrated by choosing the eigenstates of $\hat{\boldsymbol{Q}}$ as the representation space. This is referred to as the *configuration space representation*. This begins by noting that products of the position operator components satisfy

$$\langle \boldsymbol{x} | \hat{Q}_k \cdots \hat{Q}_j | \boldsymbol{y} \rangle = x_k \cdots x_j \langle \boldsymbol{x} | \boldsymbol{y} \rangle = x_k \cdots x_j \, \delta^n(\boldsymbol{x} - \boldsymbol{y}). \quad (6.91)$$

Next, the completeness of the momentum eigenstates (6.58) and the inner product (6.66) are used to find the configuration space representation of the components of the momentum operator,

$$\begin{aligned}\langle \boldsymbol{x} \mid \hat{P}_j \mid \boldsymbol{y} \rangle &= \int \mathrm{d}^n p \, \langle \boldsymbol{x} \mid \hat{P}_j \mid \boldsymbol{p} \rangle \langle \boldsymbol{p} \mid \boldsymbol{y} \rangle = \int \mathrm{d}^n p \, p_j \langle \boldsymbol{x} \mid \boldsymbol{p} \rangle \langle \boldsymbol{p} \mid \boldsymbol{y} \rangle \\ &= \int \frac{\mathrm{d}^n p}{(2\pi\hbar)^n} p_j \, e^{i\boldsymbol{p}\cdot(\boldsymbol{x}-\boldsymbol{y})/\hbar} = -i\hbar \frac{\partial}{\partial x_j} \int \frac{\mathrm{d}^n p}{(2\pi\hbar)^n} e^{i\boldsymbol{p}\cdot(\boldsymbol{x}-\boldsymbol{y})/\hbar} \\ &= -i\hbar \frac{\partial}{\partial x_j} \delta^n(\boldsymbol{x} - \boldsymbol{y}).\end{aligned} \tag{6.92}$$

This result can be extended to powers of momentum, since it is straightforward to repeat the technique of (6.92) to show that

$$\langle \boldsymbol{x} \mid \hat{\boldsymbol{P}} \cdot \hat{\boldsymbol{P}} \mid \boldsymbol{y} \rangle = \langle \boldsymbol{x} \mid \hat{P}^2 \mid \boldsymbol{y} \rangle = -\hbar^2 \nabla_x^2 \, \delta^n(\boldsymbol{x} - \boldsymbol{y}), \tag{6.93}$$

where ∇_x^2 is the Laplacian acting on the x-coordinates of the Dirac delta.

For the case that the classical Hamiltonian has the form

$$H = \frac{p^2}{2m} + V(\boldsymbol{x}), \tag{6.94}$$

it follows that the Hamiltonian appearing in the Schrödinger equation (6.70) can be placed into a configuration space representation using (6.91), (6.93), and the definition of the wave function (6.60),

$$\begin{aligned}\langle \boldsymbol{x} \mid \hat{H}(\hat{\boldsymbol{P}}, \hat{\boldsymbol{Q}}) \mid \Psi, t \rangle &= \int \mathrm{d}^n y \, \langle \boldsymbol{x} \mid \left(\frac{1}{2m} \hat{P}^2 + V(\hat{\boldsymbol{Q}}) \right) \mid \boldsymbol{y} \rangle \langle \boldsymbol{y} \mid \Psi, t \rangle \\ &= \left(-\frac{\hbar^2}{2m} \nabla_x^2 + V(\boldsymbol{x}) \right) \int \mathrm{d}^n y \, \delta^n(\boldsymbol{x} - \boldsymbol{y}) \Psi(\boldsymbol{y}, t) \\ &= \left(-\frac{\hbar^2}{2m} \nabla^2 + V(\boldsymbol{x}) \right) \Psi(\boldsymbol{x}, t).\end{aligned} \tag{6.95}$$

Using the time-independence of $\langle \boldsymbol{x} \mid$ gives

$$\langle \boldsymbol{x} \mid i\hbar \frac{\partial}{\partial t} \mid \Psi, t \rangle = i\hbar \frac{\partial}{\partial t} \langle \boldsymbol{x} \mid \Psi, t \rangle = i\hbar \frac{\partial}{\partial t} \Psi(\boldsymbol{x}, t). \tag{6.96}$$

Combining (6.99) and (6.96) gives

$$\begin{aligned}\left(-\frac{\hbar^2}{2m} \nabla^2 + V(\boldsymbol{x}) \right) \Psi(\boldsymbol{x}, t) &= \langle \boldsymbol{x} \mid \hat{H}(\hat{\boldsymbol{P}}, \hat{\boldsymbol{Q}}) \mid \Psi, t \rangle \\ &= \langle \boldsymbol{x} \mid i\hbar \frac{\partial}{\partial t} \mid \Psi, t \rangle = i\hbar \frac{\partial}{\partial t} \Psi(\boldsymbol{x}, t),\end{aligned} \tag{6.97}$$

which reproduces the wave mechanical version of the Schrödinger equation (4.11). Wave mechanics is therefore the configuration space representation of the Hilbert space version of quantum mechanics.

Similar results are obtained for the choice of a *momentum space representation*. Once again this begins by noting that the product of momentum operator components satisfies

$$\langle \boldsymbol{p} \mid \hat{P}_j \cdots \hat{P}_k \mid \boldsymbol{q} \rangle = (p_j \cdots p_k) \langle \boldsymbol{p} \mid \boldsymbol{q} \rangle = (p_j \cdots p_k)\, \delta^n(\boldsymbol{p} - \boldsymbol{q}). \tag{6.98}$$

The completeness of the position eigenstates (6.56) and the inner product (6.66) gives the momentum space representation of the components of the position operator,

$$\begin{aligned} \langle \boldsymbol{p} \mid \hat{Q}_j \mid \boldsymbol{q} \rangle &= \int \mathrm{d}^n x \, \langle \boldsymbol{p} \mid \hat{Q}_j \mid \boldsymbol{x} \rangle \langle \boldsymbol{x} \mid \boldsymbol{q} \rangle = \int \mathrm{d}^n x \; x_j \, \langle \boldsymbol{p} \mid \boldsymbol{x} \rangle \langle \boldsymbol{x} \mid \boldsymbol{q} \rangle \\ &= \int \frac{\mathrm{d}^n x}{(2\pi\hbar)^n} \, x_j \, e^{i\boldsymbol{x}\cdot(\boldsymbol{q}-\boldsymbol{p})/\hbar} = i\hbar \frac{\partial}{\partial p_j} \int \frac{\mathrm{d}^n x}{(2\pi\hbar)^n} \, e^{-i\boldsymbol{x}\cdot(\boldsymbol{q}-\boldsymbol{p})/\hbar} \\ &= i\hbar \frac{\partial}{\partial p_j} \delta^n(\boldsymbol{p} - \boldsymbol{q}), \end{aligned} \tag{6.99}$$

where the sign property (3.77) of the Dirac delta was used. This result can be extended to powers of position, since it is straightforward to repeat the technique of (6.99) to show that

$$\langle \boldsymbol{p} \mid \hat{Q}_j \cdots \hat{Q}_k \mid \boldsymbol{q} \rangle = \left(i\hbar \frac{\partial}{\partial p_j} \right) \cdots \left(i\hbar \frac{\partial}{\partial p_k} \right) \delta^n(\boldsymbol{p} - \boldsymbol{q}). \tag{6.100}$$

The final step is to give the Hamiltonian in the Schrödinger equation (6.70) a momentum space representation. Assuming the Hamiltonian has the form (6.94) and using the momentum space wave function defined by (6.64), it follows that

$$\begin{aligned} \langle \boldsymbol{p} \mid \hat{H}(\hat{\boldsymbol{P}}, \hat{\boldsymbol{Q}}) \mid \Psi, t \rangle &= \int \mathrm{d}^n q \, \langle \boldsymbol{p} \mid \left(\frac{1}{2m} \hat{P}^2 + V(\boldsymbol{Q}) \right) \mid \boldsymbol{q} \rangle \langle \boldsymbol{q} \mid \Psi, t \rangle \\ &= \left(\frac{p^2}{2m} + V(i\hbar\nabla_p) \right) \int \mathrm{d}^n q \; \delta^n(\boldsymbol{p} - \boldsymbol{q}) \left(\frac{1}{(2\pi\hbar)^n} \tilde{\Psi}(\boldsymbol{q}, t) \right) \\ &= \frac{1}{(2\pi\hbar)^n} \left(\frac{p^2}{2m} + V(i\hbar\nabla_p) \right) \tilde{\Psi}(\boldsymbol{p}, t), \end{aligned} \tag{6.101}$$

where ∇_p is the gradient operator for momentum, given in Cartesian coordinates by

$$\nabla_p = \sum_{j=1}^{n} \boldsymbol{e}_j \frac{\partial}{\partial p_j}. \tag{6.102}$$

Using the time-independence of $\langle \boldsymbol{p} \mid$ gives

$$\langle \boldsymbol{p} \mid i\hbar \frac{\partial}{\partial t} \mid \Psi, t \rangle = i\hbar \frac{\partial}{\partial t} \langle \boldsymbol{p} \mid \Psi, t \rangle = \frac{i\hbar}{(2\pi\hbar)^n} \frac{\partial}{\partial t} \tilde{\Psi}(\boldsymbol{p}, t). \tag{6.103}$$

Combining (6.101) and (6.103) and dropping the common factor of $1/(2\pi\hbar)^n$ gives

$$\left(\frac{p^2}{2m} + V(i\hbar\nabla_p) \right) \tilde{\Psi}(\boldsymbol{p}, t) = i\hbar \frac{\partial}{\partial t} \tilde{\Psi}(\boldsymbol{p}, t), \tag{6.104}$$

which is the wave mechanics Schrödinger equation in momentum space. It is straightfoward to show that result (6.104) is the configuration space Schrödinger equation (6.97) after it has undergone a Fourier transformation and several integrations by parts.

6.2.6 An example: the linear potential in momentum space

The momentum space representation (6.104) is entirely equivalent to the configuration space representation (6.97), and the choice of representation can be made based on convenience. As an example, a particle subject to a uniform gravitational field oriented in the z-direction has the classical Hamiltonian

$$H = \frac{\boldsymbol{p}^2}{2m} + mgz. \tag{6.105}$$

Using the momentum space representation (6.104) gives the Schrödinger equation

$$\left(\frac{1}{2m}(p_x^2 + p_y^2 + p_z^2) + i\hbar mg\frac{\partial}{\partial p_z}\right)\tilde{\Psi}(\boldsymbol{p}, t) = i\hbar\frac{\partial}{\partial t}\tilde{\Psi}(\boldsymbol{p}, t), \tag{6.106}$$

where $\hat{Q}_z \to i\hbar\, \partial/\partial p_z$ in the momentum representation. Equation (6.106) can undergo separtion of variables between $\boldsymbol{p}$ and t in the same manner as the configuration space Schrödinger equation in (4.46). Setting

$$\tilde{\Psi}(\boldsymbol{p}, t) = \tilde{\psi}(\boldsymbol{p})\, e^{-iEt/\hbar}, \tag{6.107}$$

and repeating the steps subsequent to (4.46) yields the energy eigenvalue equation

$$\left(\frac{1}{2m}(p_x^2 + p_y^2 + p_z^2) + i\hbar mg\frac{\partial}{\partial p_z}\right)\tilde{\psi}(\boldsymbol{p}) = E\tilde{\psi}(\boldsymbol{p}). \tag{6.108}$$

Equation (6.108) can be separated further by writing

$$\tilde{\psi}(\boldsymbol{p}) = \tilde{R}(p_x, p_y)\tilde{Z}(p_z). \tag{6.109}$$

Inserting (6.109) into (6.108) and dividing by $\tilde{\psi}(\boldsymbol{p})$ gives

$$\frac{1}{2m}(p_x^2 + p_y^2) + \frac{1}{\tilde{Z}(p_z)}\left(i\hbar mg\frac{\partial}{\partial p_z} + \frac{p_z^2}{2m}\right)\tilde{Z}(p_z) = E. \tag{6.110}$$

The radial part of the wave function, $\tilde{R}(p_x, p_y)$, has disappeared from (6.110), consistent with it being an *arbitrary* two-dimensional normalizable function with the energy eigenvalue $(p_x^2 + p_y^2)/2m$. This is because the motion in both the x- and y-directions is free. Denoting

$$\varepsilon = E - \frac{1}{2m}(p_x^2 + p_y^2), \tag{6.111}$$

the eigenvalue equation (6.110) becomes a first order differential equation,

$$\left(i\hbar mg\frac{\partial}{\partial p_z}+\frac{p_z^2}{2m}-\varepsilon\right)\tilde{Z}(p_z)=0. \tag{6.112}$$

Solving (6.112) is facilitated by changing to the dimensionless quantities k and λ, defined by

$$p_z=(2\hbar m^2 g)^{1/3}k,\quad \varepsilon=\left(\frac{1}{2}mg^2\hbar^2\right)^{1/3}\lambda. \tag{6.113}$$

Equation (6.112) simplifies to

$$\left(i\frac{\partial}{\partial k}+k^2-\lambda\right)\tilde{Z}(k)=0. \tag{6.114}$$

The eigenvalue is now λ. The solution to (6.114) is easily seen to be a modified momentum eigenfunction,

$$\tilde{Z}_\lambda(k)=\frac{1}{\sqrt{2\pi}}\exp\left(i\frac{1}{3}k^3-ik\lambda\right). \tag{6.115}$$

These functions are orthonormal,

$$\int\frac{dk}{2\pi}\,\tilde{Z}^*_{\lambda'}(k)\tilde{Z}_\lambda(k)=\int\frac{dk}{2\pi}\,e^{ik(\lambda'-\lambda)}=\delta(\lambda'-\lambda), \tag{6.116}$$

and complete since

$$\begin{aligned}\int_{-\infty}^{\infty}d\lambda\,\tilde{Z}^*_\lambda(p)\tilde{Z}_\lambda(k)&=\int_{-\infty}^{\infty}\frac{d\lambda}{2\pi}\,e^{i(p-k)\lambda}\,e^{i\frac{1}{3}(k^3-p^3)}\\&=\delta(p-k)\,e^{i\frac{1}{3}(k^3-p^3)}=\delta(p-k).\end{aligned} \tag{6.117}$$

The factor $e^{i(k^3-p^3)}$ can be dropped since the delta function is nonzero only if $p=k$.

The energy eigenvalue λ has not been constrained in any way. It was treated as continuous in the integration (6.117) and included arbitrarily negative values. In that regard, it is important to note that the classical potential energy $U=mgz$ is *not bounded from below* unless the motion is restricted through some boundary condition. The Newtonian particle can therefore have an arbitrarily negative total energy E as long as it is located at z such that $U(z)=mgz\leqslant E$. Because it should be possible to add an arbitrary constant to the energy, the restriction $z<E/mg$ can be adjusted arbitrarily and has no meaning. This is remedied by restricting the particle to values of z such that $z\geqslant 0$ and $V(z)\geqslant 0$. In effect, the motion of the particle is bounded from below by an infinite barrier or *floor* at $z=0$. In order to place such a restriction on the quantum mechanical motion, the wave function in position representation $Z_\lambda(z)$ is required. This is obtained from the wave function (6.115),

$$
\begin{aligned}
Z_\lambda(z) &= \langle z \mid Z_\lambda \rangle = \int \mathrm{d}p_z \, \langle z \mid p_z \rangle \langle p_z \mid Z_\lambda \rangle \\
&= \frac{1}{\sqrt{2\pi\hbar}} \int \mathrm{d}p_z \, e^{ip_z z/\hbar} \tilde{Z}_\lambda(p_z) = \frac{1}{N\sqrt{2\pi\hbar}} \int \mathrm{d}k \, e^{ik(\zeta-\lambda)} e^{ik^3},
\end{aligned} \tag{6.118}
$$

where $\zeta = (2m^2g/\hbar^2)^{1/3}z$ is a dimensionless vertical position and N is a normalization factor.

Evaluating (6.118) results in an *Airy function* of the first kind, denoted $\mathrm{Ai}(\zeta - \lambda)$ and depicted in figure 6.1 for dimensionless variables. The Airy function is an oscillatory function for negative values of its argument, where $\zeta - \lambda < 0$, and has an infinite sequence of zeroes located along the z-axis at $\zeta - \lambda = -\lambda_n$. For large values of the integer n the zeroes occur at $\lambda_n \approx 2.8\, n^{2/3}$. For positive values of the argument, so that $\zeta - \lambda > 0$, the Airy function $\mathrm{Ai}(\zeta - \lambda)$ is exponentially damped. Restricting the motion quantum mechanically to satisfy $z > 0$ requires the wave function $Z(z)$ to vanish for $z \leqslant 0$. Continuity of the wave function therefore requires that $\mathrm{Ai}(-\lambda) = 0$, so that $-\lambda$ must be one of the zeroes of the Airy function. This creates a set of allowed energy eigenvalues,

$$
\varepsilon_n = \left(\frac{1}{2}mg^2\hbar^2\right)^{1/3} \lambda_n \approx 2.8\left(\frac{1}{2}n^2\, mg^2\hbar^2\right)^{1/3}. \tag{6.119}
$$

The Airy function with the argument $\zeta - \lambda_n$ therefore corresponds to the associated eigenfunction for the case $z > 0$.

This includes the classically forbidden region, $mgz - \varepsilon_n > 0$, which occurs in the exponentially damped region $\zeta - \lambda_n > 0$ in the Airy function. This follows from

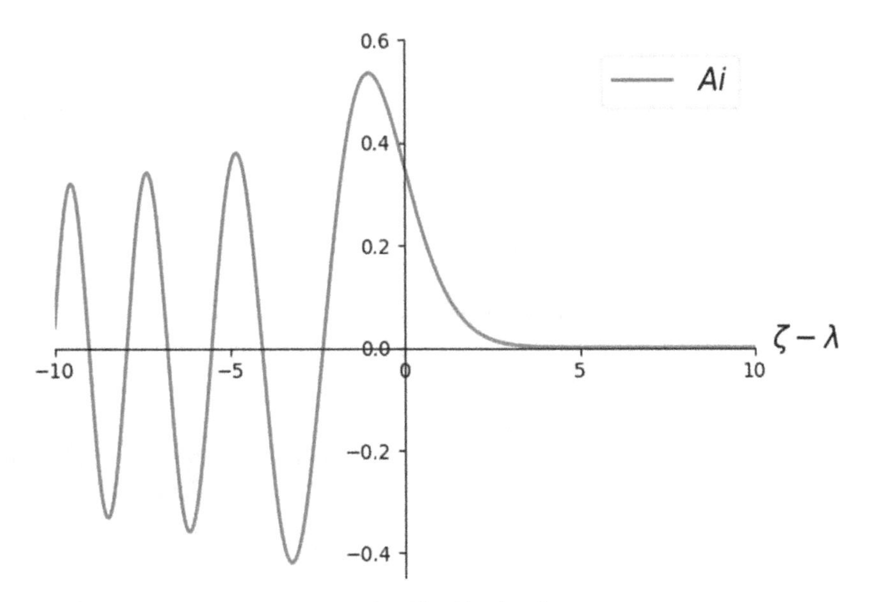

Figure 6.1. The Airy function.

$$mgz - \varepsilon_n = mg\left(\frac{\hbar^2}{2m^2g}\right)^{1/3}\zeta - \left(\frac{1}{2}mg^2\hbar^2\right)^{1/3}\lambda_n = \left(\frac{1}{2}mg^2\hbar^2\right)^{1/3}(\zeta - \lambda_n). \quad (6.120)$$

This shows that the classically forbidden region corresponds to positive values of $\zeta - \lambda_n$, which is where the Airy function is exponentially damped. The exponential damping for $mgz > \varepsilon_n$ is therefore the counterpart of barrier penetration discussed in the previous chapter. The larger the energy eigenvalue ε_n, the greater the height z that corresponds to the classically forbidden region where the wave function is exponentially damped. The restriction on motion, $z \geqslant 0$, has created a boundary condition on the wave function that *quantizes* the possible values of ε in the same manner that confining a particle to a one-dimensional well resulted in quantized energies.

6.2.7 Discrete representation spaces

Both the configuration space and momentum space representations of the Hilbert space Schrödinger equation yield continuous wave functions and differential operators for $\hat{\boldsymbol{P}}$ or $\hat{\boldsymbol{X}}$. It is also possible, and in some cases useful, to project the state of the system onto the *discrete set* of eigenstates $|\, \lambda_j \,\rangle$ of another Hermitian observable $\hat{\mathrm{O}}$. The procedure for obtaining the Schrödinger equation in the $\hat{\mathrm{O}}$ representation follows the same steps as the configuration and momentum representations. Using the assumed completeness of the $\hat{\mathrm{O}}$ eigenstates $|\, \lambda_k \,\rangle$,

$$\hat{1} = \sum_k |\, \lambda_k \,\rangle\langle\, \lambda_k \,|, \quad (6.121)$$

allows the Hilbert space Schrödinger equation to be written

$$\langle\, \lambda_j \,|\, \hat{H}(\hat{\boldsymbol{P}}, \hat{\boldsymbol{Q}}) \,|\, \Psi, t \,\rangle = \sum_j \langle\, \lambda_j \,|\, \hat{H}(\hat{\boldsymbol{P}}, \hat{\boldsymbol{Q}}) \,|\, \lambda_k \,\rangle\langle\, \lambda_k \,|\, \Psi, t \,\rangle = i\hbar\frac{\partial}{\partial t}\langle\, \lambda_j \,|\, \Psi, t \,\rangle. \quad (6.122)$$

The quantities appearing in (6.122) are denoted

$$\langle\, \lambda_j \,|\, \hat{H}(\hat{\boldsymbol{P}}, \hat{\boldsymbol{Q}}) \,|\, \lambda_k \,\rangle \equiv H^{\lambda}_{jk}, \quad \langle\, \lambda_j \,|\, \Psi, t \,\rangle \equiv \Psi^{\lambda}_j(t), \quad (6.123)$$

so that the $\hat{\mathrm{O}}$ representation of the Schrödinger equation becomes

$$\sum_k H^{\lambda}_{jk}\Psi_k(t) = i\hbar\frac{\partial}{\partial t}\Psi^{\lambda}_j(t). \quad (6.124)$$

The linear equation (6.124) is referred to as a *matrix representation* of quantum mechanics. This approach to quantum mechanics was developed by Born, Heisenberg, and Jordan prior to the Schrödinger equation. The details of such a representation depend on the choice of the observable $\hat{\mathrm{O}}$, since that will determine the eigenstates to be used. However, the mathematics of matrices and linear algebra provides powerful tools for solving (6.124). As a result, before proceeding to solve equations of the form (6.124), a brief review of matrices and linear algebra is presented.

6.3 Matrices and basic linear algebra

Matrices and linear algebra are powerful tools in analysing quantum mechanical systems. The reader familiar with these techniques may skip this section. The reader unfamiliar with these techniques will be given a brief overview.

6.3.1 Basic definitions

A matrix **A** is visualized as a rectangular array of *elements*, which can be any type of mathematical object, possessing n rows and m columns. Such an array is referred to as an $(n \times m)$-dimensional matrix and its elements are denoted by their row i and column j, so that A_{ij} is the element of **A** at row i and column j. For a square matrix the elements A_{ij} along the *diagonal* are those where $i = j$, and so these elements are referred to as the *diagonal elements*. Although matrices can have any dimensions, this book deals exclusively with square matrices **A**, where $n = m$, and row **R** and column **C** vectors, which are $(1 \times n)$- and $(n \times 1)$-dimensional, respectively. In addition, the elements of these matrices can be either real or complex numbers. These are displayed as

$$\mathbf{A} = \begin{pmatrix} A_{11} & A_{12} & \dots & A_{1n} \\ A_{21} & A_{22} & \dots & A_{2n} \\ \dots & \dots & \dots & \dots \\ A_{n1} & A_{n2} & \dots & A_{nn} \end{pmatrix}, \quad \mathbf{C} = \begin{pmatrix} C_{11} \\ C_{21} \\ \dots \\ C_{n1} \end{pmatrix}, \quad \mathbf{R} = (R_{11}\; R_{12}\; \dots\; R_{1n}), \tag{6.125}$$

where the dots '...' refer to all the rest of the sequentially numbered elements. Two matrices, **A** and **B**, are equal, written $\mathbf{A} = \mathbf{B}$, if and only if $A_{ij} = B_{ij}$ for all i and j. This requires two equal matrices to have the same dimensions and the same elements. The diagonal elements of a matrix are those for which the row number and the column number coincide, so that $A_{11}, A_{22}, \dots, A_{nn}$ are the diagonal elements of the $n \times n$ matrix A.

The sum of two matrices **A** and **B** is denoted $\mathbf{A} + \mathbf{B}$ and is defined by specifying the elements of the sum,

$$(\mathbf{A} + \mathbf{B})_{ij} = A_{ij} + B_{ij}. \tag{6.126}$$

This requires both matrices to have identical dimensions. A matrix **A** can be multiplied by a real or complex number a to obtain a new matrix,

$$\mathbf{C} = a\mathbf{A} \implies C_{ij} = aA_{ij}. \tag{6.127}$$

The additive inverse of a matrix is defined by $\mathbf{A} + (-\mathbf{A}) \equiv \mathbf{A} - \mathbf{A} = 0$, where 0 is the matrix with the same dimensions as **A** with all elements zero, referred to as the *zero matrix*. The additive inverse $-\mathbf{A}$ has the elements $-A_{ij}$. The zero matrix obeys the usual property $\mathbf{A} + 0 = \mathbf{A}$.

Similarly, the product $\mathbf{A} \cdot \mathbf{B}$ of two matrices is obtained by defining the elements of the product as

$$(\mathbf{A} \cdot \mathbf{B})_{ik} = \sum_{j=1}^{n} A_{ij} B_{jk}. \tag{6.128}$$

In order for (6.128) to be defined, $\mathbf{A}$ must have n columns and $\mathbf{B}$ must have n rows. The product $\mathbf{A} \cdot \mathbf{B}$ results in another matrix with the number of rows inherited from $\mathbf{A}$ and the number of columns inherited from $\mathbf{B}$, so that multiplying an $n \times 1$ column vector matrix on the left by an $n \times n$ matrix results in another column vector matrix with dimensions $n \times 1$. It also follows that $A \cdot 0 = 0$, where 0 is the previously defined zero matrix.

Restricting attention to square matrices allows the definition of an $n \times n$ *unit matrix*, denoted 1, such that its product with any other $n \times n$ matrix $\mathbf{A}$ gives

$$1 \cdot \mathbf{A} = \mathbf{A} \cdot 1 = \mathbf{A}. \tag{6.129}$$

The elements of the unit matrix can be written in terms of the *Kronecker delta*, denoted δ_{ij} and defined in (3.57), so that the unit matrix 1 has the elements $1_{ij} = \delta_{ij}$. The property of the Kronecker delta (3.58) shows that (6.128) gives

$$(\mathbf{A} \cdot 1)_{ik} = \sum_{j} A_{ij} \delta_{jk} = A_{ik}, \tag{6.130}$$

which verifies that $\mathbf{A} \cdot 1 = \mathbf{A}$. The matrix 1 can therefore be visualized as a $n \times n$ square array of numbers which has one for each of its *diagonal elements* and zero for all others. The unit matrix 1 plays the role of 'one' in *all matrix manipulations and operations*.

6.3.2 Matrices and vectors

Row and column matrices will be referred to as *vectors* because of their obvious resemblance to the components of a vector in an n-dimensional coordinate system. In this case the '1' in the dimension of the row and column matrices can be dropped, and the $1 \times n$ matrices are simply referred to as n-dimensional *row vectors*, while the $n \times 1$ matrices are referred to as n-dimensional *column vectors*. In n dimensions a row vector $\boldsymbol{u}$ and a column vector $\boldsymbol{v}$ can be written in terms of their components as

$$\boldsymbol{u} = (u_1 \ \ u_2 \ \ \cdots \ \ u_n), \quad \boldsymbol{v} = \begin{pmatrix} v_1 \\ v_2 \\ \vdots \\ v_n \end{pmatrix}, \tag{6.131}$$

where the u_i and v_i are referred to as the *elements* of the matrix vectors.

If $\mathbf{A}$ is an $n \times n$ square matrix and $\boldsymbol{v}$ is an $n \times 1$-dimensional column vector, the matrix product $\mathbf{A} \cdot \boldsymbol{v} = \boldsymbol{v}'$ yields another $n \times 1$-dimensional column vector denoted $\boldsymbol{v}'$. Using the rule of matrix multiplication, the elements of the new column vector $\boldsymbol{v}'$ are given by

$$v_i' = \sum_{j=1}^{n} A_{ij} v_j, \tag{6.132}$$

which is identical in form to the action of the Hamiltonian H_{jk}^{λ} on the *state vector* Ψ_j^{λ} in the matrix representation (6.124) of the Schrödinger equation. In this sense, the matrix **A** represents an action or *operation* on vectors through matrix multiplication, making contact with the idea of quantum mechanical operators. It follows that this operation is *linear* since the action of the sum of two matrices **A** and **B** on a column vector v is given by

$$(\mathbf{A} + \mathbf{B}) \cdot v = \mathbf{A} \cdot v + \mathbf{B} \cdot v, \tag{6.133}$$

giving rise to the term *linear algebra.*

6.3.3 The matrix inverse

Using the unit matrix allows the inverse of a square matrix **A**, denoted $\mathbf{A}^{-1}$, to be defined as the matrix such that

$$\mathbf{A}^{-1} \cdot \mathbf{A} = \mathbf{A} \cdot \mathbf{A}^{-1} = 1. \tag{6.134}$$

Clearly, the inverse $\mathbf{A}^{-1}$ must have the same dimensions as **A**. Since the unit matrix 1 leaves square matrices unchanged by multiplication from either the left or the right, it follows from associativity that the inverse from the left **L** is the same as the inverse from the right **R**,

$$\mathbf{L} = \mathbf{L} \cdot 1 = \mathbf{L} \cdot (\mathbf{A} \cdot \mathbf{R}) = (\mathbf{L} \cdot \mathbf{A}) \cdot \mathbf{R} = 1 \cdot \mathbf{R} = \mathbf{R}, \tag{6.135}$$

which justifies (6.134).

An example of a square matrix inverse is given in the 2×2 case by the two matrices

$$\mathbf{A} = \begin{pmatrix} 0 & 1 \\ -1 & 0 \end{pmatrix}, \quad \mathbf{B} = \begin{pmatrix} 0 & -1 \\ 1 & 0 \end{pmatrix}, \tag{6.136}$$

which have the property that

$$\mathbf{A} \cdot \mathbf{B} = \mathbf{B} \cdot \mathbf{A} = \begin{pmatrix} 1 & 0 \\ 0 & 1 \end{pmatrix} = 1 \quad \Longrightarrow \quad \mathbf{B} = \mathbf{A}^{-1}. \tag{6.137}$$

However, *not every matrix possesses an inverse.* The zero matrix 0 defined earlier is a trivial example of a matrix with no inverse. However, even non-trivial matrices may not be invertible. For example, the matrix

$$\mathbf{A} = \begin{pmatrix} 0 & 1 \\ 0 & 1 \end{pmatrix} \tag{6.138}$$

has no inverse. A general criterion for the existence of an inverse is presented later.

6.3.4 The transpose and Hermitian adjoint matrix

Given a matrix **A**, other matrices can be formed from it. The transpose of **A**, denoted $\mathbf{A}^T$, is the matrix whose elements are given by $(\mathbf{A}^T)_{ij} = A_{ji}$. In effect, the transpose of a square matrix is formed by simply interchanging rows and columns, so that the ith row of a matrix becomes the ith column of its transpose. A matrix is *symmetric* if it is equal to its transpose, $\mathbf{A} = \mathbf{A}^T$, since for such a case $A_{ij} = A_{ji}$ and the matrix is symmetric about its diagonal elements. For example, any matrix with only diagonal elements is symmetric. A matrix is *antisymmetric* if it is the additive inverse of its transpose, $\mathbf{A}^T = -\mathbf{A}$, so that $A_{ij} = -A_{ji}$. The diagonal elements of an antisymmetric matrix must vanish. The transpose of the product of two matrices **A** and **B** satisfies

$$(\mathbf{A}\cdot\mathbf{B})^{\mathrm{T}}_{ij} = (\mathbf{A}\cdot\mathbf{B})_{ji} = \sum_{k=1}^{n} A_{jk}B_{ki} = \sum_{k=1}^{n}(\mathbf{B}^T)_{ik}(\mathbf{A}^T)_{kj} = (\mathbf{B}^T\cdot\mathbf{A}^T)_{ij}, \tag{6.139}$$

so that

$$(\mathbf{A}\cdot\mathbf{B})^{\mathrm{T}} = \mathbf{B}^T\cdot\mathbf{A}^T. \tag{6.140}$$

The act of transposing a column vector v gives a row vector. This follows since a column vector has the elements v_{j1}, so that its transpose must have the elements $(v^{\mathrm{T}})_{1j} = v_{j1}$, making it an element of a row vector. Using the same definition of vector analysis, the *inner product* of two *real* column vectors $\boldsymbol{u}$ and v is given by

$$(\boldsymbol{u}, v) = \sum_{j=1}^{n} u_j v_j = \boldsymbol{u}^{\mathrm{T}}\cdot v, \tag{6.141}$$

which allows the inner product to be written in terms of matrix multiplication. If v is the column vector found by acting on the column vector $\boldsymbol{u}$ with the matrix **A**, so that $v = \mathbf{A}\cdot\boldsymbol{u}$, then it follows that

$$\begin{aligned}(v^{\mathrm{T}})_{1i} &= (\mathbf{A}\cdot\boldsymbol{u})^{\mathrm{T}}_{1i} = (\mathbf{A}\cdot\boldsymbol{u})_{i1} = \sum_{j=1}^{n} A_{ij}u_{j1} = \sum_{j=1}^{n}(\mathbf{A}^T)_{ji}(\boldsymbol{u}^{\mathrm{T}})_{1j} \\ &= \sum_{j=1}^{n}(\boldsymbol{u}^{\mathrm{T}})_{1j}(\mathbf{A}^T)_{ji} = (\boldsymbol{u}^{\mathrm{T}}\cdot\mathbf{A}^T)_{1i},\end{aligned} \tag{6.142}$$

so that

$$v = \mathbf{A}\cdot\boldsymbol{u} \Longrightarrow v^{\mathrm{T}} = \boldsymbol{u}^{\mathrm{T}}\cdot\mathbf{A}^T. \tag{6.143}$$

A matrix **A** is called *orthogonal* if its transpose is also its inverse, so that an orthogonal matrix obeys

$$\mathbf{A}^T\cdot\mathbf{A} = \mathbf{A}\cdot\mathbf{A}^{\mathrm{T}} = 1 \;\Longrightarrow\; \mathbf{A}^{\mathrm{T}} = \mathbf{A}^{-1}. \tag{6.144}$$

An example of a simple orthogonal matrix is given by the **A** appearing in (6.136), where $\mathbf{A}^{\mathrm{T}} = \mathbf{B} = \mathbf{A}^{-1}$.

If $\boldsymbol{u}' = \mathbf{A} \cdot \boldsymbol{u}$ and $\boldsymbol{v}' = \mathbf{A} \cdot \boldsymbol{v}$ are the two column vectors created by multiplying two column vectors $\boldsymbol{u}$ and $\boldsymbol{v}$ by an orthogonal matrix $\mathbf{A}$, then the (6.143) and (6.144) show that their inner product (6.141) is given by

$$(\boldsymbol{u}', \boldsymbol{v}') = (\boldsymbol{u}')^{\mathrm{T}} \cdot \boldsymbol{v}' = (\mathbf{A} \cdot \boldsymbol{u})^{\mathrm{T}} \cdot \mathbf{A} \cdot \boldsymbol{v} = \boldsymbol{u}^{\mathrm{T}} \cdot \mathbf{A}^{\mathrm{T}} \cdot \mathbf{A} \cdot \boldsymbol{v} = \boldsymbol{u}^{\mathrm{T}} \cdot 1 \cdot \boldsymbol{v} = \boldsymbol{u}^{\mathrm{T}} \cdot \boldsymbol{v} = (\boldsymbol{u}, \boldsymbol{v}). \tag{6.145}$$

The action of an orthogonal matrix therefore preserves the inner product of two real valued vectors. This is identical to the action of a rotation on two three-dimensional vectors, and so orthogonal matrices are used to represent rotations in chapter 7.

If a matrix $\mathbf{A}$ has complex elements, then it is useful to define the *Hermitian adjoint* or *complex transpose* of the matrix, denoted $\mathbf{A}^{\dagger}$, as the complex conjugate of its transpose,

$$\mathbf{A}^{\dagger} = (\mathbf{A}^{T})^{*}. \tag{6.146}$$

The elements of the Hermitian adjoint are therefore given by $(\mathbf{A}^{\dagger})_{ij} = A^{*}_{ji}$. A complex matrix $\mathbf{A}$ is called *Hermitian* if it satisfies

$$\mathbf{A} = \mathbf{A}^{\dagger} \implies A^{*}_{jk} = A_{kj}. \tag{6.147}$$

If the elements of a matrix are all real numbers, the definition of the Hermitian adjoint and the transpose are identical. A Hermitian matrix must have real numbers along the diagonal, since the complex conjugate of a diagonal element must not change the element. A matrix is *anti-Hermitian* if it satisfies $\mathbf{A}^{\dagger} = -\mathbf{A}$. Similarly, an anti-Hermitian matrix must have imaginary numbers along its diagonal since the diagonal elements must change sign under complex conjugation. The action of the Hermitian adjoint operation on products of matrices is found identically to the steps of (6.139) and (6.142), with the operation of transposition being replaced by the Hermitian adjoint. This yields

$$\begin{aligned} (\mathbf{A} \cdot \mathbf{B})^{\dagger} &= \mathbf{B}^{\dagger} \cdot \mathbf{A}^{\dagger}, \\ (\mathbf{A} \cdot \boldsymbol{v})^{\dagger} &= \boldsymbol{v}^{\dagger} \cdot \mathbf{A}^{\dagger}, \end{aligned} \tag{6.148}$$

where $\boldsymbol{v}^{\dagger}$ is the row vector that is the Hermitian adjoint of the column vector $\boldsymbol{v}$. The existence of the complex version of the inner product is central to defining the Hilbert space for *complex valued functions* used in quantum mechanics.

If the elements of the row and column vectors are complex numbers, then their inner product (3.40) defined earlier can be stated using the Hermitian adjoint. The complex inner product can be written

$$(\boldsymbol{u}, \boldsymbol{v}) = \sum_{j=2}^{n} u_j^{*} v_j = \boldsymbol{u}^{\dagger} \cdot \boldsymbol{v}. \tag{6.149}$$

Like the inner product (3.40), this yields a complex number, so that

$$(\boldsymbol{u} \cdot \boldsymbol{v})^{*} = (\boldsymbol{u}^{\dagger} \cdot \boldsymbol{v})^{\dagger} = \boldsymbol{v}^{\dagger} \cdot \boldsymbol{u} = (\boldsymbol{v}, \boldsymbol{u}). \tag{6.150}$$

This is identical to the properties of the inner product in quantum mechanics. This can be combined with the relations of (6.148) to obtain

$$(\boldsymbol{u}, (\mathbf{A} \cdot \boldsymbol{v}))^* = (\boldsymbol{u}^\dagger \cdot (\mathbf{A} \cdot \boldsymbol{v}))^* = (\mathbf{A} \cdot \boldsymbol{v})^\dagger \cdot \boldsymbol{u} = (\boldsymbol{v}^\dagger \cdot \mathbf{A}^\dagger) \cdot \boldsymbol{u} = ((\mathbf{A} \cdot \boldsymbol{v}), \boldsymbol{u}). \tag{6.151}$$

This is identical to the definition of the adjoint of an operator made in (3.154).

A matrix is *unitary* if its Hermitian adjoint is its inverse, so that

$$\mathbf{A}^\dagger \cdot \mathbf{A} = 1. \tag{6.152}$$

The unitary matrix is the extension of the orthogonal matrix to the complex case, as seen from its action on the inner product for complex vectors. If $\boldsymbol{u}' = \mathbf{A} \cdot \boldsymbol{u}$ and $\boldsymbol{v}' = \mathbf{A} \cdot \boldsymbol{v}$ are the vectors created by multiplying two vectors $\boldsymbol{u}$ and $\boldsymbol{v}$ by a unitary matrix $\mathbf{A}$, then it follows from the previous results that

$$\begin{aligned}(\boldsymbol{v}', \boldsymbol{u}') = (\boldsymbol{v}')^\dagger \cdot \boldsymbol{u}' &= (\mathbf{A} \cdot \boldsymbol{v})^\dagger \cdot \mathbf{A} \cdot \boldsymbol{u} = \boldsymbol{v}^\dagger \cdot \mathbf{A}^\dagger \cdot \mathbf{A} \cdot \boldsymbol{u} \\ &= \boldsymbol{v}^\dagger \cdot 1 \cdot \boldsymbol{u} = \boldsymbol{u}^\dagger \cdot \boldsymbol{v} = (\boldsymbol{v}, \boldsymbol{u}).\end{aligned} \tag{6.153}$$

The action of a unitary matrix therefore preserves the inner product of complex vectors.

6.3.5 Functions of matrices

Similarly, functions of a matrix act as a mapping of a matrix to another matrix. Using the substitution $1 \to \mathbf{1}$, the power series representation of a real or complex function can be adapted to the corresponding function of a square matrix. For example, given a number x the exponential function e^x has the Taylor series representation

$$e^x = \sum_{n=0}^{\infty} \frac{x^n}{n!}. \tag{6.154}$$

It follows that the exponential of the square matrix $\mathbf{A}$ multiplied by the number θ is obtained by the replacement $x \to \theta\mathbf{A}$ in the power series to obtain

$$\exp(\theta\mathbf{A}) = e^{\theta\mathbf{A}} = \sum_{n=0}^{\infty} \frac{1}{n!}\theta^n \mathbf{A}^n, \tag{6.155}$$

where $\mathbf{A}^n$ is simply the product of n copies of the matrix $\mathbf{A}$ with the understanding that $\mathbf{A}^0 = \mathbf{1}$ and $\mathbf{A}^{-n} = (\mathbf{A}^{-1})^n$. The result of (6.155) is a new matrix with the same dimension as $\mathbf{A}$. It is critical to note that $\exp \mathbf{A}$ is *not* the matrix with the exponential of the original elements of $\mathbf{A}$, i.e. $(\exp \mathbf{A})_{ij} \neq \exp A_{ij}$. As an example, the matrix

$$\mathbf{A} = \begin{pmatrix} 0 & 1 \\ 1 & 0 \end{pmatrix} \tag{6.156}$$

has the properties that if n is an even integer $\mathbf{A}^n = \mathbf{1}$, while if n is an odd integer $\mathbf{A}^n = \mathbf{A}$. Inserting these results into the power series (6.155) shows that

$$e^{\theta \mathbf{A}} = \begin{pmatrix} \cosh\theta & \sinh\theta \\ \sinh\theta & \cosh\theta \end{pmatrix}, \tag{6.157}$$

where the following Taylor series for the hyperbolic sine and cosine were used,

$$\sinh\theta = \sum_{n=0}^{\infty} \frac{\theta^{2n+1}}{(2n+1)!}, \quad \cosh\theta = \sum_{n=0}^{\infty} \frac{\theta^{2n}}{(2n)!}. \tag{6.158}$$

If a matrix is diagonal, so that its elements take the form

$$A_{jk} = \theta_j \delta_{jk}, \tag{6.159}$$

then the function of such a matrix is easily calculated. This starts by noting that the elements of an arbitrary power of the diagonal matrix take the form

$$(\mathbf{A}^m)_{jk} = \sum_{i_1,\ldots,i_m}^{n} \theta_j \delta_{ji_1} \theta_{i_1} \delta_{i_1 i_2} \cdots \theta_{i_m} \delta_{i_m k} = (\theta_j)^m \delta_{jk}, \tag{6.160}$$

which is another diagonal matrix. An arbitrary function has the Taylor series representation

$$f(x) = \sum_i a_i x^i. \tag{6.161}$$

The same function of the matrix $f(\mathbf{A})$ has the elements given by

$$f(\mathbf{A})_{jk} = \sum_i a_i (\mathbf{A}^i)_{jk} = \sum_i a_i (\theta_j)^i \delta_{jk} = \left(\sum_i a_i (\theta_j)^i \right) \delta_{jk} = f(\theta_j)\delta_{jk}. \tag{6.162}$$

The function of a diagonal matrix is therefore another diagonal matrix whose elements are the functions of the diagonal elements.

6.3.6 The Levi–Civita symbol

The *Levi–Civita symbol*, $\varepsilon^{abc\ldots}$, is an extremely important mathematical object in matrix manipulation. It was previously defined in (1.29) for the case of three superscripts. The general Levi–Civita symbol can have an arbitrary number of superscripts, but once the number of superscripts is specified then the range of the superscripts must match that number. For example, the Levi–Civita symbol ε^{ab} has two superscripts, each of which can take the values of 1 or 2. The Levi–Civita symbol $\varepsilon^{jk\ell}$ has three superscripts, each of which can take the values of 1, 2, or 3, and so on. The second property of the Levi–Civita symbol is its value, which is found by generalizing (1.29),

$$\varepsilon^{abc\ldots} = \begin{cases} +1 \text{ if } abc\ldots \text{ is an even permutation of } 1, 2, 3, \ldots \\ -1 \text{ if } abc\ldots \text{ is an odd permutation of} 1, 2, 3, \ldots. \\ 0 \quad \text{if any of the } abc\ldots \text{ are the same} \end{cases} \tag{6.163}$$

For example, in the case of two superscripts $\varepsilon^{12} = -\varepsilon^{21} = 1$ while $\varepsilon^{11} = \varepsilon^{22} = 0$.

The Levi–Civita symbol satisfies many useful identities which depend on the number of superscripts. Treating ε^{ab} as the elements of an antisymmetric 2×2 matrix immediately shows

$$\sum_{b=1}^{2} \varepsilon^{ab}\varepsilon^{bc} = -\delta_{ac}, \tag{6.164}$$

so that

$$\sum_{a,b=1}^{2} \varepsilon^{ab}\varepsilon^{ab} = \sum_{a=1}^{2} \delta_{aa} = 2 = 2!. \tag{6.165}$$

Treating $\varepsilon^{jk\ell}$ as a collection of three antisymmetric 3×3 matrices shows that

$$\sum_{i=1}^{3} \varepsilon^{ijk}\varepsilon^{i\ell m} = \delta_{j\ell}\delta_{km} - \delta_{jm}\delta_{k\ell}. \tag{6.166}$$

Result (6.166) gives

$$\sum_{i,j=1}^{3} \varepsilon^{ijk}\varepsilon^{ij\ell} = \sum_{j=1}^{3} (\delta_{jj}\delta_{k\ell} - \delta_{j\ell}\delta_{kj}) = 2\,\delta_{k\ell}, \tag{6.167}$$

from which it follows that

$$\sum_{i,j,k=1}^{3} \varepsilon^{ij\ell}\varepsilon^{ij\ell} = \sum_{\ell=1}^{3} 2\,\delta_{\ell\ell} = 2 \cdot 3 = 3!. \tag{6.168}$$

Using induction shows that

$$\sum_{j,k,\ell,\ldots=1}^{n} \varepsilon^{jk\ell\ldots}\varepsilon^{jk\ell\ldots} = n!, \tag{6.169}$$

where n is the number of indices. This follows by noting that the n-dimensional case of (6.169) is simply n copies of the $(n-1)$-dimensional case, giving $n \cdot (n-1)! = n!$.

6.3.7 The determinant

Using the Levi–Civita symbol the *determinant* of an n-dimensional square matrix $\mathbf{A}$, denoted $\det \mathbf{A}$, is defined as

$$\det \mathbf{A} = \sum_{a,b,c,\ldots=1}^{n} \varepsilon^{abc\ldots} A_{1a}A_{2b}A_{3c}\ldots, \tag{6.170}$$

where the number of superscripts n on the Levi–Civita symbol matches the dimension n of the square matrix. For example, the determinant of a 2×2 square matrix $\mathbf{A}$ is

$$\det \mathbf{A} = \sum_{a,b=1}^{2} \varepsilon^{ab} A_{1a} A_{2b} = \varepsilon^{12} A_{11} A_{22} + \varepsilon^{21} A_{12} A_{21} = A_{11} A_{22} - A_{12} A_{21}. \tag{6.171}$$

All terms in the determinant of an $n \times n$ matrix are products of n elements of the matrix. If the matrix is diagonal then the determinant is simply the product of all the diagonal elements since all other terms in (6.170) vanish. This gives $\det 1 = 1$ and $\det 0 = 0$.

The determinant has a number of important properties. The first is

$$\varepsilon^{abc\ldots} \det \mathbf{A} = \sum_{a',b',c',\ldots=1}^{n} \varepsilon^{a'b'c'\ldots} A_{aa'} A_{bb'} A_{cc'} \ldots. \tag{6.172}$$

The proof of (6.172) hinges on noting that if any of the indices in the set $ab\ldots$ are equal in the product of elements, then the right-hand side of (6.172) vanishes, just as the left-hand side does, since for such a case

$$\begin{aligned} \sum_{a',b',\ldots=1}^{n} \varepsilon^{a'b'\ldots} A_{aa'} A_{ab'} \ldots &= \sum_{b',a',\ldots=1}^{n} \varepsilon^{b'a'\ldots} A_{ab'} A_{aa'} \ldots \\ &= - \sum_{a',b',\ldots=1}^{n} \varepsilon^{a'b'\ldots} A_{aa'} A_{ab'} \ldots, \end{aligned} \tag{6.173}$$

where the freedom to relabel the indices, $a' \leftrightarrow b'$, was used in the second step since they are both summed over, while the antisymmetry of the Levi–Civita symbol was used in the third step. Since (6.173) must vanish, this shows that both sides of (6.172) vanish unless $ab\ldots$ is an odd or even permutation of the indices. The final step follows by noting that the right-hand side of (6.172) is $\det \mathbf{A}$ if $ab\ldots$ is an even permutation and is $-\det \mathbf{A}$ for an odd permutation, which matches the value of the left-hand side's Levi–Civita symbol. Result (6.172) can be rewritten by using (6.169) to obtain

$$\det \mathbf{A} = \frac{1}{n!} \sum_{a,a',b,b',c,c',\ldots=1}^{n} \varepsilon^{abc\ldots} \varepsilon^{a'b'c'\ldots} A_{aa'} A_{bb'} A_{cc'} \ldots. \tag{6.174}$$

Denoting $\mathbf{B}$ as a second n-dimensional square matrix and using (6.172) gives

$$\begin{aligned} \sum_{a,b,\ldots=1}^{n} \varepsilon^{ab\ldots} B_{1a} B_{2b} \ldots \det \mathbf{A} &= \sum_{a',b',\ldots=1}^{n} \varepsilon^{a'b'\ldots} \sum_{a=1}^{n} B_{1a} A_{aa'} \sum_{b=1}^{n} B_{2b} A_{bb'} \ldots \\ &= \sum_{a',b',\ldots=1}^{n} \varepsilon^{a'b'\ldots} (\mathbf{B} \cdot \mathbf{A})_{1a'} (\mathbf{B} \cdot \mathbf{A})_{2b'} \ldots. \end{aligned} \tag{6.175}$$

Comparing (6.175) with the definition of the determinant (6.170) and using the rule for matrix multiplication (6.128) gives a very useful result,

$$\det \mathbf{B} \det \mathbf{A} = \det(\mathbf{B} \cdot \mathbf{A}). \tag{6.176}$$

Combining (6.176) with

$$\det \mathbf{A} \det \mathbf{A}^{-1} = \det(\mathbf{A} \cdot \mathbf{A}^{-1}) = \det 1 = 1, \tag{6.177}$$

which shows that

$$\det \mathbf{A}^{-1} = \frac{1}{\det \mathbf{A}}. \tag{6.178}$$

If $\det \mathbf{A} = 0$ then $\det \mathbf{A}^{-1}$ does not exist since the number 0 has no inverse. Therefore, *if* $\det \mathbf{A} = 0$ then the matrix $\mathbf{A}$ has no inverse. An example of this is the matrix appearing in (6.138), which has a determinant of zero and therefore has no inverse. Result (6.176) is easily extended to show

$$\det(\mathbf{A} \cdot \mathbf{B} \cdots \mathbf{C}) = \det \mathbf{A} \det \mathbf{B} \cdots \det \mathbf{C}. \tag{6.179}$$

Using this result and assuming $\mathbf{A}^{-1}$ exists, replacing $\mathbf{B}$ with $\mathbf{A}^{-1} \cdot \mathbf{B}$ in (6.176) gives $\det(\mathbf{A}^{-1} \cdot \mathbf{B} \cdot \mathbf{A}) = \det \mathbf{B}$. Form (6.174) also gives $\det \mathbf{A} = \det \mathbf{A}^T$. This follows by simply substituting $A_{aa'} = A^T_{a'a}$ for all the factors and then relabeling the dummy indices $a \leftrightarrow a'$ in the sums.

6.3.8 Matrix eigenvalues and eigenvectors

An aspect of a matrix important to quantum mechanics is its set of *eigenvalues*. If $\mathbf{A}$ is a square matrix of dimension n, the number λ is an eigenvalue if there is a nonzero n-dimensional column vector v_λ such that

$$\mathbf{A} \cdot v^{(\lambda)} = \lambda\, v^{(\lambda)}. \tag{6.180}$$

The vector $v^{(\lambda)}$ is called the *eigenvector* associated with the eigenvalue λ.

For any given eigenvalue λ there is a related square matrix

$$\mathbf{B} = \mathbf{A} - \lambda 1. \tag{6.181}$$

The action of the matrix $\mathbf{B}$ on the associated eigenvector gives

$$\mathbf{B} \cdot v^{(\lambda)} = \mathbf{A} \cdot v^{(\lambda)} - \lambda 1 \cdot v^{(\lambda)} = \mathbf{A} \cdot v^{(\lambda)} - \lambda v^{(\lambda)} = 0, \tag{6.182}$$

which follows from (6.180). This means that $\mathbf{B}$ has no inverse, since if the inverse existed it would give

$$\mathbf{B}^{-1} \cdot \mathbf{B} \cdot v^{(\lambda)} = 1 \cdot v^{(\lambda)} = v^{(\lambda)}, \tag{6.183}$$

in contradiction to the result (6.182), which gives

$$\mathbf{B}^{-1} \cdot \mathbf{B} \cdot v^{(\lambda)} = \mathbf{B}^{-1} \cdot 0 = 0. \tag{6.184}$$

Therefore $\mathbf{B}$ has no inverse and its determinant must vanish. This gives the *characteristic polynomial* for the eigenvalues,

$$\det \mathbf{B} = \det(\mathbf{A} - \lambda 1) = 0. \tag{6.185}$$

For an n-dimensional square matrix (6.185) gives an nth order polynomial in λ whose solutions are the eigenvalues of the matrix. The fundamental theorem of algebra then states that an n-dimensional square matrix has at most n distinct eigenvalues.

Once the eigenvalues of a matrix are determined the general form of the eigenvectors can be determined by solving (6.180) for each given eigenvalue. The eigenvectors determined this way are not unique. Any multiple of an eigenvector, $\alpha\, v^{(\lambda)}$, is still an eigenvector since the α factor drops out of (6.180). The eigenvectors are made unique by placing a *normalization* condition,

$$(v^{(\lambda)}, v^{(\lambda)}) = v^{(\lambda)\dagger} \cdot v^{(\lambda)} = 1, \tag{6.186}$$

which fixes the overall multiplicative factor. In what follows it is assumed that all the $v^{(\lambda)}$ have been *normalized*, so that (6.186) holds for all λ.

Square Hermitian matrices play an important role in physics. The primary reason is that a Hermitian matrix $\mathbf{A}$, like a Hermitian quantum mechanical observable, has real eigenvalues. This is demonstrated by starting with the eigenvalue equation (6.180) and forming the inner product with $v^{(\lambda)}$, which gives

$$v^{(\lambda)\dagger} \cdot \mathbf{A} \cdot v^{(\lambda)} = \lambda\, v^{(\lambda)\dagger} \cdot v^{(\lambda)} = \lambda. \tag{6.187}$$

Taking the complex conjugate of (6.187) and using $\mathbf{A}^\dagger = \mathbf{A}$ gives

$$\lambda^* = v^{(\lambda)\dagger} \cdot \mathbf{A}^\dagger \cdot v^{(\lambda)} = v^{(\lambda)\dagger} \cdot \mathbf{A} \cdot v^{(\lambda)} = \lambda\, v^{(\lambda)\dagger} \cdot v^{(\lambda)} = \lambda. \tag{6.188}$$

It follows that λ must be real.

A second important property of Hermitian matrices is the orthogonality of their eigenvectors. If $v^{(\lambda)}$ and $v_{\lambda'}$ are eigenvectors of the Hermitian matrix $\mathbf{A}$ associated with different distinct eigenvalues $\lambda \neq \lambda'$, then using the real nature of the eigenvalues gives

$$\begin{aligned} \lambda\, v^{(\lambda)\dagger} \cdot v^{(\lambda')} &= (v^{(\lambda)\dagger} \cdot \mathbf{A}^\dagger) \cdot v^{(\lambda')} = v^{(\lambda)\dagger} \cdot (\mathbf{A} \cdot v^{(\lambda')}) = \lambda' v^{(\lambda)\dagger} \cdot v^{(\lambda')} \\ &\Longrightarrow (\lambda - \lambda')\, v^{(\lambda)\dagger} \cdot v^{(\lambda')} = 0. \end{aligned} \tag{6.189}$$

Result (6.189) is the matrix counterpart to the orthogonality of the eigenfunctions of a Hermitian observable demonstrated in (3.168). The n different eigenvectors of the Hermitian matrix can therefore be made orthonormal, which is written

$$(v^{(\lambda_j)}, v^{(\lambda_k)}) = v^{(\lambda_j)\dagger} \cdot v^{(\lambda_k)} = \delta_{jk}, \tag{6.190}$$

where $v^{(\lambda_j)}$ is the eigenvector associated with the jth eigenvalue λ_j.

The components of the eigenvectors of a Hermitian matrix can be used to define a projection matrix $\mathbf{P}$, whose elements are given by

$$P_{jk} = \sum_{j=1}^{n} v_j^{(\lambda_j)} v_k^{(\lambda_j)*} \Longrightarrow \mathbf{P} = \sum_{j=1}^{n} v^{(\lambda_j)} v^{(\lambda_j)\dagger}, \tag{6.191}$$

where the sum is over all n eigenvectors and $v_k^{(\lambda)}$ is the kth element of the eigenvector $v^{(\lambda)}$. The orthonormality of the eigenvectors gives $\mathbf{P}$ the property that

$$(\mathbf{P}\cdot v^{(\lambda_\ell)})_j = \sum_{k=1}^{n} P_{jk} v_k^{(\lambda_\ell)} = \sum_{i=1}^{n} v_j^{(\lambda_i)} \sum_{k=1}^{n} v_k^{(\lambda_i)*} v_k^{(\lambda_\ell)} = \sum_{i=1}^{n} v^{(\lambda_i)} \delta_{i\ell} = v_j^{(\lambda_\ell)}, \tag{6.192}$$

which can be written in matrix notation as

$$\mathbf{P}\cdot v^{(\lambda_\ell)} = \sum_{j=1}^{n} v^{(\lambda_j)} v^{(\lambda_j)\dagger} \cdot v^{(\lambda_\ell)} = \sum_{j=1}^{n} v^{(\lambda_j)} \delta_{j\ell} = v^{(\lambda_\ell)}. \tag{6.193}$$

This shows that $\mathbf{P} = 1$, so that it is the unit matrix. This is the matrix equivalent of the unit projection operator (6.121) for an observable, and reflects completeness of the observable's eigenvectors.

6.3.9 Unitary transformations

The n orthonormal column eigenvectors $v^{(\lambda_j)}$ of the matrix $\mathbf{A}$ will now be used as the columns of an n-dimensional square matrix $\mathbf{U}$, whose elements are given by

$$U_{jk} = v_j^{(\lambda_k)}, \tag{6.194}$$

where (λ_j) denotes the jth eigenvector of $\mathbf{A}$. The definition of the Hermitian adjoint gives the elements of $\mathbf{U}^\dagger$,

$$U_{\ell j}^\dagger = U_{j\ell}^* = v_j^{(\lambda_\ell)*}. \tag{6.195}$$

Matrix multiplication gives

$$(\mathbf{U}^\dagger\cdot\mathbf{U})_{\ell k} = \sum_{j=1}^{n} U_{\ell j}^\dagger U_{jk} = \sum_{j=1}^{n} v_j^{(\lambda_\ell)*} v_j^{(\lambda_k)} = v^{(\lambda_\ell)\dagger}\cdot v^{(\lambda_k)} = \delta_{\ell k} = 1_{\ell k}. \tag{6.196}$$

The orthonormality of the eigenvectors makes the matrix $\mathbf{U}$ unitary, so that $\mathbf{U}^\dagger = \mathbf{U}^{-1}$.

In addition to being unitary, the matrix defined by (6.194) can be used to transform the matrix $\mathbf{A}$ into a diagonal matrix. This is accomplished by defining the *unitary transformation* of $\mathbf{A}$ as the matrix $\mathbf{A}_\mathrm{D}$ given by

$$\mathbf{A}_\mathrm{D} = \mathbf{U}^\dagger\cdot\mathbf{A}\cdot\mathbf{U}. \tag{6.197}$$

Using the definitions (6.194) and (6.195) as well as the orthonormality (6.190) gives

$$(\mathbf{A}_\mathrm{D})_{jk} = (\mathbf{U}^\dagger\cdot\mathbf{A}\cdot\mathbf{U})_{jk} = \sum_{i,\ell=1}^{n} v_i^{(\lambda_j)*} A_{i\ell} v_\ell^{(\lambda_k)} = \lambda_k \sum_{i=1}^{n} v_i^{(\lambda_j)*} v_i^{(\lambda_k)} = \lambda_k \delta_{jk}. \tag{6.198}$$

The matrix $\mathbf{A}_\mathrm{D}$ has only diagonal elements, namely the n eigenvalues of $\mathbf{A}$,

$$\mathbf{A}_\mathrm{D} = \begin{pmatrix} \lambda_1 & 0 & \dots & 0 \\ 0 & \lambda_2 & \dots & 0 \\ \vdots & \vdots & \dots & \vdots \\ 0 & 0 & \dots & \lambda_n \end{pmatrix}. \tag{6.199}$$

Using the unitary property of $\mathbf{U}$ allows (6.197) to be rewritten as

$$\mathbf{U} \cdot \mathbf{A}_{\mathrm{D}} \cdot \mathbf{U}^{\dagger} = \mathbf{U} \cdot \mathbf{U}^{\dagger} \cdot \mathbf{A} \cdot \mathbf{U} \cdot \mathbf{U}^{\dagger} = \mathbf{A}. \tag{6.200}$$

This gives the extremely useful result that

$$\det \mathbf{A} = \det(\mathbf{U} \cdot \mathbf{A}_{\mathrm{D}} \cdot \mathbf{U}^{\dagger}) = \det(\mathbf{A}_{\mathrm{D}} \cdot \mathbf{U}^{\dagger} \cdot \mathbf{U}) = \det \mathbf{A}_{\mathrm{D}} = \prod_{i=1}^{n} \lambda_i. \tag{6.201}$$

Because $\mathbf{A}_{\mathrm{D}}$ has only diagonal elements, its determinant is simply the product of the diagonal elements. In turn, (6.201) exposes the fact that the presence of a zero eigenvalue, sometimes called a *zero mode*, prevents the existence of an inverse for the matrix $\mathbf{A}$ since $\det \mathbf{A} = 0$ for such a case. This is consistent with the eigenvalue equation, since a zero value for the eigenvalue λ gives

$$\det(\mathbf{A} - \lambda 1) = \det(\mathbf{A}) = 0. \tag{6.202}$$

The unitary transformation is an extremely useful procedure since it creates a diagonal matrix, $\mathbf{A}_{\mathrm{D}}$, which has the property that its powers are also diagonal matrices. For example, the square of $\mathbf{A}_{\mathrm{D}}$ has the elements given by

$$(\mathbf{A}_{\mathrm{D}} \cdot \mathbf{A}_{\mathrm{D}})_{jk} = \sum_{\ell=1}^{n} A_{\mathrm{D}j\ell} A_{\mathrm{D}\ell k} = \sum_{\ell=1}^{n} \lambda_j \delta_{j\ell} \lambda_\ell \delta_{\ell k} = \lambda_j^2 \, \delta_{jk}, \tag{6.203}$$

so that its diagonal elements its square are simply the squares of its eigenvalues. This is very useful in evaluating the functions of a matrix, since it is easy to see that powers of a matrix $\mathbf{A}^n$ also undergo unitary transformations into the product of n diagonal matrices by inserting factors of 1,

$$\mathbf{U}^{\dagger} \cdot \mathbf{A} \cdot \mathbf{A} \cdots \mathbf{A} \cdot \mathbf{U} = \mathbf{U}^{\dagger} \cdot \mathbf{A} \cdot \mathbf{U} \cdot \mathbf{U}^{\dagger} \cdot \mathbf{A} \cdot \mathbf{U} \cdots \mathbf{U}^{\dagger} \cdot \mathbf{A} \cdot \mathbf{U} = (\mathbf{A}_{\mathrm{D}})^n. \tag{6.204}$$

This allows a function of a matrix represented by a power series to be more easily evaluated, since

$$\mathbf{U}^{\dagger} \cdot f(\mathbf{A}) \cdot \mathbf{U} = \mathbf{U}^{\dagger} \cdot \left(\sum_{n=0}^{\infty} a_n \mathbf{A}^n \right) \cdot \mathbf{U} = \sum_{n=0}^{\infty} a_n \mathbf{A}_{\mathrm{D}}^n. \tag{6.205}$$

Result (6.162) shows that $\mathbf{A}_{\mathrm{D}}^n$ is a diagonal matrix with powers of the eigenvalues along the diagonal,

$$(\mathbf{A}_{\mathrm{D}}^n)_{jk} = \lambda_j^n \, \delta_{jk}, \tag{6.206}$$

and this gives

$$f(\mathbf{A}_{\mathrm{D}})_{jk} = \left(\sum_{n=0}^{\infty} a_n \mathbf{A}_{\mathrm{D}}^n \right)_{jk} = \left(\sum_{n=0}^{\infty} a_n \lambda_j^n \right) \delta_{jk} = f(\lambda_j) \, \delta_{jk}. \tag{6.207}$$

The function of the original matrix $\mathbf{A}$ is then found by inverting expression (6.205),

$$f(\mathbf{A}) = \mathbf{U} \cdot \left(\sum_{n=0}^{\infty} a_n \mathbf{A}_{\mathrm{D}}^n \right) \cdot \mathbf{U}^\dagger = \mathbf{U} \cdot f(\mathbf{A}_{\mathrm{D}}) \cdot \mathbf{U}^\dagger. \tag{6.208}$$

6.3.10 The trace of a matrix

The *trace* of a n-dimensional matrix, written $\mathrm{Tr}\,\mathbf{A}$, is simply the sum of the n diagonal elements,

$$\mathrm{Tr}\,\mathbf{A} = \sum_{j=1}^{n} A_{jj}. \tag{6.209}$$

The trace has the obvious property that

$$\mathrm{Tr}(\mathbf{A} + \mathbf{B}) = \mathrm{Tr}(\mathbf{A}) + \mathrm{Tr}(\mathbf{B}). \tag{6.210}$$

The trace of a Hermitian matrix also has the property that it is invariant under a unitary transformation. This follows from the fact that the trace of a product of matrices is invariant under cyclic permutations of the product,

$$\mathrm{Tr}\,(\mathbf{A} \cdot \mathbf{B} \cdot \mathbf{C}) = \sum_{i,j,k=1}^{n} A_{ij} B_{jk} C_{ki} = \sum_{i,j,k=1}^{n} C_{ki} A_{ij} B_{jk} = \mathrm{Tr}\,(\mathbf{C} \cdot \mathbf{A} \cdot \mathbf{B}). \tag{6.211}$$

Combining (6.200) and (6.211) gives

$$\mathrm{Tr}\,\mathbf{A} = \mathrm{Tr}\,(\mathbf{U} \cdot \mathbf{A}_{\mathrm{D}} \cdot \mathbf{U}^\dagger) = \mathrm{Tr}\,(\mathbf{U}^\dagger \cdot \mathbf{U} \cdot \mathbf{A}_{\mathrm{D}}) = \mathrm{Tr}\,\mathbf{A}_{\mathrm{D}} = \sum_{j=1}^{n} \lambda_j. \tag{6.212}$$

The trace of an n-dimensional matrix is the sum of its n eigenvalues. The trace of a power of the matrix $\mathbf{A}$ is given by the sum of its eigenvalues raised to that power, so that

$$\mathrm{Tr}(\mathbf{A}^k) = \sum_{j=1}^{n} (\lambda_j)^k. \tag{6.213}$$

These results combine to give a very useful theorem for an n-dimensional Hermitian matrix $\mathbf{A}$, which is that

$$\det \mathbf{A} = \exp \mathrm{Tr}\,\ln \mathbf{A}. \tag{6.214}$$

The logarithm of a matrix satisfies the usual results that $\mathbf{A} = \exp(\ln \mathbf{A})$ and $\ln(\exp \mathbf{A}) = \mathbf{A}$. It is important to note that the *specific form* of the logarithm of an arbitrary matrix $\mathbf{A}$ can be understood only as a power series in $\mathbf{A}$ since it does not have some of the logarithmic properties for numbers. This is demonstrated by considering the logarithm of the product of two *non-commuting* matrices $\mathbf{A}$ and $\mathbf{B}$, which is formally written $\ln(\mathbf{A} \cdot \mathbf{B})$. If the logarithm of the product of two matrices had the same properties as the product of two numbers, it would be true that

$$\ln(\mathbf{A} \cdot \mathbf{B}) = \ln \mathbf{A} + \ln \mathbf{B} = \ln \mathbf{B} + \ln \mathbf{A} = \ln(\mathbf{B} \cdot \mathbf{A}), \tag{6.215}$$

since the addition of the logarithm of any two matrices must be commutative. This is not true since it follows that

$$\ln(\mathbf{A}\cdot\mathbf{B}) = \ln(\mathbf{B}\cdot\mathbf{A} + [\mathbf{A}, \mathbf{B}]) \neq \ln(\mathbf{B}\cdot\mathbf{A}), \tag{6.216}$$

unless the two matrices commute.

In order to avoid contending with such issues, statement (6.214) defines the logarithm of a matrix using the power series definition of the logarithm, so that

$$\mathrm{Tr}\ln\mathbf{A} = \mathrm{Tr}\ln(1 - (1 - \mathbf{A})) = \mathrm{Tr}\left(-\sum_{k=1}^{\infty}\frac{1}{k}(1 - \mathbf{A})^k\right). \tag{6.217}$$

It follows from (6.208) and (6.211) that

$$\mathrm{Tr}\ln\mathbf{A} = \mathrm{Tr}\left(\mathbf{U}\cdot\ln\mathbf{A}_{\mathrm{D}}\cdot\mathbf{U}^{\dagger}\right) = \mathrm{Tr}\left(\mathbf{U}^{\dagger}\cdot\mathbf{U}\cdot\ln\mathbf{A}_{\mathrm{D}}\right) = \mathrm{Tr}\ln\mathbf{A}_{\mathrm{D}}. \tag{6.218}$$

Since $\ln\mathbf{A}_{\mathrm{D}}$ is a diagonal matrix with the jk element $\delta_{jk}\ln\lambda_j$, it follows that

$$\mathrm{Tr}\ln\mathbf{A} = \sum_{j=1}^{n}\ln\lambda_j. \tag{6.219}$$

This expression has an ambiguity for that case that an eigenvalue is negative since the logarithm of a negative number is undefined. Negative eigenvalues are written as

$$\lambda_j = -|\lambda_j| = e^{i\pi(2n+1)}|\lambda_j|, \tag{6.220}$$

where n is an arbitrary integer. Using this convention the exponential of the logarithm for this case is given

$$e^{\ln\lambda_j} = e^{\ln\left(e^{i\pi(2n+1)}|\lambda_j|\right)} = e^{i\pi(2n+1)+\ln|\lambda_j|} = e^{i\pi(2n+1)}|\lambda_j| = -|\lambda_j| = \lambda_j, \tag{6.221}$$

so that the original negative eigenvalue is recovered. Using this result it follows from (6.201) that

$$\exp\left(\mathrm{Tr}\ln\mathbf{A}\right) = \exp\left(\sum_{j=1}^{n}\ln\lambda_j\right) = \prod_{j=1}^{n}e^{\ln\lambda_j} = \prod_{j=1}^{n}\lambda_j = \det\mathbf{A}, \tag{6.222}$$

thereby demonstrating (6.214).

6.3.11 The Jacobi identity

Like quantum mechanical operators, two matrices may not *commute* when they are multiplied. For two square n-dimensional matrices, $\mathbf{A}$ and $\mathbf{B}$, their commutator $\mathbf{C}$ is defined in a manner identical to the operator definition (3.184),

$$\mathbf{C} = [\mathbf{A}, \mathbf{B}] \equiv \mathbf{A}\cdot\mathbf{B} - \mathbf{B}\cdot\mathbf{A}. \tag{6.223}$$

The commutator of two square matrices is also a square matrix of the same dimension. Two matrices are said to commute if $[\mathbf{A}, \mathbf{B}] = 0$. The commutator satisfies an important property known as the *Jacobi identity*. For any three square n-dimensional matrices, **A**, **B**, and **C**, the definition of the commutator gives

$$[\mathbf{A}, [\mathbf{B}, \mathbf{C}]] + [\mathbf{C}, [\mathbf{A}, \mathbf{B}]] + [\mathbf{B}, [\mathbf{C}, \mathbf{A}]] = 0. \tag{6.224}$$

The order of the three matrices in each term is a cyclic permutation of the previous term. The proof of (6.224) is straightforward, simply amounting to substituting the definition of each commutator into the appropriate expression and noticing that the twelve terms all cancel. The Jacobi identity is unique in the sense that there are no similar identities involving four or more matrices.

Similarly, the *anticommutator* $\bar{\mathbf{C}}$ of **A** and **B** is another square matrix defined as

$$\bar{\mathbf{C}} = \mathbf{A} \cdot \mathbf{B} + \mathbf{B} \cdot \mathbf{A} \equiv \{\mathbf{A}, \mathbf{B}\}. \tag{6.225}$$

The anticommutator clearly satisfies $\{\mathbf{A}, \mathbf{B}\} = \{\mathbf{B}, \mathbf{A}\}$. Two matrices are said to *anticommute* if $\{\mathbf{A}, \mathbf{B}\} = 0$. There are no Jacobi-like identities for the anticommutator as long as the elements of the matrices are standard real or complex numbers.

6.4 Matrix representations of quantum mechanics

Prior to the review of matrices and linear algebra the action of the Hamiltonian on the state was cast into matrix form in equation (6.124), so that the Schrödinger equation became

$$\sum_k H_{jk} \Psi_k(t) = i\hbar \frac{\partial}{\partial t} \Psi_k(t), \tag{6.226}$$

where the jk matrix element of the Hamiltonian is given by

$$H_{jk} = \langle \lambda_j \mid \hat{H} \mid \lambda_k \rangle, \tag{6.227}$$

while the state vector elements are given by

$$\Psi_k(t) = \langle \lambda_k \mid \Psi, t \rangle. \tag{6.228}$$

The matrix elements of the Hamiltonian and the elements of the state vector are generated using the complete set of eigenstates $\mid \lambda_j \rangle$ for a Hermitian observable $\hat{\mathrm{O}}$. Because $\hat{\mathrm{O}}$ is arbitrary this is simply referred to as a matrix representation of quantum mechanics, as opposed to the configuration or momentum representation. For another observable $\hat{\mathrm{A}}$, its matrix representation is given by a procedure identical to (6.227),

$$A_{jk} = \langle \lambda_j \mid \hat{\mathrm{A}} \mid \lambda_k \rangle. \tag{6.229}$$

For the case that the complete set of states $\mid \lambda_j \rangle$ are eigenstates of both $\hat{\mathrm{A}}$ and $\hat{\mathrm{O}}$, then this is written

$$\hat{\mathrm{A}} \mid \lambda_j \rangle = a_j \mid \lambda_j \rangle. \tag{6.230}$$

This will be the case if $\hat{A}$ and $\hat{O}$ commute. For such a case the matrix representation of $\hat{A}$ in the eigenstate basis is given by

$$A_{jk} = \langle \lambda_j \mid \hat{A} \mid \lambda_k \rangle = a_k \langle \lambda_j \mid \lambda_k \rangle = a_k \delta_{jk}, \tag{6.231}$$

so that the matrix representation of $\hat{A}$ becomes diagonal. The matrix associated with an observable $\hat{A}$ will be designated $\boldsymbol{A}$.

In some cases, converting observables into matrices allows insights and possibly an easier solution to quantum mechanical eigenvalue problems. This will be true of the spin angular momentum presented in the next chapter.

6.4.1 Indices and sums

It should be clear that the indices j and k are inherited from the quantum numbers of the basis state $\mid \lambda_j \rangle$ used to create the matrix representation, and therefore j and k may represent a collection of indices. In addition, the range of the sum in (6.226) is left unspecified, since this will depend on the eigenstates of the observable $\hat{O}$. However, each of these collections can be arranged using a particular sequence in order to turn O_{jk} into a traditional square array of elements. To clarify this statement, consider an observable whose matrix elements are obtained from the basis states of the hydrogen atom. These states are characterized by the three quantum numbers $\{n\ell m\}$, with the restriction that $\ell = 0, 1, \ldots, n - 1$ and $m = -\ell, \ell + 1, \ldots, \ell$. Using this basis, the matrix element of an observable becomes

$$\langle \Psi_{n\ell m} \mid \hat{O} \mid \Psi_{n',\ell',m'} \rangle = O_{i(n\ell m),j(n'\ell' m')}. \tag{6.232}$$

Choosing the sequence order for the quantum numbers as n, ℓ, and then m gives $O_{11} = O_{(100),(100)}$, $O_{12} = O_{(100),(200)}$, $O_{13} = O_{(100),(211)}$, $O_{14} = O_{(100),(210)}$, and $O_{15} = O_{(100),(21-1)}$. The choice is arbitrary and can be made for convenience of representation and manipulation.

6.4.2 Quantum mechanical state vectors and matrix observables

It is both useful and insightful to identify the structural similarity between linear algebra and quantum mechanics. The starting point is the state of (6.29), which is now written as

$$\mid \Psi, t \rangle = \sum_j a_j(t) \mid \Psi_j \rangle, \tag{6.233}$$

where the states $\mid \Psi_j \rangle$ are a complete basis set for the Hilbert space. It is important to note that the time development of the state has been subsumed into the time-dependent coefficients $a_j(t)$. The action of a general observable on this state is denoted using (6.34),

$$\mid \Psi', t \rangle = \hat{O} \mid \Psi, t \rangle, \tag{6.234}$$

where the resulting state can be written

$$| \Psi', t \rangle = \sum_k b_k(t) | \Psi_k \rangle. \tag{6.235}$$

The sums appearing in (6.233) and (6.235) are typically over an infinite set of states.

The general state (6.235) can be expressed in a matrix representation by evaluating the inner product with $\langle \Psi_j |$ and using the orthonormality and completeness of the basis states $| \Psi_j \rangle$. The first step is to use (6.235) to obtain

$$\langle \Psi_j | \Psi', t \rangle = \sum_k b_k(t) \langle \Psi_j | \Psi_k \rangle = \sum_k b_k(t)\, \delta_{jk} = b_j(t). \tag{6.236}$$

The second step is to use (6.234) and the completeness of the $| \Psi_j \rangle$ to obtain

$$\begin{aligned} \langle \Psi_j | \Psi', t \rangle &= \sum_k \langle \Psi_j | \hat{\mathrm{O}} | \Psi_k \rangle \langle \Psi_k | \Psi, t \rangle \\ &= \sum_k \mathrm{O}_{jk} \sum_\ell a_\ell(t)\, \langle \Psi_k | \Psi_\ell \rangle = \sum_k \mathrm{O}_{jk}(t) a_k(t). \end{aligned} \tag{6.237}$$

Equating the two results gives

$$b_j(t) = \sum_k \mathrm{O}_{jk}\, a_k(t). \tag{6.238}$$

The coefficients $a_j(t)$ and $b_j(t)$ appearing in the linear superpositions that give $| \Psi, t \rangle$ and $| \Psi', t \rangle$ are viewed as the components of the *state vectors* $\boldsymbol{a}(t)$ and $\boldsymbol{b}(t)$. Result (6.238) then shows that the state vector $\boldsymbol{b}(t)$ results from the action of the *matrix observable* $\mathbf{O}$ on the state vector $\boldsymbol{a}(t)$. In matrix notation (6.238) can therefore be written as

$$\boldsymbol{b}(t) = \mathbf{O} \cdot \boldsymbol{a}(t), \tag{6.239}$$

where the elements of the *matrix observable* $\mathbf{O}$ are defined in (6.237). In a matrix formulation of quantum mechanics, the states of quantum mechanics are replaced by *state vectors* whose components correspond to the complex coefficients of the mixed state superposition (6.29). The inner product in the matrix representation is

$$\langle \Psi', t | \Psi, t \rangle = \sum_{k,j} b_k^*(t)\, a_j(t)\, \langle \Psi_k | \Psi_j \rangle = \sum_{k,j} b_k^*(t)\, a_j(t)\, \delta_{kj} = \sum_j b_j^*(t)\, a_j(t), \tag{6.240}$$

which is identical to the inner product $(\boldsymbol{b}(t), \boldsymbol{a}(t))$ of two complex vectors defined initially by (3.40) and later extended to the linear algebra definition (6.149). If the state $| \Psi, t \rangle$ is normalized, then this means the coefficients of the state vectors satisfy

$$\langle \Psi, t | \Psi, t \rangle = \sum_j a_j^*(t)\, a_j(t) = (\boldsymbol{a}(t), \boldsymbol{a}(t)) = \boldsymbol{a}^\dagger(t) \cdot \boldsymbol{a}(t) = 1. \tag{6.241}$$

This is identical to the requirement that the *state vector* associated with a normalized quantum mechanical state also has a norm of unity. The operator observable $\hat{\mathrm{O}}$ in Dirac notation becomes a *square matrix* whose elements are found by evaluating the action of the observable on the inner product using a choice of basis states *for the matrix representation.*

The matrix Schrödinger equation (6.226) can be written by noting that $\Psi_k(t) = a_k(t)$ in the notation of (6.233). This gives the matrix version of the Schrödinger equation,

$$\mathbf{H} \cdot \boldsymbol{a}(t) = i\hbar\frac{\partial}{\partial t}\boldsymbol{a}(t), \tag{6.242}$$

where $\mathbf{H}$ is the *matrix Hamiltonian* whose elements are given by

$$H_{jk} = \langle \Psi_j \mid \hat{H} \mid \Psi_k \rangle, \tag{6.243}$$

while the state vectors are time-dependent.

Because all observables are associated with Hermitian operators, the definition (3.154) shows that their matrix elements satisfy

$$O_{jk}^* = \langle \Psi_j \mid \hat{\mathrm{O}} \mid \; \rangle \Psi_k^\dagger = \langle \Psi_k \mid \hat{\mathrm{O}}^\dagger \mid \Psi_j \rangle = \langle \Psi_k \mid \hat{\mathrm{O}} \mid \Psi_j \rangle = O_{kj}. \tag{6.244}$$

Since $O_{kj} = O_{jk}^*$, definition (6.244) shows that a Hermitian observable becomes a Hermitian matrix, $\mathbf{O}^\dagger = \mathbf{O}$. Result (6.148) shows that the Hermitian conjugate of (6.238) gives

$$\boldsymbol{b}^\dagger(t) = (\mathbf{O} \cdot \boldsymbol{a}(t))^\dagger = \boldsymbol{a}^\dagger(t) \cdot \mathbf{O}^\dagger = \boldsymbol{a}^\dagger(t) \cdot \mathbf{O}, \tag{6.245}$$

where $\boldsymbol{a}^\dagger$ is the *adjoint state vector.*

The *eigenvector* of the matrix observable $\mathbf{O}$ associated with the eigenvalue λ is denoted $\boldsymbol{a}_\lambda$ and satisfies

$$\mathbf{O} \cdot \boldsymbol{a}_\lambda = \lambda\, \boldsymbol{a}_\lambda. \tag{6.246}$$

It was shown earlier in this chapter that a Hermitian matrix has real eigenvalues, so that $\lambda^*=\lambda$, and that its eigenvectors are orthogonal,

$$\lambda' \neq \lambda \implies (\boldsymbol{a}_{\lambda'}, \boldsymbol{a}_\lambda) = 0. \tag{6.247}$$

All these structures and results are identical to those of both wave mechanics and the operator formulation of quantum mechanics, and so the matrix formulation is simply *another representation of quantum mechanics.* As mentioned earlier, the matrix representation of quantum mechanics was developed by Heisenberg, Born, and Jordan *prior* to the advent of the Schrödinger equation. In the paper presenting the equation that bears his name Schrödinger proved wave mechanics and matrix representations contained equivalent information.

6.4.3 Commuting observables

The matrix representation is visualized easily for the case that the orthonormal eigenstates $\mid \lambda_j \rangle$ of the operator $\hat{\mathrm{O}}$ are used for the representation space. The matrix elements are then given by

$$\mathrm{O}_{jk} = \langle\, \lambda_j \,|\, \hat{\mathrm{O}} \,|\, \lambda_k \,\rangle = \lambda_j \delta_{jk}, \tag{6.248}$$

where λ_j is the eigenvalue for the observable associated with the state $|\, \lambda_j \,\rangle$. For such a choice, the matrix observable $\mathbf{O}$ is represented by a *diagonal matrix* with its eigenvalues appearing along the diagonal with all other elements zero. For this case the jth component of the state vector $\boldsymbol{a}_{\lambda_n}$ associated with the nth eigenvalue λ_n has the particularly simple form $(\boldsymbol{a}_{\lambda_n})_j = \delta_{jn}$, since this gives

$$(\mathbf{O} \cdot \boldsymbol{a}_{\lambda_n})_j = \sum_k \mathrm{O}_{jk} \delta_{kn} = \sum_k \lambda_j \delta_{jk} \delta_{kn} = \lambda_j \delta_{jn} = \lambda_n \delta_{jn} = \lambda_n (\boldsymbol{a}_{\lambda_n})_j. \tag{6.249}$$

The eigenvector of the diagonal matrix is a vector with a single element, which is a one at the nth position.

If two observables commute, then they share the same eigenstates. As a result, a complete set of commuting observables (CSCO) will be represented by a set of diagonal matrices if the representation space is that of their mutual eigenstates. Since diagonal matrices commute, the matrix representation reproduces the commutativity of the CSCO. However, if the matrix version of the observable is *not diagonal* for the chosen representation space, it will not commute with any matrix observables that are diagonal. This is easily seen by using a diagonal matrix observable $\mathbf{A}$, with eigenvalues α_j and elements $A_{jk} = \alpha_j \delta_{jk}$, and commuting it with another matrix observable $\mathbf{B}$ to find,

$$([\mathbf{A}, \mathbf{B}])_{jk} = \sum_\ell (A_{j\ell} B_{\ell k} - B_{j\ell} A_{\ell k}) = \sum_\ell (\alpha_j \delta_{j\ell} B_{\ell k} - B_{j\ell} \alpha_\ell \delta_{\ell k}) = (\alpha_j - \alpha_k) B_{jk}. \tag{6.250}$$

The right-hand side of (6.250) is zero if $B_{jk} = \beta_j \delta_{jk}$, which gives

$$(\alpha_j - \alpha_k) B_{jk} = \beta_j (\alpha_j - \alpha_k) \delta_{jk} = 0. \tag{6.251}$$

As a result, $\mathbf{B}$ must also be diagonal. If two matrix observables belong to a CSCO, then if one is a diagonal matrix then the other one must also be a diagonal matrix. The two are often said to be *simultaneously diagonalizable*.

6.4.4 Diagonalizing matrix observables

Depending upon the choice of representation space, a matrix observable may be non-diagonal. One of the strengths of the matrix representation of quantum mechanics is the straightforward method (6.185) for finding the eigenvalues λ of a matrix observable $\mathbf{O}$, which is to solve the polynomial equation

$$\det(\mathbf{O} - \lambda 1) = 0. \tag{6.252}$$

Once the eigenvalues are known, the associated eigenvectors can be found by solving the eigenvalue system of equations,

$$\sum_k \mathrm{O}_{jk} a_k = \lambda a_j, \tag{6.253}$$

for each λ and applying the normalization condition,

$$\sum_j |a_j|^2 = 1. \tag{6.254}$$

Since the matrix observable is Hermitian, the eigenvectors will be orthonormal. Once the eigenvector for an eigenvalue is found, its components give the coefficients of the superposition (6.29) used to generate the matrix representation, and this yields the orthonormal eigenstates of the observable in terms of the chosen representation space basis.

The drawback to this approach is that the polynomial equation generated by (6.185) is often impossible to solve for a large matrix, much less a matrix that is infinite in dimension. As a result, the matrix representation is most useful when dealing with the reasonably small matrices that result when analysing an observable in *finite subspaces* of the Hilbert space that has been chosen for the representation space. An example of this is presented in a later section where basic perturbation theory is discussed.

6.5 Operator methods in quantum mechanics

It is sometimes useful to work directly with the operator and state formalism. This applies to both the abstract Hilbert space version and the matrix formulation of quantum mechanics. Such an approach uses the commutation relations between observables to determine their time evolution and expectation values. The unitary time evolution operator $\hat{U} = e^{-i\hat{H}t/\hbar}$ defined in (6.72) is a useful starting point for discussing operator methods, since $\hat{U}(t)$ governs the time development of the state in the Schrödinger picture and the time development of operators in the Heisenberg picture. Clearly, if the Hamiltonian $\hat{H}$ commutes with an operator $\hat{\mathrm{O}}$, then $\hat{U}(t)$ also commutes with $\hat{\mathrm{O}}$ and the Heisenberg picture operator will be time-independent,

$$\hat{\mathrm{O}}_H(t) = e^{i\hat{H}t/\hbar}\,\hat{\mathrm{O}}\,e^{-i\hat{H}t/\hbar} = e^{i\hat{H}t/\hbar}e^{-i\hat{H}t/\hbar}\hat{\mathrm{O}} = \hat{\mathrm{O}}. \tag{6.255}$$

However, if an operator does not commute with the Hamiltonian then its Heisenberg picture form will be time-dependent. It is important to understand the time-dependence of the resulting operator.

6.5.1 The Baker–Campbell–Hausdorff theorem

In general, if $\hat{\mathrm{A}}$ and $\hat{\mathrm{B}}$ are two operators that do not commute, it is useful to evaluate the operator function $\hat{\mathrm{F}}(x)$ defined by

$$\hat{\mathrm{F}}(x) = e^{x\hat{\mathrm{A}}}\hat{\mathrm{B}}e^{-x\hat{\mathrm{A}}}. \tag{6.256}$$

This begins by expanding the operator $\hat{\mathrm{F}}(x)$ in a Taylor series in x,

$$\hat{\mathrm{F}}(x) = \sum_{n=0}^{\infty}\frac{1}{n!}\hat{\mathrm{F}}_n x^n, \tag{6.257}$$

where the coefficients $\hat{F}_n$ in the series expansion of $\hat{F}(x)$ are operators determined from $\hat{A}$ and $\hat{B}$ and their commutation relation in order to make the two forms identical. Expression (6.256) is identical in structure to the Heisenberg picture operator (6.255), and so determining the function $\hat{F}(x)$ will be useful in evaluating the Heisenberg picture operators.

Applying the derivative to the first form for $\hat{F}(x)$ and combining it with the expansion form gives

$$\frac{d\hat{F}}{dx} = \hat{A}e^{x\hat{A}}\hat{B}e^{-x\hat{A}} - e^{x\hat{A}}\hat{B}e^{-x\hat{A}}\hat{A} = [\hat{A}, \hat{F}(x)] = \sum_{n=0}^{\infty}\frac{1}{n!}[\hat{A}, \hat{F}_n]x^n. \tag{6.258}$$

Applying the derivative directly to the series expansion gives

$$\frac{d\hat{F}}{dx} = \sum_{n=1}^{\infty}\frac{1}{(n-1)!}\hat{F}_n x^{n-1} = \sum_{n=0}^{\infty}\frac{1}{n!}\hat{F}_{n+1}x^n. \tag{6.259}$$

Since (6.258) and (6.259) must be equal order by order in x, an *operator recurrence relationship* is obtained,

$$\hat{F}_{n+1} = [\hat{A}, \hat{F}_n]. \tag{6.260}$$

The definition (6.256) gives

$$\hat{F}(0) = \hat{B} = \hat{F}_0. \tag{6.261}$$

All other terms in the series expansion are then immediately determined from the recurrence relationship (6.260). For example, the next two terms in the expansion are given by

$$\begin{aligned}\hat{F}_1 &= [\hat{A}, \hat{F}_0] = [\hat{A}, \hat{B}],\\ \hat{F}_2 &= [\hat{A}, \hat{F}_1] = [\hat{A}, [\hat{A}, \hat{B}]],\end{aligned} \tag{6.262}$$

with higher order terms following. The final result is

$$\begin{aligned}e^{x\hat{A}}\,\hat{B}\,e^{-x\hat{A}} &= \sum_{n=0}^{\infty}\frac{1}{n!}\hat{F}_n x^n\\ &= \frac{1}{0!}\hat{B} + \frac{x}{1!}[\hat{A}, \hat{B}] + \frac{x^2}{2!}[\hat{A}, [\hat{A}, \hat{B}]] + \frac{x^3}{3!}[\hat{A}, [\hat{A}, [\hat{A}, \hat{B}]]] + \ldots.\end{aligned} \tag{6.263}$$

Contact with the Heisenberg picture operator (6.78) is made by the choice

$$x\hat{A} = i\hat{H}t/\hbar \implies e^{-x\hat{A}} = e^{-i\hat{H}t/\hbar}, \tag{6.264}$$

which gives the time evolution operator $\hat{U}(t)$ used to define the Heisenberg picture operator. Result (6.263) can then be used to evaluate the Heisenberg picture observable. A simple choice for the Hamiltonian is

$$\hat{H} = \frac{1}{2m}\hat{\boldsymbol{P}}^2, \tag{6.265}$$

so that the time evolution operator describes a *free particle*. Simultaneously choosing $\hat{\mathrm{B}} = \hat{Q}_j$ gives the Heisenberg picture position operator for a free particle. The single commutator of importance is identical to (3.188), so that

$$\left[\hat{P}^2, \hat{Q}_j\right] = -2i\hbar\hat{P}_j, \tag{6.266}$$

which commutes with the Hamiltonian. As a result, only the first and second terms on the right-hand side of (6.263) contribute,

$$\hat{Q}_{jH}(t) = \hat{Q}_j + \frac{t}{1!}\left[\frac{i}{\hbar}\frac{\hat{P}^2}{2m}, \hat{Q}_j\right], \tag{6.267}$$

with all higher order commutators vanishing. The exact result is

$$\hat{Q}_{jH}(t) = \hat{U}^\dagger(t)\hat{Q}_j\hat{U}(t) = e^{i\hat{H}t/\hbar}\hat{Q}_j e^{-i\hat{H}t/\hbar} = \hat{Q}_j + \frac{\hat{P}_j t}{m}. \tag{6.268}$$

The result (6.268) is identical in form to the classical result,

$$x_j(t) = x_j(0) + \frac{p_j(0)\, t}{m}, \tag{6.269}$$

which is obtained by solving Hamilton's equations (1.23) for $H = p^2/2m$. Obviously, more complicated Hamiltonians lead to more complicated results for the Heisenberg picture operators.

There are several useful results that can derived from (6.263). Of chief importance in quantum mechanics is how the product of exponentiated operators $\hat{\mathrm{A}}$ and $\hat{\mathrm{B}}$ combine to form a single exponentiated operator. In order to analyse this, the product is written as

$$e^{x\hat{\mathrm{A}}}e^{x\hat{\mathrm{B}}} = e^{\hat{\mathrm{G}}(x)} \implies e^{-\hat{\mathrm{G}}(x)} = e^{-x\hat{\mathrm{B}}}e^{-x\hat{\mathrm{A}}}, \tag{6.270}$$

where the operator $\hat{\mathrm{G}}(x)$ is to be determined from $\hat{\mathrm{A}}$ and $\hat{\mathrm{B}}$. The first step is to note that

$$e^{-\hat{\mathrm{G}}(x)}\frac{\mathrm{d}}{\mathrm{d}x}e^{\hat{\mathrm{G}}(x)} = e^{-x\hat{\mathrm{B}}}e^{-x\hat{\mathrm{A}}}\frac{\mathrm{d}}{\mathrm{d}x}e^{x\hat{\mathrm{A}}}e^{x\hat{\mathrm{B}}} = e^{-x\hat{\mathrm{B}}}\hat{\mathrm{A}}e^{-x\hat{\mathrm{B}}} + \hat{\mathrm{B}} + \frac{\mathrm{d}}{\mathrm{d}x}. \tag{6.271}$$

Using (6.263) allows the right-hand side of (6.271) to be written

$$e^{-\hat{\mathrm{G}}(x)}\frac{\mathrm{d}}{\mathrm{d}x}e^{\hat{\mathrm{G}}(x)} = \frac{\mathrm{d}}{\mathrm{d}x} + \hat{\mathrm{A}} + \hat{\mathrm{B}} - \frac{x}{1!}[\hat{\mathrm{B}}, \hat{\mathrm{A}}] + \frac{x^2}{2!}[\hat{\mathrm{B}}, [\hat{\mathrm{B}}, \hat{\mathrm{A}}]] + \ldots. \tag{6.272}$$

where the alternation in the signs of the terms is brought about by the reversal of sign for x in (6.263).

The next step is to treat d/dx as the operator $\hat{B}$ in (6.263), which gives

$$e^{-\hat{G}(x)}\frac{d}{dx}e^{\hat{G}(x)} = \frac{d}{dx} - \frac{1}{1!}\left[\hat{G}(x), \frac{d}{dx}\right] + \frac{1}{2!}\left[\hat{G}(x), \left[\hat{G}(x), \frac{d}{dx}\right]\right] - .... \quad (6.273)$$

The commutators in (6.273) are given by

$$\begin{aligned}\left[\hat{G}(x), \frac{d}{dx}\right] &= \hat{G}(x)\frac{d}{dx} - \frac{d}{dx}\hat{G}(x) \\ &= \hat{G}(x)\frac{d}{dx} - \frac{d\hat{G}(x)}{dx} - \hat{G}(x)\frac{d}{dx} = -\frac{d\hat{G}(x)}{dx}.\end{aligned} \quad (6.274)$$

so that (6.273) becomes

$$e^{-\hat{G}(x)}\frac{d}{dx}e^{\hat{G}(x)} = \frac{d}{dx} + \frac{1}{1!}\frac{d\hat{G}(x)}{dx} + \frac{1}{2!}\left[\frac{d\hat{G}(x)}{dx}, \hat{G}(x)\right] + \cdots, \quad (6.275)$$

where the antisymmetric nature of the commutators allows the reordering.

The next step is to expand $\hat{G}(x)$ in a power series in x, so that

$$\hat{G}(x) = \hat{G}_1 x + \hat{G}_2 x^2 + \hat{G}_3 x^3 + \cdots. \quad (6.276)$$

This expansion is consistent with the requirement that $\hat{G}(0) = 0$ in order for (6.270) to hold. It follows that

$$\frac{d\hat{G}(x)}{dx} = \hat{G}_1 + 2\hat{G}_2 x + 3\hat{G}_3 x^2 + \cdots. \quad (6.277)$$

Substituting (6.276) and (6.277) into (6.275) generates a power series in x and the coefficients $\hat{G}_n$. The first three terms are given by

$$e^{-\hat{G}(x)}\frac{d}{dx}e^{\hat{G}(x)} = \frac{d}{dx} + \hat{G}_1 + 2\hat{G}_2 x + \left(3\hat{G}_3 - \frac{1}{2}[\hat{G}_1, \hat{G}_2]\right)x^2 + \cdots. \quad (6.278)$$

This must be equal order by order to the expansion given by (6.272). This allows a complete determination of all coefficients $\hat{G}_n$, with the first three given by

$$\begin{aligned}\hat{G}_1 &= \hat{A} + \hat{B}, \\ \hat{G}_2 &= \frac{1}{2!}[\hat{A}, \hat{B}], \\ \hat{G}_3 &= \frac{1}{3!}\left(\frac{1}{2}[[\hat{A}, \hat{B}], \hat{B}] + \frac{1}{2}[\hat{A}, [\hat{A}, \hat{B}]]\right).\end{aligned} \quad (6.279)$$

Higher order terms include higher order commutators. This result is known as the *Baker–Campbell–Hausdorff theorem.*

Result (6.279) simplifies considerably for the case that the commutator $[\hat{A}, \hat{B}]$ commutes with both $\hat{A}$ and $\hat{B}$. If this condition holds, only $\hat{G}_1$ and $\hat{G}_2$ are nonzero. For this special case, setting $x = 1$ in (6.270) gives

$$e^{\hat{A}}e^{\hat{B}} = e^{\hat{A}+\hat{B}+\frac{1}{2}[\hat{A},\hat{B}]} \implies e^{\hat{A}+\hat{B}} = e^{\hat{A}}e^{\hat{B}}e^{-\frac{1}{2}[\hat{A},\hat{B}]}. \tag{6.280}$$

This relation is also a useful first approximation to the product of the exponentiated operators.

6.5.2 The Feynman–Hellmann theorem

The Feynman–Hellmann theorem generalizes the earlier Ehrenfest result (4.32) and is useful for evaluating the expectation values of operators. It starts with the general assumption of a Hermitian operator, $\hat{O}(\lambda)$, which is a function of some parameter λ. For example, the operator could be the harmonic oscillator Hamiltonian $\hat{H}(\omega)$, which is a function of the natural frequency ω. It is assumed this Hermitian operator has a set of normalized eigenstates $|\, e_n(\lambda)\,\rangle$, which satisfy

$$\begin{aligned} \hat{O}(\lambda)|\, e_n(\lambda)\,\rangle &= e_n(\lambda)|\, e_n(\lambda)\,\rangle, \\ \langle\, e_n(\lambda)\,|\hat{O}(\lambda) &= e_n(\lambda)\langle\, e_n(\lambda)\,|, \\ \langle\, e_n(\lambda)\,|\, e_n(\lambda)\,\rangle &= 1, \end{aligned} \tag{6.281}$$

where it is assumed that eigenvalues $e(\lambda)$ of $\hat{O}(\lambda)$ may also be characterized by a dependence on the parameter λ. Such is the case with the energy eigenvalues of the harmonic oscillator Hamiltonian.

The proof of the Feynman–Hellman theorem begins by noting that the normalization condition gives

$$\frac{\partial}{\partial\lambda}(\langle\, e_n(\lambda)\,|\, e_n(\lambda)\,\rangle) = \left(\frac{\partial}{\partial\lambda}\langle\, e_n(\lambda)\,|\right)|\, e_n(\lambda)\,\rangle + \langle\, e_n(\lambda)\,|\left(\frac{\partial}{\partial\lambda}|\, e_n(\lambda)\,\rangle\right) = 0. \tag{6.282}$$

The next step is to note that the expectation value of $\hat{O}(\lambda)$ gives

$$\frac{\partial}{\partial\lambda}(\langle\, e_n(\lambda)\,|\,\hat{O}(\lambda)\,|\, e_n(\lambda)\,\rangle) = \frac{\partial e_n(\lambda)}{\partial\lambda}. \tag{6.283}$$

Combining the Leibniz rule with (6.281) and (6.282) gives

$$\begin{aligned} &\frac{\partial}{\partial\lambda}\big(\langle\, e_n(\lambda)\,|\,\hat{O}(\lambda)\,|\, e_n(\lambda)\,\rangle\big) \\ &= \left(\frac{\partial}{\partial\lambda}\langle\, e_n(\lambda)\,|\right)\hat{O}(\lambda)|\, e_n(\lambda)\,\rangle + \langle\, e_n(\lambda)\,|\hat{O}(\lambda)\left(\frac{\partial}{\partial\lambda}|\, e_n(\lambda)\,\rangle\right) + \langle\, e_n(\lambda)\,|\,\frac{\partial\hat{O}(\lambda)}{\partial\lambda}\,|\, e_n(\lambda)\,\rangle \\ &= e_n(\lambda)\left(\frac{\partial}{\partial\lambda}\langle\, e_n(\lambda)\,|\right)|\, e_n(\lambda)\,\rangle + e_n(\lambda)\langle\, e_n(\lambda)\,|\left(\frac{\partial}{\partial\lambda}|\, e_n(\lambda)\,\rangle\right) + \langle\, e_n(\lambda)\,|\,\frac{\partial\hat{O}(\lambda)}{\partial\lambda}\,|\, e_n(\lambda)\,\rangle \\ &= e_n(\lambda)\frac{\partial}{\partial\lambda}\langle\, e_n(\lambda)\,|\, e_n(\lambda)\,\rangle + \langle\, e_n(\lambda)\,|\,\frac{\partial\hat{O}(\lambda)}{\partial\lambda}\,|\, e_n(\lambda)\,\rangle \\ &= \langle\, e_n(\lambda)\,|\,\frac{\partial\hat{O}(\lambda)}{\partial\lambda}\,|\, e_n(\lambda)\,\rangle. \end{aligned} \tag{6.284}$$

Combining this with (6.283) immediately gives the Feynman–Hellmann theorem,

$$\frac{\partial e_n(\lambda)}{\partial \lambda} = \langle\, e_n(\lambda) \mid \frac{\partial \hat{O}(\lambda)}{\partial \lambda} \mid e_n(\lambda)\, \rangle. \tag{6.285}$$

For example, using the one-dimensional harmonic oscillator Hamiltonian (5.35) gives the operator

$$\frac{\partial \hat{H}(\omega)}{\partial \omega} = m\omega \hat{Q}^2. \tag{6.286}$$

Noting that the eigenvalues of $\hat{H}(\omega)$ are $E_n(\omega) = (n + \frac{1}{2})\hbar\omega$, the Feynman–Hellmann theorem gives

$$m\omega\langle\, E_n(\omega) \mid \hat{Q}^2 \mid E_n(\omega)\, \rangle = \frac{\partial E_n(\omega)}{\partial \omega} = (n + \frac{1}{2})\hbar, \tag{6.287}$$

which matches the ground state result (5.57) for $n = 0$. The Feynman–Hellman theorem often gives a simple method to find expectation values of importance.

6.5.3 The virial theorem

The classical version of the virial theorem for a single particle states that the time-averaged value of the kinetic energy K is related to the time-averaged value of the quantity $-\frac{1}{2}\boldsymbol{x} \cdot \nabla V(x)$, where $V(x)$ is the potential energy. This is written

$$\langle K \rangle - \frac{1}{2}\langle \boldsymbol{x} \cdot \nabla V(x) \rangle = 0. \tag{6.288}$$

The proof of this theorem in classical mechanics is available in the references.

The quantum mechanical counterpart is given in terms of expectation values. It begins by noting that the commutator of the Hamiltonian $\hat{H} = \hat{P}^2/2\,m + V(\hat{Q})$ with the quantity $\hat{\boldsymbol{Q}} \cdot \boldsymbol{P}$ is given by

$$[\hat{H},\, \hat{\boldsymbol{Q}} \cdot \hat{\boldsymbol{P}}] = -i\hbar\left(\frac{\hat{P}^2}{m} - \hat{\boldsymbol{Q}} \cdot \nabla V(\hat{Q})\right). \tag{6.289}$$

Demonstrating (6.289) follows from the two commutators

$$\begin{aligned} \left[\hat{P}^2,\, \hat{Q}_j\right] &= -\,2i\hbar \hat{P}_j, \\ \left[V(\hat{Q}),\, \hat{P}_j\right] &= i\hbar \frac{\partial V(\hat{Q})}{\partial Q_j}. \end{aligned} \tag{6.290}$$

These were derived in (3.189) and (3.188). Taking the expectation value of (6.289) using an energy eigenstate gives

$$\begin{aligned} \langle\, E_n \,|[\hat{H},\, \hat{\boldsymbol{Q}} \cdot \hat{\boldsymbol{P}}]|\, E_n\, \rangle &= \langle\, E_n \mid \hat{H}\hat{\boldsymbol{Q}} \cdot \hat{\boldsymbol{P}} \mid E_n\, \rangle - \langle\, E_n \mid \hat{\boldsymbol{Q}} \cdot \hat{\boldsymbol{P}}\hat{H} \mid E_n\, \rangle \\ &= (E_n - E_n)\langle\, E_n \mid \hat{\boldsymbol{Q}} \cdot \hat{\boldsymbol{P}} \mid E_n\, \rangle = 0. \end{aligned} \tag{6.291}$$

Replacing the commutator in (6.291) with the right-hand side of (6.289) and dropping the common factor of $i\hbar$ gives the quantum mechanical virial theorem,

$$\langle E_n | \frac{\hat{P}^2}{2m} | E_n \rangle - \frac{1}{2}\langle E_n | \hat{\boldsymbol{Q}} \cdot \nabla V(\hat{\boldsymbol{Q}}) | E_n \rangle = \langle \hat{K} \rangle - \frac{1}{2}\langle \hat{\boldsymbol{Q}} \cdot \nabla V(\hat{\boldsymbol{Q}}) \rangle = 0. \quad (6.292)$$

Like the Feynman–Hellmann theorem, the virial theorem (6.292) is useful for calculating operator products. Considering the case of the one-dimensional simple harmonic oscillator, it follows from result (6.287) that the expectation value of the kinetic energy in the nth energy eigenstate is given by

$$\begin{aligned} \langle \hat{K} \rangle &= \frac{1}{2}\langle E_n | \hat{Q}\frac{\partial}{\partial \hat{Q}}\left(\frac{1}{2}m\omega^2\hat{Q}^2\right) | E_n \rangle \\ &= \frac{1}{2}m\omega^2\langle E_n | \hat{Q}^2 | E_n \rangle = \langle \hat{V} \rangle = \frac{1}{2}(n + \frac{1}{2})\hbar\omega, \end{aligned} \quad (6.293)$$

which shows that the expectation value of the total energy is evenly divided between the kinetic energy and the potential energy.

6.5.4 Harmonic oscillator coherent states

It is instructive to apply operator techniques to the one-dimensional harmonic oscillator analysed in chapter 5. In Dirac notation the Schrödinger picture harmonic oscillator energy eigenstates are denoted $| n, t \rangle$ and satisfy

$$\hat{H}(\hat{P}, \hat{Q}) | n, t \rangle = (n + \frac{1}{2})\hbar\omega | n, t \rangle. \quad (6.294)$$

The annihilation and creation operators defined in (5.71) are written

$$\begin{aligned} \hat{a} &= \frac{1}{\sqrt{2}}\left(\alpha\hat{Q} + \frac{i}{\hbar\alpha}\hat{P}\right), \\ \hat{a}^\dagger &= \frac{1}{\sqrt{2}}\left(\alpha\hat{Q} - \frac{i}{\hbar\alpha}\hat{P}\right), \end{aligned} \quad (6.295)$$

where $\alpha = \sqrt{m\omega/\hbar}$. The quantization condition (6.49) maintains the commutation relation

$$[\hat{a}, \hat{a}^\dagger] = 1, \quad (6.296)$$

while the Hamiltonian is still given by

$$\hat{H} = \hbar\omega(\hat{a}^\dagger\hat{a} + \frac{1}{2}). \quad (6.297)$$

The Dirac notation Schrödinger picture ground state is written $| 0, t \rangle$ and satisfies

$$\hat{a}| 0, t \rangle = 0. \quad (6.298)$$

The Schrödinger picture higher energy or excited states are obtained by applying powers of $\hat{a}^\dagger$ to the ground state, so that

$$\begin{aligned} |\, n,\, t\, \rangle &= \frac{1}{\sqrt{n!}}\hat{a}^{\dagger n}|\, 0,\, t\, \rangle, \\ \langle\, n,\, t\, | &= \langle\, 0,\, t\, |\hat{a}^{n}\frac{1}{\sqrt{n!}}. \end{aligned} \tag{6.299}$$

These states are orthonormal,

$$\langle\, n,\, t\, |\, n',\, t\, \rangle = \delta_{nn'}. \tag{6.300}$$

Recalling that $\hat{N} = \hat{a}^\dagger\hat{a}$ is the number operator, the Heisenberg picture creation operator can be found by applying (6.263) with the identification

$$x\hat{\mathrm{A}} = i\hat{H}t\Big/\hbar = i\omega(\hat{N} + \frac{1}{2})t, \tag{6.301}$$

so that $x = i\omega t$, $\hat{\mathrm{A}} = \hat{N}$, and $\hat{\mathrm{B}} = \hat{a}^\dagger$ in (6.263). Since (5.82) gives

$$[\hat{N},\, \hat{a}^\dagger] = \hat{a}^\dagger, \tag{6.302}$$

it follows that

$$[\hat{N},\, [\dots,\, [\hat{N},\, \hat{a}^\dagger]]] = \hat{a}^\dagger. \tag{6.303}$$

Using (6.263) gives the Heisenberg picture creation operator,

$$\hat{a}_H^\dagger(t) = \hat{U}^\dagger(t)\hat{a}^\dagger\hat{U}(t) = \sum_{n=0}^{\infty}\frac{1}{n!}(i\omega t)^n\hat{a}^\dagger = \hat{a}^\dagger e^{i\omega t}. \tag{6.304}$$

Similarly, $[\hat{N},\, \hat{a}] = -\hat{a}$ gives $\hat{a}_H(t) = \hat{a}\, e^{-i\omega t}$, which is consistent with (6.304).

Result (6.280) allows the definition of a useful mixed state of the harmonic oscillator, first analysed by Glauber and known as a *coherent state*. A coherent state $|\, \lambda,\, t\, \rangle$ is defined by combining a *complex number* λ with the annihilation and creation operators,

$$|\, \lambda,\, t\, \rangle = e^{(\lambda\hat{a}^\dagger - \lambda^*\hat{a})}|\, 0,\, t\, \rangle \implies \langle\, \lambda,\, t\, | = |\, \lambda,\, t\, \rangle^\dagger = \langle\, 0,\, t\, |e^{-(\lambda\hat{a}^\dagger - \lambda^*\hat{a})}. \tag{6.305}$$

Identifying $\hat{\mathrm{A}} = \lambda\hat{a}^\dagger$ and $\hat{\mathrm{B}} = -\lambda^*\hat{a}$ shows that $[\hat{\mathrm{A}},\, \hat{\mathrm{B}}] = \lambda^*\lambda$, which commutes with both $\hat{\mathrm{A}}$ and $\hat{\mathrm{B}}$. The condition required for (6.280) holds and the coherent state can be written

$$|\, \lambda,\, t\, \rangle = e^{(\lambda\hat{a}^\dagger - \lambda^*\hat{a})}|\, 0,\, t\, \rangle = e^{\lambda\hat{a}^\dagger}e^{-\lambda^*\hat{a}}|\, 0,\, t\, \rangle e^{-\frac{1}{2}\lambda^*\lambda} = e^{\lambda\hat{a}^\dagger}|\, 0,\, t\, \rangle e^{-\frac{1}{2}|\lambda|^2}, \tag{6.306}$$

where the final step follows from using $\hat{a}|\, 0,\, t\, \rangle = 0$ and the power series for $e^{-\lambda^*\hat{a}}$.

Using the normalized energy eigenstate $|\, n,\, t\, \rangle$ of (6.299) allows (6.306) to be written

$$| \lambda, t \rangle = e^{-\frac{1}{2}|\lambda|^2} e^{\lambda \hat{a}^\dagger} | 0, t \rangle = e^{-\frac{1}{2}|\lambda|^2} \sum_{n=0}^{\infty} \frac{\lambda^n \hat{a}^{\dagger n}}{n!} | 0, t \rangle = e^{-\frac{1}{2}|\lambda|^2} \sum_{n=0}^{\infty} \frac{\lambda^n}{\sqrt{n!}} | n, t \rangle, \quad (6.307)$$

revealing the coherent state as a superposition of energy eigenstates. In effect, the coherent state is a harmonic oscillator wave packet. Combining orthonormality of the energy eigenstates with result (6.307) gives the inner product of two coherent states as

$$\begin{aligned} \langle \sigma, t \mid \lambda, t \rangle &= e^{-\frac{1}{2}|\sigma|^2 - \frac{1}{2}|\lambda|^2} \sum_{n,m=0}^{\infty} \frac{(\sigma^*)^m \lambda^n}{\sqrt{m!n!}} \langle m, t \mid n, t \rangle \\ &= e^{-\frac{1}{2}|\sigma|^2 - \frac{1}{2}|\lambda|^2} \sum_{n=0}^{\infty} \frac{(\sigma^* \lambda)^n}{n!} = e^{-\frac{1}{2}|\sigma|^2 - \frac{1}{2}|\lambda|^2 + \sigma^* \lambda}, \end{aligned} \quad (6.308)$$

demonstrating that the coherent states are *not* orthogonal although they are normalized since $\langle \lambda, t \mid \lambda, t \rangle = 1$.

The coherent state unit projection operator is defined using integration over the entire complex plane, so that $\mathrm{d}^2\lambda = \mathrm{d}\lambda_R\, \mathrm{d}\lambda_I$, which is given by $r\, \mathrm{d}r\, \mathrm{d}\theta$ in the polar representation $\lambda = re^{i\theta}$. The projection operator $\hat{\mathrm{P}}_\lambda$ is first defined, and then evaluated using (6.307) to show that

$$\begin{aligned} \hat{\mathrm{P}}_\lambda &\equiv \int \frac{\mathrm{d}^2\lambda}{\pi} | \lambda, t \rangle\langle \lambda, t | = \int_0^{2\pi} \frac{\mathrm{d}\theta}{\pi} \int_0^{\infty} \mathrm{d}r\, r\, e^{-|\lambda|^2} \sum_{m,n=0}^{\infty} \frac{\lambda^n \lambda^{*m}}{\sqrt{n!m!}} | n, t \rangle\langle m, t | \\ &= \sum_{n,m=0}^{\infty} \int_0^{\infty} \mathrm{d}r\, r \frac{r^{n+m}}{\sqrt{n!m!}} e^{-r^2} | n, t \rangle\langle m, t | \int_0^{2\pi} \frac{\mathrm{d}\theta}{\pi} e^{i(n-m)\theta} \\ &= \sum_{n=0}^{\infty} \left(\int_0^{\infty} \mathrm{d}r \frac{2}{n!} r^{2n+1} e^{-r^2} \right) | n, t \rangle\langle n, t | = \sum_{n=0}^{\infty} | n, t \rangle\langle n, t | = \hat{1}, \end{aligned} \quad (6.309)$$

where the completeness of the energy eigenstates has identified $\hat{\mathrm{P}}_\lambda$ as a unit operator. The set of coherent states are complete but not orthogonal and are referred to as *overcomplete*.

The coherent state is an *eigenstate of the annihilation operator*. This is demonstrated by first writing the action of $\hat{a}$ on the coherent state as

$$\hat{a} | \lambda, t \rangle = \hat{a}\, e^{(\lambda \hat{a}^\dagger - \lambda^* \hat{a})} | 0, t \rangle = e^{(\lambda \hat{a}^\dagger - \lambda^* \hat{a})} (e^{-(\lambda \hat{a}^\dagger - \lambda^* \hat{a})}\, \hat{a}\, e^{(\lambda \hat{a}^\dagger - \lambda^* \hat{a})}) | 0, t \rangle, \quad (6.310)$$

and then evaluating the term in parentheses using (6.263). Since $[\hat{a}, \hat{a}^\dagger] = 1$ it commutes with both $\hat{a}$ and $\hat{a}^\dagger$ and the series given by (6.263) terminates after one term, with the result

$$\hat{a} | \lambda, t \rangle = e^{(\lambda \hat{a}^\dagger - \lambda^* \hat{a})} (\hat{a} + \lambda) | 0, t \rangle = \lambda\, e^{(\lambda \hat{a}^\dagger - \lambda^* \hat{a})} | 0, t \rangle = \lambda | \lambda, t \rangle, \quad (6.311)$$

where $\hat{a} | 0, t \rangle = 0$ was used. Result (6.311) immediately gives

$$\langle \lambda, t | \hat{a}^\dagger = (\hat{a} | \lambda, t \rangle)^\dagger = (\lambda | \lambda, t \rangle)^\dagger = \langle \lambda, t | \lambda^*. \quad (6.312)$$

Results (6.311) and (6.312) can be used to find the expectation values for position and momentum associated with a coherent state by using the definitions of the annihilation and creation operators (5.71), so that

$$\begin{aligned} \hat{Q} &= \frac{1}{\alpha\sqrt{2}}(\hat{a} + \hat{a}^\dagger), \\ \hat{P} &= \frac{i\hbar\alpha}{\sqrt{2}}(\hat{a} - \hat{a}^\dagger). \end{aligned} \tag{6.313}$$

It follows that the coherent state expectation value for position is given by

$$\langle \hat{Q} \rangle_\lambda = \frac{1}{\sqrt{2}\,\alpha}\langle \lambda, t \mid (\hat{a} + \hat{a}^\dagger) \mid \lambda, t \rangle = \frac{1}{\sqrt{2}\,\alpha}(\lambda + \lambda^*), \tag{6.314}$$

with a similar result for momentum,

$$\langle \hat{P} \rangle_\lambda = \frac{i\alpha\hbar}{\sqrt{2}}\langle \lambda, t \mid (\hat{a}^\dagger - \hat{a}) \mid \lambda, t \rangle = \frac{i\alpha\hbar}{\sqrt{2}}(\lambda^* - \lambda). \tag{6.315}$$

Using the results

$$\begin{aligned} (\hat{a} + \hat{a}^\dagger)^2 &= \hat{a}^2 + \hat{a}\hat{a}^\dagger + \hat{a}^\dagger\hat{a} + \hat{a}^{\dagger 2} = \hat{a}^2 + 2\hat{a}^\dagger\hat{a} + \hat{a}^{\dagger 2} + [\hat{a}, \hat{a}^\dagger] = \hat{a}^2 + 2\hat{a}^\dagger\hat{a} + \hat{a}^{\dagger 2} + 1, \\ (\hat{a}^\dagger - \hat{a})^2 &= \hat{a}^2 - 2\hat{a}^\dagger\hat{a} + \hat{a}^{\dagger 2} - [\hat{a}, \hat{a}^\dagger] = \hat{a}^2 - 2\hat{a}^\dagger\hat{a} + \hat{a}^{\dagger 2} - 1, \end{aligned} \tag{6.316}$$

it follows that

$$\begin{aligned} \langle \hat{Q}^2 \rangle_\lambda &= \frac{1}{2\alpha^2}\langle \lambda, t \mid (\hat{a} + \hat{a}^\dagger)^2 \mid \lambda, t \rangle = \frac{1}{2\alpha^2}(\lambda + \lambda^*)^2 + \frac{1}{2\alpha^2} \\ \langle \hat{P}^2 \rangle_\lambda &= -\frac{\hbar^2\alpha^2}{2}\langle \lambda, t \mid (\hat{a}^\dagger - \hat{a})^2 \mid \lambda, t \rangle = -\frac{\hbar^2\alpha^2}{2}(\lambda - \lambda^*)^2 + \frac{\hbar^2\alpha^2}{2}. \end{aligned} \tag{6.317}$$

These results can be used in definition (4.125) to find the uncertainties in position and momentum associated with a coherent state. The results are

$$\begin{aligned} (\Delta x)^2 &= \langle \hat{X}^2 \rangle_\lambda - \langle \hat{X} \rangle_\lambda^2 = \frac{1}{2}\alpha^2 = \frac{\hbar}{2\,m\omega}, \\ (\Delta p)^2 &= \langle \hat{P}^2 \rangle_\lambda - \langle \hat{P} \rangle_\lambda^2 = \frac{1}{2}\hbar^2\alpha^2 = \frac{1}{2}\hbar m\omega. \end{aligned} \tag{6.318}$$

These results give

$$\Delta x \, \Delta p = \frac{1}{2}\hbar, \tag{6.319}$$

which is the minimum uncertainty relationship. Therefore, the coherent state is a *minimum uncertainty wave packet*. Due to their symmetric treatment of momentum and position, their minimum uncertainty, and their easily expressed completeness, coherent states and their extensions find a large number of applications in physics, including quantum field theory and quantum optics.

6.6 Matrix analysis of the Hamiltonian

It can be the case that the Hamiltonian governing a quantum mechanical system is impossible to solve for its exact eigenvalues. This often occurs when the quantum mechanical system is subject to complicated external forces such as electromagnetic fields. Rather than abandoning the analysis of such systems, it is often possible to find approximate values for the energy eigenvalues using matrix techniques.

6.6.1 Basic concepts of perturbation theory

A common application of the matrix representation is to find approximate eigenvalues of an operator Hamiltonian $\hat{H}$ using the eigenstates of a simpler operator Hamiltonian. It is often the case that the Hamiltonian can be written as a sum,

$$\hat{H} = \hat{H}_0 + \hat{H}_\mathrm{I}, \tag{6.320}$$

where $\hat{H}_0$ is a Hamiltonian whose eigenvalues E_{0n} and the complete set of associated eigenstates $|\, E_{0n} \,\rangle$ are exactly known, satisfying

$$\hat{H}_0|\, E_{0n} \,\rangle = E_{0n}|\, E_{0n} \,\rangle. \tag{6.321}$$

Examples include the harmonic oscillator or the hydrogen atom. For such a case $\hat{H}_0$ is known as the *basis Hamiltonian* and $\hat{H}_\mathrm{I}$ is referred to as the *perturbation. If the perturbation is weak*, then it is reasonable to expect that the eigenstates of the full Hamiltonian $\hat{H}$ can be written as a superposition of the eigenstates of $\hat{H}_0$. This assumption is easily made but often impossible to justify rigorously. However, this assumption often leads to a simplified analysis of the full quantum mechanical system.

The term *weak* is typically used to mean that the magnitude of the coefficients appearing in $\hat{H}_\mathrm{I}$ are small. In the perturbative approach to analysis it is often assumed that it is valid to separate the evolution operator (6.73) into two distinct factors,

$$\hat{U}(t) = e^{-i\hat{H}t/\hbar} = e^{-i(\hat{H}_0+\hat{H}_\mathrm{I})t/\hbar} \approx e^{-i\hat{H}_\mathrm{I}t/\hbar}e^{-i\hat{H}_0t/\hbar}, \tag{6.322}$$

Identifying $\hat{A} = -i\hat{H}_0t/\hbar$ and $\hat{B} = -i\hat{H}_\mathrm{I}t/\hbar$ in the BCH theorem (6.279) shows that the commutators of $\hat{H}_0$ and $\hat{H}_\mathrm{I}$ appearing in the BCH expansion will generate a power series in the coefficients of $\hat{H}_\mathrm{I}$. For a weak perturbation, the powers of these small coefficients will rapidly converge to zero, indicating that ignoring the commutators in the BCH theorem is at least approximately valid. However, this ignores the nature of the operators generated by the commutators, and there is no guarantee that they do not offset the small coefficients. It therefore must remain an *assumption* that the eigenvalues and eigenstates of $\hat{H}$ are close to those of $\hat{H}_0$.

Understanding the utility of assumption (6.322) starts by writing an eigenstate $|\, E_n, t \,\rangle$ of $\hat{H}$ at $t = 0$ as a superposition of the normalized complete eigenstates of $\hat{H}_0$, so that

$$| E_n, 0 \rangle = \sum_j a_j^n | E_{0j}, 0 \rangle, \tag{6.323}$$

where the coefficients a_j^n are to be determined. The eigenstates of $\hat{H}$ are orthonormal, and this gives

$$\langle E_m, 0 | E_n, 0 \rangle = \sum_{j,k} a_k^{m*} a_j^n \langle E_{0k}, 0 | E_{0j}, 0 \rangle = \sum_{j,k} a_k^{m*} a_j^n \delta_{jk} = \sum_j a_j^{m*} a_j^n = \delta_{mn}. \tag{6.324}$$

The superposition (6.323) is an eigenstate of $\hat{H}$, which gives

$$\hat{H} | E_n, 0 \rangle = \sum_j a_j^n \hat{H} | E_{0j}, 0 \rangle = \sum_j a_j^n (\hat{H}_0 + \hat{H}_\mathrm{I}) | E_{0j}, 0 \rangle = \sum_j a_j^n \left(E_{0j} + \hat{H}_\mathrm{I} \right) | E_{0j}, 0 \rangle. \tag{6.325}$$

Forming the inner product gives the matrix elements of the full Hamiltonian,

$$\begin{aligned} H_{mn} &= \langle E_m, 0 | \hat{H} | E_n, 0 \rangle = \sum_{j,k} a_k^{m*} a_j^n \left(E_{0j} \langle E_{0k}, 0 | E_{0j}, 0 \rangle + \langle E_{0k}, 0 | H_\mathrm{I} | E_{0j}, 0 \rangle \right) \\ &= \sum_{j,k} a_k^{m*} a_j^n \left(E_{0j}\, \delta_{kj} + V_{kj} \right), \end{aligned} \tag{6.326}$$

where

$$V_{kj} = \langle E_{0k}, 0 | H_\mathrm{I} | E_{0j}, 0 \rangle. \tag{6.327}$$

The V_{kj} can be viewed as the elements of a matrix $\mathbf{V}$ that is not diagonal. The matrix $\mathbf{V}$ is Hermitian since its elements obey

$$V_{kj}^* = \langle E_{0k}, 0 | H_\mathrm{I} | \rangle E_{0j}, 0^* = \langle E_{0j}, 0 | H_\mathrm{I}^\dagger | E_{0k}, 0 \rangle = \langle E_{0j}, 0 | H_\mathrm{I} | E_{0k}, 0 \rangle = V_{jk}. \tag{6.328}$$

However, the full Hamiltonian must be diagonal over its eigenstates, and this gives

$$H_{mn} = E_n\, \delta_{mn}, \tag{6.329}$$

which are the elements of the matrix $\mathbf{H}$. Expressions (6.326) and (6.329) must be equal.

The two equations (6.326) and (6.329) are related by first defining the matrix $\tilde{\mathbf{H}}$, which has the elements

$$\tilde{H}_{kj} = E_{0j}\, \delta_{kj} + V_{kj}. \tag{6.330}$$

The matrix $\tilde{\mathbf{H}}$ is not diagonal due to the elements V_{kj}. The next step is to choose the coefficients a_j^n of the expansion (6.323) to be the elements of the unitary matrix $\mathbf{U}$ that diagonalizes the matrix $\tilde{\mathbf{H}}$, so that

$$U_{jn} = a_j^n, \quad U_{mk}^\dagger = U_{km}^* = a_k^{m*}. \tag{6.331}$$

The orthonormality condition (6.324) immediately demonstrates unitarity,

$$(\mathbf{U}^\dagger \cdot \mathbf{U})_{mn} = \sum_k U_{mk}^\dagger U_{kn} = \sum_k a_k^{m*} a_k^n = \delta_{mn}. \tag{6.332}$$

As a result, the matrix equation (6.326) can be written

$$\mathbf{H} = \mathbf{U}^{\dagger} \cdot \tilde{\mathbf{H}} \cdot \mathbf{U}. \tag{6.333}$$

In the perturbative approach to solving the Schrödinger equation, the problem reduces to finding the energy eigenvalues from the eigenvalue equation

$$\det(\tilde{\mathbf{H}} - E\ 1) = 0, \tag{6.334}$$

where the matrix elements of $\tilde{\mathbf{H}}$ are given by (6.330). The unitary transformation $\mathbf{U}$ is then found from the eigenvectors of $\tilde{\mathbf{H}}$ by using the definition (6.194).

Of course, this approach becomes unwieldy as the dimension of the Hamiltonian matrix $\tilde{\mathbf{H}}$ becomes large. However, there are systems and perturbations for which the off-diagonal elements, V_{kj}, are suppressed for the case that $E_{0j} - E_{0k}$ becomes large. For such a case, the matrix $\tilde{\mathbf{H}}$ can be restricted in dimension to a workable size.

6.6.2 Two-state oscillations

The matrix method is readily demonstrated for a two state subspace where the matrix Hamiltonian is two-dimensional. Two-level systems can occur if there are two energy eigenvalues of $\hat{H}_0$ that are close together, such as the large n energy levels in an atom. When a weak energy perturbation is switched on, the resulting energy eigenvalues are close to those of $\hat{H}_0$, so that these two energy eigenstates will be the predominant components in the mixture that has an eigenvalue for $\hat{H}$ that is close to these two energy eigenvalues of $\hat{H}_0$. This can also arise if there is a value for j such that $V_{jk'} \ll V_{jk}$ when $k' < k$ or $k' > k + 1$, since then it justifies ignoring the matrix elements involving all other states. In particular, if there is a degeneracy in the ground state energy eigenstates of $\hat{H}_0$, these conditions come into play. It will be seen in the next chapter that the ground state of hydrogen is doubly degenerate when electron spin is taken into account. For such a case a perturbation in the form of a weak magnetic field shows how the energy levels of the ground state can be changed by a two-level analysis. Experimentally such a situation can also be realized by energetically exciting an outer electron of a multi-electron atom, such as rubidium, into a high energy Rydberg state where the next energy level is very close in value.

In the simplest possible case, the two eigenstates of $\hat{H}_0$ of importance, $|\, 0\, \rangle$ and $|\, 1\, \rangle$, are assumed to share the same energy eigenvalue E_0. This does not need to be the case, but it simplifies the presentation and will be useful in chapter 8 when formal perturbation theory is developed. The matrix Hamiltonian for these two states takes the general form

$$\tilde{\mathbf{H}} = \begin{pmatrix} E_0 + V_{00} & V_{01} \\ V_{01}^* & E_0 + V_{11} \end{pmatrix} \equiv \begin{pmatrix} E_0 + \alpha & \gamma \\ \gamma^* & E_0 + \beta \end{pmatrix}. \tag{6.335}$$

The eigenvalue equation for (6.335) is given by

$$\det(\tilde{\mathbf{H}} - E1) = E^2 - (2E_0 + \alpha + \beta)E + (E_0 + \alpha)(E_0 + \beta) - |\gamma|^2 = 0. \tag{6.336}$$

Using the quadratic formula, equation (6.336) yields the two energy eigenvalues

$$E_{\pm} = E_0 + \frac{1}{2}(\alpha + \beta) \pm \frac{1}{2}\sqrt{(\alpha - \beta)^2 + 4|\gamma|^2}\,. \tag{6.337}$$

Regardless of the details of α, β, and γ, this shows that the resulting states are no longer degenerate, and the perturbation is said to have *lifted* the energy degeneracy of the basis states. Even if $\alpha = \beta = 0$, the energy eigenvalues of $\tilde{\mathbf{H}}$ are

$$E_{\pm} = E_0 \pm |\gamma|, \tag{6.338}$$

which shows that the off-diagonal element alone is sufficient to lift the degeneracy.

For simplicity the case where $\alpha = \beta = 0$ will be considered, and this gives the energy eigenvalues $E_{\pm}$ of (6.338). For this case the normalized eigenvectors, $\boldsymbol{a}_{\pm}$, are found by solving

$$\tilde{\mathbf{H}}\cdot\boldsymbol{a}_{\pm} = E_{\pm}\,\boldsymbol{a}_{\pm}. \tag{6.339}$$

Omitting the details of the calculation, the eigenvectors are given by

$$\begin{aligned} \boldsymbol{a}_{+} &= \frac{1}{\sqrt{2}}\begin{pmatrix} 1 \\ \gamma^*/|\gamma| \end{pmatrix} = \frac{1}{\sqrt{2}}\begin{pmatrix} 1 \\ e^{-i\theta} \end{pmatrix}, \\ \boldsymbol{a}_{-} &= \frac{1}{\sqrt{2}}\begin{pmatrix} -\gamma/|\gamma| \\ 1 \end{pmatrix} = \frac{1}{\sqrt{2}}\begin{pmatrix} -e^{i\theta} \\ 1 \end{pmatrix}, \end{aligned} \tag{6.340}$$

where the polar representation $\gamma = |\gamma|e^{i\theta}$ was used. For example,

$$\tilde{\mathbf{H}}\cdot\boldsymbol{a}_{+} = \frac{1}{\sqrt{2}}\begin{pmatrix} E_0 & |\gamma|e^{i\theta} \\ |\gamma|e^{-i\theta} & E_0 \end{pmatrix}\cdot\begin{pmatrix} 1 \\ e^{-i\theta} \end{pmatrix} = (E_0 + |\gamma|)\frac{1}{\sqrt{2}}\begin{pmatrix} 1 \\ e^{-i\theta} \end{pmatrix} = (E_0 + |\gamma|)\boldsymbol{a}_{+}. \tag{6.341}$$

The two eigenvectors of (6.344) satisfy the required orthonormality relations,

$$\begin{aligned} (\boldsymbol{a}_{+}, \boldsymbol{a}_{-}) &= (\boldsymbol{a}_{+})^{\dagger}\cdot\boldsymbol{a}_{-} = 0, \\ (\boldsymbol{a}_{\pm}, \boldsymbol{a}_{\pm}) &= \boldsymbol{a}_{\pm}^{\dagger}\cdot\boldsymbol{a}_{\pm} = 1. \end{aligned} \tag{6.342}$$

Using the components of the eigenvectors as the elements of the unitary transformation that diagonalizes $\tilde{\mathbf{H}}$ gives the two eigenstates $|\,E_{\pm}, t\,\rangle$ in terms of the basis states,

$$\begin{aligned} |\,E_{+}, 0\,\rangle &= \frac{1}{\sqrt{2}}(|\,0, 0\,\rangle + e^{-i\theta}|\,1, 0\,\rangle), \\ |\,E_{-}, 0\,\rangle &= \frac{1}{\sqrt{2}}(|\,1, 0\,\rangle - e^{i\theta}|\,0, 0\,\rangle). \end{aligned} \tag{6.343}$$

These are eigenstates of the full Hamiltonian, so that their time evolution is given by

$$\begin{aligned} |\,E_{+}, t\,\rangle &= e^{-iE_{+}t/\hbar}|\,E_{+}, 0\,\rangle = \frac{1}{\sqrt{2}}e^{-iE_{+}t/\hbar}(|\,0, 0\,\rangle + e^{-i\theta}|\,1, 0\,\rangle), \\ |\,E_{-}, t\,\rangle &= e^{-iE_{-}t/\hbar}|\,E_{-}, 0\,\rangle = \frac{1}{\sqrt{2}}e^{-iE_{-}t/\hbar}(|\,1, 0\,\rangle - e^{i\theta}|\,0, 0\,\rangle). \end{aligned} \tag{6.344}$$

By virtue of the orthonormality of the two basis states $|\,0,\,0\,\rangle$ and $|\,1,\,0\,\rangle$, the two eigenstates of (6.344) remain orthonormal at all times regardless of the value for θ. The results of (6.344) are easily inverted to express the eigenstates of $\hat{H}_0$ in terms of the eigenstates of $\hat{H}$,

$$
\begin{aligned}
|\,0,\,t\,\rangle &= \frac{1}{\sqrt{2}}(|\,E_+,\,t\,\rangle - e^{-i\theta}|\,E_-,\,t\,\rangle) \\
|\,1,\,t\,\rangle &= \frac{1}{\sqrt{2}}(e^{i\theta}|\,E_+,\,t\,\rangle + |\,E_-,\,t\,\rangle).
\end{aligned}
\tag{6.345}
$$

This shows that the original basis states prior to the perturbation are now mixed states of the new eigenstates.

In the Copenhagen interpretation it is possible to have prepared the system in the basis eigenstate $|\,0,\,0\,\rangle$ at the time $t = 0$ when the perturbation H_I is switched on. The initial eigenstate then evolves in time according to (6.345). Using the orthonormality of the basis states it follows that

$$
\begin{aligned}
\langle\,0,\,0\,|\,E_+,\,t\,\rangle &= e^{-iE_+t/\hbar}\langle\,0,\,0\,|\,E_+,\,0\,\rangle = \frac{1}{\sqrt{2}}e^{-iE_+t/\hbar}, \\
\langle\,0,\,0\,|\,E_-,\,t\,\rangle &= e^{-iE_-t/\hbar}\langle\,0,\,0\,|\,E_-,\,0\,\rangle = -\frac{e^{i\theta}}{\sqrt{2}}e^{-iE_-t/\hbar}, \\
\langle\,1,\,0\,|\,E_+,\,t\,\rangle &= e^{-iE_+t/\hbar}\langle\,1,\,0\,|\,E_+,\,0\,\rangle = \frac{1}{\sqrt{2}}e^{-i\theta}e^{-iE_+t/\hbar}, \\
\langle\,1,\,0\,|\,E_-,\,t\,\rangle &= e^{-iE_-t/\hbar}\langle\,1,\,0\,|\,E_-,\,0\,\rangle = \frac{1}{\sqrt{2}}e^{-iE_-t/\hbar}.
\end{aligned}
\tag{6.346}
$$

These results provide the tools to analyse several important quantum amplitudes. These are the amplitudes for the initial state to transit into the state $|\,1,\,t\,\rangle$ at the time t or to remain in the original state $|\,0,\,t\,\rangle$ at time t. These amplitudes are found using (6.345) and (6.346), so that

$$
\begin{aligned}
T_{00} &= \langle\,0,\,0\,|\,0,\,t\,\rangle = \frac{1}{\sqrt{2}}(\langle\,0,\,0\,|\,E_+,\,t\,\rangle - e^{-i\theta}\langle\,0,\,0\,|\,E_-,\,t\,\rangle) = \frac{1}{2}(e^{-iE_+t/\hbar} + e^{-iE_-t/\hbar}), \\
T_{01} &= \langle\,0,\,0\,|\,1,\,t\,\rangle = \frac{1}{\sqrt{2}}(e^{i\theta}\langle\,0,\,0\,|\,E_+,\,t\,\rangle + \langle\,0,\,0\,|\,E_-,\,t\,\rangle) = \frac{e^{i\theta}}{2}(e^{-iE_+t/\hbar} - e^{-iE_-t/\hbar}).
\end{aligned}
\tag{6.347}
$$

These quantities are referred to as the *transition amplitudes* for the state prepared as $|\,0,\,0\,\rangle$ at $t = 0$ to *transit* to either the same state $|\,0,\,t\,\rangle$ at the time t or to the state $|\,1,\,t\,\rangle$ at the time t. This quantity is of central importance in the perturbative approach to quantum mechanical state transitions. Since it is the projection of the state $|\,0,\,0\,\rangle$ onto the state $|\,0,\,t\,\rangle$ or the state $|\,1,\,0\,\rangle$, it is interpreted as the *probability amplitude* for the perturbation to cause the transition.

In keeping with the Copenhagen interpretation, the quantity $P_{01} = |T_{01}|^2$ is the probability that the system, initially prepared in the state $|\,0,\,0\,\rangle$, will be found in the state $|\,1,\,t\,\rangle$ at the time t. Using (6.347) gives

$$P_{01} = |T_{01}|^2 = \frac{1}{2}(1 - \cos((E_+ - E_-)t/\hbar)) = \sin^2(\omega t), \tag{6.348}$$

where $\omega = (E_+ - E_-)/2\hbar = |\gamma|/\hbar$. Similarly, the probability of being in the same state $| 0, t \rangle$ at the time t is given by

$$P_{00} = |T_{00}|^2 = \frac{1}{2}(1 + \cos((E_+ - E_-)t/\hbar)) = \cos^2(\omega t), \tag{6.349}$$

where the Euler relation (3.19) was used. These results are a specific case of *Rabi oscillations*, although Rabi oscillations typically involve an oscillatory perturbation. It is straightforward to show that $P_{10} = P_{01}$ and $P_{11} = P_{00}$. In the general case where there are many final states, this probabilistic interpretation is consistent with the result that completeness gives

$$\sum_j |T_{jk}|^2 = \sum_j |\langle k, 0 | e^{-i\hat{H}T} | j, 0 \rangle|^2 = \sum_j \langle k, 0 | e^{-i\hat{H}T} | j, 0 \rangle \langle j, 0 | e^{i\hat{H}T} | k, 0 \rangle = \langle k, 0 | k, 0 \rangle = 1, \tag{6.350}$$

a property referred to as *unitarity*. In other words, the particle must be in *some state* at a later time, and so the sum of the probabilities must be one. In this simple example there are only two states under consideration and (6.348) and (6.349) clearly sum to one. For the case that T_{01} is given by (6.348), the probability of the transition from $| 0, 0 \rangle$ to $| 1, t \rangle$ is consistent with the perturbation inducing an oscillation between the two basis states with the frequency $\omega = |\gamma|/\hbar$. These exact results will be used as guidelines in understanding time-dependent perturbation theory in chapter 8.

For the case that the system is prepared in state $| 0, 0 \rangle$ at time $t = 0$ the expressions (6.345) written in terms of energy eigenstates easily give the energy expectation values

$$\begin{aligned} \langle \hat{H} \rangle &= \langle 0, t | \hat{H} | 0, t \rangle = \frac{1}{2}(E_+ + E_-), \\ \langle \hat{H}^2 \rangle &= \langle 0, t | \hat{H}^2 | 0, t \rangle = \frac{1}{2}(E_+^2 + E_-^2). \end{aligned} \tag{6.351}$$

Using (6.351) it follows that the uncertainty in energy for the state $| 0, t \rangle$ is given by

$$\begin{aligned} \Delta E &= \sqrt{\langle \hat{H}^2 \rangle - \langle \hat{H} \rangle^2} \\ &= \sqrt{\frac{1}{2}(E_+^2 + E_-^2) - \frac{1}{4}(E_+ + E_-)^2} = \frac{1}{2}(E_+ - E_-) = |\gamma|. \end{aligned} \tag{6.352}$$

The perturbation therefore induces the uncertainty in energy in the oscillating state.

Because the transition probabilities are known, it is possible to find the uncertainty in the time for measuring for the system. This begins by defining a hypothetical observable $\hat{O}$ that has the basis states as eigenstates at the time $t = 0$,

$$\hat{O}| 0, 0 \rangle = \lambda_0 | 0, 0 \rangle, \quad \hat{O}| 1, 0 \rangle = \lambda_1 | 1, 0 \rangle. \tag{6.353}$$

It is important to note that these states at a later time are no longer eigenstates of Ô since Ô no longer commutes with $\hat{H}$. This is easily seen in the matrix representation associated with the basis states,

$$\begin{aligned}[\hat{O}, \hat{H}]_{j\ell} &= \sum_{k=0}^{1}(O_{jk}H_{k\ell} - H_{jk}O_{k\ell}) \\ &= \sum_{k=0}^{1}(\lambda_j\delta_{jk}H_{k\ell} - H_{jk}\lambda_k\delta_{k\ell}) = \lambda_j H_{j\ell} - \lambda_\ell H_{j\ell} = (\lambda_j - \lambda_\ell)H_{j\ell},\end{aligned} \tag{6.354}$$

which does not vanish if $\lambda_1 \neq \lambda_0$ since H_{01} does not vanish. As a result, there will be an uncertainty in Ô after the perturbation is switched on. For the case that the system is initially prepared in the state $|\, 0, 0\,\rangle$, using the results (6.345) and (6.346) allows the expectation value of Ô to be found,

$$\begin{aligned}\langle\hat{O}\rangle(t) &= \langle\, 0, t \,|\, \hat{O} \,|\, 0, t\,\rangle = \sum_{j=0}^{1}\lambda_j\langle\, 0, t \,|\, j, 0\,\rangle\langle\, j, 0 \,|\, 0, t\,\rangle = \sum_{j=1}^{1}\lambda_j\, |\langle\, j, 0 \,|\, 0, t\,\rangle|^2 \\ &= \lambda_0 P_{00} + \lambda_1 P_{10} = \lambda_0 \cos^2(\omega t) + \lambda_1 \sin^2(\omega t).\end{aligned} \tag{6.355}$$

This gives

$$\left|\frac{\mathrm{d}}{\mathrm{d}t}\langle\hat{O}\rangle(t)\right| = \omega\, |\, \sin(2\omega t)(\lambda_1 - \lambda_0)|. \tag{6.356}$$

Similarly, the expectation value $\langle\hat{O}^2\rangle(t)$ is found to be

$$\langle\hat{O}^2\rangle(t) = \sum_{j=1}^{1}\lambda_j^2\, |\langle\, j, 0 \,|\, 0, t\,\rangle|^2 = \lambda_0^2\cos^2(\omega t) + \lambda_1^2\sin^2(\omega t). \tag{6.357}$$

Combining results (6.355) and (6.357) gives the uncertainty in Ô,

$$\Delta\hat{O} = \sqrt{\langle\hat{O}^2\rangle - \langle\hat{O}\rangle^2} = \frac{1}{2}\, |\, \sin(2\omega t)(\lambda_1 - \lambda_0)|. \tag{6.358}$$

The definition (4.145) for the time interval Δt required for measurement can be combined with (6.356) and (6.358) to give

$$\Delta\hat{O} = \left|\frac{\mathrm{d}}{\mathrm{d}t}\langle\hat{O}\rangle(t)\right| \Delta t \implies \Delta t = \frac{1}{2\omega}. \tag{6.359}$$

Using this in the uncertainty relation (4.147) gives the uncertainty in the energy as

$$\Delta E = \frac{\hbar}{2\Delta t} = \hbar\omega = |\gamma| = \frac{1}{2}(E_+ - E_-), \tag{6.360}$$

which agrees exactly with (6.352).

6.7 Electron diffraction revisited

It is instructive to apply the formal rules of quantum mechanics to electron diffraction. Along the way, some very useful results will be established.

6.7.1 The propagator

Barring the action of a measurement, the behavior of a quantum state is governed by unitary time evolution. For the case that the Hamiltonian is time-independent, the state $|\,\Psi,\, t\,\rangle$ at a later time t' is given by

$$|\,\Psi,\, t'\,\rangle = \hat{U}(t')|\,\Psi,\, 0\,\rangle = \hat{U}(t')\hat{U}^{\dagger}(t)|\,\Psi,\, t\,\rangle = \hat{U}(t'-t)|\,\Psi,\, t\,\rangle = e^{-i\hat{H}(t'-t)/\hbar}|\,\Psi,\, t\,\rangle. \quad (6.361)$$

Placing this statement into configuration representation gives

$$\Psi(\boldsymbol{x}',\, t') = \langle\, \boldsymbol{x}'\,|\,\Psi,\, t'\,\rangle = \int \mathrm{d}^n x\, \langle\, \boldsymbol{x}'\,|\,e^{-i\hat{H}(t'-t)/\hbar}\,|\,\boldsymbol{x}\,\rangle\langle\,\boldsymbol{x}\,|\,\Psi,\, t\,\rangle \equiv \int \mathrm{d}^n x\, G(\boldsymbol{x}',\, t',\, \boldsymbol{x},\, t)\Psi(\boldsymbol{x},\, t). \quad (6.362)$$

The function $G(\boldsymbol{x}',\, t',\, \boldsymbol{x},\, t)$ is defined as

$$G(\boldsymbol{x}',\, t',\, \boldsymbol{x},\, t) \equiv \langle\, \boldsymbol{x}'\,|\,e^{-i\hat{H}(t'-t)/\hbar}\,|\,\boldsymbol{x}\,\rangle, \quad (6.363)$$

and is referred to as the *propagator*. Taken on its own, it gives the probability amplitude for a particle prepared at position $\boldsymbol{x}$ at the time t to be observed at $\boldsymbol{x}'$ at the later time t'. This interpretation follows from Born's rule since (6.363) is the inner product of $|\,\boldsymbol{x}'\,\rangle$ with the time-evolved state $\hat{U}(t'-t)|\,\boldsymbol{x}\,\rangle$. In that sense, it is identical to the energy state transition elements (6.347) considered in the previous section. As a result, the transition element (6.363) serves as the basis of the *path integral* formulation of quantum mechanics discussed in chapter 8. Statement (6.362) shows that the propagator can be used to find the wave function at $\boldsymbol{x}'$ at the time t' from the wave function at $\boldsymbol{x}$ at the time t. In other words, the propagator moves or *propagates* the wave function through position and time.

If the particle is *free*, so that $\hat{H} = \hat{\boldsymbol{P}}^2/2m$, the propagator can be evaluated using the momentum projection operator (6.58) and the inner product (6.66). This reduces the propagator to n copies of a Gaussian integral that can be evaluated using Fresnel's theorem. The result is

$$\begin{aligned} G(\boldsymbol{x}',\, t',\, \boldsymbol{x},\, t) &= \int \mathrm{d}^n p\, \langle\, \boldsymbol{x}'\,|\,e^{-i\hat{\boldsymbol{P}}^2(t'-t)/2m\hbar}\,|\,\boldsymbol{p}\,\rangle\langle\,\boldsymbol{p}\,|\,\boldsymbol{x}\,\rangle \\ &= \int \frac{\mathrm{d}^n p}{(2\pi\hbar)^n}\, e^{-ip^2(t'-t)/2m\hbar} e^{i\boldsymbol{p}\cdot(\boldsymbol{x}'-\boldsymbol{x})/\hbar} \\ &= \left(\frac{m}{2\pi i\hbar(t'-t)}\right)^{n/2} \exp\left(\frac{im\,|\boldsymbol{x}'-\boldsymbol{x}|^2}{2\hbar(t'-t)}\right). \end{aligned} \quad (6.364)$$

6.7.2 Formal quantum mechanics and electron diffraction

Result (6.364) can be used to evaluate the transition element for particle diffraction depicted in figure 6.2. This is represented in two dimensions by the particle

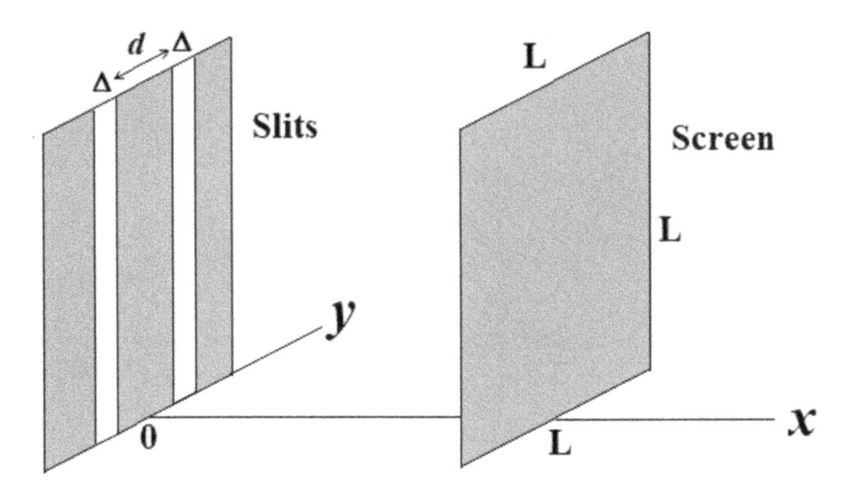

Figure 6.2. The diffraction grating.

originating from a two-slit interferometer located at $x = 0$ and aligned along the y-axis with *infinitesimal slit width* Δ and a slit separation d symmetric around $y = 0$. The screen is located at $x = L$, aligned along the y-axis, and has the area L^2, where L is arbitrarily large. The z-direction parallel to the slits will be ignored in the analysis. This is possible because the slits are considered to be infinitely long, extending from $z = -\infty$ to $z = +\infty$. The diffraction pattern formed on the screen is therefore independent of the z-coordinate.

The initial state $|\, s, t = 0 \,\rangle$ of the particle is described by a superposition of two slit states, located at $x = 0$ and written in terms of the two-dimensional position states, $|\, \boldsymbol{x} \,\rangle = |\, x = 0, y \,\rangle$, as

$$|\, s, t = 0 \,\rangle = \alpha \,|\, s_1, t = 0 \,\rangle + \beta \,|\, s_2, t = 0 \,\rangle, \tag{6.365}$$

where the two slit states appearing in (6.365) are given by a superposition of position states,

$$\begin{aligned} |\, s_1, t = 0 \,\rangle &= \int_{d/2}^{d/2+\Delta} \frac{\mathrm{d}y}{N} \,|\, x = 0, y \,\rangle, \\ |\, s_2, t = 0 \,\rangle &= \int_{-d/2-\Delta}^{-d/2} \frac{\mathrm{d}y}{N} \,|\, x = 0, y \,\rangle. \end{aligned} \tag{6.366}$$

The two slit states are orthogonal, $\langle\, s_1, t = 0 \,|\, s_2, t = 0 \,\rangle = 0$, since the domain of integration in their definition does not overlap and the position states in the definition of each slit state are therefore orthogonal. The dimensionless constant N will be fixed by normalizing the final *transition probability*. The mixed state coefficients are chosen to satisfy the usual requirement that $|\alpha|^2 + |\beta|^2 = 1$.

The *transition element* T_{fi} of interest is from the state (6.365) to a final position state on the screen, $|\, \boldsymbol{x}' \,\rangle = |\, L, y' \,\rangle$, after the time T has elapsed. The position y' is measured perpendicular to the slits from the center of the screen. For the case of a *free particle*, T_{fi} is found by using the free particle propagator (6.364) for $n = 2$, which gives

$$\begin{aligned} T_{fi} &= \langle\, \mathbf{x}' \mid e^{-i\hat{H}T/\hbar} \mid s, 0 \,\rangle \\ &= \alpha \int_{d/2}^{d/2+\Delta} \frac{\mathrm{d}y}{N} \langle\, L, y' \mid e^{-i\hat{H}T/\hbar} \mid 0, y \,\rangle + \beta \int_{-d/2-\Delta}^{-d/2} \frac{\mathrm{d}y}{N} \langle\, L, y' \mid e^{-i\hat{H}T/\hbar} \mid 0, y \,\rangle \\ &= \left(\frac{m e^{imL^2/2\hbar T}}{2\pi i\hbar NT}\right)\left(\alpha \int_{d/2}^{d/2+\Delta} \mathrm{d}y\, e^{im(y'-y)^2/2\hbar T} + \beta \int_{-d/2-\Delta}^{-d/2} \mathrm{d}y\, e^{im(y'-y)^2/2\hbar T}\right). \end{aligned} \tag{6.367}$$

Because Δ is arbitrarily small, the integrals coalesce onto the integrands multiplied by Δ and evaluated at $y = d/2$ and $y = -d/2$ respectively, so that (6.367) becomes

$$T_{fi} = \left(\frac{m\,\Delta}{2\pi i\hbar NT}\right) e^{imL^2/2\hbar T}\left(\alpha\, e^{im(y'-d/2)^2/2\hbar T} + \beta\, e^{im(y'+d/2)^2/2\hbar T}\right). \tag{6.368}$$

For the case that the slit separation d is small compared to y', expression (6.368) becomes

$$T_{fi} \approx \left(\frac{m\,\Delta}{2\pi i\hbar NT}\right) e^{im(L^2+y'^2)/2\hbar T}\left(\alpha\, e^{-imy'd/2\hbar T} + \beta e^{imy'd/2\hbar T}\right). \tag{6.369}$$

Choosing α and β to be real, the Copenhagen interpretation gives the *probability density per unit area* that the particle will strike the screen at the position y',

$$P(y') = |T_{fi}|^2 = \left(\frac{m\,\Delta}{2\pi\hbar NT}\right)^2\left(\alpha^2 + \beta^2 + 2\alpha\beta\cos\left(\frac{my'd}{\hbar T}\right)\right), \tag{6.370}$$

where the Euler relation (3.19) was used.

Result (6.370) exhibits the importance of the *quantum interference terms*, which appear proportional to $\alpha\beta$. If either α or β is zero, corresponding to only a single slit, the dependence of $P(y')$ on y' vanishes and it reduces to a constant probability. This corresponds to the wave function originating from one or the other of the two slits accompanied by a divergent uncertainty in momentum parallel to the slit due to the slit width of $\Delta \approx 0$. The case $\alpha = \beta = 1/\sqrt{2}$ corresponds to the electron passing through either slit with an equal probability, and for that situation using the identity $1 + \cos\theta = 2\cos^2(\theta/2)$ shows that the probability per unit area becomes

$$P(y') = \left(\frac{m^2\,\Delta^2}{2\pi^2\hbar^2 N^2 T^2}\right)\cos^2\left(\frac{my'd}{2\hbar T}\right). \tag{6.371}$$

Requiring the integral of (6.371) over the screen area to be unity for L arbitrarily large gives

$$\int_A \mathrm{d}y'\,\mathrm{d}z\, P(y') = \left(\frac{m^2\,\Delta^2}{2\pi^2\hbar^2 N^2 T^2}\right)\int_0^L \mathrm{d}y' \int_0^L \mathrm{d}z\,\cos^2\left(\frac{my'd}{2\hbar T}\right) \approx \frac{m^2 L^2 \Delta^2}{4\pi^2\hbar^2 T^2 N^2} = 1, \tag{6.372}$$

which fixes the normalization factor to be

$$N = \frac{mL\Delta}{2\pi\hbar T}. \tag{6.373}$$

The probability density for striking the screen, which is the probability per unit area, is therefore given by

$$P(y') = \left(\frac{2}{L^2}\right)\cos^2\left(\frac{my'd}{2\hbar T}\right). \tag{6.374}$$

It will now be shown that result (6.374) reproduces the earlier de Broglie wavelength analysis of electron diffraction presented in chapter 2. The maximum value of (6.371) occurs at those values of y' such that

$$\frac{my'd}{2\hbar T} = n\pi, \tag{6.375}$$

where n is an arbitrary integer. Denoting θ as the angle of diffraction and using $L \gg y'$ gives

$$y' = L\sin\theta. \tag{6.376}$$

The angles of maximum probability therefore satisfy

$$\frac{mLd\sin\theta}{2\hbar T} = n\pi. \tag{6.377}$$

Assuming $y' \ll L$, the velocity v of the particle *relative* to the screen is given by $v = L/T$ since it strikes the screen at the time T after passing through the interferometer, and this gives

$$d\sin\theta = n\frac{2\pi\hbar T}{mL} = n\frac{h}{m(L/T)} = n\frac{h}{mv} = n\lambda, \tag{6.378}$$

where $\lambda = h/mv$ is the de Broglie wavelength first defined in (2.54). This probability pattern is identical to the one obtained by using the de Broglie wavelength in the wave interference formula (1.79), but has been obtained from formal quantum mechanics.

References and recommended further reading

Dirac first published his abstract version of quantum mechanics in 1930. It is reprinted as

- Dirac P 1982 *The Principles of Quantum Mechanics* (Oxford: Clarendon Press)

Mathematical treatises on functional analysis relevant to physics include

- Courant R and Hilbert D 1989 *Methods of Mathematical Physics* vol I (New York: Wiley)

- Reed M and Simon B 1980 *Functional Analysis* vol I (New York: Academic Press)

Introductory quantum mechanics texts that begin by introducing Dirac notation include

- Sakurai J and Napolitano J 2020 *Modern Quantum Mechanics* (New York: Cambridge University Press)

Monographs that present linear algebra include

- Bamberg P and Sternberg S 1988 *A Course in Mathematics for Students of Physics* vol I (Cambridge: Cambridge University Press)
- Courant R and Hilbert D 1989 *Methods of Mathematical Physics* vol II (New York: Wiley)
- Byron F and Fuller R 1992 *Mathematics of Classical and Quantum Physics* (New York: Dover)
- Arfken G, Weber H and Harris F 2013 *Mathematical Methods for Physicists* 7th edn (Amsterdam: Elsevier)
- Hassani S 2013 *Mathematical Physics* 2nd edn (Berlin: Springer)

The matrix representation of quantum mechanics was developed in

- Born M, Heisenberg W and Jordan P 1925 Zur Quantenmechanik. II. *Z. Phys.* **35** 557 (https://doi.org/10.1007/BF01379806)
- Dirac P 1925 The fundamental equations of quantum mechanics *Proc. R. Soc. Lond.* A **109** 642 (https://doi.org/10.1098/rspa.1925.0150)

The demonstration of the equivalence of matrix representations and wave mechanics was given in

- Schrödinger E 1926 Quantisierung als Eigenwertproblem *Ann. Phys.* **384** 361 (https://doi.org/10.1002/andp.19263840404)

The Baker–Campbell–Hausdorff theorem was developed in

- Campbell J 1898 On a law of combination of operators (second paper) *Proc. Lond. Math. Soc.* **29** 14 (https://doi.org/10.1112/plms/s1-29.1.14)
- Baker H 1904 Alternants and continuous groups *Proc. Lond. Math. Soc.* **3** 24 (https://doi.org/10.1112/plms/s2-3.1.24)
- Hausdorff F 1906 Die symbolische Exponentialformel in der Gruppentheorie *Ber Verh Saechs Akad Wiss Leipzig* **58** 19

Feynman presented his version of the Feynman–Hellmann theorem in his senior thesis. It was published in

- Hellmann H 1937 *Einführung in die Quantenchemie* (Leipzig: Franz Deuticke)
- Feynman R 1939 Forces in molecules *Phys. Rev.* **56** 340 (https://doi.org/10.1103/PhysRev.56.340)

The classical mechanics version of the virial theorem is derived in

- Goldstein H, Poole C and Safko J 2000 *Classical Mechanics* 3rd edn (San Francisco: Addison-Wesley)

Harmonic oscillator coherent states were developed in

- Glauber R 1963 The quantum theory of optical coherence *Phys. Rev.* **130** 2529 (https://doi.org/10.1103/PhysRev.130.2529)

IOP Publishing

A Concise Introduction to Quantum Mechanics (Second Edition)

Mark S Swanson

Chapter 7

Symmetry, angular momentum, and multiparticle states

This chapter will deal with the basic properties of angular momentum and multiparticle systems. This will begin by recasting quantum observables in terms of symmetry generators. The nature of angular momentum and its relation to the generators of rotation is presented, leading to the general properties of angular momentum as well as its addition in multiparticle systems. Representing multiparticle systems is achieved using the tensor product of their respective Hilbert spaces, and tensor product states and operators are introduced and developed. These are used to understand the angular momentum of multiparticle states. Spin angular momentum is defined and developed for electrons and photons. The results are used to postulate the relationship between spin angular momentum and statistics for multiparticle systems. The chapter concludes with the ideas of quantum entanglement and decoherence.

7.1 Quantum mechanical observables as symmetry generators

While applications of quantum mechanics have so far focused on a complete set of commuting observables and their eigenvalue and eigenstate pairs as generating measurable quantities, there is another way to view the observables of quantum mechanics.

7.1.1 Time translation

The starting point is the evolution operator (6.72) associated with the Hamiltonian $\hat{H}$. For the case that $\partial\hat{H}/\partial t = 0$ the evolution operator takes the form

$$\hat{U}(t) = e^{-i\hat{H}t/\hbar}, \tag{7.1}$$

and its action creates a *time translation* of the Schrödinger picture state,

$$| \Psi, t \rangle = \hat{U}(t) | \Psi, 0 \rangle. \tag{7.2}$$

If an observable Ô commutes with the Hamiltonian $\hat{H}$, then it follows that its expectation value is independent of time. This is easy to see in the Heisenberg picture, where

$$\hat{O}_H(t) = \hat{U}^\dagger(t)\, \hat{O}\, \hat{U}(t) = \hat{O}\, \hat{U}^\dagger(t)\, \hat{U}(t) = \hat{O} = \hat{O}_H(0). \tag{7.3}$$

The Hamiltonian that appears in $\hat{U}(t)$ is therefore the *generator of time translations.* If the expectation values of a set of observables are constant in time, then time translation is said to be a *symmetry* of the system. Although time is a parameter in both classical and quantum mechanics, energy and time are said to be *canonically conjugate* in this respect and appear together in the Schrödinger equation and in the Heisenberg uncertainty principle.

7.1.2 Position and momentum translation

A second important canonically conjugate pair of observables are position $\hat{\boldsymbol{Q}}$ and momentum $\hat{\boldsymbol{P}}$. The momentum operator can be used to define the position translation operator,

$$\hat{T}(\boldsymbol{x}) = e^{i\hat{\boldsymbol{P}}\cdot\boldsymbol{x}/\hbar}, \tag{7.4}$$

where the position vector $\boldsymbol{x}$ is a parameter. The action of $\hat{T}(\boldsymbol{x})$ can be understood by considering its action on the momentum eigenstate $| \boldsymbol{p} \rangle$ in the position representation, which gives

$$\langle \boldsymbol{y} \mid \hat{T}(\boldsymbol{x}) \mid \boldsymbol{p} \rangle = \langle \boldsymbol{y} \mid e^{i\hat{\boldsymbol{P}}\cdot\boldsymbol{x}/\hbar} \mid \boldsymbol{p} \rangle = e^{i\boldsymbol{p}\cdot\boldsymbol{x}/\hbar} \langle \boldsymbol{y} \mid \boldsymbol{p} \rangle = \frac{1}{(2\pi\hbar)^{n/2}} e^{i\boldsymbol{p}\cdot(\boldsymbol{y}+\boldsymbol{x})/\hbar}, \tag{7.5}$$

where result (6.67) for $\langle \boldsymbol{y} \mid \boldsymbol{p} \rangle$ was used. Result (7.5) shows that

$$\langle \boldsymbol{y} \mid \hat{T}(\boldsymbol{x}) = \langle \boldsymbol{y} + \boldsymbol{x} \mid, \tag{7.6}$$

which in turn gives

$$\hat{T}^\dagger(\boldsymbol{x}) | \boldsymbol{y} \rangle = | \boldsymbol{x} + \boldsymbol{y} \rangle. \tag{7.7}$$

This shows that momentum is the *generator of position translation.* Similarly, the operator $\hat{W}(\boldsymbol{k})$, defined by

$$\hat{W}(\boldsymbol{k}) = e^{i\hat{\boldsymbol{Q}}\cdot\boldsymbol{k}/\hbar}, \tag{7.8}$$

can be used to show that position is the *generator of momentum translation,*

$$\hat{W}(\boldsymbol{k}) | \boldsymbol{p} \rangle = | \boldsymbol{p} + \boldsymbol{k} \rangle. \tag{7.9}$$

If an observable Ô commutes with the momentum operator $\hat{\boldsymbol{P}}$, then it follows that

$$\hat{T}(\boldsymbol{a}) \hat{O} \hat{T}^\dagger(\boldsymbol{a}) = e^{i\hat{\boldsymbol{P}}\cdot\boldsymbol{a}/\hbar} \hat{O} e^{-i\hat{\boldsymbol{P}}\cdot\boldsymbol{a}/\hbar} = \hat{O}. \tag{7.10}$$

The configuration space representation of Ô then has the property that

$$\langle\, \boldsymbol{x} \mid \hat{O} \mid \boldsymbol{y} \,\rangle = \langle\, \boldsymbol{x} \mid \hat{T}(\boldsymbol{a})\hat{O}\hat{T}^{\dagger}(\boldsymbol{a}) \mid \boldsymbol{y} \,\rangle = \langle\, \boldsymbol{x}+\boldsymbol{a} \mid \hat{O} \mid \boldsymbol{y}+\boldsymbol{a} \,\rangle. \tag{7.11}$$

The configuration space matrix elements of Ô are said to have the symmetry property of *translational invariance*. This means that the operator or observable has the same configuration space representation everywhere in the space of interest. A trivial example is the momentum itself, which has the property

$$\langle\, \boldsymbol{x}+\boldsymbol{a} \mid \hat{P}_j \mid \boldsymbol{y}+\boldsymbol{a} \,\rangle = -i\hbar\frac{\partial}{\partial(x_j+a_j)}\delta^n((\boldsymbol{x}+\boldsymbol{a})-(\boldsymbol{y}+\boldsymbol{a})) = -i\hbar\frac{\partial}{\partial x_j}\delta^n(\boldsymbol{x}-\boldsymbol{y}) \tag{7.12}$$

where the final step used the translational invariance of the derivative,

$$\frac{\partial}{\partial x_j} = \sum_{k=1}^{n}\frac{\partial(x_k+a_k)}{\partial x_j}\frac{\partial}{\partial(x_k+a_k)} = \sum_{k=1}^{n}\delta_{jk}\frac{\partial}{\partial(x_k+a_k)} = \frac{\partial}{\partial(x_j+a_j)}. \tag{7.13}$$

All three of the operators, $\hat{U}(t)$, $\hat{V}(\boldsymbol{x})$, and $\hat{W}(\boldsymbol{k})$, are *unitary* since their adjoint is also their inverse. As a result, all three preserve the inner product when applied to an arbitrary quantum mechanical state, e.g.

$$\langle\, \Psi' \mid \Psi' \,\rangle \equiv \langle\, \Psi \mid \hat{W}^{\dagger}(\boldsymbol{k})\hat{W}(\boldsymbol{k}) \mid \Psi \,\rangle = \langle\, \Psi \mid \Psi \,\rangle. \tag{7.14}$$

In the terminology of group theory, these three unitary operators are *abelian*, which means that each individual operator has the property

$$\hat{V}(\boldsymbol{x})\hat{V}(\boldsymbol{y}) = \hat{V}(\boldsymbol{y})\hat{V}(\boldsymbol{x}) = \hat{V}(\boldsymbol{x}+\boldsymbol{y}). \tag{7.15}$$

However, these operators do not commute among themselves. Because $[\hat{Q}_j, \hat{P}_k] = i\hbar\delta_{jk}$ commutes with both $\hat{\boldsymbol{Q}}$ and $\hat{\boldsymbol{P}}$, the simplified Baker-Campbell-Hausdorff (BCH) theorem (6.280) can be applied to $\hat{V}$ and $\hat{W}$ to show that

$$\hat{V}(\boldsymbol{x})\hat{W}(\boldsymbol{p}) = e^{i\hat{\boldsymbol{P}}\cdot\boldsymbol{x}/\hbar}e^{i\hat{\boldsymbol{Q}}\cdot\boldsymbol{p}/\hbar} = e^{i\hat{\boldsymbol{Q}}\cdot\boldsymbol{p}/\hbar}e^{i\hat{\boldsymbol{P}}\cdot\boldsymbol{x}/\hbar}e^{-[\boldsymbol{x}\cdot\hat{\boldsymbol{P}},\, \boldsymbol{p}\cdot\hat{\boldsymbol{Q}}]/\hbar^2} = \hat{W}(\boldsymbol{p})\hat{V}(\boldsymbol{x})e^{i\boldsymbol{p}\cdot\boldsymbol{x}/\hbar}. \tag{7.16}$$

Nothing specific regarding the commutators of $\hat{U}(t)$ with $\hat{V}(\boldsymbol{x})$ and $\hat{W}(\boldsymbol{p})$ can be stated until the exact form for $\hat{H}$ is known. It is very useful to note that the generators of all three translations are important quantum mechanical observables. This property of quantum mechanics extends to all observables since they are all associated with an *action on the state of the system*. In that regard, orbital angular momentum and its generalizations generate *rotations of the system*. In order to understand rotations it is necessary to understand the rotation group.

7.2 Rotation group theory

One of the most important observables in a three-dimensional system is angular momentum. The components of a particle's orbital angular momentum are represented in configuration space by

$$\hat{L}_j = (\boldsymbol{r} \times \hat{\boldsymbol{p}})_j = -i\hbar(\boldsymbol{r} \times \nabla)_j. \tag{7.17}$$

In the wave mechanical analysis of the hydrogen atom in chapter 5, the stationary states were characterized by the eigenvalues of $\hat{\mathrm{L}}^2$ and $\hat{\mathrm{L}}_z$. These were given respectively by $\ell(\ell + 1)\hbar^2$ and $m\hbar$, where the quantum numbers ℓ and m are integers that obey the relations $\ell \geqslant 0$ and $|m| \leqslant \ell$. The components of orbital angular momentum satisfy the commutation relations

$$[\hat{L}_j, \hat{L}_k] = i\hbar \sum_{\ell} \varepsilon^{jk\ell} \hat{L}_\ell, \tag{7.18}$$

where $\varepsilon^{jk\ell}$ is the Levi–Civita symbol (6.163) in three dimensions. These commutation relations define the algebra of the *rotation group*, and this algebra is the key to understanding the quantum mechanical nature of all angular momentum.

The concept of group theory has its origins in the classification of coordinate transformations and their relation to differential equations. This section provides a very brief and incomplete introduction to this rich subject, motivating it with the study of finite matrix transformations. Consideration is limited to analysing rotations in three dimensions and using these to define what is called a *continuous Lie group*.

7.2.1 Rotations around the z-axis

The starting point is the matrix representation of a transformation corresponding to the rotation of a coordinate system in three dimensions. The three coordinates of a point can be viewed as the components of a position vector from the origin to the point. Since rotation is a type of coordinate transformation, it will be accompanied by a change in the components of the vector. The position vector $\boldsymbol{x}$ can be written as a matrix vector with the three coordinates x, y, and z of the point as its elements,

$$\boldsymbol{x} = \begin{pmatrix} x \\ y \\ z \end{pmatrix}. \tag{7.19}$$

Actively rotating the coordinate system through an angle θ counterclockwise around the z-axis causes the x and y components of the point's position vector to change. This is depicted in figure 7.1. Using perpendicular projection, the value of x' for the point P is given by the segment OC = OA + AB + BC. However, simple trigonometry yields OA = ODcosθ = $x \cos\theta$ and AB + BC = PBsinθ + BDsinθ = (PB + BD)sin θ = $y \sin\theta$. This gives

$$x' = x\cos\theta + y\sin\theta. \tag{7.20}$$

A similar analysis shows that

$$\begin{aligned} y' &= y\cos\theta - x\sin\theta, \\ z' &= z. \end{aligned} \tag{7.21}$$

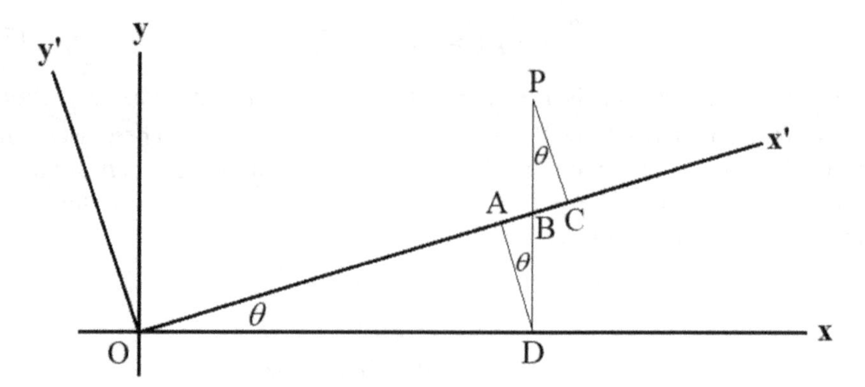

Figure 7.1. A rotation around the z-axis.

This transformation can be written in the form of a matrix $\mathbf{R}_z(\theta)$ acting on the column vector $\boldsymbol{x}$,

$$\boldsymbol{x}' = \begin{pmatrix} x' \\ y' \\ z' \end{pmatrix} = \begin{pmatrix} \cos\theta & \sin\theta & 0 \\ -\sin\theta & \cos\theta & 0 \\ 0 & 0 & 1 \end{pmatrix} \cdot \begin{pmatrix} x \\ y \\ z \end{pmatrix} \equiv \mathbf{R}_z(\theta) \cdot \boldsymbol{x}. \tag{7.22}$$

Using the rules of matrix multiplication (6.178) shows that (7.22) reproduces the results of the rotation.

The matrix $\mathbf{R}_z(\theta)$ has a number of important properties. The first is that it is parameterized by the single *continuous* variable θ in such a way that $\mathbf{R}_z(0) = \mathbf{1}$. Second, using basic trigonometric identities for $\cos(\theta_1 + \theta_2)$ and $\sin(\theta_1 + \theta_2)$ shows that the product of $\mathbf{R}_z$ for two different angles gives

$$\begin{aligned} \mathbf{R}_z(\theta_1) \cdot \mathbf{R}_z(\theta_2) &= \begin{pmatrix} \cos\theta_1 & \sin\theta_1 & 0 \\ -\sin\theta_1 & \cos\theta_2 & 0 \\ 0 & 0 & 1 \end{pmatrix} \cdot \begin{pmatrix} \cos\theta_2 & \sin\theta_2 & 0 \\ -\sin\theta_2 & \cos\theta_2 & 0 \\ 0 & 0 & 1 \end{pmatrix} \\ &= \begin{pmatrix} \cos(\theta_1+\theta_2) & \sin(\theta_1+\theta_2) & 0 \\ -\sin(\theta_1+\theta_2) & \cos(\theta_1+\theta_2) & 0 \\ 0 & 0 & 1 \end{pmatrix} = \mathbf{R}_z(\theta_1+\theta_2). \end{aligned} \tag{7.23}$$

The product is also a matrix with a form *identical* to the two matrices forming the product. These are two basic properties of a *continuous group*, which is a set of mathematical objects, in this case matrices, characterized by continuous parameters and closed under multiplication or *composition*. In other words, multiplying any two matrices in this group gives a matrix which is also a member of the group, i.e. a matrix that shares the same properties as the other members of the group.

Another basic property is the existence of an inverse for each transformation. Result (7.23) shows that

$$\mathbf{R}_z(-\theta) \cdot \mathbf{R}_z(\theta) = \mathbf{1}, \tag{7.24}$$

so that $\mathbf{R}_z(-\theta) = \mathbf{R}^{-1}(\theta)$. In this particular case any pair of matrices of the form $\mathbf{R}_z(\theta)$ satisfies

$$\mathbf{R}_z(\theta_1) \cdot \mathbf{R}_z(\theta_2) = \mathbf{R}_z(\theta_2) \cdot \mathbf{R}_z(\theta_1), \tag{7.25}$$

showing that they commute. This is reflected in their product, which depends only on the sum of the two parameter angles. Because they commute, the matrices of the form $\mathbf{R}_z(\theta)$ form an *abelian group* since the order of their composition does not matter. Most continuous or *Lie groups* do not satisfy this property and are referred to as *nonabelian.*

The matrices $\mathbf{R}_z(\theta)$ are orthogonal since $\mathbf{R}^{\mathrm{T}}(\theta) = \mathbf{R}(-\theta) = \mathbf{R}^{-1}(\theta)$, and likewise have the *special* property $\det(\mathbf{R}_z(\theta)) = \cos^2\theta + \sin^2\theta = 1$. Result (6.140) shows that the product of two orthogonal matrices is also an orthogonal matrix, since $(\mathbf{A} \cdot \mathbf{B})^{\mathrm{T}} = \mathbf{B}^T \cdot \mathbf{A}^T$, which is the inverse of $\mathbf{A} \cdot \mathbf{B}$. The product of two matrices with determinant one is also a matrix with determinant one since (6.176) gives $\det(\mathbf{A} \cdot \mathbf{B}) = \det\mathbf{A}\det\mathbf{B} = 1$. As a result, grouping together *all* 3×3 orthogonal matrices with determinant one has the property that the product of any two of these matrices will also be a member of that same *group* of matrices. This particular group of matrices is referred to as SO(3), which stands for *special orthogonal* 3×3 *matrices.*

However, $\mathbf{R}_z(\theta)$ does not exhaust all possible 3×3 orthogonal matrices with determinant $+1$ since it represents only rotations around the z-axis, and rotations around the x and y axes are also possible. The matrices of the form $\mathbf{R}_z(\theta)$ therefore constitute an abelian *subgroup* of SO(3), since they are closed under multiplication but do not exhaust all possible members of SO(3). Further investigation into SO(3) is required to understand its full range of parameters and its structure. Because these matrices represent the action of a rotation on the coordinates of a point, the column vector $\boldsymbol{x}$ is referred to as the *representation space*, and the matrices $\mathbf{R}_z$ give a *representation* of the subgroup for that choice of representation space. This is identical to the use of representation spaces in quantum mechanics discussed in chapter 6. While all this terminology may seem unnecessary for such a simple case, it will be useful when these concepts are generalized to other representation spaces.

7.2.2 Rotations in three dimensions

The first step in understanding the full nature of SO(3) is to examine the behavior of the subgroup $\mathbf{R}_z(\theta)$ in the neighborhood of $\mathbf{1}$. The continuous nature of the parameter θ allows an expansion of $\mathbf{R}_z(\theta)$ in a Taylor series around $\mathbf{1}$, and to $O(\theta)$ this gives

$$\mathbf{R}_z(\theta) \approx \mathbf{1} + \theta \frac{\partial \mathbf{R}_z}{\partial \theta}\bigg|_{\theta=0} + \cdots. \tag{7.26}$$

Using the definition (7.22) gives the constant matrix

$$\mathbf{T}_z \equiv \frac{\partial \mathbf{R}_z}{\partial \theta}\bigg|_{\theta=0} = \begin{pmatrix} 0 & 1 & 0 \\ -1 & 0 & 0 \\ 0 & 0 & 0 \end{pmatrix}. \tag{7.27}$$

The matrix $\mathbf{T}_z$ is referred to as a *generator* since it generates $\mathbf{R}_z(\theta)$ for *finite values* of θ. The demonstration of this starts by creating an infinitesimal parameter θ/N, where N is an arbitrarily large integer. For such a case, the infinitesimal transformation becomes $\mathbf{1} + (\theta/N)\mathbf{T}_z$ with the limit $N \to \infty$ understood. The form of the transformation for the finite value θ is generated through N applications of the infinitesimal expression, giving

$$\mathbf{R}_z(\theta) = \lim_{N\to\infty}\left(\mathbf{1} + \frac{\theta}{N}\mathbf{T}_z\right)^N = \exp(\theta\mathbf{T}_z). \tag{7.28}$$

The emergence of the exponential is verified by taking the logarithm of both sides and using the power series representation $\ln(1 + x) = x - x^2/2 + \cdots$ to find

$$\ln \mathbf{R}_z(\theta) = \lim_{N\to\infty} N \ln\left(\mathbf{1} + \frac{\theta}{N}\mathbf{T}_z\right) = \lim_{N\to\infty}\left(\theta\mathbf{T}_z - \frac{\theta^2}{2N}\mathbf{T}_z^2 + \cdots\right) = \theta\mathbf{T}_z, \tag{7.29}$$

which is the property of an exponential. This can be demonstrated for the special case under consideration by combining the exponential power series (6.155) with the results that even and odd powers of $\mathbf{T}_z$ are given by

$$\mathbf{T}_z^{2n} = (-1)^n\begin{pmatrix}1 & 0 & 0\\ 0 & 1 & 0\\ 0 & 0 & 0\end{pmatrix}, \quad \mathbf{T}_z^{2n+1} = (-1)^n\begin{pmatrix}0 & 1 & 0\\ -1 & 0 & 0\\ 0 & 0 & 0\end{pmatrix}. \tag{7.30}$$

The next step is to use the power series representations

$$\cos\theta = \sum_{n=0}^{\infty}\frac{(-1)^n}{(2n)!}\theta^{2n}, \quad \sin\theta = \sum_{n=0}^{\infty}\frac{(-1)^n}{(2n+1)!}\theta^{2n+1}, \tag{7.31}$$

which appear as the elements of the exponential series for (7.28). The final result is the previous form for $\mathbf{R}_z(\theta)$ that appears in (7.22).

Arbitrary rotations about the other two axes are given by $\mathbf{R}_x(\alpha)$ and $\mathbf{R}_y(\beta)$ and can be found in an identical manner, with the result that

$$\mathbf{R}_x(\alpha) = \begin{pmatrix}1 & 0 & 0\\ 0 & \cos\alpha & \sin\alpha\\ 0 & -\sin\alpha & \cos\alpha\end{pmatrix}, \quad \mathbf{R}_y(\beta) = \begin{pmatrix}\cos\beta & 0 & -\sin\beta\\ 0 & 1 & 0\\ \sin\beta & 0 & \cos\beta\end{pmatrix}. \tag{7.32}$$

Following the same procedure as (7.26) yields two new generators, which are given by

$$\mathbf{T}_x = \begin{pmatrix}0 & 0 & 0\\ 0 & 0 & 1\\ 0 & -1 & 0\end{pmatrix}, \quad \mathbf{T}_y = \begin{pmatrix}0 & 0 & -1\\ 0 & 0 & 0\\ 1 & 0 & 0\end{pmatrix}. \tag{7.33}$$

Denoting x, y, and z as 1, 2, and 3, the elements of the three matrices can be written $(\mathbf{T}_a)_{bc} = \varepsilon^{abc}$, where ε^{abc} is the Levi–Civita symbol (6.163). A full three-dimensional rotation can then be viewed as a combination of rotations around all three axes.

Denoting three angles as θ^a, (7.28) is generalized to a general three-dimensional rotation in terms of the three generators of (7.27) and (7.33) and the three angles θ^a as

$$\mathbf{R}(\theta) = \exp\left(\sum_{a=1}^{3}\theta^a\mathbf{T}_a\right). \tag{7.34}$$

The 3×3 matrix of (7.34) is understood to act on the representation space $\boldsymbol{x}$ and gives a representation of the *rotation group*.

For infinitesimal rotations (7.34) becomes $\mathbf{R}(\theta) \approx \mathbf{1} + \sum_a \theta^a\mathbf{T}_a$, and its action on the representation space is

$$x_i{}' = x_i + \sum_{k,j=1}^{3} \theta^k(\mathbf{T}_k)_{ij}x_j = x_i + \sum_{k,j=1}^{3} \varepsilon^{kij}\theta^k x_j = x_i - (\boldsymbol{\theta} \times \boldsymbol{x})_i. \tag{7.35}$$

For infinitesimal rotations about the x and y axes the commutator $[\mathbf{R}_x(\alpha), \mathbf{R}_y(\beta)]$ yields

$$[\mathbf{1} + \alpha\mathbf{T}_x, \mathbf{1} + \beta\mathbf{T}_y] = \alpha\beta[\mathbf{T}_x, \mathbf{T}_y] = -\alpha\beta\, \mathbf{T}_z, \tag{7.36}$$

where straightforward matrix multiplication results in the appearance of $\mathbf{T}_z$. This result shows two things: first, the order in which rotations are performed produces different results; and second, the generators themselves form a closed set of mathematical objects under multiplication. Because the generators $\mathbf{T}_a$ do not commute, they are the generators of what is referred to as *a nonabelian group*.

7.2.3 Properties of SO(3)

These results emerged by considering rotations around all three possible axes. It is important to confirm that these are the specific properties of SO(3). The act of rotating a coordinate system does not change the length of a position vector, and this means that a general rotation matrix $\mathbf{R}$, characterized by three angles, cannot change the inner product of a vector with itself. Using $\boldsymbol{x}' = \mathbf{R} \cdot \boldsymbol{x}$, this requirement is stated mathematically as

$$\boldsymbol{x}'^{\mathrm{T}} \cdot \boldsymbol{x}' = \boldsymbol{x}^{\mathrm{T}} \cdot \mathbf{R}^{\mathrm{T}} \cdot \mathbf{R} \cdot \boldsymbol{x} = \boldsymbol{x}^{\mathrm{T}} \cdot \boldsymbol{x}. \tag{7.37}$$

As anticipated with (6.145), this requires that $\mathbf{R}$ is an orthogonal matrix since it must satisfy $\mathbf{R}^{\mathrm{T}} \cdot \mathbf{R} = 1$.

In three dimensions and for real values, this means that the elements of $\mathbf{R}$ must obey the set of equations given by

$$(\mathbf{R}^{\mathrm{T}} \cdot \mathbf{R})_{ij} = \sum_{k=1}^{3}(\mathbf{R}^{\mathrm{T}})_{ik}(\mathbf{R})_{kj} = \sum_{k=1}^{3} R_{ki}R_{kj} = \delta_{ij}. \tag{7.38}$$

Three of the requirements of (7.38) are symmetric under interchange of i and j. For example, $i = 2$ and $j = 3$ gives the same equation as $i = 3$ and $j = 2$. This means that (7.38) gives a total of $3 \times 3 - 3 = 6$ *unique* equations. In turn, this means that the nine elements of the matrix $\mathbf{R}$ are determined entirely by $9 - 6 = 3$ free parameters. Once any three elements are specified, all nine elements of $\mathbf{R}$ are uniquely determined through the six equations required for orthogonality. The three angles of rotation around the three possible rotational axes are therefore sufficient to determine the matrix $\mathbf{R}$. Each of the three generators $\mathbf{T}_a$ has the transpose property that $\mathbf{T}_a^T = -\mathbf{T}_a$. It follows from the power series representation of (7.34) that its transpose is given by

$$\mathbf{R}^T = \exp\left(\sum_{a=1}^{3}\theta^a\mathbf{T}_a^T\right) = \exp\left(-\sum_{a=1}^{3}\theta^a\mathbf{T}_a\right) = \mathbf{R}^{-1}, \tag{7.39}$$

where the property of the matrix exponential, $(e^{\mathbf{A}})^{-1} = e^{-\mathbf{A}}$, was used. As a result, (7.34) is an orthogonal 3×3 matrix with three parameters, as it must be. The generators also have the property that $\mathrm{Tr}(\mathbf{T}_a) = 0$. Therefore, theorem (6.222) shows that (7.34) gives

$$\det \mathbf{R} = \exp\mathrm{Tr}(\ln \mathbf{R}) = \exp\mathrm{Tr}(\sum_a \theta^a\mathbf{T}_a) = \exp(0) = 1. \tag{7.40}$$

Because it is an orthogonal 3×3 matrix with three free parameters and a determinant of one, the matrix (7.34) provides a representation for a general member of SO(3).

In the case of SO(3) the generators satisfy the commutation relation

$$[\mathbf{T}^a, \mathbf{T}^b] = -\sum_{c=1}^{3}\varepsilon^{abc}\mathbf{T}^c \implies [(-i\mathbf{T}^a), (-i\mathbf{T}^b)] = i\sum_{c=1}^{3}\varepsilon^{abc}(-i\mathbf{T}^c), \tag{7.41}$$

where the Levi–Civita symbol (6.163) in three dimensions has appeared. The matrices $\mathbf{H}^a \equiv -i\mathbf{T}^a$ are now Hermitian, and this results in an overall factor of i appearing on the right-hand side of (7.41). This is consistent with the Hermitian adjoint of the commutator for Hermitian generators, since it obeys

$$[\mathbf{H}^a, \mathbf{H}^b]^\dagger = [\mathbf{H}^{b\dagger}, \mathbf{H}^{a\dagger}] = [\mathbf{H}^b, \mathbf{H}^a] = -[\mathbf{H}^a, \mathbf{H}^b]. \tag{7.42}$$

Using Hermitian generators allows the basic relation (7.41) to be written

$$[\mathbf{H}^a, \mathbf{H}^b] = i\sum_{c=1}^{3}\varepsilon^{abc}(\mathbf{H}^c). \tag{7.43}$$

While the Hermitian version of the matrix generators derived from three-dimensional rotations satisfy (7.43), there are other mathematical objects that also satisfy (7.43). The orbital angular momentum operator $\hat{\boldsymbol{L}}$ in a position representation is given by the Hermitian operator

$$\hat{\boldsymbol{L}} = -i\hbar\, \boldsymbol{x} \times \nabla, \tag{7.44}$$

so that

$$\hat{L}_j = -i\hbar \sum_{k,\ell} \varepsilon^{jk\ell}\, x_k \frac{\partial}{\partial x_\ell}. \tag{7.45}$$

Since x_j and $\partial/\partial x_j$ do not commute, the components of this operator also satisfy

$$[\hat{L}_j, \hat{L}_k] = i\hbar \sum_{\ell=1}^{3} \varepsilon^{jk\ell} \hat{L}_\ell. \tag{7.46}$$

For example, choosing $j = 1$ and $k = 2$ gives

$$[\hat{L}_x, \hat{L}_y] = -\hbar^2\left[y\frac{\partial}{\partial z} - z\frac{\partial}{\partial y}, z\frac{\partial}{\partial x} - x\frac{\partial}{\partial z}\right] = i\hbar\left(-i\hbar\left(x\frac{\partial}{\partial y} - y\frac{\partial}{\partial x}\right)\right) = i\hbar\hat{L}_z. \tag{7.47}$$

As a result, (7.43) can be viewed as the definition of a *Lie algebra*, and does not require the objects obeying that algebra to be matrices. The matrices discussed so far are a *representation* of the SO(3) *algebra* for the case that it acts on the three-dimensional representation space $\boldsymbol{x}$.

7.3 Rotations and quantum mechanics

Just as (7.5) corresponds to the spatial translation operator through the action of the momentum operator, the same approach can be applied to angular momentum.

7.3.1 Angular momentum as the generator of rotations

A *rotation* of the wave function Ψ through the angle α around the z-axis is given by the action of the operator

$$\hat{R}_z(\alpha) = e^{i\alpha\hat{L}_z/\hbar}. \tag{7.48}$$

This is especially easy to see in a spherical coordinate representation, since (5.134) showed that

$$\hat{L}_z = -i\hbar\frac{\partial}{\partial\phi}. \tag{7.49}$$

The action of the power series for $\hat{R}_z(\alpha)$ on a function in spherical coordinates gives

$$\hat{R}_z(\alpha)f(r, \theta, \phi) = e^{\alpha\,\partial/\partial\phi}f(r, \theta, \phi) = \sum_{j=0}^{\infty}\frac{\alpha^n}{n!}\frac{\partial^n}{\partial\phi^n}f(r, \theta, \phi) = f(r, \theta, \phi + \alpha), \tag{7.50}$$

by virtue of this reproducing the Taylor series expansion of $f(r, \theta, \phi + \alpha)$. As a result, $\hat{L}_z$ is the *generator of rotations* around the z-axis.

However, angular momentum is a vector operator, and an arbitrary finite rotation requires the action of all three angular momentum operators or *generators*, so that a general spatial rotation operator is written

$$\hat{\boldsymbol{R}}(\boldsymbol{\alpha}) = e^{i\boldsymbol{\alpha}\cdot\hat{\boldsymbol{L}}/\hbar}. \tag{7.51}$$

The angle vector $\boldsymbol{\alpha}$ represents the angular rotation around all three axes required for a general rotation. Since the components of the angular momentum do no commute among themselves, $\hat{\boldsymbol{R}}(\boldsymbol{\alpha})$ is a member of a *nonabelian group*, identified in the previous section as the *rotation group* for the specific case that the *representation space* for the rotation is the set of spatial vectors. A choice of representation space determines the specific form for the generators $\hat{\boldsymbol{L}}$ appearing in (7.51). This simply means that $\hat{L}_z$ is the differential operator (7.49) if the representation space is a scalar function of spherical coordinates, while $\hat{L}_z$ is the matrix of (7.22) if the representation space is a position vector. Regardless of the representation space chosen, the operator $\hat{\boldsymbol{R}}(\boldsymbol{\alpha})$ is a unitary operator, and therefore its action on a state preserves the inner product. If

$$\hat{\boldsymbol{R}}(\boldsymbol{\alpha})|\,\Psi,\, t\,\rangle = |\,\Psi,\, t\,\rangle, \tag{7.52}$$

then the state $|\,\Psi,\, t\,\rangle$ is said to be *rotationally symmetric*. If an operator $\hat{\mathrm{O}}$ satisfies

$$\hat{\boldsymbol{R}}^{\dagger}(\boldsymbol{\alpha})\,\hat{\mathrm{O}}\,\hat{\boldsymbol{R}}(\boldsymbol{\alpha}) = \hat{\mathrm{O}}, \tag{7.53}$$

then the operator $\hat{\mathrm{O}}$ is said to be *rotationally symmetric*.

In quantum mechanics there are many spaces that can be used to give a *representation* to the angular momentum operators, which are now identified as the generators of the *rotation group*. However, all representations of angular momentum must share the same *algebra* as the orbital angular momentum of (7.46),

$$[\hat{J}_j,\, \hat{J}_k] = i\hbar\sum_{\ell=1}^{3}\varepsilon^{jk\ell}\hat{J}_\ell, \tag{7.54}$$

where $\hat{J}_j$ now stands for a generalized angular momentum operator. For example, a generalized angular momentum could be the total orbital angular momentum of a multiparticle system, or it could be the sum of the orbital angular momentum and the *intrinsic angular momentum* of a particle. The intrinsic angular momentum of a particle is referred to as *spin*, in analogy to a spinning macroscopic object.

7.3.2 Spin angular momentum for the photon

A good starting point for understanding spin is the behavior of a light wave under a rotation of coordinates. In the case of a light wave, the vector potential $\mathbf{A}(\boldsymbol{x}, t)$ is the three-dimensional representation space for the action of the rotation. An arbitrary rotation $\hat{\boldsymbol{R}}(\boldsymbol{\alpha})$ acts on *both* the position vector $\boldsymbol{x}$ that defines the location under consideration *and* on the components of the three-dimensional vector potential $\mathbf{A}$ at that location, so that

$$\hat{\boldsymbol{R}}(\boldsymbol{\alpha})\,\mathbf{A}(\boldsymbol{x},\, t) = \mathbf{A}'(\boldsymbol{x}',\, t). \tag{7.55}$$

Because the representation space is a vector valued *function*, the rotation of $\boldsymbol{x}$ into $\boldsymbol{x}'$ is represented by using the orbital angular momentum operators $\hat{\boldsymbol{L}}$ defined in (7.44). This gives the position rotation operator

$$\hat{R}_x(\alpha) = e^{i\alpha\cdot\hat{L}/\hbar}. \tag{7.56}$$

This operator acts on the spatial argument x of the vector potential function to rotate it onto x'.

The rotation of the components of A is obtained by treating them as the components of a column vector and using the matrix generators (7.27) and (7.33) multiplied by a factor of $-i\hbar$ to define the *spin angular momentum operator* $\hat{S}$. To make this explicitly clear, the matrices $\mathbf{T}_a$ of (7.27) and (7.33) are used to define the Hermitian matrix components of $\hat{S}$, so that

$$\hat{\mathbf{S}}_a = -i\hbar\mathbf{T}_a. \tag{7.57}$$

The rotation operator for the vector potential components is then given by

$$\hat{R}_S(\alpha) = e^{i\alpha\cdot S/\hbar}, \tag{7.58}$$

which gives a three-dimensional rotation matrix. The components of $\mathbf{A}$ are then rotated using the matrix equation

$$\hat{R}_S(\alpha)\cdot\mathbf{A}(x) = \mathbf{A}'(x). \tag{7.59}$$

It is important to note that $\hat{R}_S$ and $\hat{R}_x$ commute since they act on different aspects of the vector potential function. It is common notation to write

$$\hat{R}(\alpha) = \hat{R}_S(\alpha)\,\hat{R}_x(\alpha) = e^{i\alpha\cdot(\hat{L}+\hat{S})/\hbar} = e^{i\alpha\cdot\hat{J}/\hbar}, \tag{7.60}$$

where

$$\hat{J} = \hat{L} + \hat{S}. \tag{7.61}$$

The operator $\hat{J}$ is referred to as the *total angular momentum operator* for a particle with *spin*, with the understanding that $[\hat{L}_j, \hat{S}_k] = 0$ because they act to rotate different aspects of the vector potential.

The need to include the action of the operator $\hat{S}$ represents the fact that the light wave may possess an intrinsic angular momentum. This intrinsic angular momentum is apparent in the case of *circular polarization* for the light wave. As it passes a point, the vector $\mathbf{A}$ for circular polarization will rotate around the direction of propagation in either a left or right-handed sense, and this rotation reflects an *intrinsic* angular momentum present in the light wave. Because the light wave consists of photons, the three matrices of the vector operator $\hat{S}$ therefore represent the observable known as the *spin angular momentum* for the photon. This identification is possible since these matrices obey the defining rotation algebra,

$$[\hat{S}_j, \hat{S}_k] = i\hbar\sum_{\ell=1}^{3}\varepsilon^{jk\ell}\hat{S}_\ell, \tag{7.62}$$

and therefore serve as the generators of rotations on the components of the light wave described by $\mathbf{A}(x, t)$.

7.3.3 Spin angular momentum eigenvalues for the photon

Because the photon spin operator $\boldsymbol{S}$ obeys the algebra (7.62), identical to the orbital angular momentum operator of the hydrogen atom analysed in (5.135), it is possible to include $\hat{S}_z$ and $\hat{\boldsymbol{S}} \cdot \hat{\boldsymbol{S}} = \hat{S}^2$ in the CSCO. In particular, (7.27) gives the z-component of the spin operator for the photon,

$$\hat{\mathbf{S}}_z = -i\hbar \mathbf{T}_z = \begin{pmatrix} 0 & -i\hbar & 0 \\ i\hbar & 0 & 0 \\ 0 & 0 & 0 \end{pmatrix}. \tag{7.63}$$

The eigenvalues s_z of $\hat{S}_z$ for a photon can then be found using (7.63) and solving the characteristic equation,

$$\det(\hat{S}_z - s_z \mathbf{1}) = -s_z^3 + \hbar^2 s_z = 0. \tag{7.64}$$

There are therefore three eigenvalues of $\hat{S}_z$, given by the set $s_z = \{\hbar, 0, -\hbar\}$.

In the case of a massless photon, the eigenvalue $s_z = 0$ is forbidden if the associated light wave is moving in the z-direction. This follows by using (7.63) to show that the eigenvector of $\hat{S}_z$ with the eigenvalue of zero is given by a vector potential $\mathbf{A}_0$ with the general form

$$\hat{\mathbf{S}}_z \cdot \mathbf{A}_0 = 0 \implies \mathbf{A}_0 = \begin{pmatrix} 0 \\ 0 \\ A_z \end{pmatrix}. \tag{7.65}$$

However, the wave vector $\boldsymbol{k}$ and the vector potential $\mathbf{A}$ of the electromagnetic wave (1.72) must satisfy the transversality condition, $\boldsymbol{k} \cdot \mathbf{A} = 0$. A vector potential of the form $\mathbf{A}_0$ is therefore forbidden for the case that the wave vector $\boldsymbol{k}$ lies along the z-direction since $\boldsymbol{k} \cdot \mathbf{A}_0 = k_z A_z$. As a result, $\hat{S}_z$ can have only the eigenvalues $\pm\hbar$ for a wave traveling in the z-direction since the associated orthonormal eigenvectors of (7.63) for the eigenvalues $\pm\hbar$ are given by

$$\mathbf{A}_\pm = \frac{1}{\sqrt{2}} \begin{pmatrix} 1 \\ \pm i \\ 0 \end{pmatrix}. \tag{7.66}$$

These two eigenvectors satisfy $\mathbf{A}_+^\dagger \cdot \mathbf{A}_- = 0$. For a general particle of momentum $\boldsymbol{p}$, the quantity $\boldsymbol{p} \cdot \boldsymbol{S}/p$, the projection of $\boldsymbol{S}$ along the momentum, is referred to as the *helicity* of the particle. The methods of quantum field theory show that the helicity *for a massless particle* cannot be zero, in agreement with the result obtained here for the photon.

Using the forms $\hat{S}_x$ and $\hat{S}_y$ obtained from (7.33) as the other two components of photon spin gives

$$\hat{S}^2 = \hat{S}_x^2 + \hat{S}_y^2 + \hat{S}_z^2 = \begin{pmatrix} 0 & 0 & 0 \\ 0 & \hbar^2 & 0 \\ 0 & 0 & \hbar^2 \end{pmatrix} + \begin{pmatrix} \hbar^2 & 0 & 0 \\ 0 & 0 & 0 \\ 0 & 0 & \hbar^2 \end{pmatrix} + \begin{pmatrix} \hbar^2 & 0 & 0 \\ 0 & \hbar^2 & 0 \\ 0 & 0 & 0 \end{pmatrix} = 2\hbar^2 \mathbf{1}, \tag{7.67}$$

showing that $\hat{S}^2$ is a multiple of the unit matrix. These results for photon spin are identical to the $\ell = 1$ case of orbital angular momentum in the hydrogen atom. In the hydrogen atom case, results (5.177) and (5.178) show that the total orbital angular momentum $\hat{L}^2$ has the eigenvalues $\ell(\ell + 1)\hbar^2$ and, once ℓ is specified, the possible eigenvalues of $\hat{L}_z$ are $(\ell\,\hbar, (\ell - 1)\hbar, \dots, -\ell\,\hbar)$. By comparison, the photon is referred to as a *spin one particle* since its spin eigenvalues match those corresponding to $\ell = 1$ in the hydrogen atom. Because of these similarities it is important to analyse the rotation algebra to determine the general form of angular momentum and the possible spin quantum numbers.

7.4 General angular momentum

The first step in analysing a general angular momentum, denoted $\hat{\boldsymbol{J}}$, is to pick the operators $\hat{J}^2$ and $\hat{J}_z$ as the two observables of interest. Because the components of $\hat{\boldsymbol{J}}$ are assumed to satisfy the rotation algebra, these two operators commute and so they share common eigenstates. The orthonormalized eigenstates are designated as $|\,\lambda, m\,\rangle$, and satisfy

$$\begin{aligned} \hat{J}^2|\,\lambda, m\,\rangle &= \lambda\hbar^2\,|\,\lambda, m\,\rangle \\ \hat{J}_z|\,\lambda, m\,\rangle &= m\hbar\,|\,\lambda, m\,\rangle. \end{aligned} \tag{7.68}$$

Initially, the eigenvalues λ and m are considered unknowns and unrelated, with the restriction that $\lambda \geqslant 0$ since it is the eigenvalue of $\hat{J}^2$. The angular momentum operator algebra can now be used to determine these eigenvalues as well as the relationship between λ and m.

7.4.1 Properties of the rotation algebra

The process begins by defining two non-Hermitian operators

$$\begin{aligned} \hat{J}_+ &= \hat{J}_x + i\hat{J}_y, \\ \hat{J}_- &= \hat{J}_x - i\hat{J}_y, \end{aligned} \tag{7.69}$$

which have the property that $\hat{J}_+^\dagger = \hat{J}_-$ and $\hat{J}_-^\dagger = \hat{J}_+$. The rotation group algebra shows that their products satisfy

$$\hat{J}_+\hat{J}_- = \hat{J}_x^2 + \hat{J}_y^2 - i(\hat{J}_x\hat{J}_y - \hat{J}_y\hat{J}_x) = \hat{J}^2 - \hat{J}_z^2 + \hbar\hat{J}_z, \tag{7.70}$$

$$\hat{J}_-\hat{J}_+ = \hat{J}_x^2 + \hat{J}_y^2 + i(\hat{J}_x\hat{J}_y - \hat{J}_y\hat{J}_x) = \hat{J}^2 - \hat{J}_z^2 - \hbar\hat{J}_z. \tag{7.71}$$

Using these two results gives

$$[\hat{J}_+, \hat{J}_-] = \hat{J}_+\hat{J}_- - \hat{J}_-\hat{J}_+ = 2\hbar\hat{J}_z. \tag{7.72}$$

The rotation algebra also gives

$$[\hat{J}_z, \hat{J}_+] = [\hat{J}_z, \hat{J}_x] + i[\hat{J}_z, \hat{J}_y] = i\hbar\hat{J}_y + \hbar\hat{J}_x = \hbar\hat{J}_+, \tag{7.73}$$

$$[\hat{J}_z, \hat{J}_-] = [\hat{J}_z, \hat{J}_x] - i[\hat{J}_z, \hat{J}_y] = i\hbar\hat{J}_y - \hbar\hat{J}_x = -\hbar\hat{J}_-. \tag{7.74}$$

Since $[\hat{J}^2, \hat{J}_k] = 0$ for all three components $\hat{J}_k$, it follows that $[\hat{J}^2, \hat{J}_\pm] = 0$.

The action of $\hat{J}_+$ on the state $|\, \lambda, m \,\rangle$ is designated as

$$\hat{J}_+|\, \lambda, m \,\rangle = |\, \lambda_+, m_+ \,\rangle, \tag{7.75}$$

where λ_+ and m_+ are to be determined. The adjoint state is given by

$$\langle\, \lambda_+, m_+ \,| = (\hat{J}_+|\, \lambda, m \,\rangle)^\dagger = \langle\, \lambda, m \,|\hat{J}_+^\dagger = \langle\, \lambda, m \,|\hat{J}_-. \tag{7.76}$$

The inner product of this state is then given by

$$\langle\, \lambda_+, m_+ \,|\, \lambda_+, m_+ \,\rangle = \langle\, \lambda, m \,|\, \hat{J}_-\hat{J}_+ \,|\, \lambda, m \,\rangle. \tag{7.77}$$

Using the positive-definite nature of the inner product and the form (7.71) for $\hat{J}_-\hat{J}_+$ gives

$$0 \leqslant \langle\, \lambda_+, m_+ \,|\, \lambda_+, m_+ \,\rangle = \langle\, \lambda, m \,|\, (\hat{J}^2 - \hat{J}_z^2 - \hbar\hat{J}_z) \,|\, \lambda, m \,\rangle = (\lambda - m(m+1))\hbar^2, \tag{7.78}$$

which yields the inequality $\lambda \geqslant m(m+1)$. In an identical manner, the action of $\hat{J}_-$ is written

$$\hat{J}_-|\, \lambda, m \,\rangle = |\, \lambda_-, m_- \,\rangle, \tag{7.79}$$

so that its adjoint state is given by

$$\langle\, \lambda_-, m_- \,| = (\hat{J}_-|\, \lambda, m \,\rangle)^\dagger = \langle\, \lambda, m \,|\hat{J}_+. \tag{7.80}$$

The positive-definite nature of the inner product and the form (7.70) for $\hat{J}_+\hat{J}_-$ gives

$$0 \leqslant \langle\, \lambda_-, m_- \,|\, \lambda_-, m_- \,\rangle = \langle\, \lambda, m \,|\, (\hat{J}^2 - \hat{J}_z^2 + \hbar\hat{J}_z) \,|\, \lambda, m \,\rangle = (\lambda - m(m-1))\hbar^2, \tag{7.81}$$

yielding the second inequality $\lambda \geqslant m(m-1)$. The sum of the two inequalities gives the important result

$$\lambda - m(m+1) + \lambda - m(m-1) \geqslant 0 \implies \lambda \geqslant m^2. \tag{7.82}$$

For a given value of λ, which must be positive since it is the eigenvalue of $\hat{J}^2$, this inequality requires m to lie in the range

$$-\sqrt{\lambda} \leqslant m \leqslant +\sqrt{\lambda}. \tag{7.83}$$

7.4.2 Raising and lowering operators and their action

Because $\hat{J}^2$ commutes with $\hat{J}_\pm$, it follows that

$$\hat{J}^2|\,\lambda_\pm,\, m_\pm\,\rangle = \hat{J}^2\hat{J}_\pm|\,\lambda,\, m\,\rangle = \hat{J}_\pm\hat{J}^2|\,\lambda,\, m\,\rangle = \lambda\hbar^2\hat{J}_\pm|\,\lambda,\, m\,\rangle = \lambda\hbar^2|\,\lambda_\pm,\, m_\pm\,\rangle, \quad (7.84)$$

which shows that $\lambda_\pm = \lambda$ and the operators $\hat{J}_\pm$ have no effect on the eigenvalues of $\hat{J}^2$. As a result, the states defined in (7.75) and (7.79) can be written

$$|\,\lambda_\pm,\, m_\pm\,\rangle = |\,\lambda,\, m_\pm\,\rangle, \quad (7.85)$$

since only the eigenvalue m is changed by $\hat{J}_\pm$.

Applying the commutator (7.73) to the state $|\,\lambda,\, m\,\rangle$ gives

$$[\hat{J}_z,\, \hat{J}_+]\,|\,\lambda,\, m\,\rangle = \hbar\hat{J}_+|\,\lambda,\, m\,\rangle = \hbar|\,\lambda,\, m_+\,\rangle. \quad (7.86)$$

The action of the commutator on the left-hand side of (7.86) is given by

$$[\hat{J}_z,\, \hat{J}_+]\,|\,\lambda,\, m\,\rangle = \hat{J}_z\hat{J}_z\hat{J}_+\,|\,\lambda,\, m\,\rangle - \hat{J}_+\hat{J}_z\,|\,\lambda,\, m\,\rangle = \hat{J}_z\,|\,\lambda,\, m_+\,\rangle - m\hbar|\,\lambda,\, m_+\,\rangle. \quad (7.87)$$

Equating (7.86) and (7.87) therefore gives

$$\hat{J}_z|\,\lambda,\, m_+\,\rangle = (m+1)\hbar\,|\,\lambda,\, m_+\,\rangle \implies \hat{J}_z\hat{J}_+|\,\lambda,\, m\,\rangle = (m+1)\hbar\,\hat{J}_+|\,\lambda,\, m\,\rangle. \quad (7.88)$$

The operator $\hat{J}_+$ is therefore referred to as a *raising operator* since it acts to raise the eigenvalue of $\hat{J}_z$ from m to $m + 1$. This means that the state defined in (7.75) and restricted by (7.85) must be given by

$$|\,\lambda,\, m_+\,\rangle = \hat{J}_+|\,\lambda,\, m\,\rangle = C_+(\lambda,\, m)|\,\lambda,\, m+1\,\rangle, \quad (7.89)$$

where $C_+(\lambda, m)$ is a constant to be determined. However, if $m + 1 \geqslant \sqrt{\lambda}$, then the inequality $\lambda \geqslant m^2$ would be violated, so that

$$m + 1 \geqslant \sqrt{\lambda} \implies \hat{J}_+|\,\lambda,\, m\,\rangle = 0. \quad (7.90)$$

Similarly, applying the commutator (7.74) to the state $|\,\lambda,\, m\,\rangle$ gives

$$[\hat{J}_z,\, \hat{J}_-]\,|\,\lambda,\, m\,\rangle = -\hbar\hat{J}_-|\,\lambda,\, m\,\rangle \implies \hat{J}_z|\,\lambda,\, m_-\,\rangle = (m-1)\hbar\,|\,\lambda,\, m_-\,\rangle. \quad (7.91)$$

The operator $\hat{J}_-$ is referred to as a *lowering operator* since it lowers the eigenvalue of $\hat{J}_z$ from m to $m - 1$. As a result,

$$|\,\lambda,\, m_-\,\rangle = \hat{J}_-|\,\lambda,\, m\,\rangle = C_-(\lambda,\, m)|\,\lambda,\, m-1\,\rangle, \quad (7.92)$$

where $C_-(\lambda, m)$ is a constant to be determined. It follows that if $m - 1 \leqslant -\sqrt{\lambda}$, then the inequality $\lambda \geqslant m^2$ would be violated. As a result,

$$m - 1 \leqslant -\sqrt{\lambda} \implies \hat{J}_-|\,\lambda,\, m\,\rangle = 0. \quad (7.93)$$

7.4.3 The eigenvalues of $\hat{J}^2$ and $\hat{J}_z$

The two results (7.90) and (7.93) show that there is a maximum and a minimum value for m in order for a non-zero state to exist, and these are designated j and j', respectively. These two values satisfy

$$\hat{J}_+|\,\lambda, j\,\rangle = 0, \quad \hat{J}_-|\,\lambda, j'\,\rangle = 0. \tag{7.94}$$

Combining this with (7.71) gives

$$0 = \hat{J}_-\hat{J}_+|\,\lambda, j\,\rangle = (\hat{J}^2 - \hat{J}_z^2 - \hbar\hat{J}_z)|\,\lambda, j\,\rangle = (\lambda - j^2 - j)\hbar^2|\,\lambda, j\,\rangle, \tag{7.95}$$

so that

$$\lambda = j(j+1), \tag{7.96}$$

where j is the maximum value for m. Similarly, (7.70) gives

$$0 = \hat{J}_+\hat{J}_-|\,\lambda, j'\,\rangle = (\hat{J}^2 - \hat{J}_z^2 + \hbar\hat{J}_z)|\,\lambda, j'\,\rangle = (\lambda - j'^2 + j')\hbar^2|\,\lambda, j'\,\rangle, \tag{7.97}$$

so that

$$\lambda = j'(j'-1), \tag{7.98}$$

where j' is the minimum value for m.

Since the two results (7.96) and (7.98) for λ must be equal, it follows that

$$j(j+1) = j'(j'-1). \tag{7.99}$$

This equation has two solutions, $j' = -j$ and $j' = j + 1$. The latter solution must be discarded since j' would be greater than j, in contradiction of the assumption that j is the maximum value. The eigenstates consistent with the algebra of the rotation group are therefore written $|\,j, m\,\rangle$, and satisfy

$$\hat{J}^2|\,j, m\,\rangle = j(j+1)\hbar^2|\,j, m\,\rangle, \quad \hat{J}_z|\,j, m\,\rangle = m\hbar|\,j, m\,\rangle. \tag{7.100}$$

where $m = j$ is the maximum value of m and $m = -j$ is the minimum value of m. Because $\hat{J}^2$ and $\hat{J}_z$ are Hermitian operators, the eigenstates of (7.100) will be orthonormal,

$$\langle\,j, m\,|\,j', m'\,\rangle = \delta_{jj'}\delta_{mm'}. \tag{7.101}$$

The angular momentum states are also assumed to be complete. Because the allowed values of m are limited by j, this completeness is expressed as

$$\sum_{m=-j}^{j}\sum_{j=0}^{\infty}|\,j, m\,\rangle\langle\,j, m\,| = \hat{\mathbf{1}}. \tag{7.102}$$

Using (7.101) shows that (7.102) is idempotent. An extremely important result is obtained by using the fact that the state $|\,j, -j\,\rangle$ must result from an integer power of $\hat{J}_-$ being applied to the state $|\,j, j\,\rangle$. If that integer power is n, then it must be related to j according to $n = 2j$, since $2j$ integer steps are required to go from j to $-j$. Therefore, the *allowable values of j* are determined from the sequence of positive integers n according to $j = \frac{1}{2}n$, and are therefore given by

$$j = 0, \frac{1}{2}, 1, \frac{3}{2}, \dots. \tag{7.103}$$

For consistency with the correspondence principle *all quantum mechanical angular momentum* must be governed by these general results, so that *all* angular momentum is quantized and the eigenstates are governed by the relations (7.100). All the results for angular momentum so far has been consistent with this idea. The hydrogen atom exhibited only integer values of $j = \ell$ for its orbital angular momentum along with the relation $-\ell \leqslant m \leqslant \ell$. The photon spin angular momentum determined in the previous section was associated with the single value $j = 1$ with the $\hat{s}_z$ values associated with $-1 \leqslant m \leqslant 1$, although the case $m = 0$ was excluded due to the transversality of the electromagnetic wave.

For a general angular momentum state, the actions of $J_{\pm}$ can be written

$$J_+|\, j, m\,\rangle = C_+(j, m)|\, j, m + 1\,\rangle, \qquad J_-|\, j, m\,\rangle = C_-(j, m)|\, j, m - 1\,\rangle. \tag{7.104}$$

Results (7.95) and (7.97) can now be adapted to determine the two coefficients $C_{\pm}$ in (7.104). Using

$$\langle\, j, m\,| J_- = \langle\, j, m\,| J_+^{\dagger} = (J_+|\, j, m\,\rangle)^{\dagger} = (C_+(j, m)\,|\, j, m + 1\,\rangle)^{\dagger} = C_+^*(j, m)\,\langle\, j, m + 1\,|, \tag{7.105}$$

along with a similar result for J_- in combination with the orthonormality of the angular momentum eigenstates, it follows that

$$|C_+(j, m)|^2 = \langle\, j, m\,|\, \hat{J}_-\hat{J}_+\,|\, j, m\,\rangle = (j(j + 1) - m(m + 1))\hbar^2, \tag{7.106}$$

$$|C_-(j, m)|^2 = \langle\, j, m\,|\, \hat{J}_+\hat{J}_-\,|\, j, m\,\rangle = (j(j + 1) - m(m - 1))\hbar^2. \tag{7.107}$$

These two results will be used in a later section when angular momentum is added quantum mechanically.

7.4.4 Angular momentum and uncertainty

The x-component of $\hat{\boldsymbol{J}}$ can be written

$$\hat{J}_x = \frac{1}{2}(\hat{J}_+ + \hat{J}_-), \tag{7.108}$$

which immediately gives the expectation value in an arbitrary angular momentum state as

$$\langle \hat{J}_x \rangle = \frac{1}{2}\langle\, j, m\,|\,(\hat{J}_+ + \hat{J}_-)\,|\, j, m\,\rangle = \frac{1}{2}C_+(j, m)\langle\, j, m\,|\, j, m + 1\,\rangle + \frac{1}{2}C_-(j, m)\langle\, j, m\,|\, j, m - 1\,\rangle = 0. \tag{7.109}$$

Similarly, $\hat{J}_x^2$ is given by

$$\hat{J}_x^2 = \frac{1}{4}(\hat{J}_+^2 + \hat{J}_-^2 + \hat{J}_+\hat{J}_- + \hat{J}_-\hat{J}_+). \tag{7.110}$$

The first two terms have the property that

$$\langle j, m \mid \hat{J}_+^2 \mid j, m \rangle = \langle j, m \mid \hat{J}_-^2 \mid j, m \rangle = 0. \tag{7.111}$$

Combining this with (7.106) and (7.107) shows that

$$\langle \hat{J}_x^2 \rangle = \frac{1}{4}\langle j, m \mid \hat{J}_-\hat{J}_+ \mid j, m \rangle + \frac{1}{4}\langle j, m \mid \hat{J}_+\hat{J}_- \mid j, m \rangle = \frac{1}{2}(j(j+1) - m^2)\hbar^2. \tag{7.112}$$

These results give the uncertainty in $\hat{J}_x$,

$$\Delta\hat{J}_x = \sqrt{\langle \hat{J}_x^2 \rangle - \langle \hat{J}_x \rangle^2} = \hbar\sqrt{\frac{1}{2}j(j+1) - \frac{1}{2}m^2} = \sqrt{\frac{1}{2}\langle \hat{J}^2 \rangle - \frac{1}{2}\langle \hat{J}_z^2 \rangle} = \sqrt{\frac{1}{2}\langle \hat{J}_x^2 \rangle + \frac{1}{2}\langle \hat{J}_y^2 \rangle}. \tag{7.113}$$

The minimum uncertainty is $\hbar\sqrt{j/2}$, which occurs when $m = \pm j$, corresponding to the two largest values for $\hat{J}_z$. Squaring both sides of (7.113) gives

$$\langle \hat{J}_x^2 \rangle = \langle \hat{J}_y^2 \rangle = \frac{1}{2}(j(j+1) - m^2)\hbar^2, \tag{7.114}$$

which also follows from using $\hat{J}_y = -\frac{1}{2}i(\hat{J}_+ - \hat{J}_-)$ and repeating the analysis for $\hat{J}_x$. Coupling this with $\langle \hat{J}_x \rangle = \langle \hat{J}_y \rangle = 0$ shows that the projection of $\hat{\boldsymbol{J}}$ onto the x–y plane can point in any direction with *equal probability*. Quantum mechanics yields no correlation between the two remaining components of $\hat{\boldsymbol{J}}$.

In an early attempt at an alternative version of quantum mechanics the eigenvalues of $\hat{J}_x$ and $\hat{J}_y$ were postulated to exist simultaneously with those of $\hat{J}_z$, so that the vector $\hat{\boldsymbol{J}}$ is well-defined. However, two of its components remain *unknown* subsequent to preparation. The values of J_x and J_y can then play the role of *hidden variables*, and their values are constrained to satisfy conservation of angular momentum if the system is isolated. In the hidden variable approach it is the random nature of the hidden variable that gives rise to the statistical behavior characterizing quantum mechanical measurements. The assumption of hidden angular momentum variables results in additional correlations for measurements of angular momentum in an isolated two-particle system. These additional correlations stem from *Bell's theorem*. The observation of these correlations would enable such a hidden variable theory to be distinguished from standard quantum mechanics, and in fact would render standard quantum mechanics incorrect. However, experimental results have consistently contradicted the predictions of hidden variable theories by violating Bell's theorem. This is discussed in more detail subsequent to understanding electron spin and developing the rules for addition of angular momentum.

7.5 The Stern–Gerlach experiment and electron spin

The Stern–Gerlach experiment placed a stream of *electrically neutral* silver atoms into an *inhomogeneous magnetic field.* Since the atoms were neutral they did not perform the helical motion for a charge q in a magnetic field that was found in chapter 5. Instead, the stream split into *two distinct spatial parts* as a result of interaction with the magnetic field. The Stern–Gerlach experiment predates the Schrödinger equation, and so the Bohr model provided the current theoretical understanding of the energy levels of the electrons. It is to be recalled that the Bohr model assumed that the orbital angular momentum was quantized in integer values. The experiment was intended to test for the expected integer values of a net orbital angular momentum for the electrons in the silver atom, which was referred to as *spatial quantization*. However, the splitting of the beam into *two* distinct branches revealed that the silver atom possessed only two distinct values for its net angular momentum.

7.5.1 The electron spin dipole moment

In order to explain the outcome of the experiment it is necessary to postulate that, despite the atom being electrically neutral, the electrons in the silver atoms possess an intrinsic angular momentum that results in an interaction with the magnetic field. In this regard, the atomic number of silver is 47, so that there is an odd number of electrons present in the neutral silver atom. If the electrons in the silver atoms each possess an intrinsic angular momentum or spin $\hat{\boldsymbol{S}}$, then each electron of charge q would have a magnetic dipole moment. In analogy to the dipole moment (5.253) of circular motion, it is written as

$$\boldsymbol{M}_{\mathrm{s}} = \kappa \frac{q}{mc} \hat{\boldsymbol{S}}, \tag{7.115}$$

where κ was postulated as an unknown constant. This magnetic dipole moment is the spin counterpart to the orbital moment (5.253), which results from circular motion,

$$\boldsymbol{M}_{\mathrm{L}} = \frac{q}{2mc} \hat{\boldsymbol{L}}. \tag{7.116}$$

This was defined in (5.252) and used to analyse the energy levels of a charge moving in a uniform magnetic field. It is present with an electron for the same basic reason that a classical electrically charged spinning sphere possesses a magnetic dipole moment. However, unlike the orbital dipole moment (7.116) the spin magnetic dipole moment derived using a classical argument results in a value for $\mathbf{M}_{\mathrm{s}}$ that is incorrect by a factor of 2. The value of κ was unknown at the time, but has subsequently been determined both theoretically and experimentally to great accuracy as $\kappa \approx 1.001\,16$. This makes the factor in (7.115) almost exactly a factor of 2 larger than that of the orbital magnetic dipole moment (7.116), and so κ will be treated as unity in what follows. The explanation for this difference between the spin and orbital dipole moment can be obtained by linearizing the *Dirac equation*, which is the relativistically correct equation governing the electron and will not be presented in this text. However, it can also be obtained by linearizing the Schrödinger equation and using the properties of electron spin, which is to be developed in what follows. The interested reader is recommended to consult the references.

By analogy to the Hamiltonian term (5.252), the energy of a dipole in a magnetic field, assumed to be oriented in the z-direction, is

$$U_d(\boldsymbol{x}) = -\mathbf{M} \cdot \mathbf{B}(\boldsymbol{x}) = -\frac{q}{mc} B_z(\boldsymbol{x}) \hat{S}_z. \tag{7.117}$$

Since the magnetic field is inhomogeneous, this potential energy corresponds to a force on each electron in the atom given by

$$\boldsymbol{F}(\boldsymbol{x}) = -\nabla U_d(\boldsymbol{x}) = \left(\frac{q}{mc} \nabla B_z(\boldsymbol{x}) \right) \hat{S}_z. \tag{7.118}$$

The net spin of the electrons in the z-direction is the sum of the spins in the z-direction of the individual electrons. The experimental outcome dictates the idea that pairs of electrons in the silver atom have equal and opposite spins in the z-direction, but since there is an odd number of electrons the outermost electron is *unpaired.* This unpaired electron creates the dipole moment of the silver atom responsible for the beam splitting. The two distinct pieces of the beam then indicate that the outermost electron has only *two distinct spin states* along the axis of the magnetic field.

While unexpected, the Stern–Gerlach experiment demonstrated that the electron has a quantized half-integer spin angular momentum. The upshot of all subsequent experiments that involve electron spin is that there are only two possible values for $\hat{S}_z$ *for each electron*. From the analysis of the previous section, this is consistent with the value $j = \frac{1}{2}$, so that $\hat{S}_z$ has the eigenvalues $\frac{1}{2}\hbar$ and $-\frac{1}{2}\hbar$, while $\hat{S}^2$ has the eigenvalue $j(j + 1)\hbar^2 = \frac{3}{4}\hbar^2$. The electron is therefore referred to as a *spin one-half* particle. The electron spin operators share the eigenstates $| s, m_s \rangle$, where $s = \frac{1}{2}$ and $m_s = \pm\frac{1}{2}$. These states can be viewed as a two-component complex column vector, known as a *spinor*, and it is important to investigate the rotational behavior of spinors.

7.5.2 The group SU(2)

The spinor representation of the rotation group involves the *special unitary matrices of order* 2, denoted SU(2). These are represented by the set of complex valued 2×2 matrices that are both unitary and special, so that they have a determinant of one. The product of two such matrices, $\mathbf{A}$ and $\mathbf{B}$, is another unitary matrix with determinant one. This follows since

$$\begin{aligned}(\mathbf{A}\cdot\mathbf{B})^{\dagger} &= \mathbf{B}^{\dagger}\cdot\mathbf{A}^{\dagger} = (\mathbf{A}\cdot\mathbf{B})^{-1},\\ \det(\mathbf{A}\cdot\mathbf{B}) &= \det\mathbf{A}\det\mathbf{B} = 1.\end{aligned} \tag{7.119}$$

Therefore, this set of matrices is closed under multiplication and therefore constitutes the group known as SU(2) or special unitary group of 2×2 matrices.

An arbitrary 2×2 complex valued matrix has eight values since the four elements of the matrix can be complex, corresponding to both a real and imaginary number. For complex elements the unitarity requirement,

$$(\mathbf{A}^{\dagger}\cdot\mathbf{A})_{\ell j} = \sum_{k=1}^{2} A^{\dagger}_{\ell k}A_{kj} = \sum_{k=1}^{2} A^{*}_{k\ell}A_{kj} = \delta_{\ell j}, \tag{7.120}$$

gives four distinct equations. This follows from the fact that the two diagonal equations, where $\ell = j$ in the unitarity requirement, give a real number on the left-hand side,

$$\sum_{k=1}^{2} A^{*}_{kj}A_{kj} = A^{*}_{1j}A_{1j} + A^{*}_{2j}A_{2j} = |A_{1j}|^2 + |A_{2j}|^2 = 1, \tag{7.121}$$

and therefore correspond to only two equations for the real parts of the elements that must be satisfied. For the two non-diagonal cases, where $\ell \neq j$, the unitarity requirement takes the form of a complex number,

$$\sum_{k=1}^{2} A^{*}_{k\ell}A_{kj} = A^{*}_{1\ell}A_{1j} + A^{*}_{2\ell}A_{2j} = \alpha + \beta i = 0, \tag{7.122}$$

which gives two equations, $\alpha = 0$ and $\beta = 0$. However, interchanging ℓ and j results in

$$\sum_{k=1}^{2} A^{*}_{kj}A_{k\ell} = \sum_{k=1}^{2} (A^{*}_{k\ell}A_{kj})^{*} = (\alpha + \beta i)^{*} = \alpha - \beta i = 0, \tag{7.123}$$

which is the same equation as (7.122). As a result, there are only two equations from the off-diagonal elements of the unitarity requirement. The total number of unique equations obtained from the unitarity requirement is therefore four.

The special determinant requirement gives

$$\det\mathbf{A} = \gamma + \delta i = 1. \tag{7.124}$$

However, the unitarity requirement has already given

$$\det(\mathbf{A}^{\dagger}\mathbf{A}) = \det\mathbf{A}^{\dagger}\det\mathbf{A} = (\det\mathbf{A}^{T})^{*}\det\mathbf{A} = (\det\mathbf{A})^{*}\det\mathbf{A} = \gamma^2 + \delta^2 = \det\mathbf{1} = 1. \tag{7.125}$$

The requirement that the determinant is unity gives only one unique equation, $\delta = 0$, while the unitarity requirement reduces to $\gamma^2 = 1$, like SO(3). The matrices of SU(2) therefore have three free parameters, which is the same as the matrices of SO(3).

The three free parameters of the group SU(2) can be mapped into the three free parameters of the rotation group SO(3). Using two complex numbers, α and β, a general member of SU(2) can be written as the 2×2 matrix

$$\mathbf{U} = \begin{pmatrix} \alpha & \beta \\ -\beta^* & \alpha^* \end{pmatrix}. \tag{7.126}$$

The unitarity requirement gives

$$\mathbf{U}^\dagger \cdot \mathbf{U} = \begin{pmatrix} \alpha^* & -\beta \\ \beta^* & \alpha \end{pmatrix} \cdot \begin{pmatrix} \alpha & \beta \\ -\beta^* & \alpha^* \end{pmatrix} = \begin{pmatrix} \alpha^*\alpha + \beta^*\beta & 0 \\ 0 & \alpha^*\alpha + \beta^*\beta \end{pmatrix} = \begin{pmatrix} 1 & 0 \\ 0 & 1 \end{pmatrix}. \tag{7.127}$$

The determinant requirement gives

$$\det \mathbf{U} = \det \begin{pmatrix} \alpha & \beta \\ -\beta^* & \alpha^* \end{pmatrix} = \alpha^*\alpha + \beta^*\beta = 1. \tag{7.128}$$

Both requirements are satisfied by enforcing the condition

$$\alpha^*\alpha + \beta^*\beta = 1. \tag{7.129}$$

This is a real valued equation, and so it places one constraint on the four parameters of α and β to reduce them to three free parameters, consistent with SU(2).

Using the polar form $\alpha = ae^{i\theta}$, this constraint becomes

$$a^2 = 1 - \beta_{\mathrm{R}}^2 - \beta_{\mathrm{I}}^2, \tag{7.130}$$

where β_{R} and β_{I} are the real and imaginary parts of β. Both β_{R} and β_{I} can be treated as infinitesimal parameters so that (7.130) reduces to $a \approx 1$. For the infinitesimal case the matrix (7.127) is given by

$$\mathbf{U} = \begin{pmatrix} e^{i\theta} & \beta_{\mathrm{R}} + i\beta_{\mathrm{I}} \\ -\beta_{\mathrm{R}} + i\beta_{\mathrm{I}} & e^{-i\theta} \end{pmatrix}. \tag{7.131}$$

The generators for SU(2) can now be found by taking the derivative of the matrix (7.131). Employing the same procedure (7.27) that was used for SO(3), expanding the matrix (7.131) in a Taylor series around zero, gives the SU(2) generators,

$$\begin{aligned} \boldsymbol{\tau}^1 &= \frac{1}{2}\frac{\partial}{\partial \beta_{\mathrm{I}}}\mathbf{U} = \frac{1}{2} i\boldsymbol{\sigma}^1 = \frac{1}{2} i \begin{pmatrix} 0 & 1 \\ 1 & 0 \end{pmatrix}, \\ \boldsymbol{\tau}^2 &= \frac{1}{2}\frac{\partial}{\partial \beta_r}\mathbf{U} = \frac{1}{2} i\boldsymbol{\sigma}^2 = \frac{1}{2} i \begin{pmatrix} 0 & -i \\ i & 0 \end{pmatrix}, \\ \boldsymbol{\tau}^3 &= \frac{1}{2}\frac{\partial}{\partial \theta}\mathbf{U} = \frac{1}{2} i\boldsymbol{\sigma}^3 = \frac{1}{2} i \begin{pmatrix} 1 & 0 \\ 0 & -1 \end{pmatrix}, \end{aligned} \tag{7.132}$$

where the factor of 1/2 is required to conform with the algebra of the SO(3) group. The $\boldsymbol{\sigma}^a$ matrices defined by (7.132) are known as the *Pauli spin matrices*. The generators $\boldsymbol{\tau}^a = \frac{1}{2}i\boldsymbol{\sigma}^a$ give

$$[\boldsymbol{\tau}^1, \boldsymbol{\tau}^2] = -\frac{1}{4}\begin{pmatrix} 0 & 1 \\ 1 & 0 \end{pmatrix} \cdot \begin{pmatrix} 0 & -i \\ i & 0 \end{pmatrix} + \frac{1}{4}\begin{pmatrix} 0 & -i \\ i & 0 \end{pmatrix} \cdot \begin{pmatrix} 0 & 1 \\ 1 & 0 \end{pmatrix} = -\frac{1}{2}i\begin{pmatrix} 1 & 0 \\ 0 & -1 \end{pmatrix} = -\boldsymbol{\tau}^3. \quad (7.133)$$

Calculation of the other commutators shows that the $\boldsymbol{\tau}^a$ matrices satisfy

$$[\boldsymbol{\tau}^a, \boldsymbol{\tau}^b] = -\sum_{c=1}^{3} \varepsilon^{abc}\boldsymbol{\tau}^c, \quad (7.134)$$

which is the same Lie algebra (7.41) that was found for the generators $\mathbf{T}_a$ of SO(3). It is straightforward to verify that the Pauli matrices also satisfy the anticommutation relation

$$\frac{1}{2}\{\boldsymbol{\sigma}^a, \boldsymbol{\sigma}^b\} = \delta_{ab}\,\mathbf{1}, \quad (7.135)$$

which is not a property of the SO(3) generators.

7.5.3 Spinor representations of rotations

Because it reproduces the algebra of SO(3), SU(2) gives what is known as a *spinor representation* of rotations. The representation space will be the set of two-dimensional complex vectors, written

$$v = \begin{pmatrix} v_1 \\ v_2 \end{pmatrix}, \quad (7.136)$$

and known as *spinors*. The 2×2 matrices of SU(2) act on v to rotate it through the angles θ^a. This is given by

$$v' = \mathbf{U} \cdot v, \quad (7.137)$$

where, analogously to SO(3),

$$\mathbf{U} = \exp\left(\sum_{a=1}^{3} \theta^a \boldsymbol{\tau}^a\right) = \exp\left(i\frac{1}{2}\sum_{a=1}^{3} \theta^a \boldsymbol{\sigma}^a\right). \quad (7.138)$$

This transformation is unitary since the definitions (7.132) show that $\boldsymbol{\tau}^{a\dagger} = -\boldsymbol{\tau}^a$, equivalently $\boldsymbol{\sigma}^{a\dagger} = \boldsymbol{\sigma}^a$, and therefore

$$\mathbf{U}^\dagger = \exp\left(\sum_{a=1}^{3} \theta^a \boldsymbol{\tau}^{a\dagger}\right) = \exp\left(-\sum_{a=1}^{3} \theta^a \boldsymbol{\tau}^a\right) = \mathbf{U}^{-1}. \quad (7.139)$$

The definitions (7.132) of the Pauli spin matrices have the property $\mathrm{Tr}(\boldsymbol{\sigma}^a) = 0$, and so theorem (6.222) again shows that

$$\det \mathbf{U} = \exp\mathrm{Tr}\ln\mathbf{U} = \exp\mathrm{Tr}\left(\sum_{a=1}^{3}\theta^a \boldsymbol{\tau}^a\right) = \exp(0) = 1. \tag{7.140}$$

Unitarity insures that the transformed spinor satisfies

$$\boldsymbol{v}'^{\dagger} \cdot \boldsymbol{v}' = \boldsymbol{v}^{\dagger} \cdot \mathbf{U}^{\dagger} \cdot \mathbf{U} \cdot \boldsymbol{v} = \boldsymbol{v}^{\dagger} \cdot \boldsymbol{v}, \tag{7.141}$$

so that the magnitude of the spinor is maintained. Although $\boldsymbol{v}$ is complex, the bilinear $\boldsymbol{v}^{\dagger} \cdot \boldsymbol{v}$ is a real number and therefore observable.

The transformation $\mathbf{U}$ defined by (7.138) can be associated with a physical rotation of a vector in three-dimensional space in the following way. If $\boldsymbol{A}$ is a vector in three-dimensional space with the components A_j, the matrix $\boldsymbol{A\!\!\!/}$ is defined using the components of $\mathbf{A}$ and the Pauli spin matrices,

$$\boldsymbol{A\!\!\!/} \equiv \sum_{j=1}^{3} A_j \boldsymbol{\sigma}^j. \tag{7.142}$$

This matrix has the property that its determinant is given by

$$\det\boldsymbol{A\!\!\!/} = \det\begin{pmatrix} A_3 & A_1 - iA_2 \\ A_1 + iA_2 & -A_3 \end{pmatrix} = -\sum_{j=1}^{3} A_j A_j = -\boldsymbol{A} \cdot \boldsymbol{A} = -A^2. \tag{7.143}$$

Since det/$\mathbf{A}$ depends only on the magnitude of the vector, it does not change under a rotation. The rotated version of the matrix $\boldsymbol{A\!\!\!/}$ is defined as

$$\boldsymbol{A\!\!\!/}' = \mathbf{U} \cdot \boldsymbol{A\!\!\!/} \cdot \mathbf{U}^{\dagger}. \tag{7.144}$$

The transformation (7.144) represents a rotation since result (6.176) combined with the unitarity of the matrix $\mathbf{U}$ defined by (7.138) gives

$$-A'^{\,2} = \det\boldsymbol{A\!\!\!/}' = \det(\mathbf{U} \cdot \boldsymbol{A\!\!\!/} \cdot \mathbf{U}^{\dagger}) = \det\boldsymbol{A\!\!\!/} = -A^2, \tag{7.145}$$

showing that the magnitude of the vector $\boldsymbol{A}$ is preserved. Since the transformation $\mathbf{U}$ preserves the magnitude of the vector $\boldsymbol{A}$, it therefore acts as a rotation of the vector $\boldsymbol{A}$.

For example, a spinor rotation around the z-axis by the angle θ is performed using

$$\mathbf{U}_3(\theta) = \exp\left(\frac{1}{2} i\theta\boldsymbol{\sigma}^3\right) = \sum_{n=0}^{\infty} \frac{(i\theta\boldsymbol{\sigma}^3)^n}{2^n n!}. \tag{7.146}$$

For n an integer, $(\boldsymbol{\sigma}^3)^{2n} = \mathbf{1}$ and $(\boldsymbol{\sigma}^3)^{2n+1} = \boldsymbol{\sigma}^3$, and the infinite series for $\mathbf{U}_3(\theta)$ sums to

$$\mathbf{U}_3(\theta) = \begin{pmatrix} e^{\frac{1}{2}i\theta} & 0 \\ 0 & e^{-\frac{1}{2}i\theta} \end{pmatrix} \Longrightarrow \mathbf{U}_3^{\dagger}(\theta) = \begin{pmatrix} e^{-\frac{1}{2}i\theta} & 0 \\ 0 & e^{\frac{1}{2}i\theta} \end{pmatrix}. \tag{7.147}$$

Performing the spinor transformation (7.144) gives

$$\boldsymbol{\mathcal{A}}' = \begin{pmatrix} A_3' & (A_1' - iA_2') \\ (A_1' + iA_2') & -A_3' \end{pmatrix} = \mathbf{U} \cdot \boldsymbol{\mathcal{A}} \cdot \mathbf{U}^\dagger = \begin{pmatrix} A_3 & (A_1 - iA_2)e^{i\theta} \\ (A_1 + iA_2)e^{-i\theta} & -A_3 \end{pmatrix}. \quad (7.148)$$

Using the Euler relation $e^{i\theta} = \cos\theta + i\sin\theta$ shows that the elements of the transformed matrix $\boldsymbol{\mathcal{A}}'$ are given by

$$\begin{aligned} A_3' &= A_3, \\ A_1' + iA_2' &= (A_1\cos\theta + A_2\sin\theta) + i(A_2\cos\theta - A_1\sin\theta). \end{aligned} \quad (7.149)$$

Since the real and imaginary parts must be equal separately, the results (7.20) and (7.21) for the rotation of a vector around the z-axis have appeared. Therefore the transformation (7.147) represents the same rotation around the z-axis. The quantity $v^\dagger \cdot \boldsymbol{\mathcal{A}} \cdot v$ is invariant under rotations due to the unitarity of the transformation matrix, and so it corresponds to a scalar or rotationally invariant quantity.

There is an unusual aspect to the spinor version of rotations, which is apparent when the angle of rotation $\theta = 2\pi$ is considered. For that case the transformation matrix (7.147) is given by $\mathbf{U}_3(2\pi) = -\mathbf{1}$, so that under a rotation of 2π the spinor changes sign, $v \to -v$. No physically observable quantity can change sign under a rotation of 2π, since it would mean that it must be zero. Like the wave function, a spinor does not correspond to a directly observable physical quantity. However, the bilinear $v^\dagger \cdot v$ is a real number and does not change sign under a rotation of 2π. As a result, it can be used to define the observable probability density associated with a *spinor particle*. The spinor version of a vector $\boldsymbol{\mathcal{A}}$ does not change sign under a rotation of 2π, and therefore does not face this dilemma of interpretation. Although rotations can be given a representation using SU(2), its representation space of two-dimensional complex spinors is fundamentally different from the representation space of three-dimensional real vectors used for SO(3). It was not possible to smoothly connect the SO(3) transformation of (7.34) to an improper rotation $-\mathbf{1}$ since such a transformation has a determinant of -1 in the 3×3 case. However, $-\mathbf{1}$ is a member of SU(2) since it has a determinant of $+1$ in the 2×2 case. In effect, the transformations of SU(2) map the interval $[0, 4\pi]$ into two copies of the same rotation for the vector $\boldsymbol{\mathcal{A}}$. Such a representation of rotations is referred to as *unfaithful* since it is two to one. However, because SU(2) has the same Lie algebra as the rotation group, SU(2) is said to be the *universal covering group* for rotations in three dimensions and is sometimes referred to as Spin(3).

7.5.4 Electron spin angular momentum

These results can now be adapted to the quantum mechanics of electron spin. The electron has two spin eigenstates, and these can be used to define a two-dimensional matrix representation for the electron spin operators, with elements defined by $\hat{\mathbf{s}}_{j\, m_s m_s'} = \langle\, s, m_s \mid \hat{S}_j \mid s, m_s' \,\rangle$, so that they are given explicitly by

$$\hat{\mathbf{s}}_x = \frac{\hbar}{2}\begin{pmatrix} 0 & 1 \\ 1 & 0 \end{pmatrix}, \quad \hat{\mathbf{s}}_y = \frac{\hbar}{2}\begin{pmatrix} 0 & -i \\ i & 0 \end{pmatrix}, \quad \hat{\mathbf{s}}_z = \frac{\hbar}{2}\begin{pmatrix} 1 & 0 \\ 0 & -1 \end{pmatrix}. \quad (7.150)$$

The Pauli spin matrices $\hat{\boldsymbol{\sigma}}_j$, defined in (7.132) for the spinor representation of the rotation group, allow the electron spin matrices of (7.150) to be written

$$\hat{\mathbf{s}}_j = \frac{1}{2}\hbar\hat{\boldsymbol{\sigma}}_j. \tag{7.151}$$

The spin matrices satisfy the required Hermitian version of the angular momentum algebra (7.46),

$$[\hat{\mathbf{s}}_j, \hat{\mathbf{s}}_k] = i\hbar\sum_{\ell=1}^{3}\varepsilon^{jk\ell}\hat{\mathbf{s}}_\ell, \tag{7.152}$$

as well as the relation

$$\hat{\mathbf{s}}^2 = \sum_{j=1}^{3}\hat{\mathbf{s}}_j^2 = \frac{3}{4}\hbar^2\mathbf{1}. \tag{7.153}$$

As a result, the spin matrices $\hat{\mathbf{s}}_j$ represent a quantum mechanical angular momentum operator. Since $\hat{\mathbf{s}}_z$ is a diagonal matrix, it is apparent its eigenvalues are the expected $\pm\frac{1}{2}\hbar$, allowing the electrons in atoms to have the required spin $\frac{1}{2}$ values.

7.5.5 Electron spinor wave functions

To recap, the Pauli spin matrices serve as the generators of the group SU(2), which shares the same algebra as the rotation group. The members of SU(2), defined by the matrix (7.138), act on two-dimensional complex vectors, referred to as *spinors*, which provide the two components of the electron *spinor wave function*, $\Psi_a(\boldsymbol{x}, t)$. The operators $\hat{\mathbf{s}}_z$ and $\hat{\mathbf{s}}^2$ can be added to the CSCO. The general electron spinor wave function is therefore a linear superposition of the eigenspinors of $\hat{\mathbf{s}}_z$, which take the form

$$\Psi_1(\boldsymbol{x}, t) = \begin{pmatrix}1\\0\end{pmatrix}\Psi_\uparrow(\boldsymbol{x}, t), \quad \Psi_2(\boldsymbol{x}, t) = \begin{pmatrix}0\\1\end{pmatrix}\Psi_\downarrow(\boldsymbol{x}, t). \tag{7.154}$$

The notation is chosen to reflect the fact that $\Psi_\uparrow$ is the positive or *spin up z*-component and $\Psi_\downarrow$ is the negative or *spin down z*-component of the electron wave function. When solving the Schrödinger equation for the electron, it is now understood that the spinor form for Ψ should be used, with the result that the Schrödinger equation for *both* components must be solved. Doing so allows spin-dependent effects to be included in the Hamiltonian.

As an example, the Hamiltonian for a charge in an magnetic field can be adapted to reflect the spinor nature of the electron by amending the magnetic dipole energy term (5.252) to read

$$U_\mathrm{d}(\boldsymbol{x}) = -\frac{q}{2\,mc}B_z(\boldsymbol{x})\hat{L}_z\mathbf{1} + \kappa q B_z(\boldsymbol{x})\,\hat{\mathbf{s}}_z, \tag{7.155}$$

where **1** is the two-dimensional unit matrix and $\hat{L}_z$ is the z-component of the *orbital angular momentum*, which acts on both components of the spinor. The electron wave function provides a *spinor representation space* of the rotation algebra, just as the electromagnetic vector potential provides a *vector representation space* of the rotation algebra. As pointed out earlier, the spinor representation has the unusual property that it *changes sign* under a rotation of 2π. If the spinor wave function was directly physically measurable, this would immediately require it to be zero. However, it is the probability density associated with the spinor wave function that is measurable, and it is given by the inner product

$$\mathcal{P}(\boldsymbol{x}, t) = \Psi^{\dagger}(\boldsymbol{x}, t) \cdot \Psi(\boldsymbol{x}, t) = \sum_{a=1}^{2} \Psi_a^{*}(\boldsymbol{x}, t)\Psi_a(\boldsymbol{x}, t). \tag{7.156}$$

The probability density found using a spinor wave function is therefore unchanged by a rotation of 2π, and the structure of quantum mechanics allows the electron to be modeled as a *spinor particle*.

7.5.6 Electron spin, chemical properties, and the Pauli exclusion principle

The additional quantum number represented by the eigenvalue of the intrinsic spin z-component $\hat{s}_z$ must be included in *all electron states*. For example, the eigenstates of the electron in the hydrogen atom must be written $| \, n \; \ell \; m_\ell \; m_s \, \rangle$, where $m_s = \pm\frac{1}{2}$ represents the two possibilities for the z-component of the electron spin. The quantum number m_ℓ represents the z-component of the orbital angular momentum, and the restrictions on the values of ℓ and m_ℓ found earlier by solving the Schrödinger equation or by using the algebra of the rotation group still apply. As a result, $\ell < n$ and $-\ell \leqslant m_\ell \leqslant \ell$. The presence of electron spin has the effect of doubling the number of possible eigenstates for the electron in the hydrogen atom. For example, the ground state, given by the principle quantum number $n = 1$, has only $\ell = m_\ell = 0$ possible. However, there are two possible values for m_s, and so the ground state of hydrogen is doubly degenerate. The first excited state, with the principle quantum number $n = 2$, can have $\ell = 0$ and $\ell = 1$, with the latter having $m_\ell = \{-1, 0, 1\}$ possible. Each of those states can have $m_s = \pm\frac{1}{2}$, and so the first excited state has eight possible electron states available.

The pattern just uncovered reflects the chemical properties of the first ten elements on the periodic table. Covalent chemical bonding is characterized by atoms sharing their outermost electrons. For example, the elements of hydrogen and lithium both have the chemical valence of +1 and behave similarly in chemical reactions. If lithium is similar to hydrogen in terms of its energy levels, they would both have a single electron in the outermost energy level or *shell*, but *only if* two of the three electrons in a neutral lithium atom are *filling* the $n = 1$ energy level. Similarly, the elements of helium, with atomic number 2, and neon, with atomic number 10, are chemically inert with valence 0. An explanation for their chemically inert behavior is that helium and neon have an energy shell structure similar to hydrogen *and* they have a fully occupied outer energy level. This would render the atom inert since the energy required to share an electron with another atom and chemically bond would be prohibitive. However, this explanation works only if each possible state is restricted to having a *single electron* occupying it. Otherwise, all electrons in a neutral unexcited atom would eventually radiate its excess energy away and be found in the lowest possible energy state.

This led Pauli to postulate the *Pauli exclusion principle*, which states that an electron will not share *all* its quantum numbers with another electron in the same atom. In quantum mechanical terms this means that, if an electron is found in the state with the quantum numbers n, ℓ, m_ℓ, and m_s, no other electron in that atom will be found with those same quantum numbers. Its state must have at least one of its quantum numbers with a value different from *any other electron in the atom.* Subsequent developments in relativistic quantum field theory showed that *all spin one-half particles* in nature are required to obey the Pauli exclusion principle with all other particles of the same type. For example, the mu meson or *muon*, is considerably more massive than an electron, but in every other way it is identical to the electron. Since it has spin one-half, it also obeys the Pauli exclusion principle with regard to other muons. In order to state this principle quantum mechanically, it is necessary to formulate *multiparticle states.*

7.6 Multiparticle states and statistics

Underlying quantum mechanics is the observational fact that there are fundamental building blocks to matter, often referred to as *elementary particles.* For example, each electron is identical to each other electron in the sense that all electrons have the same intrinsic mass, electric charge, and spin, and will behave identically to all other electrons *experimentally.* The identification of the types of elementary particles present in nature has been a major area of effort in the reductionist approach to physics. The identical nature of fundamental particles has ramifications for their behavior in multiparticle systems. This aspect of identical particles is often referred to as statistics, since multiparticle systems play a central role in statistical mechanics and condensed matter theory.

7.6.1 Tensor product Hilbert spaces

In the early discussion of wave functions for two-particle systems in chapter 5, it was stated that a two-particle wave function would be written $\Psi(\boldsymbol{x}_1, \boldsymbol{x}_2, t)$. In addition, if the wave function is describing a system with two indistinguishable identical elementary particles, the wave function was required to be either symmetric (5.97) or antisymmetric (5.100) under the exchange of $\boldsymbol{x}_1$ and $\boldsymbol{x}_2$. The infinitesimal probability,

$$dP = |\Psi(\boldsymbol{x}_1, \boldsymbol{x}_2, t)|^2 \, d^n x_1 \, d^n x_2, \tag{7.157}$$

is invariant under the exchange $\boldsymbol{x}_1 \leftrightarrow \boldsymbol{x}_2$ for both of these options.

If the two particles are *free of each other* and do not directly interact with each other in any way, then the Hamiltonian will not depend on the relative position

$x_2 - x_1$. For such a case, the total wave function is separable in the sense that, for identical particles,

$$\Psi_{\pm}(\mathbf{x}_1, \mathbf{x}_2, t) = \psi_1(\mathbf{x}_1, t)\psi_2(\mathbf{x}_2, t) \pm \psi_1(\mathbf{x}_2, t)\psi_2(\mathbf{x}_1, t). \tag{7.158}$$

This is not the case in the hydrogen atom, although the wave function is still separable in terms of the center of mass position $\mathbf{R}$ defined by (5.107) and the reduced mass position $\mathbf{r}$ defined by (5.106). It is also possible that interactions between the two particles can be ignored as a first approximation. An example is multi-electron atoms, where the basic energy shell structure of single electron hydrogen atom serves as a guideline for the states available to the electrons. Clearly, two electrons are not free of each other since their electric charge creates the repulsive Coulomb potential that depends on their position, but it is initially ignored in analyzing multi-electron atoms by assuming that their Coulomb repulsion does not alter the *general structure* of the hydrogen energy levels. Separability means that differential operators will operate on only those wave functions that match their argument. An example is the momentum of the particle at position $\mathbf{x}_1$, which is given by

$$-i\hbar\nabla_{x_1}\Psi(\mathbf{x}_1, \mathbf{x}_2, t) = (-i\hbar\nabla_{x_1}\psi_1(\mathbf{x}, t))\psi_2(\mathbf{x}_2, t), \tag{7.159}$$

which receives no contribution from $\psi_2(\mathbf{x}_2, t)$.

In Dirac notation, a separable state combines the component one-particle states into what is called a *tensor product*. For example, if $|\,\Psi_1, t\,\rangle$ and $|\,\Psi_2, t\,\rangle$ are two possible one-particle states, then a system where there are two independent particles present could be found in the state

$$|\,\Psi_1, t\,\rangle \otimes |\,\Psi_2, t\,\rangle \equiv |\,\Psi_1, \Psi_2, t\,\rangle, \tag{7.160}$$

where the notation suppressing the tensor product symbol $\otimes$ is almost universal in physics. In this case, the first particle is in the state $|\,\Psi_1, t\,\rangle$ and the second particle is in the state $|\,\Psi_2, t\,\rangle$. Each particle state exists in its own Hilbert space, but both are present in the full state of the two-particle system, whose Hilbert space is the tensor product of the two individual particle Hilbert spaces.

The inner product of the two-particle tensor product states $|\,\Psi_1', \Psi_2', t\,\rangle$ and $|\,\Psi_1, \Psi_2, t\,\rangle$ is given by

$$\langle\,\Psi_1', \Psi_2', t\,|\,\Psi_1, \Psi_2, t\,\rangle = \langle\,\Psi_1', t\,| \otimes \langle\,\Psi_2', t\,|)(|\,\Psi_1, t\,\rangle \otimes |\,\Psi_2, t\,\rangle) = \langle\,\Psi_1', t\,|\,\Psi_1, t\,\rangle\langle\,\Psi_2', t\,|\,\Psi_2, t\,\rangle. \tag{7.161}$$

This idea has been implicitly deployed already, since three-dimensional position states can be defined as a tensor product of three one-dimensional position states,

$$|\,\mathbf{x}\,\rangle = |\,x, y, z\,\rangle = |\,x\,\rangle \otimes |\,y\,\rangle \otimes |\,z\,\rangle, \tag{7.162}$$

so that each degree of spatial freedom corresponds to a Hilbert space. The inner product of two position states is given by (7.161), so that

$$\langle\,\mathbf{x}_1\,|\,\mathbf{x}_2\,\rangle = \langle\,x_1\,|\,x_2\,\rangle\langle\,y_1\,|\,y_2\,\rangle\langle\,z_1\,|\,z_2\,\rangle = \delta(x_1 - x_2)\,\delta(y_1 - y_2)\,\delta(z_1 - z_2) = \delta^3(\mathbf{x}_1 - \mathbf{x}_2). \tag{7.163}$$

The tensor product state can be given a configuration space representation using the inner product with the tensor product state

$$| x_1, x_2 \rangle = | x_1 \rangle \otimes | x_2 \rangle \implies | x_1, x_2 \rangle^\dagger = \langle x_1 | \otimes \langle x_2 | = \langle x_1, x_2 |. \tag{7.164}$$

Using the inner product recipe (7.161) for a tensor product state yields

$$\begin{aligned} \langle x_1, x_2 | \Psi_1, \Psi_2, t \rangle &= (\langle x_1 | \otimes \langle x_2 |)(| \Psi_1, t \rangle \otimes | \Psi_2, t \rangle) \\ &= \langle x_1 | \Psi_1, t \rangle \langle x_2 | \Psi_2, t \rangle = \Psi_1(x_1, t)\Psi_2(x_2, t), \end{aligned} \tag{7.165}$$

demonstrating that the tensor product state maps into a direct product of wave functions under the action of the tensor product configuration space.

Operators can also form tensor products, so that $\hat{A} \otimes \hat{B}$ is an operator that acts on the tensor product state according to

$$(\hat{A} \otimes \hat{B}) (| \Psi_1, t \rangle \otimes | \Psi_2, t \rangle) = \hat{A} | \Psi_1, t \rangle \otimes \hat{B} | \Psi_2, t \rangle, \tag{7.166}$$

where $\hat{A}$ acts on the first particle state and $\hat{B}$ acts on the second particle state. If the two states are eigenstates of both $\hat{A}$ and $\hat{B}$ with the eigenvalues a_1 and b_2, then

$$(\hat{A} \otimes \hat{B}) (| \Psi_1, t \rangle \otimes | \Psi_2, t \rangle) = a_1 | \Psi_1, t \rangle \otimes b_2 | \Psi_2, t \rangle = a_1 b_2 (| \Psi_1, t \rangle \otimes | \Psi_2, t \rangle), \tag{7.167}$$

which represents the *product* of two operators acting on *different* particles.

This operator formalism has already been implicitly employed. For example, the z-component of the three-dimensional position operator should have been written as

$$\hat{Q}_z = \hat{\mathbf{1}} \otimes \hat{\mathbf{1}} \otimes \hat{Q}, \tag{7.168}$$

where $\hat{Q}$ is the one-dimensional position operator and $\hat{\mathbf{1}}$ is the *unit operator*. Its action on the tensor position state $| \boldsymbol{x} \rangle = | x \rangle \otimes | y \rangle \otimes | z \rangle$ is given by

$$\hat{Q}_z | \boldsymbol{x} \rangle = \hat{\mathbf{1}} | x \rangle \otimes \hat{\mathbf{1}} | y \rangle \otimes \hat{Q} | z \rangle = z \, (| x \rangle \otimes | y \rangle \otimes | z \rangle) = z \, | \boldsymbol{x} \rangle, \tag{7.169}$$

which is the result that was assumed throughout previous presentations.

The sum of two operators, $\hat{\text{A}}$ and $\hat{\text{B}}$, that act separately on each of the two-particle states is understood in tensor terms as

$$\hat{\text{A}} + \hat{\text{B}} = \hat{\text{A}} \otimes \hat{\mathbf{1}} + \hat{\mathbf{1}} \otimes \hat{\text{B}}, \tag{7.170}$$

so that (7.166) and $\hat{\mathbf{1}} | \Psi_{1,2}, t \rangle = | \Psi_{1,2}, t \rangle$ gives

$$(\hat{\text{A}} + \hat{\text{B}}) | \Psi_1, \Psi_2, t \rangle = \hat{\text{A}} | \Psi_1, t \rangle \otimes | \Psi_2, t \rangle + | \Psi_1, t \rangle \otimes \hat{\text{B}} | \Psi_2, t \rangle. \tag{7.171}$$

In the event the states are eigenstates of $\hat{\text{A}}$ and $\hat{\text{B}}$ as in (7.167), the expected result is obtained,

$$(\hat{A} + \hat{B}) | \Psi_1, \Psi_2, t \rangle = (a_1 + b_2) | \Psi_1, \Psi_2, t \rangle. \tag{7.172}$$

Needless to say, tensor product notation is often cumbersome, and so it is typically left implicit when there is no ambiguity, as in the case of $|\, \boldsymbol{x}\, \rangle$, $\hat{Q}_z$, and $\hat{A} + \hat{B}$. Using these definitions shows that the commutator of a tensor product of operators satisfies

$$\begin{aligned}[\hat{A} \otimes \hat{B}, \hat{C} \otimes \hat{D}] &= \hat{A}\hat{C} \otimes \hat{B}\hat{D} - \hat{C}\hat{A} \otimes \hat{D}\hat{B} = (\hat{A}\hat{C} \otimes \hat{B}\hat{D} - \hat{C}\hat{A} \otimes \hat{B}\hat{D}) + (\hat{C}\hat{A} \otimes \hat{B}\hat{D} - \hat{C}\hat{A} \otimes \hat{D}\hat{B}) \\ &= [\hat{A}, \hat{C}] \otimes \hat{B}\hat{D} + \hat{C}\hat{A} \otimes [\hat{B}, \hat{D}],\end{aligned} \tag{7.173}$$

so that the commutator of the tensor product operator vanishes *only if both commutators vanish.*

As a final useful tensor operation, it is possible to give tensor products of operators a matrix representation by the tensor product version of (6.229). Using the same basis states for both operators gives the tensor product representation space,

$$|\, \lambda_j, \lambda_{j'} \rangle = |\, \lambda_j \rangle \otimes |\, \lambda_{j'} \rangle. \tag{7.174}$$

The completeness of these states gives a generalization of the projection operator onto the $|\, \lambda_j, \lambda_{j'} \rangle$ basis,

$$\sum_{j,j'} |\, \lambda_j, \lambda_{j'} \rangle\langle \lambda_j, \lambda_{j'} \,| = \sum_j |\, \lambda_j \rangle\langle \lambda_j \,| \otimes \sum_{j'} |\, \lambda_{j'} \rangle\langle \lambda_{j'} \,| = \hat{\mathbf{1}} \otimes \hat{\mathbf{1}} = \hat{\mathbf{1}}. \tag{7.175}$$

The generalization of the single particle observable matrix element (6.229) to the tensor product operator $\hat{A} \otimes \hat{B}$ is given by

$$(\mathbf{A} \otimes \mathbf{B})_{jk,j'k'} = \langle \lambda_j, \lambda_{j'} \,|\, \hat{A} \otimes \hat{B} \,|\, \lambda_k, \lambda_{k'} \rangle = \langle \lambda_j \,|\, \hat{A} \,|\, \lambda_k \rangle\langle \lambda_{j'} \,|\, \hat{B} \,|\, \lambda_{k'} \rangle = A_{jk} B_{j'k'}. \tag{7.176}$$

This allows the trace to be taken for the individual matrices, which is written

$$\mathrm{Tr}_A(\mathbf{A} \otimes \mathbf{B}) = (\mathrm{Tr}\, \mathbf{A})\mathbf{B} = \left(\sum_j A_{jj} \right) \mathbf{B}, \tag{7.177}$$

with a similar notation for the trace of **B**,

$$\mathrm{Tr}_B(\mathbf{A} \otimes \mathbf{B}) = \mathbf{A}(\mathrm{Tr}\, \mathbf{B}) = \left(\sum_j B_{jj} \right) \mathbf{A}. \tag{7.178}$$

This will be useful in understanding the nature of two-particle interactions.

7.6.2 Observables and statistics: fermions and bosons

The choice of a CSCO for a system of particles must reflect observations made on *all the particles* in the system. For example, the total energy or total angular momentum of the system may be chosen as an observable. However, once the CSCO *for the system* is chosen, it follows that this CSCO must be constructed from the associated observables for the *individual particles* in the system. It is a simple observation that this means the individual particles in the *system* cannot have different choices for a CSCO. For example, the observable $\hat{O}$ of a two-particle system is obtained from the corresponding observable $\hat{\text{O}}$ for a single particle system, and it is written as the tensor product

$$\hat{\mathcal{O}} = \hat{\mathbf{1}} \otimes \hat{\text{O}} + \hat{\text{O}} \otimes \hat{\mathbf{1}}. \tag{7.179}$$

This definition ensures that the contributions of both states are included in the value of the *system observable,* so that $\hat{\mathcal{O}}$ yields the action of the total observable for the system. This form for a system observable *assumes* that the system observable is the *sum* of the corresponding observable for each particle in the system. Since this is an underlying assumption of Newtonian mechanics for systems of non-interacting particles, it can be viewed as an extension of the basic assumptions of quantum mechanics.

As before, in Dirac notation the separable two-particle state is written

$$|\, \Psi, t \,\rangle = |\, \Psi_1, \Psi_2, t \,\rangle \equiv |\, \Psi_1, t \,\rangle \otimes |\, \Psi_2, t \,\rangle, \tag{7.180}$$

so that $|\, \Psi, t \,\rangle$ is a tensor product state made from two one-particle states $|\, \Psi_1, t \,\rangle$ and $|\, \Psi_2, t \,\rangle$. However, if the system consists of two *identical* particles, there is an additional quantum mechanical aspect of the tensor product state that must be taken into account. This is revealed by introducing the operator $\hat{\text{P}}_{12}$ that *permutes* the two particles, so that its action on the two-particle state is

$$\hat{\text{P}}_{12}|\, \Psi_1, \Psi_2, t \,\rangle = \alpha|\, \Psi_2, \Psi_1, t \,\rangle, \tag{7.181}$$

where α is a constant to be determined. This operation is understood as exchanging the *quantum numbers* of the two particles, so that the state of particle one becomes the state of particle two and *vice versa.* It should be obvious that this operation is possible only if the two particles are identical, since only then do they have the same mass, electric charge, spin, and so forth. Applying $\hat{\text{P}}_{12}$ again yields

$$(\hat{\text{P}}_{12})^2|\, \Psi_1, \Psi_2, t \,\rangle = \alpha\hat{\text{P}}_{12}|\, \Psi_2, \Psi_1, t \,\rangle = \alpha^2|\, \Psi_1, \Psi_2, t \,\rangle. \tag{7.182}$$

Demanding that $\hat{\text{P}}_{12}$ is idempotent requires $\alpha^2 = 1$, which allows $\alpha = \pm 1$.

If the two-component states $|\, \Psi_1, t \,\rangle$ and $|\, \Psi_2, t \,\rangle$ are both eigenstates of an observable $\hat{\text{O}}$ with the eigenvalues λ_1 and λ_2, respectively, then the action of the total observable $\hat{\mathcal{O}}$ defined by (7.179) on the two-particle tensor product state satisfies

$$\hat{\mathcal{O}}|\, \Psi_1, \Psi_2, t \,\rangle = \hat{\text{O}}|\, \Psi_1, t \,\rangle \otimes |\, \Psi_2, t \,\rangle + |\, \Psi_1, t \,\rangle \otimes \hat{\text{O}}|\, \Psi_2, t \,\rangle = (\lambda_1 + \lambda_2)|\, \Psi_1, \Psi_2, t \,\rangle. \tag{7.183}$$

The permuted state yields

$$\hat{\mathcal{O}}|\, \Psi_2, \Psi_1, t \,\rangle = \hat{\text{O}}|\, \Psi_2, t \,\rangle \otimes |\, \Psi_1, t \,\rangle + |\, \Psi_2, t \,\rangle \otimes \hat{\text{O}}|\, \Psi_1, t \,\rangle = (\lambda_2 + \lambda_1)|\, \Psi_2, \Psi_1, t \,\rangle, \tag{7.184}$$

which is identical to (7.183) since the eigenvalues are simply additive. This establishes that the two operators $\hat{\mathcal{O}}$ and $\hat{\text{P}}_{12}$ commute, since the order in which they are applied to the two-particle state is irrelevant.

In the Copenhagen interpretation of measurement, this means that the permuted state is experimentally *indistinguishable* from the original state *if the total value of the observable for a system is being measured.* This is similar to the case of a mixed state discussed at the end of chapter 3, where it was argued that the constituent states of a mixed state are indistinguishable if they share eigenvalues. Since $\hat{P}_{12}$ commutes with all observables in the CSCO, it is also a member of the CSCO for a system of two identical particles. This has the important result that the eigenstates of the CSCO must also be the eigenstates of $\hat{P}_{12}$. Since the allowed eigenvalues of $\hat{P}_{12}$ were earlier established to be $\alpha = \pm 1$, it follows that the two eigenstates of $\hat{P}_{12}$ are given, up to an irrelevant phase factor $e^{i\theta}$, by

$$| \Psi, t \rangle_A = \frac{1}{\sqrt{2}} | \Psi_1, \Psi_2, t \rangle - \frac{1}{\sqrt{2}} | \Psi_2, \Psi_1, t \rangle, \tag{7.185}$$

$$| \Psi, t \rangle_S = \frac{1}{\sqrt{2}} | \Psi_1, \Psi_2, t \rangle + \frac{1}{\sqrt{2}} | \Psi_2, \Psi_1, t \rangle. \tag{7.186}$$

These clearly obey

$$\begin{aligned} \hat{P}_{12} | \Psi, t \rangle_A &= \frac{1}{\sqrt{2}} | \Psi_2, \Psi_1, t \rangle - \frac{1}{\sqrt{2}} | \Psi_1, \Psi_2, t \rangle = -| \Psi, t \rangle_A \\ \hat{P}_{12} | \Psi, t \rangle_S &= \frac{1}{\sqrt{2}} | \Psi_2, \Psi_1, t \rangle + \frac{1}{\sqrt{2}} | \Psi_1, \Psi_2, t \rangle = +| \Psi, t \rangle_S. \end{aligned} \tag{7.187}$$

The two states of (7.185) and (7.186) are orthonormal, and therefore represent the two possible choices for the eigenstates of a system of two identical particles. They are also remain eigenstates of $\hat{\mathcal{O}}$ with the total eigenvalue $\lambda_1 + \lambda_2$.

The antisymmetric eigenstate (7.185) yields a probability density in configuration space given by

$$\mathcal{P}_A(\boldsymbol{x}_1, \boldsymbol{x}_2, t) = \frac{1}{2} | \Psi_1(\boldsymbol{x}_1, t)\Psi_2(\boldsymbol{x}_2, t) - \Psi_2(\boldsymbol{x}_1, t)\Psi_1(\boldsymbol{x}_2, t)|^2. \tag{7.188}$$

Result (7.188) is unchanged by $\Psi_1(\boldsymbol{x}_j, t) \leftrightarrow \Psi_2(\boldsymbol{x}_j, t)$ for both values of j, showing that the two particles are indistinguishable. The two wave functions, Ψ_1 and Ψ_2, are characterized by a set of quantum numbers, and if the quantum numbers are the same, then $\Psi_2(\boldsymbol{x}, t) = \Psi_1(\boldsymbol{x}, t)$ and the probability density (7.188) *vanishes everywhere*. This property applies to states containing more than two particles.

This property allows a mathematical expression of the Pauli exclusion principle discussed in the previous section. The methods of quantum field theory show that identical *fermions*, which are characterized by a *half-integral spin quantum number*, must have multiparticle states that are *antisymmetric*, like the eigenstate (7.185). Since it is not possible to form an antisymmetric wave function if any two states are identical, the probability that two identical fermions share identical quantum numbers is therefore zero. For example, the ground state of a hydrogen-like atom is characterized by two possible sets of quantum numbers, $n = 1$, $\ell = 0$, $m_\ell = 0$, and $m_s = \pm\frac{1}{2}$. If two electrons are in the ground state, they must have a two-particle state that is antisymmetric, and the only quantum number that can be different is the z-component of spin. This identifies the states in expression (7.185) as

$$
\begin{aligned}
|\,\Psi_1, t\,\rangle &= \left|\, n=1, \ell=0, m_\ell=0, m_s=\frac{1}{2}, t \right\rangle, \\
|\,\Psi_2, t\,\rangle &= \left|\, n=1, \ell=0, m_\ell=0, m_s=-\frac{1}{2}, t \right\rangle.
\end{aligned} \tag{7.189}
$$

As a result, once two electrons are in the ground state, there is no available set of quantum numbers *in the ground state* that is different from the two states that are already *occupied*, and a third electron must therefore occupy a higher energy state. Similarly, if spin one-half fermions are placed in the one-dimensional well, only two may reside in each available energy level. If there are $2n$ fermions in the box, the lowest energy configuration available to the system is the case where all energy levels up to and including $E_n = n^2\pi^2\hbar^2/2\,mL^2$ are occupied by two fermions. The energy E_n is referred to as the *Fermi energy*.

A similar result from quantum field theory shows that two identical *bosons*, which are characterized by an *integral spin quantum number*, must have a two-particle state that is *symmetric*, as in (7.186). As a result, the probability density for a pair of identical bosons is given by

$$
\mathcal{P}_S(x_1, x_2, t) = \frac{1}{2}\,|\,\Psi_1(x_1, t)\Psi_2(x_2, t) + \Psi_2(x_1, t)\Psi_1(x_2, t)|^2. \tag{7.190}
$$

Result (7.190) does *not* vanish if the two particles share identical quantum numbers.

These results can be extended to multiparticle states and are the essence of the connection between *spin and statistics*, with extremely important ramifications for statistical mechanics. These fundamental properties are what allows a *Bose–Einstein condensate*, which is the situation where all the bosons in a gas are sharing the ground state energy of the system. Similarly, since photons are spin one particles, any number of them may share the same momentum in a light wave. Since the proof of spin and statistics lies outside the scope of this text, it will made into the fourth postulate of quantum mechanics.

> **Postulate four**: The state for identical particles must be antisymmetric under the interchange of the quantum numbers for any two particles if the particles have half-integer spin and must be symmetric under the interchange of the quantum numbers of any two particles if the particles have integer spin. The set of all half-integer spin particles are known generically as *fermions*, while the set of all integer spin particles are known generically as *bosons*.

7.6.3 Antisymmetric multiparticle wave functions

A totally antisymmetric N-particle state can be constructed using the Levi–Civita symbol defined by (6.163). Using the fact that the N-dimensional Levi–Civita symbol has $N!$ permutations, the state is given by

$$| \Psi, t \rangle_A = \frac{1}{\sqrt{N!}} \sum_{j_1,\ldots,j_N=1}^{N} \varepsilon^{j_1 \ldots j_N} | \Psi_{j_1}, \ldots, \Psi_{j_N}, t \rangle. \tag{7.191}$$

The antisymmetric nature of the Levi–Civita symbol guarantees that the state (7.191) changes sign under interchange of any pair $j_\ell \leftrightarrow j_m$. The factor $1/\sqrt{N!}$ provides the normalization factor for the sum. Performing the inner product with $| \boldsymbol{x}_1, \ldots, \boldsymbol{x}_N \rangle$ gives the totally antisymmetric N-particle wave function,

$$\Psi_A(\boldsymbol{x}_1, \ldots, \boldsymbol{x}_N, t) = \frac{1}{\sqrt{N!}} \sum_{j_1,\ldots,j_N=1}^{N} \varepsilon^{j_1 \ldots j_N} \Psi_{j_1}(\boldsymbol{x}_1, t) \cdots \Psi_{j_N}(\boldsymbol{x}_N, t). \tag{7.192}$$

This can be expressed in matrix terms by defining the square matrix $\boldsymbol{\Psi}$ whose elements are given by $(\boldsymbol{\Psi})_{jk} = \Psi_j(\boldsymbol{x}_k, t)$. The definition of the determinant (6.170) shows that (7.192) can be written as

$$\Psi_A = \frac{\det \boldsymbol{\Psi}}{\sqrt{N!}}, \tag{7.193}$$

and is referred to as the *Slater determinant.*

7.7 Angular momentum addition

It is possible for a quantum mechanical state to manifest two forms of angular momentum. A particularly simple example is that of a single particle with both orbital angular momentum and spin angular momentum, such as an electron in an excited state of the hydrogen atom. Another simple example is a two-electron system, where both electrons have spin angular momentum.

7.7.1 The uncoupled representation

If there are two particles with angular momentum present and the individual angular momenta are part of the CSCO, the eigenstates are given by a tensor product of two angular momentum eigenstates, so that

$$| j_1, j_2, m_1, m_2 \rangle = | j_1, m_1 \rangle \otimes | j_2, m_2 \rangle. \tag{7.194}$$

The two angular momentum operators are also tensor products, $\hat{\boldsymbol{J}}_1 \otimes \hat{\mathbf{1}}$ and $\hat{\mathbf{1}} \otimes \hat{\boldsymbol{J}}_2$, where the two states in the tensor product obey the eigenvalue equations (7.100) for $\hat{\boldsymbol{J}}_1$ and $\hat{\boldsymbol{J}}_2$, respectively. This form for the tensor product state assumes that the CSCO consists of

$$\begin{aligned} \hat{J}_1^2 &\equiv \hat{J}^2 \otimes \hat{\mathbf{1}}, \\ \hat{J}_2^2 &\equiv \hat{\mathbf{1}} \otimes \hat{J}^2, \\ \hat{J}_{1z} &\equiv \hat{J}_z \otimes \hat{\mathbf{1}}, \\ \hat{J}_{2z} &\equiv \hat{\mathbf{1}} \otimes \hat{J}_z. \end{aligned} \tag{7.195}$$

Including (7.195) in the CSCO gives rise to the *uncoupled representation* of the system's angular momentum.

Since the four operators of (7.195) are Hermitian, the states of (7.194) are orthonormal,

$$\langle j_1, j_2, m_1, m_2 \,|\, j_1', j_2', m_1', m_2' \rangle = \delta_{j_1 j_1'} \delta_{j_2 j_2'} \delta_{m_1 m_1'} \delta_{m_2 m_2'}. \tag{7.196}$$

It will also be useful to employ the completeness of the uncoupled representation states, which is expressed as the tensor product of two copies of the single particle completeness (7.102),

$$\begin{aligned} &\sum_{m_1=-j_1}^{j_1} \sum_{j_1=0}^{\infty} \sum_{m_2=-j_2}^{j_2} \sum_{j_2=0}^{\infty} | j_1, j_2, m_1, m_2 \rangle\langle j_1, j_2, m_1, m_2 | \\ &= \sum_{m_1=-j_1}^{j_1} \sum_{j_1=0}^{\infty} | j_1, m_1 \rangle\langle j_1, m_1 | \otimes \sum_{m_2=-j_2}^{j_2} \sum_{j_2=0}^{\infty} | j_2, m_2 \rangle\langle j_2, m_2 | \\ &= \hat{\mathbf{1}} \otimes \hat{\mathbf{1}} = \hat{\mathbf{1}}. \end{aligned} \tag{7.197}$$

When there is no chance of confusion, this will be written

$$\sum_{m_1, m_2} \sum_{j_1, j_2} | j_1, j_2, m_1, m_2 \rangle\langle j_1, j_2, m_1, m_2 | = \hat{\mathbf{1}}, \tag{7.198}$$

where the limits on m_1 and m_2 are understood.

7.7.2 The coupled representation

A second choice for CSCO is possible for this case. It starts by choosing the *total angular momentum*,

$$\hat{\boldsymbol{J}}_{\mathrm{T}} = \hat{\boldsymbol{J}} \otimes \hat{\mathbf{1}} + \hat{\mathbf{1}} \otimes \hat{\boldsymbol{J}} = \hat{\boldsymbol{J}}_1 + \hat{\boldsymbol{J}}_2. \tag{7.199}$$

Applying the tensor product commutator (7.173) and using $[\hat{\mathbf{1}}, \hat{\boldsymbol{J}}] = 0$ shows that the components of $\hat{\boldsymbol{J}}_{\mathrm{T}}$ satisfy

$$\begin{aligned} [\hat{J}_{\mathrm{T}j}, \hat{J}_{\mathrm{T}k}] &= [\hat{J}_j, \hat{J}_k] \otimes \hat{\mathbf{1}} + \hat{\mathbf{1}} \otimes [\hat{J}_j, \hat{J}_k] = i\hbar \sum_{\ell=1}^{3} \varepsilon^{jk\ell} (\hat{J}_\ell \otimes \hat{\mathbf{1}} + \hat{\mathbf{1}} \otimes \hat{J}_\ell) \\ &= i\hbar \sum_{\ell=1}^{3} \varepsilon^{jk\ell} (\hat{J}_{1\ell} + \hat{J}_{2\ell}) = i\hbar \sum_{\ell=1}^{3} \varepsilon^{jk\ell} \hat{J}_{\mathrm{T}\ell}, \end{aligned} \tag{7.200}$$

which is the usual angular momentum algebra. The definition (7.199) of $\hat{\boldsymbol{J}}_{\mathrm{T}}$ and the rules of tensor product operator manipulation show that

$$\hat{\boldsymbol{J}}_{\mathrm{T}}^2 = \hat{\boldsymbol{J}}_1^{\,2} + \hat{\boldsymbol{J}}_2^{\,2} + 2\sum_{k=1}^{3} \hat{J}_{1k} \hat{J}_{2k} = \hat{\boldsymbol{J}}_1^{\,2} + \hat{\boldsymbol{J}}_2^{\,2} + \hat{J}_{1+} \hat{J}_{2-} + \hat{J}_{1-} \hat{J}_{2+} + 2\hat{J}_{1z} \hat{J}_{2z}, \tag{7.201}$$

where

$$\hat{J}_+ = \hat{J}_x + i\hat{J}_y,$$
$$\hat{J}_- = \hat{J}_x - i\hat{J}_y. \tag{7.202}$$

It is the presence of the raising and lowering operators in (7.201) that prevent the simple tensor product state of (7.194) from being an eigenstate of $\hat{J}_T^2$.

However, the rotation algebra gives

$$[\hat{J}_1^2, \hat{J}_{1j}] = [\hat{J}_2^2, \hat{J}_{2j}] = 0, \tag{7.203}$$

which holds for all j. It then follows that

$$[\hat{J}_T^2, \hat{J}_1^2] = [\hat{J}_T^2, J_2^2] = 0. \tag{7.204}$$

The angular momentum algebra shows that the summation term appearing in the first line of (7.201) has the commutators with $\hat{J}_{1z}$ and $\hat{J}_{2z}$ given by

$$\sum_{k=1}^{3}[\hat{J}_{1z}, \hat{J}_{1k}\hat{J}_{2k}] = \sum_{k=1}^{3}[\hat{J}_z, \hat{J}_k] \otimes \hat{J}_k = i\hbar\hat{J}_y \otimes \hat{J}_x - i\hbar\hat{J}_x \otimes \hat{J}_y = i\hbar(\hat{J}_{1y}\hat{J}_{2x} - \hat{J}_{1x}\hat{J}_{2y}), \tag{7.205}$$

$$\sum_{k=1}^{3}[\hat{J}_{2z}, \hat{J}_{1k}\hat{J}_{2k}] = \sum_{k=1}^{3}\hat{J}_k \otimes [\hat{J}_z, \hat{J}_k] = i\hbar\hat{J}_x \otimes \hat{J}_y - i\hbar\hat{J}_y \otimes \hat{J}_x = i\hbar(\hat{J}_{1x}\hat{J}_{2y} - \hat{J}_{1y}\hat{J}_{2x}). \tag{7.206}$$

This shows that the sum $\hat{J}_{Tz} = \hat{J}_{1z} + \hat{J}_{2z}$ satisfies

$$[\hat{J}_T^2, \hat{J}_{Tz}] = 0, \tag{7.207}$$

but that the individual terms that make up $\hat{J}_{Tz}$ do not separately commute with $\hat{J}_T^2$. Therefore, a second CSCO including $\hat{\boldsymbol{J}}_T$ consists of the set given by

$$\begin{aligned}
\hat{J}_T^2 &= \hat{J}_1^2 + \hat{J}_2^2 + 2\hat{\boldsymbol{J}}_1 \cdot \hat{\boldsymbol{J}}_2 = \hat{J}^2 \otimes \hat{\mathbf{1}} + \hat{\mathbf{1}} \otimes \hat{J}^2 + 2\sum_{k=1}^{3}\hat{J}_k \otimes \hat{J}_k,\\
\hat{J}_{Tz} &= \hat{J}_{1z} + \hat{J}_{2z} = \hat{J}_z \otimes \hat{\mathbf{1}} + \hat{\mathbf{1}} \otimes \hat{J}_z,\\
\hat{J}_1^2 &= \hat{J}^2 \otimes \hat{\mathbf{1}},\\
\hat{J}_2^2 &= \hat{\mathbf{1}} \otimes \hat{J}^2.
\end{aligned} \tag{7.208}$$

Including (7.208) in the CSCO gives rise to the *coupled representation* of the system's angular momentum.

Because $\hat{\boldsymbol{J}}_T$ satisfies the angular momentum algebra (7.200), result (7.100) shows that the eigenstates associated with the second CSCO can be written $|\, j_1, j_2, J, M\,\rangle$ and satisfy

$$\begin{aligned}
\hat{J}_T^2|\, j_1, j_2, J, M\,\rangle &= J(J+1)\hbar^2|\, j_1, j_2, J, M\,\rangle,\\
\hat{J}_{Tz}|\, j_1, j_2, J, M\,\rangle &= M\hbar\,|\, j_1, j_2, J, M\,\rangle.
\end{aligned} \tag{7.209}$$

From the previous results (7.100) the quantum number J is a positive integer or half-integer, while M satisfies $-J \leqslant M \leqslant J$ in integer steps. Since *both* sets of observables contain $\hat{J}^2 \otimes \hat{\mathbf{1}} = \hat{J}_1^2$ and $\hat{\mathbf{1}} \otimes \hat{J}^2 = \hat{J}_2^2$, their eigenvalues must be the same for both cases,

$$\begin{aligned} \hat{J}_1^2 | j_1, j_2, J, M \rangle &= j_1(j_1 + 1)\hbar^2 | j_1, j_2, J, M \rangle, \\ \hat{J}_2^2 | j_1, j_2, J, M \rangle &= j_2(j_2 + 1)\hbar^2 | j_1, j_2, J, M \rangle. \end{aligned} \tag{7.210}$$

Because both the coupled and uncoupled representations are eigenstates of $\hat{J}_1^2$ and $\hat{J}_2^2$, it follows that the two representations are orthonormal for their eigenvalues, which is written

$$\langle j_1', j_2', m_1, m_2 | j_1, j_2, J, M \rangle = \delta_{j_1 j_1'} \delta_{j_2 j_2'} \langle j_1, j_2, m_1, m_2 | j_1, j_2, J, M \rangle. \tag{7.211}$$

Further analysis is required to find the relationship of J and M to the other four eigenvalues, j_1, j_2, m_1, and m_2.

7.7.3 The coupled representation eigenvalues

The natural question arises as to how the quantum numbers J, M, j_1, and j_2 for the second CSCO, referred to as the *coupled representation*, are related to the quantum numbers j_1, j_2, m_1, and m_2 for the first CSCO, referred to as the *uncoupled representation*. A directly related question pertains to the allowed values that J, and therefore M, can assume, given the possible values of j_1 and j_2 that are shared by both representations. This is the central question of how the individual angular momenta add to create the total angular momentum, and it is complicated by the quantum nature of the angular momenta, which allows only the magnitude and z-component of the individual angular momenta to be known. The exact orientation of the angular momentum vectors is therefore unknown.

The relationship between the two representations can be analysed by fixing the values of j_1 and j_2 and finding the states of the coupled representation in terms of the states of the uncoupled representation. Using the completeness of the uncoupled representation completeness (7.198) and the partial orthonormality (7.211) gives

$$\begin{aligned} | j_1, j_2, J, M \rangle &= \sum_{j_1', j_2', m_1, m_2} \langle j_1', j_2', m_1, m_2 | j_1, j_2, J, M \rangle | j_1', j_2', m_1, m_2 \rangle \\ &= \sum_{j_1', j_2', m_1, m_2} \delta_{j_1 j_1'} \delta_{j_2 j_2'} \langle j_1', j_2', m_1, m_2 | j_1, j_2, J, M \rangle | j_1', j_2', m_1, m_2 \rangle \\ &= \sum_{m_1 = -j_1}^{j_1} \sum_{m_2 = -j_2}^{j_2} (\langle j_1, j_2, m_1, m_2 | j_1, j_2, J, M \rangle) | j_1, j_2, m_1, m_2 \rangle. \end{aligned} \tag{7.212}$$

The *Clebsch–Gordan coefficients* are defined as

$$C_{m_1, m_2}^{J,M} = \langle j_1, j_2, m_1, m_2 | j_1, j_2, J, M \rangle. \tag{7.213}$$

The symbol $C^{J,M}_{m_1,\, m_2}$ should be understood to depend on j_1 and j_2, which are the two eigenvalues shared by both the uncoupled and coupled representations. The coupled representation states are then written

$$|\, j_1, j_2, J, M\, \rangle = \sum_{m_1, m_2} C^{J,M}_{m_1,\, m_2} |\, j_1, j_2, m_1, m_2\, \rangle, \tag{7.214}$$

where it is implicit that the sum over m_1 ranges from $-j_1$ to j_1 and the sum over m_2 ranges from $-j_2$ to j_2.

Before proceeding it is useful to note that the coupled representation states satisfy orthonormality,

$$\langle\, j_1, j_2, J, M\, |\, j_1, j_2, J', M'\, \rangle = \delta_{JJ'}\delta_{MM'}, \tag{7.215}$$

which follows from the Hermitian nature of $\hat{J}_{\mathrm{T}}^2$ and $\hat{J}_{\mathrm{T}z}$. Using the completeness of the uncoupled representation (7.198) and the partial orthonormality (7.211) as well as the definition (7.213) of the Clebsch–Gordan coefficients gives

$$\begin{aligned}\langle\, j_1, j_2, J, M\, |\, j_1, j_2, J', M'\, \rangle &= \sum_{m_1, m_2} \sum_{j_1', j_2'} \langle\, j_1, j_2, J, M\, |\, j_1', j_2', m_1, m_2\, \rangle \langle\, j_1', j_2', m_1, m_2\, |\, j_1, j_2, J', M'\, \rangle \\ &= \sum_{m_1, m_2} \left(C^{J,M}_{m_1,\, m_2}\right)^* C^{J',M'}_{m_1,\, m_2} = \delta_{JJ'}\delta_{MM'},\end{aligned} \tag{7.216}$$

which shows that the Clebsch–Gordan coefficients are orthonormal. In effect, the Clebsch–Gordan coefficients are the elements of orthogonal eigenvectors in the angular momentum space.

Applying $\hat{J}_{\mathrm{T}z}$ to both sides of (7.212) gives

$$\begin{aligned}\hat{J}_{\mathrm{T}z} |\, j_1, j_2, J, M\, \rangle = M\hbar |\, j_1, j_2, J, M\, \rangle &= \sum_{m_1, m_2} C^{J,M}_{m_1,\, m_2} M\hbar |\, j_1, j_2, m_1, m_2\, \rangle \\ &= \sum_{m_1, m_2} C^{J,M}_{m_1,\, m_2} (\hat{J}_{1z} |\, j_1, m_1\, \rangle \otimes |\, j_2, m_2\, \rangle + |\, j_1, m_1\, \rangle \otimes \hat{J}_{2z} |\, j_2, m_2\, \rangle) \\ &= \sum_{m_1, m_2} (m_1 + m_2)\hbar\, C^{J,M}_{m_1,\, m_2} |\, j_1, j_2, m_1, m_2\, \rangle.\end{aligned} \tag{7.217}$$

Result (7.217) shows that $C^{J,M}_{m_1,\, m_2} = 0$ unless $M = m_1 + m_2$. This result reflects the fact that the z-components of both angular momenta simply add to give the total z-component. The rotation algebra also requires that $C^{J,M}_{m_1,\, m_2} = 0$ if $M < -J$ or $M > J$.

The next step is to infer the allowed values of J. The maximum value of m_1 is j_1, while the maximum value of M is J. Setting $m_1 = j_1$ and $M = J$ requires that

$$M = J = m_1 + m_2 = j_1 + m_2, \tag{7.218}$$

which yields the requirement that a non-zero Clebsch–Gordan coefficient can occur only if

$$m_2 = J - j_1. \tag{7.219}$$

However, m_2 can range between the values of $-j_2$ to j_2 in integer steps, so that $-j_2 \leqslant m_2 \leqslant j_1$. This gives the inequality

$$-j_2 \leqslant J - j_1 \leqslant j_2 \implies j_1 - j_2 \leqslant J \leqslant j_1 + j_2. \tag{7.220}$$

Similarly, the maximum value of m_2 is j_2, and choosing $M = J$ requires

$$J = M = m_1 + m_2 = m_1 + j_2, \tag{7.221}$$

so that a non-zero Clebsch–Gordan coefficient requires $m_1 = J - j_2$. Since m_1 must satisfy $-j_1 \leqslant m_1 \leqslant j_1$, this gives the inequality

$$-j_1 \leqslant J - j_2 \leqslant j_1 \implies j_2 - j_1 \leqslant J \leqslant j_1 + j_2. \tag{7.222}$$

The two inequalities for J given by (7.220) and (7.222) must hold simultaneously, so that combining them gives

$$|j_2 - j_1| \leqslant J \leqslant j_1 + j_2. \tag{7.223}$$

The possible eigenvalues J for the total angular momentum lie in a range of values from $|j_2 - j_1|$ to $j_1 + j_2$ in *integer steps*. Once a possible value for J is selected, the allowed values of M will lie in the range $-J \leqslant M \leqslant J$ in *integer steps*.

This result can be visualized in the following way. The total angular momentum $\boldsymbol{J}_{\mathrm{T}}$ is the sum of two angular momenta, $\boldsymbol{J}_1$ and $\boldsymbol{J}_2$. The sum will depend upon the orientation of the two vectors. If these were classical angular momenta, then their addition would have the following property. If the two were parallel then the sum would have the magnitude $J_1 + J_2$, which is the sum of their magnitudes. If the two were antiparallel, then the magnitude of their sum would be $|J_2 - J_1|$, the absolute value of the difference in their magnitudes. In the quantum mechanical case this classical picture is complicated by the inability to determine if the two angular momenta are parallel. Nevertheless, the sum will have a similar range of angular momentum values between $j_1 + j_2$ and $|j_2 - j_1|$. It is to be remembered that j_1 and j_2 are *not* the magnitudes of the two angular momenta. Instead, $j_1(j_1 + 1)\hbar^2$ and $j_2(j_2 + 1)\hbar^2$ are the magnitudes squared of the respective angular momenta. Nevertheless, it is common to refer to $J = j_1 + j_2$ as the *parallel case*, while $J = |j_2 - j_1|$ is referred to as the *antiparallel case*. Result (7.223) shows that the completeness of the coupled representation states is written

$$\sum_{M=-J}^{J} \sum_{J=|j_2-j_1|}^{j_1+j_2} \sum_{j_1, j_2} |j_1, j_2, J, M\rangle\langle j_1, j_2, J, M| = \hat{\mathbf{1}}. \tag{7.224}$$

An example is the ground state of neutral helium, where there are two electrons with no orbital angular momentum. In that case there is only electron spin angular momentum, so that $j_1 = j_2 = \frac{1}{2}$. The total angular momentum J is constrained to satisfy (7.223), which gives $0 \leqslant J \leqslant 1$, so that either $J = 0$ or $J = 1$. For the $J = 0$ case only $M = 0$ is possible, while for the $J = 1$ case M can be one of $\{-1, 0, 1\}$. There are therefore a total of four possible states in the coupled representation, matching in number the four possible tensor product states of the uncoupled representation. In terms of the rotation group this is written as $2 \otimes 2 = 3 \oplus 1$, which means that the tensor product space of two two-dimensional spinors decomposes into the *direct sum* of a triplet and a singlet. Clarifying this statement requires an explicit construction of the Clebsch–Gordan coefficients.

7.7.4 Clebsch–Gordan coefficients

In order to develop a recurrence relation for the Clebsch–Gordan coefficients, the raising and lowering operators, given by

$$\hat{J}_{\mathrm{T}\pm} = \hat{J}_{\mathrm{T}x} \pm i\hat{J}_{\mathrm{T}y} = \hat{J}_{\pm} \otimes \hat{\mathbf{1}} + \hat{\mathbf{1}} \otimes \hat{J}_{\pm}, \tag{7.225}$$

are applied to (7.212). Their action on the states was determined in (7.89) and (7.92). For compactness of notation, the symbols j_1 and j_2 will be suppressed in the states, so that

$$|\, j_1, j_2, m_1, m_2 \,\rangle \to |\, m_1, m_2 \,\rangle, \quad |\, j_1, j_2, J, M \,\rangle \to |\, J, M \,\rangle. \tag{7.226}$$

Using the result

$$\hat{J}_{\mathrm{T}\pm}|\, J, M \,\rangle = C_{\pm}(J, M)|\, J, M \pm 1 \,\rangle, \tag{7.227}$$

where the symbols $C_{\pm}(J, M)$ were defined earlier and explicitly determined in (7.106) and (7.107), the result is

$$\begin{aligned} \hat{J}_{\mathrm{T}\pm}|\, J, M \,\rangle &= C_{\pm}(J, M)|\, J, M \pm 1 \,\rangle \\ &= \sum_{m_1', m_2'} C^{J,M}_{m_1', m_2'} \left(\hat{J}_{\pm}|\, m_1' \,\rangle \otimes |\, m_2' \,\rangle + |\, m_1' \,\rangle \otimes \hat{J}_{\pm}|\, m_2' \,\rangle \right) \\ &= \sum_{m_1', m_2'} C^{J,M}_{m_1', m_2'} \big(C_{\pm}(j_1, m_1')|\, m_1' \pm 1, m_2' \,\rangle + C_{\pm}(j_2, m_2')|\, m_1', m_2' \pm 1 \,\rangle \big). \end{aligned} \tag{7.228}$$

Forming the inner product of $|\, m_1, m_2 \,\rangle$ with (7.228), using the individual inner products

$$\langle\, m_1, m_2 \,|\, J, M \pm 1 \,\rangle = C^{J,M\pm 1}_{m_1, m_2}, \tag{7.229}$$

$$\langle\, m_1, m_2 \,|\, m_1' \pm 1, m_2' \,\rangle = \delta_{m_1 \mp 1, m_1'} \delta_{m_2 m_2'}, \tag{7.230}$$

$$\langle\, m_1, m_2 \,|\, m_1', m_2' \pm 1 \,\rangle = \delta_{m_1 m_1'} \delta_{m_2 \mp 1, m_2'}, \tag{7.231}$$

and performing the sum over m_1' and m_2' yields a pair of recurrence relations for the Clebsch–Gordan coefficients,

$$C^{J,M\pm 1}_{m_1, m_2} C_{\pm}(J, M) = C^{J,M}_{m_1 \mp 1, m_2} C_{\pm}(j_1, m_1 \mp 1) + C^{J,M}_{m_1, m_2 \mp 1} C_{\pm}(j_2, m_2 \mp 1). \tag{7.232}$$

Using (7.106) and (7.107) allows these recurrence relations to be written explicitly as

$$C^{J,M+1}_{m_1,\,m_2}\sqrt{J(J+1)-M(M+1)} = C^{J,M}_{m_1-1,\,m_2}\sqrt{j_1(j_1+1)-m_1(m_1-1)} + C^{J,M}_{m_1,\,m_2-1}\sqrt{j_2(j_2+1)-m_2(m_2-1)}\,, \tag{7.233}$$

$$C^{J,M-1}_{m_1,\,m_2}\sqrt{J(J+1)-M(M-1)} = C^{J,M}_{m_1+1,\,m_2}\sqrt{j_1(j_1+1)-m_1(m_1+1)} + C^{J,M}_{m_1,\,m_2+1}\sqrt{j_2(j_2+1)-m_2(m_2+1)}\,. \tag{7.234}$$

The strategy for using (7.233) and (7.234) is first to eliminate all zero Clebsch–Gordan coefficients and then solve for the remaining Clebsch–Gordan coefficients in terms of a single Clebsch–Gordan coefficient. Once this is complete the value of the single Clebsch–Gordan coefficient can be determined by the orthonormality requirement (7.216),

$$\langle\, J,\, M \mid J,\, M\,\rangle = \sum_{m_1=-j_1}^{j_1}\sum_{m_2=-j_2}^{j_2} |C^{J,M}_{m_1,\,m_2}|^2 = 1, \tag{7.235}$$

where all the non-zero $C^{J,M}_{m_1,\,m_2}$ are expressed in terms of the single coefficient. As an example, the maximal values for M and m_1 are chosen, so that $M = J$ and $m_1 = j_1$. In order to give a non-zero Clebsch–Gordan coefficient, it must be that

$$m_2 = M - m_1 = J - j_1. \tag{7.236}$$

Therefore, $C^{J,J}_{j_1,\,J-j_1}$ will be chosen as the single coefficient to be determined by the normalization condition. Since choosing $m_1 = j_1$ means that $j_1(j_1+1) - m_1(m_1+1) = 0$, relationship (7.234) gives

$$C^{J,J-1}_{j_1,\,J-j_1-1}\sqrt{2J} = C^{J,J}_{j_1,\,J-j_1}\sqrt{j_2(j_2+1)-(J-j_1-1)(J-j_1)}\,, \tag{7.237}$$

which expresses $C^{J,J-1}_{j_1,\,J-j_1-1}$ in terms of $C^{J,J}_{j_1,\,J-j_1}$. Choosing $M = J - 1$ in (7.233) gives

$$C^{J,J}_{j_1,\,J-j_1}\sqrt{2J} = C^{J,J-1}_{j_1-1,\,J-j_1}\sqrt{2j_1} + C^{J,J-1}_{j_1,\,J-j_1-1}\sqrt{j_2(j_2+1)-(J-j_1)(J-j_1-1)}\,, \tag{7.238}$$

which allows $C^{J,J-1}_{j_1-1,\,J-j_1}$ to be expressed in terms of $C^{J,J}_{j_1,\,J-j_1}$. This iterative process can be continued until all possible Clebsch–Gordan coefficients are expressed in terms of $C^{J,J}_{j_1,\,J-j_1}$, which is then determined by the normalization condition (7.235).

7.7.5 Spin-$\frac{1}{2}$ case

This process becomes algebraically tedious if there are a large number of uncoupled states and therefore many non-zero coefficients. However, for the relatively simpler case where j_1 is arbitrary but $j_2 = \frac{1}{2}$ and $m_2 = \pm\frac{1}{2}$, the results are fairly compact,

$$C^{j_1\pm\frac{1}{2},M}_{m_1,\,\frac{1}{2}} = \pm\sqrt{\frac{j_1 \pm M + \frac{1}{2}}{2j_1+1}}\,, \qquad C^{j_1\pm\frac{1}{2},M}_{m_1,\,-\frac{1}{2}} = \sqrt{\frac{j_1 \mp M + \frac{1}{2}}{2j_1+1}}\,. \tag{7.239}$$

It is important to remember that the $\hat{J}_{Tz}$ eigenvalue M is the sum $m_1 \pm \frac{1}{2}$. For the case that $m_2 = \frac{1}{2}$, the eigenvalue m_1 appearing in the uncoupled representation state is given by

$$m_1 = M - \frac{1}{2}, \tag{7.240}$$

while for the case $m_2 = -\frac{1}{2}$, the eigenvalue m_1 appearing in the uncoupled representation state is

$$m_1 = M + \frac{1}{2}. \tag{7.241}$$

These results can be applied to the earlier case of two free electrons, where $j_1 = j_2 = \frac{1}{2}$. For such a case $J = j_1 \pm \frac{1}{2}$ can assume the two values $J = 1$ and $J = 0$. For the case that the positive sign is chosen, giving $J = 1$, the formulas of (7.239) for the $M = 1$ case give

$$C^{1,1}_{\frac{1}{2},\frac{1}{2}} = 1, \tag{7.242}$$

with all others vanishing, while for the $M = -1$ case they give

$$C^{1,-1}_{-\frac{1}{2},-\frac{1}{2}} = 1 \tag{7.243}$$

with all others vanishing. However, for the case that $J = 1$ and $M = 0$, it must be that $m_1 = -m_2$, and therefore these formulas yield

$$C^{1,0}_{-\frac{1}{2},\frac{1}{2}} = C^{1,0}_{\frac{1}{2},-\frac{1}{2}} = \frac{1}{\sqrt{2}}, \tag{7.244}$$

with all others vanishing. Finally, for the case that $J = 0$ and therefore $M = 0$, the negative sign in (7.239) must be chosen, along with the condition $m_1 = -m_2$. The formulas yield

$$C^{0,0}_{\frac{1}{2},-\frac{1}{2}} = -C^{0,0}_{-\frac{1}{2},\frac{1}{2}} = \frac{1}{\sqrt{2}}, \tag{7.245}$$

with all others vanishing. The final result for the two-electron total angular momentum states are the *singlet*,

$$|0, 0\rangle = \frac{1}{\sqrt{2}}\left|\frac{1}{2}, -\frac{1}{2}\right\rangle - \frac{1}{\sqrt{2}}\left|-\frac{1}{2}, \frac{1}{2}\right\rangle, \tag{7.246}$$

and the vector or *triplet* set of coupled representation states,

$$|1, 1\rangle = \left|\frac{1}{2}, \frac{1}{2}\right\rangle, \quad |1, 0\rangle = \frac{1}{\sqrt{2}}\left|\frac{1}{2}, -\frac{1}{2}\right\rangle + \frac{1}{\sqrt{2}}\left|-\frac{1}{2}, \frac{1}{2}\right\rangle, \quad |1, -1\rangle = \left|-\frac{1}{2}, -\frac{1}{2}\right\rangle. \tag{7.247}$$

The singlet state (7.246) is antisymmetric under the interchange $m_1 \leftrightarrow m_2$, so it can give the antisymmetric two-electron wave function required by (7.185). This means that the occupied ground state of neutral helium has two electrons with zero total spin angular momentum as well as zero orbital angular momentum. Similarly, the triplet states of (7.247) are symmetric under the interchange $m_1 \leftrightarrow m_2$, which is consistent with the integer value of angular momentum.

It is worth noting that the states of (7.246) and (7.247) are explicit eigenstates of the tensor product operator $\hat{J}_T^2$ given by (7.201), as well as

$$\hat{J}_{Tz} = \hat{J}_z \otimes \hat{\mathbf{1}} + \hat{\mathbf{1}} \otimes \hat{J}_z. \tag{7.248}$$

For example, since

$$\hat{J}_+ \otimes \hat{J}_- \left| \frac{1}{2}, \frac{1}{2} \right\rangle = \hat{J}_- \otimes \hat{J}_+ \left| \frac{1}{2}, \frac{1}{2} \right\rangle = 0, \tag{7.249}$$

applying the formula (7.201) for $\hat{J}_T^2$ to the state $|\, 1, 1 \rangle$ defined by (7.247) gives

$$\begin{aligned} \hat{J}_T^2 |\, 1, 1 \rangle = \hat{J}_T^2 \left| \frac{1}{2}, \frac{1}{2} \right\rangle &= (j_1(j_1 + 1) + j_2(j_2 + 1) + 2m_1 m_2)\hbar^2 \left| \frac{1}{2}, \frac{1}{2} \right\rangle \\ &= \left(\frac{3}{4} + \frac{3}{4} + \frac{1}{2} \right)\hbar^2 \left| \frac{1}{2}, \frac{1}{2} \right\rangle = 1(1 + 1)\hbar^2 \left| \frac{1}{2}, \frac{1}{2} \right\rangle = J(J + 1)\hbar^2 \left| \frac{1}{2}, \frac{1}{2} \right\rangle. \end{aligned} \tag{7.250}$$

Since

$$\begin{aligned} (\hat{J}_{1+}\hat{J}_{2-}) \left| -\frac{1}{2}, \frac{1}{2} \right\rangle &= (\hat{J}_+ \otimes \hat{J}_-) \left(\left| -\frac{1}{2} \right\rangle \otimes \left| \frac{1}{2} \right\rangle \right) = \hbar^2 \left(\left| \frac{1}{2} \right\rangle \otimes \left| -\frac{1}{2} \right\rangle \right) = \hbar^2 \left| \frac{1}{2}, -\frac{1}{2} \right\rangle, \\ (\hat{J}_{1-}\hat{J}_{2+}) \left| \frac{1}{2}, -\frac{1}{2} \right\rangle &= (\hat{J}_- \otimes \hat{J}_+) \left(\left| \frac{1}{2} \right\rangle \otimes \left| -\frac{1}{2} \right\rangle \right) = \hbar^2 \left(\left| -\frac{1}{2} \right\rangle \otimes \left| \frac{1}{2} \right\rangle \right) = \hbar^2 \left| -\frac{1}{2}, \frac{1}{2} \right\rangle, \end{aligned} \tag{7.251}$$

it also follows that the symmetric $M = 0$ state in (7.247) obeys

$$\left(\hat{J}_{1+}\hat{J}_{2-} + \hat{J}_{1-}\hat{J}_{2+} \right) |\, 1, 0 \rangle = \left(\hat{J}_{1+}\hat{J}_{2-} + \hat{J}_{1-}\hat{J}_{2+} \right) \frac{1}{\sqrt{2}} \left(\left| -\frac{1}{2}, \frac{1}{2} \right\rangle + \left| \frac{1}{2}, -\frac{1}{2} \right\rangle \right) = \hbar^2 |\, 1, 0 \rangle. \tag{7.252}$$

Similarly,

$$(\hat{J}_{1z}\hat{J}_{2z}) |\, 1, 0 \rangle = (\hat{J}_z \otimes \hat{J}_z) \frac{1}{\sqrt{2}} \left(\left| -\frac{1}{2}, \frac{1}{2} \right\rangle + \left| \frac{1}{2}, -\frac{1}{2} \right\rangle \right) = -\frac{1}{4}\hbar^2 |\, 1, 0 \rangle. \tag{7.253}$$

The result of applying the operator $\hat{J}_T^2$ given by (7.201) to the symmetric $M = 0$ state is given by

$$\begin{aligned} \hat{J}_T^2 |\, 1, 0 \rangle &= \left(j_1(j_1 + 1) + j_2(j_2 + 1) + 1 - \frac{1}{2} \right)\hbar^2 |\, 1, 0 \rangle = \left(\frac{3}{4} + \frac{3}{4} + \frac{1}{2} \right)\hbar^2 |\, 1, 0 \rangle \\ &= 2\hbar^2 |\, 1, 0 \rangle = J(J + 1)\hbar^2 |\, 1, 0 \rangle. \end{aligned} \tag{7.254}$$

These results can be extended to the singlet state of (7.246) by identical methods.

These tensor product states can be understood using the matrix representation of the electron spin angular momentum given by the Pauli spin matrices. The tensor product two-particle spinor wave function is obtained from two single particle spinor wave functions of the form (7.154),

$$\Psi = \psi_1 \otimes \psi_2 = \begin{pmatrix} \psi_{1\uparrow}\psi_{2\uparrow} \\ \psi_{1\downarrow}\psi_{2\uparrow} \\ \psi_{1\uparrow}\psi_{2\downarrow} \\ \psi_{1\downarrow}\psi_{2\downarrow} \end{pmatrix}. \tag{7.255}$$

The total spin operators are obtained using the tensor product (7.166) of the Pauli spin matrices $\hat{\sigma}_j$ of (7.132) and the 2×2 unit matrix $\mathbf{1}$, referred to as the $2 \otimes 2$ representation of the rotation group. For example, $\hat{S}_x$ is given by

$$\begin{aligned} \hat{S}_{\mathrm{T}x} &= \frac{1}{2}\hbar(\mathbf{1} \otimes \hat{\sigma}_x + \hat{\sigma}_x \otimes \mathbf{1}) = \frac{1}{2}\hbar\begin{pmatrix} 1 & 0 \\ 0 & 1 \end{pmatrix} \otimes \begin{pmatrix} 0 & 1 \\ 1 & 0 \end{pmatrix} + \frac{1}{2}\hbar\begin{pmatrix} 0 & 1 \\ 1 & 0 \end{pmatrix} \otimes \begin{pmatrix} 1 & 0 \\ 0 & 1 \end{pmatrix} \\ &= \frac{1}{2}\hbar\begin{pmatrix} 0 & 0 & 1 & 0 \\ 0 & 0 & 0 & 1 \\ 1 & 0 & 0 & 0 \\ 0 & 1 & 0 & 0 \end{pmatrix} + \frac{1}{2}\hbar\begin{pmatrix} 0 & 1 & 0 & 0 \\ 1 & 0 & 0 & 0 \\ 0 & 0 & 0 & 1 \\ 0 & 0 & 1 & 0 \end{pmatrix} = \frac{1}{2}\hbar\begin{pmatrix} 0 & 1 & 1 & 0 \\ 1 & 0 & 0 & 1 \\ 1 & 0 & 0 & 1 \\ 0 & 1 & 1 & 0 \end{pmatrix}. \end{aligned} \tag{7.256}$$

The other two are obtained similarly, with the final results being

$$\hat{S}_{\mathrm{T}y} = \frac{1}{2}\hbar\begin{pmatrix} 0 & -i & -i & 0 \\ i & 0 & 0 & -i \\ i & 0 & 0 & -i \\ 0 & i & i & 0 \end{pmatrix}, \quad \hat{S}_{\mathrm{T}z} = \hbar\begin{pmatrix} 1 & 0 & 0 & 0 \\ 0 & 0 & 0 & 0 \\ 0 & 0 & 0 & 0 \\ 0 & 0 & 0 & -1 \end{pmatrix}. \tag{7.257}$$

The matrix $S_{\mathrm{T}z}$ is diagonal and its eigenvalues are clearly $\{\hbar, 0, 0, -\hbar\}$, consistent with the addition rule determined earlier.

The total spin squared is not diagonal, but given by

$$\hat{S}_{\mathrm{T}}^2 = \hat{S}_{\mathrm{T}x}^2 + \hat{S}_{\mathrm{T}y}^2 + \hat{S}_{\mathrm{T}z}^2 = \hbar^2\begin{pmatrix} 2 & 0 & 0 & 0 \\ 0 & 1 & 1 & 0 \\ 0 & 1 & 1 & 0 \\ 0 & 0 & 0 & 2 \end{pmatrix}. \tag{7.258}$$

The eigenvalues of (7.258) are determined by the characteristic polynomial

$$\det(\hat{S}_{\mathrm{T}}^2 - \lambda\mathbf{1}) = -(\lambda - 2)^3\lambda = 0. \tag{7.259}$$

This gives the expected result that there are three eigenvalues of $\lambda = 2$, which matches $J(J + 1)$ for the case that $J = 1$. There is one eigenvalue of $\lambda = 0$, which matches $J(J + 1)$ for that case that $J = 0$. Direct multiplication by the matrix of (7.258) shows that the normalized tensor product eigenspinor Ψ_0 associated with the $J = 0$ eigenvalue of (7.258) is given by

$$\hat{S}_{\mathrm{T}}^2 \cdot \Psi_0 = \frac{\hbar^2}{\sqrt{2}}\begin{pmatrix} 2 & 0 & 0 & 0 \\ 0 & 1 & 1 & 0 \\ 0 & 1 & 1 & 0 \\ 0 & 0 & 0 & 2 \end{pmatrix} \cdot \begin{pmatrix} 0 \\ -1 \\ 1 \\ 0 \end{pmatrix} = 0. \tag{7.260}$$

Forming the inner product of the Ψ_0 used in (7.260) with the representation space spinor (7.255) shows that (7.260) corresponds to the antisymmetric $J = 0$ singlet state of (7.246) found earlier, only now in tensor product wave function form,

$$\Psi = \frac{1}{\sqrt{2}}(\psi_{1\downarrow}\psi_{2\uparrow} - \psi_{1\uparrow}\psi_{2\downarrow}). \tag{7.261}$$

The $2 \otimes 2$ tensor product space of spinors can be rewritten as

$$\Psi = \begin{pmatrix} \psi_{1\uparrow}\psi_{2\uparrow} \\ \frac{1}{\sqrt{2}}(\psi_{1\downarrow}\psi_{2\uparrow} + \psi_{1\uparrow}\psi_{2\downarrow}) \\ \psi_{1\downarrow}\psi_{2\downarrow} \\ \frac{1}{\sqrt{2}}(\psi_{1\downarrow}\psi_{2\uparrow} - \psi_{1\uparrow}\psi_{2\downarrow}) \end{pmatrix}, \tag{7.262}$$

and can be viewed as a $3 \oplus 1$ *direct sum* spinor. The top three components transform under rotations using the 3×3 matrix representation (7.63) associated with *vectors* and used for photon spin, while the bottom singlet transforms as a *scalar* under rotations. This demonstrates that spin one-half particles can combine to form spin one and spin zero particles. This is precisely the mechanism of the *quark model*, where quantum mechanical states of two spin one-half fermionic quarks can behave as bosonic particles.

7.8 Multiparticle states, entanglement, and decoherence

Several of the more unusual aspects of quantum mechanics occur in systems comprised of several subsystems. These include quantum entanglement as well as the suppression of quantum effects due to interaction of the system with its *environment*. Both the entangled two-particle system and the particle plus environment are modeled as a mixture of tensor states. The density matrix is a useful tool for describing such phenomena.

7.8.1 The density matrix

Consideration begins with a one-particle system that is prepared in a single normalized Schrödinger picture quantum state denoted $|\,\Psi, t\,\rangle$. This state may be a single eigenstate of some observable $\hat{\mathrm{O}}$, or it may be a superposition of a complete set of the Schrödinger picture eigenstates $|\,\lambda_j, t\,\rangle$ associated with $\hat{\mathrm{O}}$, as in expression (3.267),

$$|\,\Psi, t\,\rangle = \sum_j c_j |\,\lambda_j, t\,\rangle. \tag{7.263}$$

Such a state is commonly referred to as a *pure state*, signifying that this state is the unique *purely* quantum state resulting from preparation. The state $|\,\Psi, t\,\rangle$ is used to define the *density operator* $\hat{\rho}(t)$ as a projection operator,

$$\hat{\rho}(t) = | \Psi, t \rangle\langle \Psi, t |. \tag{7.264}$$

The density operator is Hermitian since

$$\hat{\rho}^{\dagger}(t) = (| \Psi, t \rangle\langle \Psi, t |)^{\dagger} = | \Psi, t \rangle\langle \Psi, t | = \hat{\rho}(t). \tag{7.265}$$

The density operator acts upon another normalized state $| \varphi, t \rangle$ to give its projection onto the state $| \Psi, t \rangle$,

$$\hat{\rho}(t) | \varphi, t \rangle = \langle \Psi, t | \varphi, t \rangle | \Psi, t \rangle = a(t) | \Psi, t \rangle, \tag{7.266}$$

where $|a(t)|^2 \leqslant 1$. Due to the normalization of the state $| \Psi, t \rangle$ the density operator defined by using a pure quantum state in (7.264) satisfies the idempotency requirement,

$$\hat{\rho}^2(t) = | \Psi, t \rangle\langle \Psi, t | \Psi, t \rangle\langle \Psi, t | = | \Psi, t \rangle\langle \Psi, t | = \hat{\rho}(t). \tag{7.267}$$

Using the Schrödinger equation gives the time-development of $\hat{\rho}(t)$,

$$\begin{aligned} i\hbar\frac{\mathrm{d}}{\mathrm{d}t}\hat{\rho}(t) &= \left(i\hbar\frac{\mathrm{d}}{\mathrm{d}t}| \Psi, t \rangle\right)\langle \Psi, t | + \langle \Psi, t |\left(i\hbar\frac{\mathrm{d}}{\mathrm{d}t}\langle \Psi, t |\right) \\ &= \hat{H}| \Psi, t \rangle\langle \Psi, t | - | \Psi, t \rangle\langle \Psi, t |\hat{H} = [\hat{H}, \hat{\rho}(t)], \end{aligned} \tag{7.268}$$

which is known as the *von Neumann equation.* It resembles the Heisenberg picture evolution of an operator (6.81), although it has the opposite sign for the commutator. While the two states $| \Psi, t \rangle$ and $e^{i\theta}| \Psi, t \rangle$ give rise to the same expectation values, the density operator is unchanged under a phase transformation,

$$| \Psi, t \rangle\langle \Psi, t | \rightarrow e^{i\theta}| \Psi, t \rangle\langle \Psi, t |e^{-i\theta} = | \Psi, t \rangle\langle \Psi, t |, \tag{7.269}$$

and so it is unique.

The matrix representation of the density operator, denoted $\boldsymbol{\rho}$, is referred to as the *density matrix*. Like all matrix representations of operators, it uses a complete set $| \lambda_j \rangle$ of eigenstates to define the matrix elements of the representation,

$$\rho_{jk}(t) = \langle \lambda_j | \hat{\rho}(t) | \lambda_k \rangle. \tag{7.270}$$

It follows that the trace of $\boldsymbol{\rho}(t)$ is unity,

$$\begin{aligned} \mathrm{Tr}\, \boldsymbol{\rho}(t) &= \sum_j \rho_{jj}(t) = \sum_j \langle \lambda_j | \hat{\rho}(t) | \lambda_j \rangle = \sum_j \langle \lambda_j | \Psi, t \rangle\langle \Psi, t | \lambda_j \rangle \\ &= \sum_j \langle \Psi, t | \lambda_j \rangle\langle \lambda_j | \Psi, t \rangle = \langle \Psi, t | \Psi, t \rangle = 1. \end{aligned} \tag{7.271}$$

It has the property that the expectation value for an arbitrary observable $\hat{\mathrm{O}}$ in the state $| \Psi, t \rangle$ is given by

$$\begin{aligned} \langle \hat{\mathrm{O}} \rangle(t) &= \langle \Psi, t | \hat{\mathrm{O}} | \Psi, t \rangle = \sum_{j,k} \langle \Psi, t | \lambda_j \rangle\langle \lambda_j | \hat{\mathrm{O}} | \lambda_k \rangle\langle \lambda_k | \Psi, t \rangle \\ &= \sum_{j,k} \langle \lambda_k | \Psi, t \rangle\langle \Psi, t | \lambda_j \rangle\langle \lambda_j | \hat{\mathrm{O}} | \lambda_k \rangle = \sum_{j,k} \rho_{kj}(t) O_{jk} = \mathrm{Tr}\, \boldsymbol{\rho}(t) \cdot \mathbf{O}, \end{aligned} \tag{7.272}$$

where $\mathbf{O}$ denotes the matrix for the observable $\hat{\mathrm{O}}$ in the representation $|\,\lambda_j\,\rangle$, while $\boldsymbol{\rho}(t)$ is the *density matrix* in the same representation. This result is independent of the choice of complete eigenstates used to define the matrix representations of $\hat{\rho}(t)$ and $\hat{\mathrm{O}}$.

Result (7.272) can be applied to a projection operator onto the state $|\,\varphi_n\,\rangle$. The projection operator is defined by

$$\hat{\mathrm{P}}_n = |\,\varphi_n\,\rangle\langle\,\varphi_n\,|, \tag{7.273}$$

where $|\,\varphi_n\,\rangle$ is an eigenstate of some observable $\hat{\mathrm{O}}$. This operator has the matrix elements in the $|\,\lambda_j\,\rangle$ representation space given by

$$P_{njk} = \langle\,\lambda_j\,|\,\hat{\mathrm{P}}_n\,|\,\lambda_k\,\rangle = \langle\,\lambda_j\,|\,\varphi_n\,\rangle\langle\,\varphi_n\,|\,\lambda_k\,\rangle. \tag{7.274}$$

Using the completeness of the $|\,\lambda_j\,\rangle$ representation space gives

$$\begin{aligned}\mathrm{Tr}\,\boldsymbol{\rho}(t)\cdot\mathbf{P}_n &= \sum_{j,k}\rho_{kj}(t)P_{njk} = \sum_{j,k}\underbrace{\langle\,\lambda_k\,|\,\Psi,\,t\,\rangle\langle\,\Psi,\,t\,|\,\lambda_j\,\rangle}_{\rho_{kj}}\underbrace{\langle\,\lambda_j\,|\,\varphi_n\,\rangle\langle\,\varphi_n\,|\,\lambda_k\,\rangle}_{P_{njk}}\\ &= \left(\sum_k\langle\,\varphi_n\,|\,\lambda_k\,\rangle\langle\,\lambda_k\,|\,\Psi,\,t\,\rangle\right)\left(\sum_j\langle\,\Psi,\,t\,|\,\lambda_j\,\rangle\langle\,\lambda_j\,|\,\varphi_n\,\rangle\right)\\ &= \langle\,\varphi_n\,|\,\Psi,\,t\,\rangle\langle\,\Psi,\,t\,|\,\varphi_n\,\rangle = |\langle\,\Psi,\,t\,|\,\varphi_n\,\rangle|^2 = |a_n(t)|^2,\end{aligned} \tag{7.275}$$

where $|a_n(t)|^2$ is the probability given by the Born rule (3.243) for observing the state $|\,\Psi,\,t\,\rangle$ with the eigenvalues of $|\,\varphi_n\,\rangle$. For the case that the states $|\,\varphi_n\,\rangle$ are complete, it follows that

$$\sum_n \mathrm{Tr}\,\rho(t)\cdot\mathbf{P}_n = \sum_n|\langle\,\Psi,\,t\,|\,\varphi_n\,\rangle|^2 = \langle\,\Psi,\,t\,|\,\Psi,\,t\,\rangle = 1. \tag{7.276}$$

The technique of reducing the trace into a sum over complete states is a staple of evaluating the density matrix and its products. Another example is the projection operator onto the position state $|\,\boldsymbol{x}\,\rangle$,

$$\hat{X} = |\,\boldsymbol{x}\,\rangle\langle\,\boldsymbol{x}\,|. \tag{7.277}$$

For this case, the matrix representation uses position space, and the trace of the product is given by

$$\begin{aligned}\mathrm{Tr}\,\boldsymbol{\rho}(t)\cdot\mathbf{X} &= \int_V \mathrm{d}^n y\,\langle\,\boldsymbol{y}\,|\,\Psi,\,t\,\rangle\langle\,\Psi,\,t\,|\,\boldsymbol{x}\,\rangle\langle\,\boldsymbol{x}\,|\,\boldsymbol{y}\,\rangle\\ &= \int_V \mathrm{d}^n y\,\langle\,\boldsymbol{y}\,|\,\Psi,\,t\,\rangle\langle\,\Psi,\,t\,|\,\boldsymbol{x}\,\rangle\delta^n(\boldsymbol{x}-\boldsymbol{y}) = |\Psi(\boldsymbol{x},\,t)|^2,\end{aligned} \tag{7.278}$$

which is the probability density for observing the particle at the position $\boldsymbol{x}$. It is important to note that the trace over a continuous representation space such as $|\,\boldsymbol{x}\,\rangle$ is implemented by integration.

So far the density matrix approach has simply provided an alternative way to find expectation values and probabilities in quantum mechanics. However, it provides a valuable tool for dealing with both multiparticle states and issues related to the preparation of states. In the case of preparing the state for analysis, it is possible that there is *not* a unique quantum state $| \Psi, t \rangle$ that is available to the observer on a consistent basis. The apparatus or the physical process preparing the system may produce *different* pure quantum states of the general form (7.263). If these states are not experimentally distinguishable, then the criteria of (3.268) show that they are not necessarily orthogonal. However, it is possible that the observer knows the relative probability p_j for each possible pure state prepared for analysis. If there are N possible states that can be prepared, then the probabilities of each respective state must sum to one, so that

$$\sum_{j=1}^{N} p_j = 1. \tag{7.279}$$

If the set $\{ | \Psi_j, t \rangle, \ldots \}$ constitutes the normalized pure states, then the density operator is redefined as

$$\hat{\rho}(t) = \sum_{j=1}^{N} p_j | \Psi_j, t \rangle\langle \Psi_j, t |. \tag{7.280}$$

It is stressed that the states appearing in (7.280) are not necessarily orthonormal. It is also possible, depending on the preparation process creating this mixture of pure states, that N is greater than or less than the dimension of the Hilbert space to which the pure states $| \Psi_j, t \rangle$ belong. It should be noted that if $p_j = \delta_{jn}$, so that $p_n = 1$ with all others vanishing, then (7.280) reduces to the previous density operator (7.264) for the pure state $| \Psi_n, t \rangle$. Because the probabilities p_j are introduced *externally*, they are not given by the Born rule applied to some quantum state of the form (3.244). Their presence in the preparation of the system *reduces* the amount of information available to the observer since the result of a measurement may not distinguish which wave function in (7.280) resulted in the eigenvalue that was measured. In that sense (7.280) is often referred to as a *statistical ensemble* of pure states. Because it combines external probabilities and more than one pure state, (7.280) is referred to as the *mixed state density operator.*

The mixed state density matrix $\rho(t)$ has the property that its trace over a complete set of states $| \lambda_j, t \rangle$ is given by

$$\begin{aligned} \operatorname{Tr} \rho(t) &= \sum_{k} \langle \lambda_k, t | \hat{\rho}(t) | \lambda_k, t \rangle = \sum_{j=1}^{N} \sum_{k} p_j \langle \Psi_j, t | \lambda_k \rangle\langle \lambda_k, t | \Psi_j, t \rangle \\ &= \sum_{j=1}^{N} p_j \langle \Psi_j, t | \Psi_j, t \rangle = \sum_{j=1}^{N} p_j = 1, \end{aligned} \tag{7.281}$$

which is the same as (7.271). However, it is no longer idempotent, since

$$\hat{\rho}^2(t) = \sum_{j,k=1}^{N} \left(p_j p_k \langle \Psi_j, t | \Psi_k, t \rangle \right) | \Psi_j, t \rangle\langle \Psi_k, t | \neq \hat{\rho}(t). \tag{7.282}$$

Even if the states are orthogonal, so that $\langle \Psi_j, t \mid \Psi_k, t \rangle = \delta_{jk}$, the mixed density operator squared becomes

$$\hat{\rho}^2(t) = \sum_{j=1}^{N} p_j^2 \mid \Psi_j, t \rangle\langle \Psi_j, t \mid \neq \hat{\rho}(t), \tag{7.283}$$

with the inequality holding as long as $p_j < 1$ for the states. The trace of (7.282) over a complete set $\mid \lambda_\ell, t \rangle$ gives

$$\mathrm{Tr}\,\boldsymbol{\rho} \cdot \boldsymbol{\rho} = \sum_{j,k=1}^{N} \sum_{\ell} p_j p_k \langle \Psi_j, t \mid \Psi_k, t \rangle\langle \Psi_k, t \mid \lambda_\ell, t \rangle\langle \lambda_\ell, t \mid \Psi_j, t \rangle = \sum_{j,k=1}^{N} p_j p_k \, |\langle \Psi_j, t \mid \Psi_k, t \rangle|^2. \tag{7.284}$$

Combining the inequality for the inner product,

$$0 \leqslant |\langle \Psi_j, t \mid \Psi_k, t \rangle|^2 \leqslant 1, \tag{7.285}$$

with the positivity of the probabilities p_j in (7.284) gives

$$0 \leqslant \mathrm{Tr}\,\boldsymbol{\rho} \cdot \boldsymbol{\rho} \leqslant \sum_{j,k=1}^{N} p_j p_k = \sum_{j=1}^{N} p_j \sum_{k=1}^{N} p_k = 1. \tag{7.286}$$

Expression (7.284) is often referred to as the *purity of the mixed state density operator*. It can equal one only if all but one of the probabilities vanish, so that $p_j = \delta_{jn}$. As a result, its deviation from one is used to quantify how far the system preparation is from a pure quantum mechanical state.

The mixed state density matrix gives a variant of the expectation value (7.272) for a pure state,

$$\begin{aligned} \mathrm{Tr}\,\boldsymbol{\rho}(t) \cdot \mathbf{A} &= \sum_{k,\ell} \sum_{j=1}^{N} p_j \langle \lambda_\ell \mid \Psi_j, t \rangle\langle \Psi_j, t \mid \lambda_k \rangle\langle \lambda_k \mid \hat{\mathrm{A}} \mid \lambda_\ell \rangle \\ &= \sum_{\ell} \sum_{j=1}^{N} p_j \langle \Psi_j, t \mid \hat{\mathrm{A}} \mid \lambda_\ell \rangle\langle \lambda_\ell \mid \Psi_j, t \rangle = \sum_{j=1}^{N} p_j \langle \Psi_j, t \mid \hat{\mathrm{A}} \mid \Psi_j, t \rangle \\ &= \sum_{j=1}^{N} p_j \langle \hat{\mathrm{A}} \rangle_j(t), \end{aligned} \tag{7.287}$$

where $\langle \hat{\mathrm{A}} \rangle_j(t)$ is the expectation value of the observable in the pure state $\mid \Psi_j, t \rangle$. For the case that the system is prepared with the probability p_j in the state $\mid \Psi_j, t \rangle$, (7.287) gives the statistical average of many measurements of $\hat{\mathrm{A}}$, and so it is the extension of the expectation value $\langle \hat{\mathrm{A}} \rangle$ to the case where the system is prepared in different quantum states weighted by known probabilities.

It is possible to find the eigenstates of the mixed state density operator $\hat{\rho}(t)$. Demonstrating this begins by writing the pure states in (7.280) as superpositions of orthonormal states $\mid \lambda_j, t \rangle$,

$$| \Psi_j, t \rangle = \sum_{\ell} c_{j\ell} | \lambda_\ell, t \rangle. \tag{7.288}$$

Inserting these superpositions into (7.280) gives

$$\hat{\rho}(t) = \sum_{\ell,k} \sum_{j=1}^{N} \left(p_j c_{j\ell}^* c_{jk}\right) | \lambda_\ell, t \rangle \langle \lambda_k, t | = \sum_{\ell,k} | \lambda_\ell, t \rangle M_{\ell k} \langle \lambda_k, t |, \tag{7.289}$$

where the matrix $\mathbf{M}$ has the elements

$$M_{\ell k} = \sum_{j=1}^{N} \left(p_j c_{j\ell}^* c_{jk}\right), \tag{7.290}$$

and is the same dimension as the Hilbert space to which the pure states belong. Because the probabilities p_j are real, the elements of $\mathbf{M}$ satisfy

$$M_{\ell k}^* = \sum_{j=1}^{N} \left(p_j c_{jk}^* c_{j\ell}\right) = M_{k\ell}, \tag{7.291}$$

so that definition (6.147) shows $\mathbf{M}$ is a Hermitian matrix. It will therefore have real eigenvalues, denoted e_j, and orthonormal eigenvectors. Result (6.197) shows that the matrix $\mathbf{M}$ can therefore be diagonalized by using a unitary transformation $\mathbf{U}$ constructed from its eigenvectors. Since $\mathbf{U} \cdot \mathbf{U}^\dagger = \mathbf{1}$, expression (7.289) can be written

$$\begin{aligned} \hat{\rho}(t) &= \sum_{i,j,k,\ell,m,n} \left(| \lambda_\ell, t \rangle U_{\ell i}\right)\left(U_{im}^\dagger M_{mn} U_{nj}\right)\left(U_{jk}^\dagger \langle \lambda_k, t |\right) \\ &= \sum_{i,j,k,\ell} \left(| \lambda_\ell, t \rangle U_{\ell i}\right)\left(e_i \delta_{ij}\right)\left(U_{jk}^\dagger \langle \lambda_k, t |\right) = \sum_j e_j | \rho_j, t \rangle \langle \rho_j, t |, \end{aligned} \tag{7.292}$$

where the orthonormal eigenstates of $\hat{\rho}(t)$ are given by the action of the unitary matrix,

$$\begin{aligned} &| \rho_\ell, t \rangle = \sum_i U_{i\ell} | \lambda_i, t \rangle, \quad \langle \rho_j, t | = \sum_k U_{jk}^\dagger \langle \lambda_k, t | \\ &\Longrightarrow \langle \rho_j, t | \rho_\ell, t \rangle = \sum_{i,k} U_{i\ell} U_{jk}^\dagger \langle \lambda_k, t | \lambda_i, t \rangle = \sum_{i,k} U_{i\ell} U_{jk}^\dagger \delta_{ik} = \sum_k U_{jk}^\dagger U_{k\ell} = \delta_{j\ell}. \end{aligned} \tag{7.293}$$

Because the transformation is unitary the resulting states $| \rho_j, t \rangle$ are orthonormal and will be assumed to be complete. As a result, the trace of the mixed density matrix can be evaluated using its eigenstates. Since the trace of $\hat{\rho}(t)$ is one, this gives

$$\mathrm{Tr}\, \boldsymbol{\rho}(t) = \sum_n \langle \rho_n, t | \sum_{j=1}^{N} p_j | \Psi_j, t \rangle \langle \Psi_j, t | \rho_n, t \rangle = \sum_n e_n \langle \rho_n, t | \rho_n, t \rangle = \sum_n e_n = 1, \tag{7.294}$$

so that the eigenvalues of $\hat{\rho}(t)$ must sum to unity.

7.8.2 Multiparticle states and entanglement

One of the more unusual properties of quantum mechanics is the possibility of what is referred to as *quantum entanglement*. For simplicity of presentation, discussion and analysis will focus on two-particle tensor product states.

The first step is to generalize Born's rule to the two-particle state by finding the corresponding probability density (3.239) using the generalization of the characteristic function (3.238) associated with the observable given by the tensor product

$$\hat{\mathrm{O}} = \hat{\mathrm{A}} \otimes \hat{\mathrm{B}}. \tag{7.295}$$

It is important to note that $\hat{\mathrm{A}}$ and $\hat{\mathrm{B}}$ do not need to commute since they act on separate Hilbert spaces. Using the eigenstates of $\hat{\mathrm{A}}$ and $\hat{\mathrm{B}}$, denoted $|\, e_j \,\rangle$ and $|\, e'_k \,\rangle$, respectively, in the tensor product matrix element (7.176) gives

$$\langle\, e_j, e'_\ell \,|\, \hat{\mathrm{A}} \otimes \hat{\mathrm{B}} \,|\, e_k, e'_n \,\rangle = \langle\, e_j \,|\, \hat{\mathrm{A}} \,|\, e_k \,\rangle\langle\, e'_\ell \,|\, \hat{\mathrm{B}} \,|\, e'_n \,\rangle = e_j\delta_{jk}\, e'_\ell\delta_{\ell n}, \tag{7.296}$$

where e_j and e'_k are eigenvalues of $\hat{\mathrm{A}}$ and $\hat{\mathrm{B}}$, respectively. The characteristic function (3.238) for the two-particle state is identical in form to the one-particle state,

$$\chi(\lambda) = \langle e^{-i\lambda\hat{\mathrm{O}}}\rangle = \langle\, \Psi, t \,|\, e^{-i\lambda\hat{\mathrm{O}}} \,|\, \Psi, t \,\rangle, \tag{7.297}$$

where $|\, \Psi, t \,\rangle$ is the state of the system. Using the completeness of the tensor product states (7.175) gives

$$\chi(\lambda) = \sum_{j,k}\langle\, \Psi, t \,|\, e^{-i\lambda\hat{\mathrm{O}}} \,|\, e_j, e'_k \,\rangle\langle\, e_j, e'_k \,|\, \Psi, t \,\rangle = \sum_{j,k} e^{-i\lambda e_j e'_k}|\langle\, \Psi, t \,|\, e_j, e'_k \,\rangle|^2. \tag{7.298}$$

The probability density for measuring the value m for the two-particle observable $\hat{\mathrm{O}}$ is given by

$$\begin{aligned} \mathcal{P}_{\hat{\mathrm{O}}}(m, t) &= \int_{-\infty}^{\infty} \frac{\mathrm{d}\lambda}{2\pi}\, e^{i\lambda m}\chi(\lambda) = \sum_{j,k}\int_{-\infty}^{\infty} \frac{\mathrm{d}\lambda}{2\pi}\, e^{i\lambda(m - e_j e'_k)}|\langle\, \Psi, t \,|\, e_j, e'_k \,\rangle|^2 \\ &= \sum_{j,k}\delta(m - e_j e'_k)\, |\langle\, \Psi, t \,|\, e_j, e'_k \,\rangle|^2. \end{aligned} \tag{7.299}$$

This shows that a single measurement of the tensor product observable $\hat{\mathrm{O}}$ for the two-particle system will result in a product of two eigenvalues, $e_j e'_k$, with the relative probability

$$P(e_j e'_k, t) = |\langle\, \Psi, t \,|\, e_j, e'_k \,\rangle|^2. \tag{7.300}$$

This is the generalization of Born's rule to the two-particle state.

It is useful to understand first when a system is *not* entangled. An *unentangled* two-particle state is given by the tensor product state

$$|\, \Psi, t \,\rangle = |\, \psi, t \,\rangle \otimes |\, \varphi, t \,\rangle, \tag{7.301}$$

where $| \psi, t \rangle$ and $| \varphi, t \rangle$ are normalized states for the respective particles. This two-particle state has the property that

$$\langle \Psi, t | \Psi, t \rangle = \langle \psi, t | \psi, t \rangle\langle \varphi, t | \varphi, t \rangle = 1. \tag{7.302}$$

Born's rule (7.300) gives the probability of observing $e_j e_k'$ for $\hat{O}$ as

$$P(e_j e_k', t) = |\langle \Psi, t | e_j, e_k' \rangle|^2 = |\langle \psi, t | e_j \rangle\langle \varphi, t | e_k' \rangle|^2 = |\langle \psi, t | e_j \rangle|^2 \, |\langle \varphi, t | e_k' \rangle|^2. \tag{7.303}$$

The probability has factorized into the *product* of two one-particle probabilities for observing the two states with the eigenvalues e_j and e_k', respectively. These observations are therefore uncorrelated, showing the quantum mechanical state (7.301) describes two particles whose measurements are independent of each other. The completeness of the two sets of eigenstates shows that the sum of all the probabilities is given by

$$\begin{aligned} \sum_{j,k} P(e_j e_k', t) &= \sum_j |\langle \psi, t | e_j \rangle|^2 \cdot \sum_k |\langle \varphi, t | e_k' \rangle|^2 \\ &= \sum_j \langle \psi, t | e_j \rangle\langle e_j | \psi, t \rangle \cdot \sum_k \langle \varphi, t | e_k' \rangle\langle e_k' | \varphi, t \rangle \\ &= \langle \psi, t | \psi, t \rangle\langle \varphi, t | \varphi, t \rangle = 1, \end{aligned} \tag{7.304}$$

satisfying the requirement for unit probability.

Next, a simple *entangled state* is constructed from two eigenstates of each of the operators $\hat{A}$ and $\hat{B}$, given explicitly by

$$| \Psi, t \rangle = \frac{1}{\sqrt{2}}(| e_1, t \rangle \otimes | e_2', t \rangle - | e_2, t \rangle \otimes | e_1', t \rangle). \tag{7.305}$$

Since $\langle e_1, t | e_2, t \rangle = \langle e_1', t | e_2', t \rangle = 0$, the state $| \Psi, t \rangle$ satisfies

$$\begin{aligned} \langle \Psi, t | \Psi, t \rangle &= \frac{1}{2}(\langle e_1, t | \otimes \langle e_2', t | - \langle e_2, t | \otimes \langle e_1', t |)(| e_1, t \rangle \otimes | e_2', t \rangle - | e_2, t \rangle \otimes | e_1', t \rangle) \\ &= \frac{1}{2}(\langle e_1, t | e_1, t \rangle\langle e_2', t | e_2', t \rangle + \langle e_2, t | e_2, t \rangle\langle e_1', t | e_1', t \rangle) = 1. \end{aligned} \tag{7.306}$$

The relative probabilities of measurement are found using Born's rule, so that

$$P(e_j e_k', t) = |\langle \Psi, t | e_j, e_k' \rangle|^2 = \frac{1}{2}\left|\langle e_1, t | e_j \rangle\langle e_2', t | e_k' \rangle - \langle e_2, t | e_j \rangle\langle e_1', t | e_k' \rangle\right|^2 \tag{7.307}$$

This can be simplified by treating $\hat{A}$ and $\hat{B}$ as members of a CSCO including the Hamiltonian. For such a case the inner products become

$$\begin{aligned} \langle e_\ell, t | e_j \rangle &= \delta_{\ell j}\, e^{iE_\ell t/\hbar}, \\ \langle e_n', t | e_k' \rangle &= \delta_{nk}\, e^{iE_n' t/\hbar}, \end{aligned} \tag{7.308}$$

where E_ℓ and E_n' are the energies of the respective eigenstates. Using $(\delta_{ij})^2 = \delta_{ij}$, the probabilities of (7.307) then simplify to

$$P(e_j e_k', t) = \frac{1}{2}\delta_{1j}\delta_{2k} + \frac{1}{2}\delta_{2j}\delta_{1k}. \tag{7.309}$$

If the eigenvalue e_1 is measured for $\hat{\mathrm{A}}$ acting on the first particle, then only the eigenvalue e_2' will be measured for $\hat{\mathrm{B}}$ acting on the second particle. Similarly, if the eigenvalue e_2 is measured for $\hat{\mathrm{A}}$ acting on the first particle, then only the eigenvalue e_1' will be measured for $\hat{\mathrm{B}}$ acting on the second particle. The result of a measurement on one particle is completely correlated to the result of a measurement on the second particle. The two particles are said to be *entangled.*

The entangled state (7.305) can be given a position representation, so that the wave function is given by

$$\Psi(x, x', t) = (\langle x | \otimes \langle x' |) | \Psi, t \rangle = \frac{1}{\sqrt{2}} \psi_1(x, t) \varphi_2(x', t) - \frac{1}{\sqrt{2}} \psi_2(x, t) \varphi_1(x', t), \tag{7.310}$$

where

$$\psi_j(\boldsymbol{x}, t) = \langle \boldsymbol{x} \,|\, e_j, t \rangle, \quad \varphi_k(\boldsymbol{x}', t) = \langle \boldsymbol{x}' \,|\, e_k', t \rangle. \tag{7.311}$$

The entanglement of the two-particle state persists *regardless* of the value of $|\boldsymbol{x} - \boldsymbol{x}'|$. This means that if a system is prepared in an entangled state, presumably through a local interaction of the two particles, then that state of entanglement will persist regardless of how far the two particles become separated. This is similar to the spreading of a wave packet as time passes.

A conceptual analogy of entanglement may be of use in understanding this phenomenon, as well as demonstrating the nature of quantum mechanical preparation and measurement. In this analogy there are two observers, known as A and B, and each receives a sealed envelope from person C. Person C has *prepared* these two envelopes secretly by flipping a coin. In the event that heads occurred, C placed a black slip of paper in *both* envelopes, while if tails occurred C placed a white slip of paper in *both* envelopes. The two envelopes represent the two quantum mechanical particles and the black or white slips give a simple two value representation of the possible state of the two particles, modeling their properties such as spin angular momenta. Until an envelope is opened or, in quantum mechanical terms, *a measurement on the particle occurs*, the state of the envelope is *unknown* to both A and B, since there is a fifty percent probability the envelope contains either color. However, by agreement both observers are aware that the preparation by C has entangled the two envelopes. If one envelope contains a black slip, then so does the other, with the same kind of entanglement applying to a white slip. It was the act of *preparation* by C, playing the role of the experimental apparatus, that created the entanglement of the two envelopes by interacting with them *locally*. However, no matter how far the two envelopes are separated spatially, the moment that A opens the envelope and examines the slip of paper, A will *know* the content of B's envelope to absolute certainty.

In the Copenhagen interpretation a measurement on the first particle will cause its wave function to collapse onto one of the two eigenstates present in the mixed state, and any subsequent measurement on the second particle, regardless of its distance from the first measurement, will yield the entangled value since its wave function must also have collapsed. Thus, a measurement of one particle immediately reveals the outcome of a measurement on the second particle, regardless of the separation between the two, allowing the first observer to know more about the second particle than the second observer *prior* to measurement. However, the observer of the first particle cannot communicate this knowledge to the observer of the second particle any faster than the speed of light, and so there is no violation of classical causality. Nevertheless, this result brought significant discomfort to physicists in the early days of quantum mechanics. In particular, Einstein famously referred to it as 'spooky action at a distance,' and felt it violated the prohibition that special relativity placed against superluminal transmission of any kind and created the paradox of the first observer knowing more about the second particle than the second observer. As a result, this aspect of entanglement is often referred to as the *Einstein–Podolsky–Rosen (EPR) paradox* after the first authors to draw attention to this phenomenon.

The apparent violation of causality contributed to the efforts to develop alternative formulations of quantum mechanics. In particular, it led to the idea of hidden variables in quantum systems, where they were used to explain the correlation of the two particles. This was discussed briefly earlier in the context of angular momentum addition. The next section examines this idea in much greater detail. To set the stage for this discussion, a hidden variable in the simple analogy with envelopes and slips of paper would be the presence of some aspect of the envelopes that would correlate to the color of the slip of paper. An example would be if the person preparing the slips put them into the envelope aligned in different directions depending on the color of the slip of paper chosen. This aspect of the slip is not observable prior to opening the envelope, yet it correlates exactly to the color of the slip of paper. Upon opening their envelopes both observers observe the color of the slip determined by this variable and there is no paradox of wave function collapse.

7.8.3 Hidden variables and Bell's theorem

The singlet two-particle spin state present in (7.262) can be written in tensor notation as

$$|\Psi_s\rangle = \frac{1}{\sqrt{2}}\left(\left|\frac{1}{2}\right\rangle \otimes \left|-\frac{1}{2}\right\rangle - \left|-\frac{1}{2}\right\rangle \otimes \left|\frac{1}{2}\right\rangle\right) \equiv \frac{1}{\sqrt{2}}\left(\left|\frac{1}{2}, -\frac{1}{2}\right\rangle - \left|-\frac{1}{2}, \frac{1}{2}\right\rangle\right), \quad (7.312)$$

where an overall minus sign has been added to the state (7.262). This is a much studied quantum state since it can be created experimentally through the decay of spinless particles such as the neutral π^0 meson. The neutral π^0 meson most often decays into two photons, but occasionally decays into an electron and a positron. The positron is the antiparticle to the electron, and so it has the same mass and spin but is oppositely charged. Both are spin $\frac{1}{2}$ particles. The resulting two-particle state must have zero angular momentum in the rest frame of the spinless π^0 meson, and so the result of the decay must be the singlet state (7.312).

The singlet state (7.312) has the property that

$$\langle \Psi_s | \hat{S}_i \otimes \hat{S}_j | \Psi_s \rangle = -\frac{\hbar^2}{4}\delta_{ij}, \quad (7.313)$$

where $\hat{S}_i$ is the Cartesian component of the spin opoerator. The proof of (7.313) begins by using the action of the spin raising and lowering operators defined identically to (7.86) and (7.91),

$$\begin{aligned}\hat{S}_+\left|\frac{1}{2}\right\rangle &= (\hat{S}_x + i\hat{S}_y)\left|\frac{1}{2}\right\rangle = 0,\\ \hat{S}_-\left|\frac{1}{2}\right\rangle &= (\hat{S}_x - i\hat{S}_y)\left|\frac{1}{2}\right\rangle = \hbar\left|-\frac{1}{2}\right\rangle.\end{aligned} \tag{7.314}$$

Adding and subtracting these two equations gives

$$\begin{aligned}\hat{S}_x\left|\frac{1}{2}\right\rangle &= \frac{\hbar}{2}\left|-\frac{1}{2}\right\rangle\\ \hat{S}_y\left|\frac{1}{2}\right\rangle &= i\frac{\hbar}{2}\left|-\frac{1}{2}\right\rangle.\end{aligned} \tag{7.315}$$

Similarly, using the state $|-\frac{1}{2}\rangle$ yields a second set of similar expressions. The action of the spin operators on the two spin states are summarized as

$$\begin{aligned}\hat{S}_x\left|\frac{1}{2}\right\rangle &= \frac{\hbar}{2}\left|-\frac{1}{2}\right\rangle, & \hat{S}_x\left|-\frac{1}{2}\right\rangle &= \frac{\hbar}{2}\left|\frac{1}{2}\right\rangle\\ \hat{S}_y\left|\frac{1}{2}\right\rangle &= i\frac{\hbar}{2}\left|-\frac{1}{2}\right\rangle, & \hat{S}_y\left|-\frac{1}{2}\right\rangle &= -i\frac{\hbar}{2}\left|\frac{1}{2}\right\rangle\\ \hat{S}_z\left|\frac{1}{2}\right\rangle &= \frac{\hbar}{2}\left|\frac{1}{2}\right\rangle, & \hat{S}_z\left|-\frac{1}{2}\right\rangle &= -\frac{\hbar}{2}\left|\frac{1}{2}\right\rangle.\end{aligned} \tag{7.316}$$

Verifying (7.313) can then be done on a case by case basis. Using the orthonormality of the states, two examples are given by

$$\begin{aligned}\langle \Psi_S \mid \hat{S}_x \otimes \hat{S}_y \mid \Psi_S \rangle &= \frac{1}{2}\left(\left\langle \frac{1}{2}, -\frac{1}{2}\right| - \left\langle -\frac{1}{2}, \frac{1}{2}\right|\right)(\hat{S}_x \otimes \hat{S}_y)\left(\left|\frac{1}{2}, -\frac{1}{2}\right\rangle - \left|-\frac{1}{2}, \frac{1}{2}\right\rangle\right)\\ &= \frac{\hbar^2}{8}\left(\left\langle \frac{1}{2}, -\frac{1}{2}\right| - \left\langle -\frac{1}{2}, \frac{1}{2}\right|\right)\left(-i\left|-\frac{1}{2}, \frac{1}{2}\right\rangle - i\left|\frac{1}{2}, -\frac{1}{2}\right\rangle\right)\\ &= -i\frac{\hbar^2}{8}\left\langle \frac{1}{2}, -\frac{1}{2}\middle|\frac{1}{2}, -\frac{1}{2}\right\rangle + i\frac{\hbar^2}{8}\left\langle -\frac{1}{2}, \frac{1}{2}\middle| -\frac{1}{2}, \frac{1}{2}\right\rangle = 0\end{aligned} \tag{7.317}$$

and

$$\begin{aligned}\langle \Psi_S \mid \hat{S}_x \otimes \hat{S}_x \mid \Psi_S \rangle &= \frac{1}{2}\left(\left\langle \frac{1}{2}, -\frac{1}{2}\right| - \left\langle -\frac{1}{2}, \frac{1}{2}\right|\right)(\hat{S}_x \otimes \hat{S}_x)\left(\left|\frac{1}{2}, -\frac{1}{2}\right\rangle - \left|-\frac{1}{2}, \frac{1}{2}\right\rangle\right)\\ &= \frac{\hbar^2}{8}\left(\left\langle \frac{1}{2}, -\frac{1}{2}\right| - \left\langle -\frac{1}{2}, \frac{1}{2}\right|\right)\left(\left|-\frac{1}{2}, \frac{1}{2}\right\rangle - \left|\frac{1}{2}, -\frac{1}{2}\right\rangle\right)\\ &= -\frac{\hbar^2}{8}\left\langle \frac{1}{2}, -\frac{1}{2}\middle|\frac{1}{2}, -\frac{1}{2}\right\rangle - \frac{\hbar^2}{8}\left\langle -\frac{1}{2}, \frac{1}{2}\middle| -\frac{1}{2}, \frac{1}{2}\right\rangle = -\frac{\hbar^2}{4}.\end{aligned} \tag{7.318}$$

All other cases in (7.313) follow similarly.

The next step is to project the two spin operators onto two *arbitrary* spatial unit vectors, $\hat{e}_a$ and $\hat{e}_b$, so that

$$\hat{S}_a = \hat{e}_a \cdot \hat{\boldsymbol{S}} = \sum_{j=1}^{3} e_{aj}\hat{S}_j, \quad \hat{S}_b = \hat{e}_b \cdot \hat{\boldsymbol{S}} = \sum_{k=1}^{3} e_{bk}\hat{S}_k. \tag{7.319}$$

These two spin projections will be interpreted as the measurements of two Stern–Gerlach devices on the spin orientation of the electron and positron emitted from the π^0 meson decay. These devices are oriented at arbitrary angles to each other, which is represented by the choice of arbitrary unit vectors for the projections. The quantum mechanical result (7.313) for the expectation value can now be used to find the expectation value for the measurement of the two spins projected along the arbitrary spatial directions. This is given by

$$\langle \hat{S}_a \otimes \hat{S}_b \rangle = \sum_{j,k=1}^{3} e_{aj}e_{bk}\langle \Psi_s \mid \hat{S}_j \otimes \hat{S}_k \mid \psi_s \rangle = -\frac{\hbar^2}{4}\sum_{j,k=1}^{3} e_{aj}e_{bk}\delta_{jk} = -\frac{\hbar^2}{4}\sum_{j=1}^{3} e_{aj}e_{bj} = -\frac{\hbar^2}{4}\hat{e}_a \cdot \hat{e}_b. \tag{7.320}$$

Bell compared this result to the effect that a general hidden variable approach would have in measuring the electron and positron spins. This begins by noting that the result (7.320) reflects the statistical nature of the singlet state (7.312), which is a mixed state of both spin up and down for the two particles. In an effort to avoid the conceptual problems surrounding wave function collapse onto one of the states as a result of measurement, it is possible to postulate that the spin state of each particle is determined by some local hidden variable, denoted by σ. This hidden variable is fixed by some random process during the preparation of the state, in this case the decay of the parent π^0 meson. The value of each spin projection for the two particles is measured after they are separated, and is given by a function of the direction of the projection vector and the value of the hidden variable. This is written in general form for the electron–positron pair as

$$\langle \hat{S}_a \otimes \hat{S}_b \rangle_\sigma = \left(\frac{\hbar}{2}E(\hat{e}_a, \sigma)\right) \times \left(\frac{\hbar}{2}P(\hat{e}_b, \sigma)\right), \tag{7.321}$$

where, for consistency with the Stern–Gerlach experiment, the two functions can take only the values

$$E(\hat{e}_a, \sigma) = \pm 1, \quad P(\hat{e}_b, \sigma) = \pm 1. \tag{7.322}$$

These functions describe the spin state of each particle in a manner determined by the value of the hidden variable. In order for angular momentum to be conserved, the two functions must have the property

$$E(\hat{e}_a, \sigma) = -P(\hat{e}_a, \sigma), \tag{7.323}$$

since then the two detectors are parallel and must give equal and opposite results for the spin of the electron and the positron.

However, the value of the hidden variable may be determined randomly during the preparation process for the state. This reflects the fact that the spin orientation of the two particles varies from decay to decay. The probability of a specific value of σ being created is then written as

$$P(\sigma) = \rho(\sigma)\, \mathrm{d}\sigma, \tag{7.324}$$

where $\rho(\sigma)$ is a *positive-definite* probability density. This probability must be normalized, so that

$$\int \rho(\sigma)\, \mathrm{d}\sigma = 1. \tag{7.325}$$

The result of many measurements, by definition the expectation value of the product of the two spins, would then be given by

$$\begin{aligned} \langle \hat{S}_a \otimes \hat{S}_b \rangle &= \int \mathrm{d}\sigma\, \rho(\sigma)\, \langle \hat{S}_a \otimes \hat{S}_b \rangle_\sigma = \frac{\hbar^2}{4} \int \mathrm{d}\sigma\, \rho(\sigma)\, E(\hat{e}_a, \sigma)\, P(\hat{e}_b, \sigma) \\ &= -\frac{\hbar^2}{4} \int \mathrm{d}\sigma\, \rho(\sigma)\, E(\hat{e}_a, \sigma)\, E(\hat{e}_b, \sigma), \end{aligned} \tag{7.326}$$

where the last step used the constraint of (7.323). This result gives

$$\langle \hat{S}_a \otimes \hat{S}_b \rangle - \langle \hat{S}_a \otimes \hat{S}_c \rangle = -\frac{\hbar^2}{4} \int \mathrm{d}\sigma\, \rho(\sigma)\, (E(\hat{e}_a, \sigma)\, E(\hat{e}_b, \sigma) - E(\hat{e}_a, \sigma)\, E(\hat{e}_c, \sigma)). \tag{7.327}$$

Property (7.322) gives $E^2(\hat{e}_b, \sigma) = 1$, so that inserting this factor of one into the second term of (7.327) gives

$$\langle \hat{S}_a \otimes \hat{S}_b \rangle - \langle \hat{S}_a \otimes \hat{S}_c \rangle = -\frac{\hbar^2}{4} \int \mathrm{d}\sigma\, \rho(\sigma)\, (1 - E(\hat{e}_b, \sigma)\, E(\hat{e}_c, \sigma)) E(\hat{e}_a, \sigma)\, E(\hat{e}_b, \sigma). \tag{7.328}$$

Property (7.322) shows that the products of E appearing in (7.328) are either 1 or -1. Coupling this with the positive-definite nature of $\rho(\sigma)$ shows that the first factor in (7.328) has the property

$$\rho(\sigma)\, (1 - E(\hat{e}_b, \sigma)\, E(\hat{e}_c, \sigma)) \geqslant 0. \tag{7.329}$$

However, the second factor in (7.328) has the property that

$$E(\hat{e}_a, \sigma)\, E(\hat{e}_b, \sigma) = \pm 1, \tag{7.330}$$

so that as σ is integrated there may be values of σ for which the second factor changes sign, thereby *decreasing* the *absolute value* of the integral. This means that the absolute value of the left-hand side is bounded by the integral

$$|\langle \hat{S}_a \otimes \hat{S}_b \rangle - \langle \hat{S}_a \otimes \hat{S}_c \rangle| \leqslant \frac{\hbar^2}{4} \int \mathrm{d}\sigma\, \rho(\sigma)\, (1 - E(\hat{e}_b, \sigma)\, E(\hat{e}_c, \sigma)). \tag{7.331}$$

Using the normalization of the probability density (7.325) and the expression for the expectation value (7.326) gives

$$|\langle \hat{S}_a \otimes \hat{S}_b \rangle - \langle \hat{S}_a \otimes \hat{S}_c \rangle| \leqslant \frac{\hbar^2}{4} + \langle \hat{S}_b \otimes \hat{S}_c \rangle, \tag{7.332}$$

which is known as *Bell's inequality*.

To reiterate, if the measurement of the spin state is determined by the value of a local hidden variable σ, the expectation values of the singlet state must satisfy the inequality (7.332) for *all choices* of the detector orientations $\hat{e}_a$, $\hat{e}_b$, and $\hat{e}_c$. It is easy to see that this inequality is not satisfied by the quantum mechanical result (7.320). For the simple case that $\hat{e}_a \cdot \hat{e}_b = \sqrt{2}/2$, $\hat{e}_b \cdot \hat{e}_c = \sqrt{2}/2$, and $\hat{e}_a \cdot \hat{e}_c = 0$, corresponding to detector directions lying in a plane at 45° angles, the quantum mechanical result (7.320) gives

$$|\langle \hat{S}_a \otimes \hat{S}_b \rangle - \langle \hat{S}_a \otimes \hat{S}_c \rangle| = \frac{\sqrt{2}}{2}\frac{\hbar^2}{4} > \frac{\hbar^2}{4}\left(1 - \frac{\sqrt{2}}{2}\right) = \frac{\hbar^2}{4} + \langle \hat{S}_b \otimes \hat{S}_c \rangle. \tag{7.333}$$

There are only two options: either the formulation of quantum mechanics used to find (7.320) is incorrect *or* local hidden variables are not a viable alternative to wave function collapse. A growing body of experimental evidence supports the latter choice and quantum mechanics is therefore understood to have a *nonlocal nature*. Over the years subsequent to its theoretical identification, the phenomenon of quantum entanglement has received significant experimental confirmation, culminating in the Nobel Prize in Physics in 2022. Such entangled spin states also play an important role in the theory of quantum computing. Unfortunately, the subject of quantum computing lies outside the purview of this text. A reference is provided at the end of this chapter.

7.8.4 Entanglement and environmental decoherence

Important properties of entanglement are revealed by examining the associated density matrix. It is useful first to examine the outcome of doing so for an unentangled state. The density operator for the unentangled state (7.301) is given by

$$\hat{\rho}(t) = |\, \Psi, t\, \rangle\langle\, \Psi, t\, | = |\, \psi, t\, \rangle\langle\, \psi, t\, | \otimes |\, \varphi, t\, \rangle\langle\, \varphi, t\, |. \tag{7.334}$$

This density operator can be written

$$\hat{\rho}(t) = \hat{\rho}_1(t) \otimes \hat{\rho}_2(t), \tag{7.335}$$

so that it factorizes into the respective one-particle density operators. By virtue of (7.302) this density operator is both idempotent and has a trace of one,

$$\mathrm{Tr}\, \rho(t) = \sum_{j,k} \langle\, e_j, e_k' \,|\, \Psi, t\, \rangle\langle\, \Psi, t \,|\, e_j, e_k'\, \rangle = \langle\, \Psi, t \,|\, \Psi, t\, \rangle = 1. \tag{7.336}$$

The expectation value of an observable $\hat{\mathrm{A}}$ can be measured *for the first particle* by using the operator $\hat{\mathrm{A}}_1 = \hat{\mathrm{A}} \otimes \hat{\mathbf{1}}$, which gives

$$\hat{\rho}(t)\hat{\mathrm{A}}_1 = |\, \psi, t\, \rangle\langle\, \psi, t\, |\hat{\mathrm{A}} \otimes |\, \varphi, t\, \rangle\langle\, \varphi, t\, |\hat{\mathbf{1}} = |\, \psi, t\, \rangle\langle\, \psi, t\, |\hat{\mathrm{A}} \otimes |\, \varphi, t\, \rangle\langle\, \varphi, t\, |. \tag{7.337}$$

This operator has the trace over the complete states $| e_j \rangle$ and $| e'_k \rangle$ given by

$$\begin{aligned} \mathrm{Tr}\,\hat{\rho}(t)\hat{\mathrm{A}}_1 &= \sum_j \langle e_j \,|\, \psi, t \rangle\langle \psi, t \,|\, \hat{\mathrm{A}} \,|\, e_j \rangle \cdot \sum_k \langle e'_k \,|\, \varphi, t \rangle\langle \varphi, t \,|\, e'_k \rangle \\ &= \sum_j \langle \psi, t \,|\, \hat{\mathrm{A}} \,|\, e_j \rangle\langle e_j \,|\, \psi, t \rangle \cdot \sum_k \langle \varphi, t \,|\, e'_k \rangle\langle e'_k \,|\, \varphi, t \rangle \\ &= \langle \psi, t \,|\, \hat{\mathrm{A}} \,|\, \psi, t \rangle\langle \varphi, t \,|\, \varphi, t \rangle = \langle \psi, t \,|\, \hat{\mathrm{A}} \,|\, \psi, t \rangle, \end{aligned} \tag{7.338}$$

which is the same result that is obtained by evaluating the expectation value ignoring the second particle and using only the state $| \psi, t \rangle$. It shows no interference between the two particles while measuring the properties of the first particle. In this regard, if measurements are limited to the first particle, the density matrix can be *reduced* by performing the trace over the second particle state, so that the *reduced density matrix* is given by

$$\mathrm{Tr}_2\,\hat{\rho}(t) = | \psi, t \rangle \sum_j \langle e'_j \,|\, \varphi, t \rangle\langle \varphi, t \,|\, e'_j \rangle\langle \psi, t \,| = \langle \varphi, t \,|\, \varphi, t \rangle | \psi, t \rangle\langle \psi, t \,| = | \psi, t \rangle\langle \psi, t \,|. \tag{7.339}$$

The reduced density operator exhibits no vestige of the second particle state $| \varphi, t \rangle$. All information available from measurements of the first particle is contained in the reduced density operator (7.339).

However, if the particle pair is entangled this will not be the case. This is demonstrated by considering the density matrix for an entangled state of the form

$$| \Psi, t \rangle = \frac{1}{\sqrt{2}} | \psi_1, t \rangle \otimes | \varphi_1, t \rangle + \frac{1}{\sqrt{2}} | \psi_2, t \rangle \otimes | \varphi_2, t \rangle. \tag{7.340}$$

It is important to clarify the states appearing in (7.340). For the sake of simplification, it will be assumed that the two normalized states of the first particle, $| \psi_1, t \rangle$ and $| \psi_2, t \rangle$, are orthogonal, so that $\langle \psi_1, t \,|\, \psi_2, t \rangle = 0$. However, no such assumption regarding the normalized states of the second particle, $| \varphi_1, t \rangle$ and $| \varphi_2, t \rangle$, will be made. Dropping the zero inner products, these relaxed assumptions still yield

$$\langle \Psi, t \,|\, \Psi, t \rangle = \frac{1}{2}(\langle \psi_1, t \,|\, \psi_1, t \rangle\langle \varphi_1, t \,|\, \varphi_1, t \rangle + \langle \psi_2, t \,|\, \psi_2, t \rangle\langle \varphi_2, t \,|\, \varphi_2, t \rangle) = 1. \tag{7.341}$$

The density operator is given by

$$\begin{aligned} \hat{\rho}(t) &= | \Psi, t \rangle\langle \Psi, t \,| = \frac{1}{2}(| \psi_1, t \rangle \otimes | \varphi_1, t \rangle + | \psi_2, t \rangle \otimes | \varphi_2, t \rangle)(\langle \psi_1, t \,| \otimes \langle \varphi_1, t \,| + \langle \psi_2, t \,| \otimes \langle \varphi_2, t \,|) \\ &= \frac{1}{2}(| \psi_1, t \rangle\langle \psi_1, t \,| \otimes | \varphi_1, t \rangle\langle \varphi_1, t \,| + | \psi_1, t \rangle\langle \psi_2, t \,| \otimes | \varphi_1, t \rangle\langle \varphi_2, t \,| \\ &\quad + | \psi_2, t \rangle\langle \psi_1, t \,| \otimes | \varphi_2, t \rangle\langle \varphi_1, t \,| + | \psi_2, t \rangle\langle \psi_2, t \,| \otimes | \varphi_2, t \rangle\langle \varphi_2, t \,|). \end{aligned} \tag{7.342}$$

Performing the trace over the second particle states using the normalization of the states $| \varphi_j, t \rangle$ gives

$$\begin{aligned} \mathrm{Tr}_2\,\hat{\rho}(t) = \frac{1}{2}(&| \psi_1, t \rangle\langle \psi_1, t \,| + \langle \varphi_1, t \,|\, \varphi_2, t \rangle \,| \psi_1, t \rangle\langle \psi_2, t \,| \\ &+ \langle \varphi_2, t \,|\, \varphi_1, t \rangle \,| \psi_2, t \rangle\langle \psi_1, t \,| + | \psi_2, t \rangle\langle \psi_2, t \,|). \end{aligned} \tag{7.343}$$

There is now a vestige of the second particle states in the form of their inner product $\langle \varphi_1, t \mid \varphi_2, t \rangle$. However, using the assumed orthonormality of the $\mid \psi_j, t \rangle$ states the trace of the resulting reduced density operator is easily found to be

$$\mathrm{Tr}\,(\mathrm{Tr}_2\hat{\rho}(t)) = \frac{1}{2}(\langle \psi_1, t \mid \psi_1, t \rangle + \langle \psi_2, t \mid \psi_2, t \rangle) = 1, \tag{7.344}$$

which is the necessary condition for the density operator.

There are two cases of interest in considering (7.343). The first case is $\mid \varphi_1, t \rangle = \mid \varphi_2, t \rangle = \mid \varphi, t \rangle$. For this case, the entangled state (7.340) becomes the unentangled state (7.301),

$$\mid \Psi, t \rangle = \frac{1}{\sqrt{2}}(\mid \psi_1, t \rangle + \mid \psi_2, t \rangle) \otimes \mid \varphi, t \rangle \;\equiv\; \mid \psi, t \rangle \otimes \mid \varphi, t \rangle. \tag{7.345}$$

In addition, for this case the reduced density operator (7.343) can be written

$$\mathrm{Tr}_2\,\hat{\rho}(t) = \frac{1}{2}(\mid \psi_1, t \rangle + \mid \psi_2, t \rangle)(\langle \psi_1, t \mid + \langle \psi_2, t \mid) = \mid \psi, t \rangle\langle \psi, t \mid. \tag{7.346}$$

The density operator has become the density operator of a *pure state*. As such, the expectation value of observables for the first particle state will take the form

$$\begin{aligned}\langle \hat{\mathrm{A}} \rangle(t) &= \mathrm{Tr}\,(\mathrm{Tr}_2\hat{\rho}(t))\hat{\mathrm{A}} \\ &= \frac{1}{2}(\langle \psi_1, t \mid \hat{\mathrm{A}} \mid \psi_1, t \rangle + \langle \psi_1, t \mid \hat{\mathrm{A}} \mid \psi_2, t \rangle + \langle \psi_2, t \mid \hat{\mathrm{A}} \mid \psi_1, t \rangle + \langle \psi_2, t \mid \hat{\mathrm{A}} \mid \psi_2, t \rangle).\end{aligned} \tag{7.347}$$

Choosing $\hat{\mathrm{A}} = \mid \boldsymbol{x} \rangle\langle \boldsymbol{x} \mid$ gives

$$\mathrm{Tr}\,(\mathrm{Tr}_2\hat{\rho}(t))\mid \boldsymbol{x} \rangle\langle \boldsymbol{x} \mid = \frac{1}{2}\left(|\psi_1(\boldsymbol{x}, t)|^2 + \psi_1^*(\boldsymbol{x}, t)\psi_2(\boldsymbol{x}, t) + \psi_2^*(\boldsymbol{x}, t)\psi_1(\boldsymbol{x}, t) + |\psi_2(\boldsymbol{x}, t)|^2\right). \tag{7.348}$$

Comparing this with result (7.278) shows that the quantum interference terms of a pure state are present. If $\hat{A}$ is not a member of the CSCO, then the cross-terms will not necessarily vanish. Assuming that the Hamiltonian is a member of the CSCO, then the two states $\mid \psi_j \rangle$ are eigenstates of the Hamiltonian, and the cross-terms in (7.348) are given by

$$\langle \psi_j, t \mid \hat{A} \mid \psi_k, t \rangle = \langle \psi_j, 0 \mid e^{i\hat{H}t/\hbar}\hat{A}\, e^{-i\hat{H}t/\hbar} \mid \psi_k, 0 \rangle = \langle \psi_j, 0 \mid \hat{A} \mid \psi_k, 0 \rangle e^{i(E_j - E_k)t/\hbar}. \tag{7.349}$$

The expectation value for the observable $\hat{A}$ will therefore exhibit oscillatory behavior in time when measured in the non-entangled state (7.345).

The second case is $\langle \varphi_1, t \mid \varphi_2, t \rangle = 0$. For this case the reduced density operator becomes

$$\mathrm{Tr}_2\,\hat{\rho}(t) = \frac{1}{2}\mid \psi_1, t \rangle\langle \psi_1, t \mid + \frac{1}{2}\mid \psi_2, t \rangle\langle \psi_2, t \mid, \tag{7.350}$$

which is the form (7.280) for a *mixed state density operator*. For such a density operator, the expectation value of the observable $\hat{\mathrm{A}}$ for the first particle is given by

$$\langle \hat{A} \rangle(t) = \mathrm{Tr}\,(\mathrm{Tr}_2\hat{\rho}(t))\hat{A} = \frac{1}{2}\langle \psi_1, t \mid \hat{A} \mid \psi_1, t \rangle + \frac{1}{2}\langle \psi_2, t \mid \hat{A} \mid \psi_2, t \rangle. \tag{7.351}$$

Again assuming that the $| \psi_j \rangle$ are eigenstates of the Hamiltonian, it follows that

$$\begin{aligned} \langle \hat{A} \rangle(t) &= \frac{1}{2}\langle \psi_1, 0 \mid e^{i\hat{H}t/\hbar}\hat{A}\, e^{-i\hat{H}t/\hbar} \mid \psi_1, 0 \rangle + \frac{1}{2}\langle \psi_2, 0 \mid e^{i\hat{H}t/\hbar}\hat{A}\, e^{-i\hat{H}t/\hbar} \mid \psi_2, 0 \rangle \\ &= \frac{1}{2}\langle \psi_1, 0 \mid \hat{A} \mid \psi_1, 0 \rangle + \frac{1}{2}\langle \psi_2, 0 \mid \hat{A} \mid \psi_2, 0 \rangle, \end{aligned} \tag{7.352}$$

which exhibits no oscillatory behavior for the expectation value of $\hat{A}$. This also follows from

$$\mathrm{Tr}\,(\mathrm{Tr}_2\hat{\rho}(t)) \mid \boldsymbol{x} \rangle\langle \boldsymbol{x} \mid = \frac{1}{2}|\psi_1(\boldsymbol{x}, t)|^2 + \frac{1}{2}|\psi_2(\boldsymbol{x}, t)|^2, \tag{7.353}$$

showing that the quantum interference terms have vanished. The vanishing of the quantum interference terms in the expectation value is referred to as *quantum decoherence* since it exhibits the *classical addition of probabilities* first discussed in (3.247). Entanglement can be accompanied by decoherence, ranging from its total absence in (7.346) to complete decoherence and the mixed state density operator exhibited in (7.350).

Quantum decoherence and entanglement have been adapted to model the quantum to classical transition through interactions of the quantum system with its environment. One of the hallmarks of physics is the restriction of application and theory to *isolated systems* that is almost always made. These are idealized systems which are closed, interacting only among the constituents of the system. Of course, this ignores the unavoidable interaction of the system with its *environment*, which is understood to be the set of objects in the larger world around the system under consideration. An example of this at the classical level is modeling the motion of the solar system using only the two-body interactions between each planet and the Sun while ignoring the effects of the gravitational interaction of the planets. At the quantum level this restriction to isolated systems manifests itself by postulating the preparation of an initial state and then ignoring the effect of interaction with its environment may have on its time evolution. Applications of decoherence include the presence of the environment and its effect on the evolution of the system.

A version of decoherence can be modeled by choosing the initial state of a one-particle and environment system to be given by (7.345),

$$| \Psi, t = 0 \rangle = \frac{1}{\sqrt{2}}(| \psi_1, t = 0 \rangle + | \psi_2, t = 0 \rangle) \otimes | \varphi, t = 0 \rangle, \tag{7.354}$$

where $| \varphi, t = 0 \rangle$ represents the initial state of the environment. This environment may include a detector with which the system is interacting or a collection of many particles interacting with the particle. Initially, the state (7.354) is unentangled since the state of the environment is the same for both the constituent states $| \psi_{1,\,2}, t \rangle$ of the particle state. Since this state has the form (7.345), result (7.348) shows that quantum interference will be present in measurements made on the particle. However, as time passes the interaction of the particle with its environment can lead to a change in the state of the environment,

$$|\,\psi_j,\,0\,\rangle \otimes |\,\varphi,\,0\,\rangle \to |\,\psi_j',\,t\,\rangle \otimes |\,\varphi_j,\,t\,\rangle = e^{-i\hat{H}t/\hbar}\Big(|\,\psi_j,\,0\,\rangle \otimes |\,\varphi,\,0\,\rangle\Big). \quad (7.355)$$

This assumes that each constituent state of the initial particle state does not disappear and that no new states appear. The states $|\,\varphi_j,\,t\,\rangle$ are referred to as *pointer states* since they mimic the action of a detector. This is discussed again in the next section.

A simple interaction that does not change the system states but modifies the initial environmental state is given by

$$\hat{V} = |\,\psi_1,\,0\,\rangle\langle\,\psi_1,\,0\,| \otimes \hat{V}_1 + |\,\psi_2,\,0\,\rangle\langle\,\psi_2,\,0\,| \otimes \hat{V}_2, \quad (7.356)$$

where $\hat{V}_1$ and $\hat{V}_2$ are two different potentials acting on the environment created by the two different system states. It modifies the state (7.354) according to

$$\hat{V}|\,\Psi,\,0\,\rangle = \frac{1}{\sqrt{2}}(|\,\psi_1,\,0\,\rangle \otimes \hat{V}_1|\,\varphi,\,0\,\rangle + |\,\psi_2,\,0\,\rangle \otimes \hat{V}_2|\,\varphi,\,0\,\rangle). \quad (7.357)$$

This interaction causes the environmental state to evolve depending on the system state to which it is coupled, while the system state is unchanged. Such an interaction is suited to modeling the elastic scattering of a light environmental particle from a heavier system particle. Such a scattering process will change the environmental state while leaving the system states unchanged. This is discussed again in chapter 8 where the time evolution operator for a potential is developed.

The unitary time evolution created by the Hamiltonian maintains the normalization of the constituent states. As a result, the state of (7.354) becomes entangled with its environment as time passes,

$$|\,\Psi,\,0\,\rangle \to \frac{1}{\sqrt{2}}(|\,\psi_1,\,t\,\rangle \otimes |\,\varphi_1,\,t\,\rangle + |\,\psi_2,\,t\,\rangle \otimes |\,\varphi_2,\,t\,\rangle). \quad (7.358)$$

For the case that

$$\langle\,\varphi_1,\,t\,|\,\varphi_2,\,t\,\rangle \to 0, \quad (7.359)$$

the pure quantum state of (7.354) decoheres into a mixed state density operator with classical properties in the sense that quantum interference terms are suppressed. Detailed analysis indicates that environmentally induced decoherence occurs very rapidly in multiparticle systems, which helps to explain why quantum interference effects play a much smaller role as the scale of the system is increased.

7.8.5 Detection and decoherence

Decoherence techniques can also be applied to electron diffraction in the case that a detector monitors the slits through which the electron may pass. The initial electron states are the slit states $|\,s_1\,\rangle$ and $|\,s_2\,\rangle$ defined in (6.366). In this case, the environmental states $| \varphi_1 \rangle$ and $| \varphi_2 \rangle$ represent the device placed at the slits to detect the passage of the electron. The device, although much more complicated in scope than a single electron, has only two states. The first is $| \varphi_1 \rangle$, which represents the electron passing through the first slit and triggering the detector. The second is $| \varphi_2 \rangle$, which represents the electron passing through the second slit and triggering the detector. As a result, the slit states act as pointer states for the electron. The electron and device are described by the entangled tensor product state,

$$| \Psi, t = 0 \rangle = \alpha | s_1, t = 0 \rangle \otimes | \varphi_1, t = 0 \rangle + \beta | s_2, t = 0 \rangle \otimes | \varphi_2, t = 0 \rangle, \quad (7.360)$$

where $|\alpha|^2 + |\beta|^2 = 1$ ensures the normalization of the state (7.360). This state reflects the correct limits on α and β. If $\alpha = 0$ then there is no first slit and the detector state must be $| \varphi_2 \rangle$, while if $\beta = 0$ there is no second slit and the detector state must be $| \varphi_1 \rangle$. In order for these pointer states to be distinguishable, the two states of the detector must be orthogonal, so that $\langle \varphi_1, t | \varphi_2, t \rangle = 0$. Performing the trace over the states $| \varphi_j \rangle$ using their orthogonality gives the reduced density operator

$$\mathrm{Tr}_2\, \hat{\rho}(t) = |\alpha|^2 | s_1, t \rangle\langle s_1, t | + |\beta|^2 | s_2, t \rangle\langle s_2, t |, \quad (7.361)$$

which is a mixed state density operator with the probabilities $p_1 = |\alpha|^2$ and $p_2 = |\beta|^2$. This density operator is equivalent to a statistical ensemble of prepared states such that the electron passes through either slit one or slit two with the relative probabilities p_1 and p_2. For such a case, there is no interference between the wave function originating from both slits since it is known for each member of the statistical ensemble which slit the electron passed through. This shows that the act of acquiring knowledge about which slit the electron passed through by detecting its passage results in a destruction of the quantum interference terms. Once again, this phenomenon has been observed experimentally. However, if the detector malfunctions and is triggered by the passage of the electron through either slit to report only the first slit, then setting $| \varphi_2 \rangle = | \varphi_1 \rangle$ in (7.360) shows that the quantum interference terms return. The interaction of the particle with the detector is the essence of an observation. This shows that the state of the particle collapses onto one of the slit states as a result of its detection during its passage through the two slit device.

References and recommended further reading

Modern group theory has its origins in the work of Lie:

- Lie S 1888 *Theorie der Transformationsgruppen* (Leipzig: Teubner)

Monographs for physicists that focus on Lie groups and Lie algebras include

- Wybourne B 1974 *Classical Groups for Physicists* (New York: Wiley)
- Hamermesh M 1989 *Group Theory and Its Application to Physical Problems* (New York: Dover)
- Sternberg S 1994 *Group Theory and Physics* (Cambridge: Cambridge University Press)

Stern and Gerlach presented the results of their experiment in

- Gerlach W and Stern O 1922 Der experimentelle Nachweis des magnetischen Moments des Silberatoms *Z. Phys.* **8** 110 (https://doi.org/10.1007/BF01329580)

The correct electron spin dipole moment is derived by linearizing the Schrödinger equation in

- Levy-Leblond J 1967 Nonrelativistic particles and wave equations *Comm. Math. Phys.* **6** 286 (https://doi.org/10.1007/BF01646020)
- Greiner W 2001 *Quantum Mechanics: An Introduction* 4th edn (New York: Springer)

Pauli first employed the spinor representation for electron spin in

- Pauli W 1927 Zur Quantenmechanik des magnetischen Elektrons *Z. Phys.* **43** 601 (https://doi.org/10.1007/BF01397326)

Pauli proposed what is now known as the Paul exclusion principle in

- Pauli W 1925 Über den Zusammenhang des Abschlusses der Elektronengruppen im Atom mit der Komplexstruktur der Spektren *Z. Phys.* **31** 765 (https://doi.org/10.1007/BF02980631)

The addition of quantum mechanical angular momentum is presented in

- Schiff L 1968 *Quantum Mechanics* 3rd edn (New York: McGraw-Hill)
- Schwabl F 2007 *Quantum Mechanics* 4th end (New York: Springer)
- Griffiths D and Schroeter D 2018 *Introduction to Quantum Mechanics* 3rd edn (New York: Cambridge University Press)
- Cohen–Tannoudji C, Diu B and Laloë F 2019 *Quantum Mechanics* vols 1 and 2, 2nd edn (New York: Wiley)
- Sakurai J and Napolitano J 2020 *Modern Quantum Mechanics* (New York: Cambridge University Press)

The density matrix was first discussed in

- von Neumann J 1927 Wahrscheinlichkeitstheoretischer Aufbau der Quantenmechanik *Nachricht. Gesellsch. Wissensch. Göttingen, Math.-Phys.* Klasse 1 **1927** 245–72 (http://eudml.org/doc/59230)

The EPR critique of quantum mechanics based on entanglement was published in

- Einstein A, Podolsky B and Rosen N 1935 Can quantum-mechanical description of physical reality be considered complete? *Phys. Rev.* **47** 777 (https://doi.org/10.1103/PhysRev.47.777)

Bell published his theorem regarding hidden variable theories in

- Bell J 1964 On the Einstein Podolsky Rosen paradox *Physics* **1** 195 (https://doi.org/10.1103/PhysicsPhysiqueFizika.1.195)

Bell's theorem on hidden variables is discussed in

- Greiner W 2001 *Quantum Mechanics: An Introduction* 4th edn (New York: Springer)
- Griffiths D and Schroeter D 2018 *Introduction to Quantum Mechanics* 3rd edn (New York: Cambridge University Press)

Mixed states and entanglement and their relation to quantum computing is presented in

- Nielsen M and Chuang I 2011 *Quantum Computation and Quantum Information* (Cambridge: Cambridge University Press)

The systematic investigation of quantum decoherence began with

- Zeh H 1970 On the interpretation of measurement in quantum theory *Found. Phys.* **1** 69 (https://doi.org/10.1007/BF00708656)
- Joos E and Zeh H 1985 The emergence of classical properties through interaction with the environment *Z. Phys.* B **59** 223 (https://doi.org/10.1007/BF01725541)

A thorough presentation of quantum decoherence is found in

- Schlosshauer M 2007 *Decoherence and the Quantum-to-Classical Transition* (Berlin: Springer)

IOP Publishing

A Concise Introduction to Quantum Mechanics (Second Edition)

Mark S Swanson

Chapter 8

Approximation techniques

So far attention has been limited to exactly solvable quantum systems such as the harmonic oscillator and the hydrogen atom. While this has provided important insights into the nature of the atomic world, there are many physical systems that cannot be solved exactly. This chapter will present some general techniques that can be employed to obtain the *approximate* properties of such systems. These are based on using the properties of systems that are assumed to be *close in nature* to the system under consideration.

8.1 Stationary perturbation theory

In the same article where Schrödinger proposed the equation named after him, he adapted work by Rayleigh on sound waves to develop perturbation theory. For that reason it is sometimes referred to as *Rayleigh–Ritz perturbation theory*. In the case under consideration, the physical system is governed by a Hamiltonian of the general form

$$\hat{H} = \hat{H}_0 + \lambda \hat{V}, \tag{8.1}$$

where $\hat{H}_0$ is an exactly solvable time-independent Hamiltonian and λ is a *coupling constant* characterizing the strength of the *time-independent perturbation* $\hat{V}$. The coupling constant λ can also be viewed as a bookkeeping parameter by characterizing the dependence of eigenvalues and eigenstates in terms of powers of λ. Because the Hamiltonian has no explicit time-dependence, the approximation method to be developed is referred to as *stationary perturbation theory*. It is assumed that the eigenstates of $\hat{H}_0$, denoted $|E_n^{(0)}\rangle$, are both known and characterized by a set of integers that enumerate the eigenvalues $E_n^{(0)}$. As an example, the harmonic oscillator eigenstates are characterized by a single integer n. The *unperturbed states* satisfy

$$\hat{H}_0 | E_n^{(0)} \rangle = E_n^{(0)} | E_n^{(0)} \rangle. \tag{8.2}$$

doi:10.1088/978-0-7503-5663-3ch8

It will be assumed in this section that these states are nondegenerate, so that $E_n^{(0)} = E_j^{(0)}$ only if the set of integers represented by n satisfies $n = j$. This holds for the harmonic oscillator, but not for the eigenstates of the Coulomb potential found for the hydrogen atom. In that case, the states are characterized by the set of quantum numbers $\{n, \ell, m_\ell, m_s\}$, but differ in energy only if the principal quantum number n is different. It is assumed that the *unperturbed states* are complete, so that

$$\sum_n | E_n^{(0)} \rangle\langle E_n^{(0)} | = \hat{\mathbf{1}}. \tag{8.3}$$

The unperturbed states are assumed to form a complete basis for the states of the full Hamiltonian $\hat{H}$, denoted $| E_n \rangle$, which satisfy

$$\hat{H}| E_n \rangle = E_n| E_n \rangle, \tag{8.4}$$

and are also complete,

$$\sum_n | E_n \rangle\langle E_n | = \hat{\mathbf{1}}. \tag{8.5}$$

It is important to note that both sets of states, $| E_n^{(0)} \rangle$ and $| E_n \rangle$, are *assumed* to be indexed by the same set of integers n. This is often justified by arguing that, if the perturbation coupling constant λ is weak, then the two sets of states should be similar in nature. There is no guarantee that this is true.

8.1.1 Perturbation expansions

The next step is to expand both the energy eigenvalues E_n and eigenstates $| E_n \rangle$ in a power series in the coupling constant, so that the energy eigenvalues and eigenstates can be written

$$E_n = E_n^{(0)} + \lambda E_n^{(1)} + \lambda^2 E_n^{(2)} + \cdots = \sum_{j=0}^{\infty} \lambda^j E_n^{(j)}, \tag{8.6}$$

$$| E_n \rangle = | E_n^{(0)} \rangle + \lambda| E_n^{(1)} \rangle + \lambda^2| E_n^{(2)} \rangle + \cdots = \sum_{j=0}^{\infty} \lambda^j| E_n^{(j)} \rangle. \tag{8.7}$$

It is important to bear in mind that $| E_n^{(j)} \rangle$ is not an eigenstate of either $\hat{H}_0$ or $\hat{V}$, but rather the jth contribution to the eigenstate $| E_n \rangle$. The two series expansions of (8.7) are assumed to converge, so that the $\lambda \to 0$ limit is well-defined and reduces to the unperturbed situation. The Schrödinger equation becomes

$$\hat{H}| E_n \rangle = \left(\hat{H}_0 + \lambda\hat{V}\right)\left(\sum_{j=0}^{\infty} \lambda^j| E_n^{(j)} \rangle\right) = E_n| E_n \rangle = \left(\sum_{k=0}^{\infty} \lambda^k E_n^{(k)}\right)\left(\sum_{j=0}^{\infty} \lambda^j| E_n^{(j)} \rangle\right). \tag{8.8}$$

Expression (8.8) must hold for each power of λ. Expanding (8.8) and equating terms of the same power of λ on the left- and right-hand sides gives an infinite sequence of equations,

$$\lambda^0 \Longrightarrow \hat{H}_0 | E_n^{(0)} \rangle = E_n^{(0)} | E_n^{(0)} \rangle, \tag{8.9}$$

$$\lambda^1 \Longrightarrow \hat{V} | E_n^{(0)} \rangle + \hat{H}_0 | E_n^{(1)} \rangle = E_n^{(1)} | E_n^{(0)} \rangle + E_n^{(0)} | E_n^{(1)} \rangle, \tag{8.10}$$

$$\lambda^2 \Longrightarrow \hat{V} | E_n^{(1)} \rangle + \hat{H}_0 | E_n^{(2)} \rangle = E_n^{(2)} | E_n^{(0)} \rangle + E_n^{(1)} | E_n^{(1)} \rangle + E_n^{(0)} | E_n^{(2)} \rangle, \tag{8.11}$$
$$\vdots\ .$$

The zeroth order equation (8.9) holds by assumption, so that the goal is to solve the higher order terms. It is important to note that this process is equivalent to diagonalizing the matrix whose elements are given by

$$H_{\ell n} = \langle E_\ell^{(0)} | (\hat{H}_0 + \lambda \hat{V}) | E_n^{(0)} \rangle = E_\ell^{(0)} \delta_{\ell n} + \langle E_\ell^{(0)} | \lambda \hat{V} | E_n^{(0)} \rangle. \tag{8.12}$$

This was done earlier for the much simpler case of the 2×2 matrix Hamiltonian of (6.326). The expansion terms given by (8.10) and (8.11) are an alternative means to generate approximations of the unitary matrix and eigenvalues that diagonalize the infinite-dimensional matrix Hamiltonian of (8.12).

8.1.2 Nondegenerate perturbation theory

The first order equation (8.10) is solved by forming the inner product with $\langle E_\ell^{(0)} |$. Denoting the relevant unknown inner product as

$$a_{\ell n}^{(1)} = \langle E_\ell^{(0)} | E_n^{(1)} \rangle, \tag{8.13}$$

equation (8.10) becomes

$$\begin{aligned} &\langle E_\ell^{(0)} | \hat{V} | E_n^{(0)} \rangle + \langle E_\ell^{(0)} | \hat{H}_0 | E_n^{(0)} \rangle = E_n^{(1)} \langle E_\ell^{(0)} | E_n^{(0)} \rangle + E_n^{(0)} \langle E_\ell^{(0)} | E_n^{(1)} \rangle \\ \Longrightarrow &\langle E_\ell^{(0)} | \hat{V} | E_n^{(0)} \rangle + E_\ell^{(0)} a_{\ell n}^{(1)} = E_n^{(1)} \delta_{\ell n} + E_n^{(0)} a_{\ell n}^{(1)}. \end{aligned} \tag{8.14}$$

Choosing the case $\ell = n$ immediately gives the first order correction to the energy eigenvalue,

$$E_n^{(1)} = \langle E_n^{(0)} | \hat{V} | E_n^{(0)} \rangle, \tag{8.15}$$

where the right-hand side is given in terms of known quantities.

As an example, it is straightforward to calculate the correction to the ground state energy of the simple harmonic oscillator brought about by a quartic perturbation,

$$\hat{V} = \frac{1}{4} \lambda \hat{Q}^4. \tag{8.16}$$

This is given by putting (8.15) into a configuration representation and using the ground state wave function (5.88) for the simple harmonic oscillator,

$$\lambda E_0^{(1)} = \langle E_0^{(0)} \mid \frac{1}{4}\lambda \hat{Q}^4 \mid E_0^{(0)} \rangle = \frac{1}{4}\lambda \int_{-\infty}^{\infty} \mathrm{d}x\, \mathrm{d}y\, \langle E_0^{(0)} \mid x \rangle \langle x \mid \hat{Q}^4 \mid y \rangle \langle y \mid E_0^{(0)} \rangle$$
$$= \frac{1}{4}\lambda \int_{-\infty}^{\infty} \mathrm{d}x\, \psi_0^*(x) x^4 \psi_0(x) = \frac{\lambda\alpha}{4\sqrt{\pi}} \int_{-\infty}^{\infty} \mathrm{d}x\, x^4 e^{-\alpha^2 x^2} = \frac{3\lambda}{16\alpha^4}, \tag{8.17}$$

where $\alpha^2 = m\omega/\hbar$. The energy correction is $O(\hbar^2)$, far smaller than the ground state energy of the unperturbed harmonic oscillator for typical values of m and ω.

Considering the $\ell \neq n$ case of (8.14) gives

$$\left(E_n^{(0)} - E_\ell^{(0)}\right) a_{\ell n}^{(1)} = \langle E_\ell^{(0)} \mid \hat{V} \mid E_n^{(0)} \rangle. \tag{8.18}$$

This yields $a_{\ell n}^{(1)}$ as long as $E_n^{(0)} - E_\ell^{(0)} \neq 0$, which requires the absence of degeneracy in the energy eigenvalues of $\hat{H}_0$. This is the case for the one-dimensional harmonic oscillator, but does not hold if the hydrogen atom solutions are chosen for $E_n^{(0)}$. The case of energy degeneracy will be treated in the next section. It then follows that, for $\ell \neq n$,

$$a_{\ell n}^{(1)} = \frac{\langle E_\ell^{(0)} \mid \hat{V} \mid E_n^{(0)} \rangle}{\left(E_n^{(0)} - E_\ell^{(0)}\right)}, \tag{8.19}$$

which gives $\mid E_n^{(1)} \rangle$,

$$\mid E_n^{(1)} \rangle = \sum_\ell \mid E_\ell^{(0)} \rangle \langle E_\ell^{(0)} \mid E_n^{(1)} \rangle = \sum_\ell a_{\ell n}^{(1)} \mid E_\ell^{(0)} \rangle = \sum_{\ell \neq n} \frac{\langle E_\ell^{(0)} \mid \hat{V} \mid E_n^{(0)} \rangle}{\left(E_n^{(0)} - E_\ell^{(0)}\right)} \mid E_\ell^{(0)} \rangle + a_{nn}^{(1)} \mid E_n^{(0)} \rangle. \tag{8.20}$$

Depending on the form of $\hat{V}$, the corrections to the energy eigenstates are a mixture of the unperturbed eigenstates $\mid E_\ell^{(0)} \rangle$.

In the absence of degeneracy all terms in the summation can in principle be calculated, so that the last remaining first order unknown is $a_{nn}^{(1)}$. This is determined by normalizing the state $\mid E_n \rangle$ to $O(\lambda)$. To first order in λ, the inner product for the state $\mid E_n \rangle$ is given by

$$\langle E_n \mid E_n \rangle = \left(\langle E_n^{(0)} \mid + \lambda \langle E_n^{(1)} \mid + \cdots\right)\left(\mid E_n^{(0)} \rangle + \lambda \mid E_n^{(1)} \rangle + \cdots\right)$$
$$\approx \langle E_n^{(0)} \mid E_n^{(0)} \rangle + \lambda\left(\langle E_n^{(0)} \mid E_n^{(1)} \rangle + \langle E_n^{(1)} \mid E_n^{(0)} \rangle\right) = 1 + \lambda\left(a_{nn}^{(1)} + a_{nn}^{(1)*}\right). \tag{8.21}$$

Setting $a_{nn}^{(1)} = i|\beta_n|$ will therefore normalize the state $\mid E_n \rangle$ to first order in λ. It is common to redefine the first correction to the eigenstate as

$$\mid E_n^{(1)} \rangle' = \mid E_n^{(1)} \rangle - i|\beta_n| \mid E_n^{(0)} \rangle. \tag{8.22}$$

This correction has the property that

$$\langle E_n^{(0)} \mid E_n^{(1)} \rangle' = \langle E_n^{(0)} \mid E_n^{(1)} \rangle - i|\beta_n| \langle E_n^{(0)} \mid E_n^{(0)} \rangle = a_{nn}^{(1)} - i|\beta_n| = 0. \tag{8.23}$$

As a result, the perturbative corrections to the eigenstate are assumed to satisfy the $j \neq 0$ condition

$$\langle E_n^{(0)} \mid E_n^{(j)} \rangle = a_{nn}^{(j)} = 0. \tag{8.24}$$

The solutions to the first order perturbation equation (8.10) can now be used in combination with (8.24) to solve the second order perturbation equation (8.11). Once again, taking the inner product of (8.11) with $\langle E_\ell^{(0)} \mid$ gives

$$\langle E_\ell^{(0)} \mid \hat{V} \mid E_n^{(1)} \rangle + E_\ell^{(0)} \langle E_\ell^{(0)} \mid E_n^{(2)} \rangle = E_n^{(2)} \delta_{\ell n} + E_n^{(1)} \langle E_\ell^{(0)} \mid E_n^{(1)} \rangle + E_n^{(0)} \langle E_\ell^{(0)} \mid E_n^{(2)} \rangle. \quad (8.25)$$

Choosing $\ell = n$ and using the normalization condition (8.24) gives the second order correction to the energy eigenvalue,

$$E_n^{(2)} = \langle E_n^{(0)} \mid \hat{V} \mid E_n^{(1)} \rangle = \sum_{\ell \neq n} \frac{|\langle E_\ell^{(0)} \mid \hat{V} \mid E_n^{(0)} \rangle|^2}{(E_n^{(0)} - E_\ell^{(0)})}, \quad (8.26)$$

where result (8.20) was used in the second step. Choosing $\ell \neq n$ and defining

$$a_{\ell n}^{(2)} = \langle E_\ell^{(0)} \mid E_n^{(2)} \rangle \quad (8.27)$$

gives

$$\langle E_\ell^{(0)} \mid \hat{V} \mid E_n^{(1)} \rangle = E_n^{(1)} a_{\ell n}^{(1)} + \left(E_n^{(0)} - E_\ell^{(0)}\right) a_{\ell n}^{(2)}. \quad (8.28)$$

For the case of nondegeneracy this gives the $\ell \neq n$ solution for $a_{\ell n}^{(2)}$,

$$a_{\ell n}^{(2)} = \frac{\langle E_\ell^{(0)} \mid \hat{V} \mid E_n^{(1)} \rangle - E_n^{(1)} a_{\ell n}^{(1)}}{\left(E_n^{(0)} - E_\ell^{(0)}\right)} \quad (8.29)$$

Determining further higher order contributions is tedious but straightforward.

8.1.3 Degenerate perturbation theory

For simplicity, the situation where the first N unperturbed states are energetically degenerate will be considered. More complicated situations will follow from the results obtained. This means that

$$\hat{H}_0 | E_n^{(0)} \rangle = \varepsilon | E_n^{(0)} \rangle, \quad n = 1, \dots N, \quad (8.30)$$

and so step (8.19) breaks down and $a_{\ell n}^{(1)}$ is undefined for $\ell, n = 1, \dots, N$. The solution is to define a new set of basis states in the degenerate sector, denoted $| \tilde{E}_\ell^{(0)} \rangle$ for $\ell = 1, \dots, N$, which diagonalize the full Hamiltonian,

$$\langle \tilde{E}_\ell^{(0)} \mid \hat{H} \mid \tilde{E}_n^{(0)} \rangle = \langle \tilde{E}_\ell^{(0)} \mid (\hat{H}_0 + \hat{V}) \mid \tilde{E}_n^{(0)} \rangle = \tilde{E}_\ell^{(0)} \delta_{\ell n}. \quad (8.31)$$

These states then replace the original degenerate states in expression (8.8), so that

$$| E_\ell^{(0)} \rangle \to | \tilde{E}_\ell^{(0)} \rangle, \quad \ell = (1, \dots, N). \quad (8.32)$$

This uses the standard procedure for diagonalizing the $N \times N$ matrix obtained from (8.12) by limiting consideration to the matrix elements $H_{\ell n}$ for $\ell, n = 1, \dots, N$.

In the 2×2 matrix case of (6.326) the diagonalization procedure consisted of finding the eigenvalues and eigenvectors of the Hamiltonian matrix, then applying a unitary transformation to find the eigenstates of the perturbation. There is a second and often more useful approach, which is to exploit any symmetries which the perturbation $\hat{V}$ may possess. In this case, a symmetry is defined as the existence of an observable $\hat{\mathrm{O}}$ that commutes with $\hat{V}$, so that the symmetry associated with $\hat{\mathrm{O}}$ leaves the perturbation unchanged,

$$e^{i\lambda\hat{\mathrm{O}}}\hat{V}e^{-i\lambda\hat{\mathrm{O}}} = \hat{V}. \tag{8.33}$$

Because $\hat{V}$ and $\hat{\mathrm{O}}$ commute it was shown in (3.194) that the two share a common set of eigenstates. As a result, using the eigenstates of $\hat{\mathrm{O}}$ will diagonalize the perturbation $\hat{V}$.

It is important to begin by noting that $\hat{H}_0$ remains diagonal in the new basis. This follows from the fact that the matrix representation of $\hat{H}_0$ in the unperturbed *degenerate sector* is a multiple of the identity matrix since each of the unperturbed states, by assumption, satisfies

$$H_{0\ell n} = \langle\, E_\ell^{(0)} \mid \hat{H}_0 \mid E_n^{(0)}\, \rangle = \varepsilon\delta_{\ell n} \implies \mathbf{H}_0 = \varepsilon\hat{\mathbf{1}}. \tag{8.34}$$

Since the new eigenstates are obtained from a unitary transformation $\mathbf{U}$ of the $\hat{H}_0$ eigenstates, the action of the $\hat{H}_0$ matrix gives

$$\mathbf{H}_0 \cdot \mathbf{U} = \varepsilon\hat{\mathbf{1}} \cdot \mathbf{U} = \varepsilon\mathbf{U}, \tag{8.35}$$

showing that the eigenstates of $\hat{H}$ will also be eigenstates of $\hat{H}_0$,

$$\hat{H}_0|\, \tilde{E}_\ell^{(0)}\, \rangle = \varepsilon|\, \tilde{E}_\ell^{(0)}\, \rangle. \tag{8.36}$$

In turn, this shows that the eigenvalues of the new states can be written

$$\tilde{E}_\ell^{(0)} = \varepsilon + V_\ell, \tag{8.37}$$

where V_ℓ are the eigenvalues of the $N \times N$ matrix whose elements are

$$V_{\ell n} = \langle\, E_\ell^{(0)} \mid \hat{V} \mid E_n^{(0)}\, \rangle, \tag{8.38}$$

so that

$$\langle\, \tilde{E}_\ell^{(0)} \mid \hat{V} \mid \tilde{E}_n^{(0)}\, \rangle = V_\ell\delta_{\ell n} \tag{8.39}$$

In what follows it will be *assumed* that all the eigenvalues and eigenstates can be found in the degenerate sector, and that the degeneracy in energy is *totally lifted* as a result. If the latter is not the case, then a different set of basis states that lifts the degeneracy must be chosen by diagonalizing the second order matrix appearing on the right-hand side of (8.26). This will not be considered in this text.

The same expansion (8.8) around the eigenstates of $\hat{H}_0$ that gave (8.9), (8.10), and (8.11) is now followed using the replacement (8.32). Result (8.9) is once again automatically satisfied in both the degenerate and nondegenerate sectors of the theory, where the eigenvalue $E_n^{(0)} = \varepsilon$ still appears in the degenerate sector by virtue of (8.36). As before, the first order correction to the energy *in the degenerate sector* is given by

$$E_n^{(1)} = \langle \tilde{E}_n^{(0)} \mid \hat{V} \mid \tilde{E}_n^{(0)} \rangle = V_n, \tag{8.40}$$

which gives the total energy of a state in the degenerate sector to first order as

$$E_n = E_n^{(0)} + E_n^{(1)} = \varepsilon + V_n. \tag{8.41}$$

This is consistent with the matrix result (8.37) for the degenerate sector. Similarly, for $\ell \neq n$ and $\ell, n \leqslant N$, combining (8.39) with (8.19) gives the degenerate sector result

$$a_{\ell n}^{(1)} = \frac{\langle \tilde{E}_\ell^{(0)} \mid \hat{V} \mid \tilde{E}_n^{(0)} \rangle}{\left(\tilde{E}_n^{(0)} - \tilde{E}_\ell^{(0)}\right)} = \frac{V_\ell \delta_{\ell n}}{\left(\tilde{E}_n^{(0)} - \tilde{E}_\ell^{(0)}\right)} = 0, \tag{8.42}$$

since $\ell \neq n$. Noting that normalization requires $a_{nn}^{(1)} = 0$, this shows that $a_{\ell n}^{(1)}$ receives no contribution from the states in the degenerate sector. As a result, the first order correction to states in the degenerate sector is given by

$$n \leqslant N \implies \mid E_n^{(1)} \rangle = \sum_{\ell > N} a_{\ell n}^{(1)} \mid E_\ell^{(0)} \rangle = \sum_{\ell > N} \frac{\langle E_\ell^{(0)} \mid \hat{V} \mid \tilde{E}_n^{(0)} \rangle}{\left(\tilde{E}_n^{(0)} - E_\ell^{(0)}\right)} \mid E_\ell^{(0)} \rangle. \tag{8.43}$$

Because the new basis states already include the effect of the perturbation in the degenerate sector they do not contribute to the sum appearing in (8.43). The correction to a state outside the degenerate sector, $n > N$, breaks into two sums, one from the degenerate sector and one from the nondegenerate sector,

$$n > N \implies \mid E_n^{(1)} \rangle = \sum_{\ell=1}^{N} \frac{\langle \tilde{E}_\ell^{(0)} \mid \hat{V} \mid E_n^{(0)} \rangle}{\left(E_n^{(0)} - \tilde{E}_\ell^{(0)}\right)} \mid \tilde{E}_\ell^{(0)} \rangle + \sum_{\ell > N} \frac{\langle E_\ell^{(0)} \mid \hat{V} \mid E_n^{(0)} \rangle}{\left(E_n^{(0)} - E_\ell^{(0)}\right)} \mid E_\ell^{(0)} \rangle. \tag{8.44}$$

At least to first order, all difficulties defining the perturbation series resulting from the energy degeneracy have been removed by using the eigenstates of the perturbation in the degenerate sector.

8.2 Atomic fine structure

The spin angular momentum of the electron results in the presence of the magnetic moment (7.115) first discussed in relation to the Stern–Gerlach experiment. This magnetic moment will result in an interaction with magnetic fields that are present in the atom or are externally applied. The spin–orbit coupling falls into the former category, while the Zeeman effect falls into the latter. Both of these result in a modification of the energy levels available to the atom, often referred to as the *fine structure of the atom*. Because the hydrogen states possess an energy degeneracy over the orbital and spin quantum numbers, it is necessary to employ the techniques of degenerate perturbation theory developed in the previous section. Understanding these effects begins with a review of angular momentum in the hydrogen atom that includes both the orbital angular momentum and the spin angular momentum of the electron.

8.2.1 Coupled and uncoupled representations for the hydrogen atom

The basic analysis of the hydrogen atom in chapter 5 gave the wave functions associated with the Bohr energy levels, as well as the relationship between the energy level quantum number n and the possible values of the orbital angular momentum quantum number ℓ and the quantum number m_ℓ of the z-projection of orbital angular momentum. The quantum numbers of the electron states are completed by including the electron spin s and its z-component m_s. For simplicity, the time-dependence of the Schrödinger picture states will be suppressed in what follows. The *uncoupled representation* Schrödinger picture states of the hydrogen atom are therefore denoted

$$| \Psi \rangle = | n, \ell, m_\ell, s, m_s \rangle, \tag{8.45}$$

and are eigenstates of orbital angular momentum $\hat{\boldsymbol{L}}$ and spin angular momentum $\hat{\boldsymbol{S}}$ in the following way,

$$\begin{aligned} \hat{L}^2 | n, \ell, m_\ell, s, m_s \rangle &= \ell(\ell + 1)\hbar^2 \, | n, \ell, m_\ell, s, m_s \rangle, \\ \hat{S}^2 | n, \ell, m_\ell, s, m_s \rangle &= s(s + 1)\hbar^2 \, | n, \ell, m_\ell, s, m_s \rangle, \\ \hat{L}_z | n, \ell, m_\ell, s, m_s \rangle &= m_\ell \hbar \, | n, \ell, m_\ell, s, m_s \rangle, \\ \hat{S}_z | n, \ell, m_\ell, s, m_s \rangle &= m_s \hbar \, | n, \ell, m_\ell, s, m_s \rangle, \end{aligned} \tag{8.46}$$

where $s = \frac{1}{2}$ and $m_s = \pm\frac{1}{2}$ for the electron. The quantum number ℓ satisfies $0 \leqslant \ell < n$ while m_ℓ ranges from ℓ to $-\ell$ in integer steps. In matrix notation, the analysis leading to (7.154) showed the electron spin states are represented by the two orthonormal spinors,

$$\xi_{\frac{1}{2}} = \begin{pmatrix} 1 \\ 0 \end{pmatrix}, \quad \xi_{-\frac{1}{2}} = \begin{pmatrix} 0 \\ 1 \end{pmatrix}, \tag{8.47}$$

which are the two eigenspinors of the z-component of the electron spin matrix operator $\hat{S}_z = \frac{1}{2}\hbar\sigma^3$.

The analysis of the previous chapter showed that the orbital and spin angular momenta of the electron, $\hat{\boldsymbol{L}}$ and $\hat{\boldsymbol{S}}$, can be combined into a total angular momentum $\hat{\boldsymbol{J}} = \hat{\boldsymbol{L}} + \hat{\boldsymbol{S}}$. This leads to the *coupled representation* states,

$$| \Psi \rangle = | n, \ell, s, j, m \rangle, \tag{8.48}$$

which satisfy the first two relations of (8.46). However, instead of the second two relations of (8.46) the coupled states satisfy

$$\begin{aligned} \hat{J}^2 | n, \ell, s, j, m \rangle &= j(j + 1)\hbar^2 \, | n, \ell, s, j, m \rangle, \\ \hat{J}_z | n, \ell, s, j, m \rangle &= m\hbar \, | n, \ell, s, j, m \rangle. \end{aligned} \tag{8.49}$$

In terms of the orbital and spin angular momenta quantum numbers, the possible values of j range from $\ell + s$ to $\ell - s$ in integer steps, while m ranges from j to $-j$ in integer steps. Since the spin of the electron is $s = \frac{1}{2}$, the two possible quantum numbers for j are $\ell + \frac{1}{2}$ and $\ell - \frac{1}{2}$. The coupled representation states were expressed as linear combinations of the uncoupled representation states in (7.212) using the Clebsch–Gordan coefficients.

8.2.2 Spin–orbit coupling and the Zeeman interaction

The presence of an intrinsic magnetic field in the atom arises from *relativistic effects*. This can be motivated using a classical argument that considers the electron with mass m_e in a circular orbit around the nucleus in a hydrogen atom with the angular momentum L given by

$$L = m_e v r, \tag{8.50}$$

where v is the velocity and the r is the radius of the orbit. In the *frame of the electron*, the positive point charge q of the nucleus appears to be moving in a circular orbit around the electron with the velocity v at the distance r from the electron. The positive nuclear charge can be written in terms of the electron charge q as $|q|$, and its relative motion corresponds to an electric current I given by

$$I = \frac{|q|}{T}, \tag{8.51}$$

where T is the time for one orbit, given by $T = 2\pi r/v$. This current gives rise to a magnetic field B acting on the electron that points along the same axis as the angular momentum L of the electron and is obtained from the Biot–Savart law (1.64) by treating the current as constant around a circular loop with the circumference $2\pi r$,

$$B = \frac{1}{c}\oint \frac{I\,\mathrm{d}\ell \times \boldsymbol{e}_r}{r^2} = \frac{2\pi I}{rc} = \frac{2\pi|q|}{rcT} = \frac{|q|v}{r^2 c} = \frac{|q|}{m_e r^3 c}(m_e v r) = \frac{|q|}{m_e r^3 c}L. \tag{8.52}$$

Elevating L to a vector gives

$$\boldsymbol{B} = \frac{|q|}{m_e r^3 c}\boldsymbol{L}\,. \tag{8.53}$$

It is important to note that this elementary derivation treats the nucleus as a point charge, and this is the origin of the singularity as $r \to 0$. Allowing the nucleus to be an extended charge distribution yields an expression that gives $\boldsymbol{B} = 0$ if $\boldsymbol{L} = 0$ regardless of the value of r. Combining the magnetic field of (8.53) with the electron spin dipole moment $\hat{\boldsymbol{M}}_s$ given by (7.115) *tentatively* identifies the interaction Hamiltonian,

$$\hat{V} = -\hat{\boldsymbol{M}}_s \cdot \hat{\boldsymbol{B}} = -\frac{q\boldsymbol{S}}{m_e c} \cdot \frac{|q|\boldsymbol{L}}{m_e r^3 c} = \left(\frac{e^2}{m_e^2 r^3 c^2}\right)\hat{\boldsymbol{S}} \cdot \hat{\boldsymbol{L}}, \tag{8.54}$$

where the negative sign of the electron charge was removed by defining $q = -e$.

However, there is a flawed assumption in this elementary derivation, which is that the electron is in an *inertial frame*. The electron is actually in an accelerated frame of reference and this causes the electron reference frame to precess around the axis of rotation, a phenomenon referred to as *Wigner rotation*. This reduces the effective current and thereby the magnetic field in the frame of the electron. The upshot of a more careful analysis using special relativity is that (8.54) must be modified by a factor of $\frac{1}{2}$, known as the *Thomas precession factor*. The corrected version of (8.54) is therefore given by

$$\hat{H}_{\text{so}} = \left(\frac{e^2}{2m_e^2 r^3 c^2}\right)\hat{\boldsymbol{S}} \cdot \hat{\boldsymbol{L}}, \tag{8.55}$$

which is known as the *spin–orbit coupling*. The reader interested in the relativistic derivation should consult the references.

If a *uniform* external magnetic field $\boldsymbol{B}_o$ is applied to an atom, then both the orbital and the spin dipole moment of the electron interact with it. This gives the Zeeman interaction,

$$\hat{H}_{\text{Z}} = -\hat{\boldsymbol{M}}_{\text{L}} \cdot \boldsymbol{B}_o - \hat{\boldsymbol{M}}_{\text{s}} \cdot \boldsymbol{B}_o = \frac{e}{2m_e c}\hat{\boldsymbol{L}} \cdot \boldsymbol{B}_o + \frac{e}{m_e c}\hat{\boldsymbol{S}} \cdot \boldsymbol{B}_o = \frac{e}{2m_e c}\left(\hat{\boldsymbol{L}} + 2\hat{\boldsymbol{S}}\right) \cdot \boldsymbol{B}_o. \tag{8.56}$$

For the case that the z-axis is chosen to be directed along the magnetic field, the Zeeman interaction becomes

$$\hat{H}_{\text{Z}} = \frac{eB_o}{2m_e c}\left(\hat{L}_z + 2\hat{S}_z\right). \tag{8.57}$$

It is useful to define the *Bohr magneton* as

$$\mu_{\text{B}} = \frac{e\hbar}{2m_e c}. \tag{8.58}$$

Inserting the values of $m_e = 9.1 \times 10^{-28}$ g, $e = 4.8 \times 10^{-10}$ esu, $\hbar = 1.0 \times 10^{-27}$ erg s, and $c = 3 \times 10^{10}$ cm s^{-1} shows that Bohr magneton has a value of $\mu_{\text{B}} \approx 9 \times 10^{-21}$ erg Gauss^{-1} $\approx 6 \times 10^{-5}$ eV T^{-1} if the magnetic field is measured in tesla. This allows the Zeeman interaction to be written

$$\hat{H}_{\text{Z}} = \frac{\mu_{\text{B}} B_o}{\hbar}\left(\hat{L}_z + 2\hat{S}_z\right). \tag{8.59}$$

It should be noted that both the spin–orbit and Zeeman terms rely upon relativistic effects for their correct strengths.

Combining the magnetic interaction derived in (5.240) with the spin–orbit and Zeeman interaction terms shows that the spin of the electron results in an altered form of the hydrogen atom Hamiltonian originally given by (5.115). The full Hamiltonian in the center of mass coordinate system is now given by

$$H = \underbrace{\frac{1}{2m_e}\hat{p}^2 - \frac{e^2}{r} + \left(\frac{e^2}{2m_e^2 r^3 c^2}\right)\hat{\boldsymbol{S}} \cdot \hat{\boldsymbol{L}}}_{\text{Internal interactions}} + \underbrace{\frac{eB_o}{2m_e c}\left(\hat{L}_z + 2\hat{S}_z\right) + \frac{1}{8}m_e\left(\frac{eB_o}{m_e c}\right)^2\left(x^2 + y^2\right)}_{\text{External magnetic field interactions}}. \tag{8.60}$$

The terms labeled as *external magnetic field interactions* vanish if $B_o = 0$, while the terms labeled as *internal interactions* do not. In the original analysis of the hydrogen atom, the spin of the electron was ignored and so the spin–orbit term was not considered. This was a good approximation since the spin–orbit term is proportional to $\hbar^2$ due to the product of the two angular momenta, and so can be ignored as small compared to the Coulomb interaction. Considering the relative strengths of the two external terms allows the second term proportional to $x^2 + y^2$ to be discarded even for very strong magnetic fields. This is due to the factor of the Bohr radius squared, $x^2 + y^2 \approx 10^{-17}$ cm^2, and the mass of the electron, $m_e \approx 9 \times 10^{-28}$ gram. Even for a magnetic field with the strength of a tesla, the second term is suppressed by a factor of $\approx 10^{-6}$ compared to the first term. As a result, the Hamiltonian of (8.60) can then be written as the original Coulomb Hamiltonian (5.115), denoted $\hat{H}_0$, with two perturbations, the spin–orbit coupling and the Zeeman term,

$$\hat{H} = \hat{H}_0 + f(r)\,\hat{\boldsymbol{S}} \cdot \hat{\boldsymbol{L}} + \omega(\hat{L}_z + 2\hat{S}_z), \tag{8.61}$$

where

$$\omega = \frac{eB_0}{2m_e c}, \quad f(r) = \frac{e^2}{2m_e^2 r^3 c^2}. \tag{8.62}$$

8.2.3 Diagonalizing the spin–orbit interaction

First order perturbation theory can be applied to the spin–orbit coupling in order to determine the alterations of the Bohr energies associated with the unperturbed Coulomb Hamiltonian $\hat{H}_0$. The original Coulomb interaction states are degenerate, possessing a $2n^2$ degeneracy in terms of the principle quantum number n. As a result, the spin–orbit coupling must be diagonalized. This requires finding the states that are eigenstates of the perturbation.

The spin–orbit coupling is present even when the external field is absent. Setting $\omega = 0$ leaves only the spin–orbit coupling, which is written as

$$\hat{H}_{so} = f(r)\,\hat{\boldsymbol{S}} \cdot \hat{\boldsymbol{L}}. \tag{8.63}$$

This perturbation is not diagonal in the uncoupled representation since the uncoupled representation has eigenstates only of $\hat{L}_z$ and $\hat{S}_z$, while $\hat{\boldsymbol{S}} \cdot \hat{\boldsymbol{L}}$ contains all of the components. This can be remedied by using

$$\hat{J}^2 = (\hat{\boldsymbol{L}} + \hat{\boldsymbol{S}}) \cdot (\hat{\boldsymbol{L}} + \hat{\boldsymbol{S}}) = \hat{L}^2 + \hat{S}^2 + 2\hat{\boldsymbol{S}} \cdot \hat{\boldsymbol{L}}, \tag{8.64}$$

so that

$$\hat{H}_{so} = \frac{1}{2}f(r)\,(\hat{J}^2 - \hat{L}^2 - \hat{S}^2). \tag{8.65}$$

The operators appearing in (8.65) are diagonal in the coupled representation (8.49). This follows from

$$\begin{aligned}\hat{H}_{\rm so}|\, n, \ell, s, j, m \,\rangle &= \frac{1}{2}f(r)\left(\hat{J}^2 - \hat{L}^2 - \hat{S}^2\right)|\, n, \ell, s, j, m \,\rangle \\ &= \frac{1}{2}\hbar^2 f(r)\big(j(j+1) - \ell(\ell+1) - s(s+1)\big)|\, n, \ell, s, j, m \,\rangle.\end{aligned} \tag{8.66}$$

The matrix elements in the coupled representation are therefore given by

$$\begin{aligned}&\langle\, n', \ell', s, j', m' \mid \hat{H}_{\rm so} \mid n, \ell, s, j, m \,\rangle \\ &= \frac{1}{2}\hbar^2\big(j(j+1) - \ell(\ell+1) - s(s+1)\big)\langle\, n', \ell', s, j', m' \mid f(r) \mid n, \ell, s, j, m \,\rangle,\end{aligned} \tag{8.67}$$

where the spin s of the electron is assumed to be the same in both states. The remaining matrix element in (8.67) can be evaluated by using the Clebsch–Gordan expansion (7.212) to write the coupled states in terms of the uncoupled representation,

$$|\, n, \ell, s, j, m \,\rangle = \sum_{m_\ell, m_s} C^{j,m}_{m_\ell,\, m_s}|\, n, \ell, m_\ell, s, m_s \,\rangle. \tag{8.68}$$

In the configuration representation the bound state hydrogen wave functions of (5.202) give

$$\langle\, \boldsymbol{x} \mid n, \ell, s, j, m \,\rangle = \sum_{m_\ell, m_s} C^{j,m}_{m_\ell,\, m_s}\langle\, \boldsymbol{x} \mid n, \ell, m_\ell, s, m_s \,\rangle = \sum_{m_\ell, m_s} C^{j,m}_{m_\ell,\, m_s} R_{n\ell}(r) Y_{\ell m_\ell}(\theta, \phi)\xi_{m_s}, \tag{8.69}$$

where the ξ_{m_s} are the electron eigenspinors defined by (8.47). It follows from the orthonormality of the spherical harmonics and the eigenspinors that

$$\begin{aligned}&\langle\, n', \ell', s, j', m' \mid f(\hat{r}) \mid n, \ell, s, j, m \,\rangle = \int {\rm d}^3 r\, f(r)\langle\, n', \ell', s, j', m' \mid \boldsymbol{x} \,\rangle\langle\, \boldsymbol{x} \mid n, \ell, s, j, m \,\rangle \\ &= \sum_{m_\ell, m_s, m_{\ell'}, m_{s'}} \left(C^{j',m'}_{m_{\ell'},\, m_{s'}}\right)^* C^{j,m}_{m_\ell,\, m_s}\xi^\dagger_{m_{s'}} \cdot \xi_{m_s} \int {\rm d}r\, r^2 f(r) R_{n'\ell'}(r) R_{n\ell}(r) \int {\rm d}\Omega\, Y_{\ell' m_{\ell'}}(\theta, \phi) Y_{\ell m_\ell}(\theta, \phi) \\ &= \sum_{m_\ell, m_s, m_{\ell'}, m_{s'}} \left(C^{j',m'}_{m_{\ell'},\, m_{s'}}\right)^* C^{j,m}_{m_\ell,\, m_s} \int {\rm d}r\, r^2 f(r) R_{n'\ell'}(r) R_{n\ell}(r)\, \delta_{\ell'\ell}\, \delta_{m_{\ell'} m_\ell}\, \delta_{m_{s'} m_s} \\ &= \int {\rm d}r\, r^2 f(r) R_{n'\ell}(r) R_{n\ell}(r)\, \delta_{\ell'\ell}\left(\sum_{m_\ell, m_s}\left(C^{j',m'}_{m_\ell,\, m_s}\right)^* C^{j,m}_{m_\ell,\, m_s}\right) \\ &= \left(\int {\rm d}r\, r^2 f(r) R_{n'\ell}(r) R_{n\ell}(r)\right)\delta_{\ell'\ell}\, \delta_{j'j}\, \delta_{m'm},\end{aligned} \tag{8.70}$$

where the last step used the orthonormality of the Clebsch–Gordan coefficients demonstrated in (7.216).

Result (8.70) shows that the spin–orbit coupling is diagonal in the energetically degenerate angular momentum subspace if the coupled representation is used. Each value of n has had its degeneracy lifted. Setting $n' = n$ in (8.70) and returning it to

(8.67) gives the first order energy correction (8.40) to the nth Bohr energy due to the spin–orbit coupling in the diagonalized subspace,

$$E_n^{(1)} = \big(j(j+1) - \ell(\ell+1) - s(s+1)\big)\hbar^2\left(\int \mathrm{d}r\, r^2 \frac{1}{2} f(r) R_{n\ell}^2(r)\right). \tag{8.71}$$

Using $f(r)$ from (8.62) the quantity $Q_{n\ell}$ is defined as

$$Q_{n\ell} \equiv \left(\int \mathrm{d}r\, r^2 \frac{1}{2} f(r) R_{n\ell}^2(r)\right) = \frac{e^2}{4m_e^2 c^2}\int_0^\infty \frac{\mathrm{d}r}{r} R_{n\ell}^2(r) = \frac{e^2}{4m_e^2 c^2}\left\langle \frac{1}{r^3}\right\rangle. \tag{8.72}$$

Applying the Feynman–Hellmann theorem (6.285) and the virial theorem (6.292) to the radial part of the Coulomb Hamiltonian and its eigenvalues it can be shown that

$$\left\langle \frac{1}{r^3}\right\rangle = \frac{m_e^3 e^6}{\hbar^6 n^3 \ell(\ell + \frac{1}{2})(\ell + 1)}, \tag{8.73}$$

which is a specific case of what is known as *Kramer's relation.* The calculation is lengthy and available in the references. Combining (8.73) with (8.72) gives

$$Q_{n\ell} = \frac{m_e e^8}{4\hbar^6 c^2 n^3 \ell(\ell + \frac{1}{2})(\ell + 1)}. \tag{8.74}$$

Using (8.74) in (8.71) yields the final expression for the fine structure modification of the Bohr energies brought about by the spin–orbit coupling,

$$E_n^{(1)} = \big(j(j+1) - \ell(\ell+1) - s(s+1)\big)\hbar^2 Q_{n\ell} = \frac{m_e e^8 \big(j(j+1) - \ell(\ell+1) - s(s+1)\big)}{4\hbar^4 c^2 n^3 \ell(\ell + \frac{1}{2})(\ell + 1)}. \tag{8.75}$$

Expression (8.75) appears to be singular for $\ell = 0$. However, the spin quantum is $s = \frac{1}{2}$, so that setting $j = \ell + \frac{1}{2}$ and taking the limit $\ell \to 0$ gives the finite result

$$E_n^{(1)} = \frac{m_e e^8}{2\hbar^4 c^2 n^3}. \tag{8.76}$$

Since the expectation values of $\boldsymbol{L}^2$ and $\boldsymbol{L}_z$ vanish for $\ell = 0$, this contribution should be $Q_{n0} = 0$. As mentioned earlier, the origin of this problem is the treatment of the nucleus as a point charge. This was explicit in the assumptions made in applying the Biot–Savart formula in the elementary derivation of the internal magnetic field (8.53). The $1/r^3$ term resulting from the Coulomb potential for a point charge can be replaced by

$$\frac{e^2}{r^3} \to -\frac{1}{r}\frac{\partial V(r)}{\partial r}, \tag{8.77}$$

in such a way that that the potential is spatially distributed and is not singular as $r \to 0$. For such a case, it follows that $Q_{n0} = 0$. As a result, in what follows it will be assumed that $Q_{n0} = 0$. The interested reader is recommended to consult the references.

Since this is being applied to the hydrogen atom it follows that $s = \frac{1}{2}$ and the possible values of j are $j = \ell + s = \ell + \frac{1}{2}$ and $j = \ell - s = \ell - \frac{1}{2}$. The first case gives

$$E_n^{(1)} = \ell\hbar^2 Q_{n\ell}, \tag{8.78}$$

while for the second case

$$E_n^{(1)} = -(\ell + 1)\hbar^2 Q_{n\ell}. \tag{8.79}$$

As an example, the $n = 2$ energy level corrections can have either $\ell = 0$ and $\ell = 1$, and this gives the four corrections,

$$E_n^{(1)} = \{-2\hbar^2 Q_{21},\ 0,\ 0,\ \hbar^2 Q_{21}\}, \tag{8.80}$$

where Q_{20} was set to zero. This shows that the spin–orbit coupling splits the $\ell = 1$ state into three states separated in total energy by $\Delta E = 3\hbar^2 Q_{21}$. Result (8.74) gives the magnitude of the spin–orbit coupling for the second Bohr energy level. Inserting the values of $m_e = 9.1 \times 10^{-28}$ g, $e = 4.8 \times 10^{-10}$ esu, $\hbar = 1.05 \times 10^{-27}$ erg s, and $c = 3.0 \times 10^{10}$ cm s^{-1} for the case $n = 2$ and $\ell = 1$ gives

$$\Delta E = 3\hbar^2 Q_{21} \approx 7.3 \times 10^{-17}\,\text{erg} \approx 4.5 \times 10^{-5}\,\text{eV}. \tag{8.81}$$

While this correction is significantly smaller than the second Bohr energy level, it is sufficient to create an observable change in the frequency of emission by the electron transition from E_2 to E_1.

8.2.4 The Zeeman effect

If an external magnetic field is applied to the hydrogen, then in addition to the spin–orbit coupling there is the Zeeman interaction $\hat{H}_Z$ given by (8.59). The Zeeman term is comparable in strength to the spin–orbit coupling for magnetic fields less than 0.1 tesla. For stronger field strengths, the Zeeman interaction dominates and a different analysis will be required.

Because the spin–orbit coupling is diagonalized in the coupled representation, finding the contribution of the Zeeman interaction to the first order energy correction for the case of a weak field requires evaluating its matrix elements in the coupled representation. The Zeeman interaction can then be written

$$H_Z = \omega(\hat{L}_z + 2\hat{S}_z) = \omega(\hat{J}_z + \hat{S}_z), \tag{8.82}$$

where $\omega = \mu_B B_o/\hbar$. The operator $\omega\hat{J}_z$ is already diagonal in the coupled representation, so that its contribution to the energy correction is given by

$$\langle\, n, \ell, s, j, m \mid \omega\hat{J}_z \mid n, \ell, s, j, m \,\rangle = m\hbar\omega. \tag{8.83}$$

This is independent of the value for j as along as $-j \leqslant m \leqslant j$.

However, the operator $\omega\hat{S}_z$ is not diagonal. Evaluating the contribution of the electron spin operator requires writing the relevant coupled representation states in terms of the uncoupled representation states. Because the electron is spin-$\frac{1}{2}$, this uses the Clebsch–Gordan coefficients of (7.239) for the two possible total angular momentum eigenvalues $j = \ell + \frac{1}{2}$ and $j = \ell - \frac{1}{2}$. Using the coefficients and the rule that $m_\ell = m - m_s$, the states are given by

$$
\begin{aligned}
| n, \ell, s, j = \ell - \frac{1}{2}, m \rangle &= -\sqrt{\frac{\ell + \frac{1}{2} - m}{2\ell + 1}} \, | m_\ell = m - \frac{1}{2}, \frac{1}{2} \rangle + \sqrt{\frac{\ell + \frac{1}{2} + m}{2\ell + 1}} \, | m_\ell = m + \frac{1}{2}, -\frac{1}{2} \rangle \\
| n, \ell, s, j = \ell + \frac{1}{2}, m \rangle &= \sqrt{\frac{\ell + \frac{1}{2} + m}{2\ell + 1}} \, | m_\ell = m - \frac{1}{2}, \frac{1}{2} \rangle + \sqrt{\frac{\ell + \frac{1}{2} - m}{2\ell + 1}} \, | m_\ell = m + \frac{1}{2}, -\frac{1}{2} \rangle .
\end{aligned}
\tag{8.84}
$$

For the first case that $j = \ell - \frac{1}{2}$, the values of m range from $-\ell + \frac{1}{2}$ to $\ell - \frac{1}{2}$, while for the second case that $j = \ell + \frac{1}{2}$, the values of m range from $-\ell - \frac{1}{2}$ to $\ell + \frac{1}{2}$. It should be noted that the two states of (8.84) are orthonormal.

Using the states of (8.84) and the orthonormality of the uncoupled representation states gives the following matrix elements for $\omega\hat{S}_z$,

$$
\begin{aligned}
\langle n, \ell, s, \ell - \frac{1}{2}, m \,|\, \omega\hat{S}_z \,|\, n, \ell, s, \ell - \frac{1}{2}, m \rangle &= \frac{1}{2}\hbar\omega \left(\frac{\ell + \frac{1}{2} - m}{2\ell + 1} \right) - \frac{1}{2}\hbar\omega \left(\frac{\ell + \frac{1}{2} + m}{2\ell + 1} \right) = -\frac{m\hbar\omega}{2\ell + 1}, \\
\langle n, \ell, s, \ell + \frac{1}{2}, m \,|\, \omega\hat{S}_z \,|\, n, \ell, s, \ell + \frac{1}{2}, m \rangle &= \frac{1}{2}\hbar\omega \left(\frac{\ell + \frac{1}{2} + m}{2\ell + 1} \right) - \frac{1}{2}\hbar\omega \left(\frac{\ell + \frac{1}{2} - m}{2\ell + 1} \right) = \frac{m\hbar\omega}{2\ell + 1}.
\end{aligned}
\tag{8.85}
$$

Combining (8.83) and (8.85) the contribution of the Zeeman is given by

$$
E_Z^{(1)} = m\hbar\omega \pm \frac{m\hbar\omega}{2\ell + 1} = m\hbar\omega \left(\frac{2\ell + 1 \pm 1}{2\ell + 1} \right) = m\hbar\omega \left(\frac{2j + 1}{2\ell + 1} \right), \tag{8.86}
$$

where the two values of (8.85) occur since the allowed values of j are $j = \ell \pm \frac{1}{2}$.

The spin–orbit coupling and the Zeeman interaction result in the following total correction to the Bohr energy E_n,

$$
E_n^{(1)} = \hbar^2 Q_{n\ell} \big(j(j + 1) - \ell(\ell + 1) - s(s + 1) \big) + m\hbar\omega \left(\frac{2j + 1}{2\ell + 1} \right), \tag{8.87}
$$

where $Q_{n\ell}$ is given by (8.74) for $\ell \geqslant 1$. Result (8.87) is referred to as the *anomalous Zeeman effect* since its observation predated the concept of electron spin, which is crucial to understanding the spin–orbit coupling as well as the strength of the Zeeman contribution.

As an example, in the case of the second Bohr energy level E_2, the allowed value of j for $\ell = 0$ is $j = \frac{1}{2}$, so that the allowed values of m are $\frac{1}{2}$ and $-\frac{1}{2}$. The corrections to the Bohr energy are

$$E_2^{(1)} = 2\,m\hbar\omega = \pm\hbar\omega. \tag{8.88}$$

Similarly, the allowed values for j for the case that $\ell = 1$ are $j = 3/2$ and $j = 1/2$. For the $\ell = 1$ and $j = 3/2$ case, the allowed values of m are 3/2, 1/2, −1/2, and −3/2. The corrections to the Bohr energy are

$$E_2^{(1)} = \hbar^2 Q_{21} + \frac{4}{3}m\hbar\omega = \hbar^2 Q_{21} + \begin{cases} \pm\,2\hbar\omega & m = \pm 3/2 \\ \pm\,\dfrac{2}{3}\hbar\omega & m = \pm 1/2 \end{cases}. \tag{8.89}$$

Finally, for $j = 1/2$ and $\ell = 1$ the allowed values of m are 1/2 and −1/2. The corrections to the Bohr energy are

$$E_2^{(1)} = -2\hbar^2 Q_{21} + \frac{2}{3}m\hbar\omega = -2\hbar^2 Q_{21} \pm \frac{1}{3}\hbar\omega. \tag{8.90}$$

The $n = 2$ energy level contains eight states that are initially energetically degenerate. The combination of the spin–orbit and Zeeman coupling has completely lifted this degeneracy by providing eight distinct corrections to the Bohr energy level. This is depicted in figure 8.1. As a result, transitions from the second Bohr level to the ground state will exhibit eight distinct wavelengths for the emitted light in the presence of an external magnetic field.

In the event that a strong magnetic field is applied, the Zeeman term will dominate over the spin–orbit coupling. The Zeeman term is diagonal in the uncoupled representation, where

$$\langle\, n, \ell, m_\ell, s, m_s \mid \hat{H}_Z \mid n, \ell, m_\ell, s, m_s \,\rangle = \langle\, m_\ell, m_s \mid \omega(\hat{L}_z + 2\hat{S}_z) \mid m_\ell, m_s \,\rangle = \hbar\omega(m_\ell + 2m_s). \tag{8.91}$$

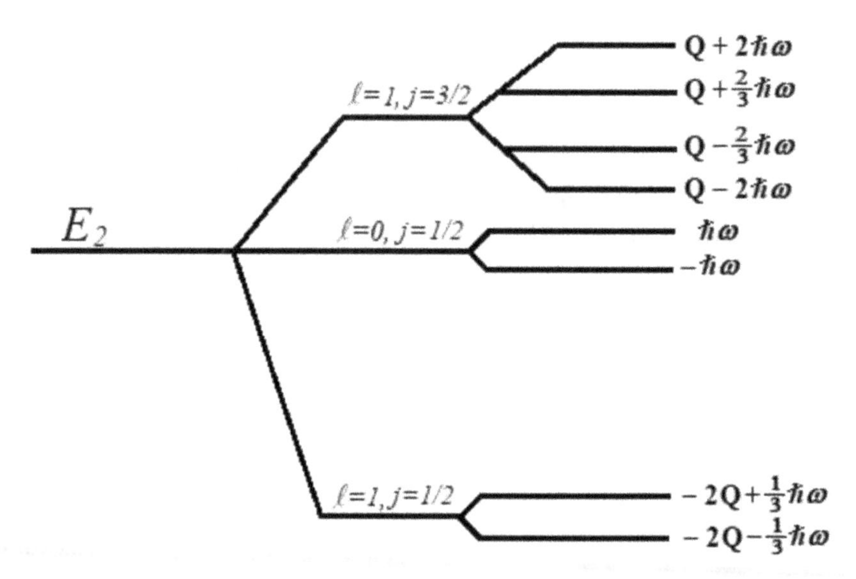

Figure 8.1. The first order corrections to the Bohr energy E_2 for a weak magnetic field.

Evaluating the spin–orbit coupling in the uncoupled state is facilitated by writing

$$\hat{\boldsymbol{S}} \cdot \hat{\boldsymbol{L}} = \hat{S}_x\hat{L}_x + \hat{S}_y\hat{L}_y + \hat{S}_z\hat{L}_z = \frac{1}{2}(\hat{S}_+\hat{L}_- + \hat{S}_-\hat{L}_+) + \hat{S}_z\hat{L}_z, \tag{8.92}$$

where $\hat{S}_\pm$ and $\hat{L}_\pm$ are the raising and lowering operators defined in (7.69). The action of the raising and lowering operators demonstrated in (7.89) and (7.92) coupled with the orthogonality of the uncoupled representation states shows that

$$\langle\, n, \ell, m_\ell, s, m_s \mid \hat{L}_\pm\hat{S}_\mp \mid n, \ell, m_\ell, s, m_s \,\rangle = 0. \tag{8.93}$$

The spin–orbit coupling therefore has the matrix element

$$\langle\, n, \ell, m_\ell, s, m_s \,|\, f(r)\hat{S} \cdot \hat{L} \,|\, n, \ell, m_\ell, s, m_s \,\rangle = \langle\, m_\ell, m_s \,|\, \hat{L}_z\hat{S}_z \,|\, m_\ell, m_s \,\rangle \int \mathrm{d}r\, r^2 f(r) = 2\hbar^2 Q_{n\ell} m_\ell m_s, \tag{8.94}$$

where $Q_{n\ell}$ is given by (8.74) for $\ell \geqslant 1$. The combination of Zeeman interaction (8.91) and the spin–orbit coupling (8.94) gives the strong magnetic field result

$$E_n^{(1)} = 2\hbar^2 Q_{n\ell} m_\ell m_s + \hbar\omega\big(m_\ell + 2m_s\big), \tag{8.95}$$

where it is assumed that the value of $Q_{n\ell}$ is smaller than the Zeeman contribution.

Again using E_2 as the example and denoting $\hbar^2 Q_{21} = Q$, the energy corrections are evaluated using (8.95). For $\ell = 1$ there are two combinations, either $m_\ell = 1$ and $m_s = -\frac{1}{2}$ or $m_\ell = -1$ and $m_s = \frac{1}{2}$, which give $E_2^{(1)} = -Q$. For $\ell = 0$ and $\ell = 1$ there are two combinations with $m_\ell = 0$ and $m_s = \pm\frac{1}{2}$ that give $E_2^{(1)} = \pm\hbar\omega$. For $\ell = 1$, the combination $m_\ell = 1$ and $m_s = \frac{1}{2}$ gives $E_2^{(1)} = Q + 2\hbar\omega$, while the combination $m_\ell = -1$ and $m_s = -\frac{1}{2}$ gives $E_2^{(1)} = Q - 2\hbar\omega$. As a result, there are only five distinct energy corrections in the presence of a strong magnetic field. This is depicted in figure 8.2 and is referred to as the *regular Zeeman effect*. Of the eight original energy states, there are three pairs of states which remain energetically degenerate, while the

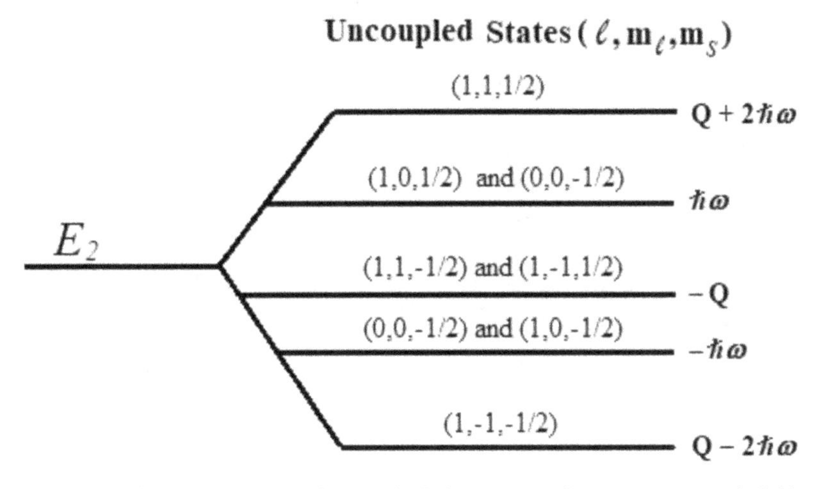

Figure 8.2. The first order corrections to the Bohr energy E_2 for a strong magnetic field.

remaining two states have had their degeneracy completely lifted. This occurs because the strong magnetic field suppresses the strength of the spin–orbit coupling to be $\pm Q$, which does not allow the energy degeneracy to be completely lifted. In that regard, it important to note that the relative strengths of the two interactions dictated the representation used in evaluating the energy corrections.

8.3 Time-dependent perturbation theory

The stationary perturbation theory developed and applied in the first part of this chapter assumes that the perturbation is time-independent. The goal of stationary perturbation theory is to construct the eigenvalues and stationary eigenstates of the full Hamiltonian.

There is a second important physical situation, which is the case that a time-dependent perturbation is applied to the system. This could include an oscillatory electromagnetic field or a sudden temporal change in the parameters of the system, such as the size of a confining potential well. Time-dependent perturbation theory treats the original system prior to the perturbation as providing the eigenstates and energies to describe the system at later times and calculates the probability that the system will transit from its original eigenstate to a new eigenstate of the original system. This is brought about by the fact that the time-dependent perturbation represents the flow of energy into or out of the system, and therefore changes the state of the original system. It assumes that the system in the presence of the perturbation can still be described in terms of the original eigenstates and eigenvalues prior to the perturbation.

8.3.1 Basic time-dependent perturbation theory

The time-dependent Hamiltonian that describes the system is given by

$$\hat{H}(t) = \hat{H}_0 + \lambda \hat{V}(t), \tag{8.96}$$

where $\hat{H}_0$ is a time-independent Hermitian Hamiltonian whose orthonormal eigenstates $|\, E_n^{(0)}\,\rangle$ at $t = 0$ are known,

$$\hat{H}_0|\, E_n^0\,\rangle = E_n^0|\, E_n^0\,\rangle, \tag{8.97}$$

and are complete,

$$\sum_n |\, E_n^0\,\rangle\langle\, E_n^0\,| = \hat{\mathbf{1}}. \tag{8.98}$$

The time evolution of the eigenstates is governed by $\hat{H}_0$, so that statement (8.98) holds at all times,

$$\begin{aligned}\sum_n |\, E_n^0,\, t\,\rangle\langle\, E_n^0,\, t\,| &= \sum_n e^{-i\hat{H}_0 t/\hbar}|\, E_n^0\,\rangle\langle\, E_n^0\,|e^{i\hat{H}_0 t/\hbar}\\ &= \sum_n e^{-iE_n^0 t/\hbar}|\, E_n^0\,\rangle\langle\, E_n^0\,|e^{iE_n^0 t/\hbar} = \sum_n |\, E_n^0\,\rangle\langle\, E_n^0\,| = \hat{\mathbf{1}}.\end{aligned} \tag{8.99}$$

The goal is to find the time-dependent solutions of the Schrödinger equation,

$$\hat{H}(t)|\,\Psi,\,t\,\rangle = \left(\hat{H}_0 + \lambda\hat{V}(t)\right)|\,\Psi,\,t\,\rangle = i\hbar\frac{\partial}{\partial t}|\,\Psi,\,t\,\rangle. \tag{8.100}$$

Using the completeness (8.99) it follows that

$$|\,\Psi,\,t\,\rangle = \sum_n \langle\, E_n^0,\,t\,|\,\Psi,\,t\,\rangle|\,E_n^0,\,t\,\rangle = \sum_n c_n(t)\,|\,E_n^0,\,t\,\rangle, \tag{8.101}$$

where the coefficients in the sum are given by

$$c_n(t) = \langle\, E_n^0,\,t\,|\,\Psi,\,t\,\rangle. \tag{8.102}$$

The coefficients $c_n(t)$ can be understood as the probability amplitude for the state to be in the nth energy eigenstate at the time t. This follows since

$$\sum_n |c_n(t)|^2 = \sum_n \langle\,\Psi,\,t\,|\,E_n^0,\,t\,\rangle\langle\,E_n^0,\,t\,|\,\Psi,\,t\,\rangle = \langle\,\Psi,\,t\,|\,\Psi,\,t\,\rangle = 1, \tag{8.103}$$

so that normalization of probability is satisfied. Substituting (8.101) into the Schrödinger equation (8.100) gives

$$\hat{H}(t)|\,\Psi,\,t\,\rangle = \sum_n c_n(t)\left(\hat{H}_0 + \lambda\hat{V}(t)\right)|\,E_n^0,\,t\,\rangle = \sum_n c_n(t)\left(E_n^0 + \lambda\hat{V}(t)\right)e^{-iE_n^0 t/\hbar}|\,E_n^0\,\rangle, \tag{8.104}$$

while

$$i\hbar\frac{\partial}{\partial t}|\,\Psi,\,t\,\rangle = \sum_n\left(i\hbar\frac{\partial c_n(t)}{\partial t} + E_n^0\right)e^{-iE_n^0 t/\hbar}|\,E_n^0\,\rangle. \tag{8.105}$$

Equating (8.104) and (8.105) gives

$$\sum_n i\hbar\frac{\partial c_n(t)}{\partial t}e^{-iE_n^0 t/\hbar}|\,E_n^0\,\rangle = \sum_n c_n(t)\,\lambda\hat{V}(t)e^{-iE_n^0 t/\hbar}|\,E_n^0\,\rangle, \tag{8.106}$$

which is exactly equivalent to the Schrödinger equation. Using the assumed orthonormality of the $\hat{H}_0$ eigenstates shows that the inner product with $\langle\, E_{\mathrm{f}}^0\,|$ gives

$$i\hbar\frac{\partial c_f(t)}{\partial t} = \lambda\sum_n c_n(t)\,e^{i(E_{\mathrm{f}}^0 - E_n^0)t/\hbar}\langle\,E_{\mathrm{f}}^0\,|\,\hat{V}(t)\,|\,E_n^0\,\rangle. \tag{8.107}$$

The next step is to expand $c_f(t)$ in a power series in λ,

$$c_f(t) = \sum_{\ell=0}^{\infty}\lambda^\ell c_f^{(\ell)}(t). \tag{8.108}$$

Inserting (8.108) into (8.107) and equating terms with equal powers of λ gives an infinite sequence of first order differential equations,

$$
\begin{aligned}
&\text{zeroth order} && i\hbar\frac{\partial}{\partial t}c_f^{(0)}(t) = 0, \\
&\text{first order} && i\hbar\frac{\partial}{\partial t}c_f^{(1)}(t) = \sum_n c_n^{(0)}(t)\, e^{i\omega_{fn}t}\langle\, E_{\mathrm{f}}^0 \mid \hat{V}(t) \mid E_n^0\, \rangle, \\
&\dots && \\
&k\text{th order} && i\hbar\frac{\partial}{\partial t}c_f^{(k)}(t) = \sum_n c_n^{(k-1)}(t)\, e^{i\omega_{fn}t}\langle\, E_{\mathrm{f}}^0 \mid \hat{V}(t) \mid E_n^0\, \rangle,
\end{aligned}
\tag{8.109}
$$

where

$$
\omega_{fn} = \frac{1}{\hbar}(E_{\mathrm{f}}^0 - E_n^0). \tag{8.110}
$$

It should be noted that $\lambda = 0$ results in only the zeroth order equation.

Solving the sequence (8.109) of first order equations requires an initial condition. To simplify finding a solution, it will be assumed that the perturbation is switched on at $t = 0$ and that the system in the *initial eigenstate* $\mid E_{\mathrm{i}}^0\, \rangle$ at that moment. This initial condition is represented by

$$
c_f^{(0)}(t = 0) = \delta_{f\mathrm{i}}. \tag{8.111}
$$

The solution to the zeroth order equation consistent with this initial condition is then given by

$$
c_f^{(0)}(t) = \delta_{f\mathrm{i}}, \tag{8.112}
$$

so that $c_f^{(0)}$ remains constant. Substituting this result into the first order equation gives

$$
\frac{\partial}{\partial t}c_f^{(1)}(t) = -\frac{i}{\hbar}\sum_n \delta_{n\mathrm{i}} e^{i\omega_{fn}t}\langle\, E_{\mathrm{f}}^0 \mid \hat{V}(t) \mid E_n^0\, \rangle = -\frac{i}{\hbar}e^{i\omega_{f\mathrm{i}}t}\langle\, E_{\mathrm{f}}^0 \mid \hat{V}(t) \mid E_{\mathrm{i}}^0\, \rangle. \tag{8.113}
$$

The initial conditions for the higher order corrections are found from (8.103) and the choice that the perturbation vanishes for $t \leqslant 0$, which requires that

$$
k > 0 \implies c_f^{(k)}(t = 0) = 0. \tag{8.114}
$$

The first order equation (8.113) can now be solved by integration. Using the boundary condition (8.114) gives

$$
c_f^{(1)}(t) = \int_0^t \mathrm{d}\tau\, \frac{\partial}{\partial \tau}c_f^{(1)}(\tau) = -\frac{i}{\hbar}\int_0^t \mathrm{d}\tau\, e^{i\omega_{f\mathrm{i}}\tau}\langle\, E_{\mathrm{f}}^0 \mid \hat{V}(\tau) \mid E_{\mathrm{i}}^0\, \rangle. \tag{8.115}
$$

The result (8.115) gives the first order probability amplitude for the transition from the state $\mid E_{\mathrm{i}}^0\, \rangle$ to the state $\mid E_{\mathrm{f}}^0\, \rangle$ in the time t,

$$
c_f(t) = c_f^{(0)}(t) + \lambda c_f^{(1)}(t) = \delta_{f\mathrm{i}} + \lambda c_f^{(1)}(t). \tag{8.116}
$$

From the assumptions that were made to derive it, such as ignoring terms higher order in λ, this amplitude is valid only if it yields a small probability.

8.3.2 The time-averaged transition probability

In a very simple case, the perturbation takes the form

$$\hat{V}(t) = \lambda \hat{V}\theta(t), \tag{8.117}$$

where $\hat{V}$ is a time-independent perturbation that is switched on at the time $t = 0$ by virtue of the step function appearing in (8.117). For such a simple case, the transition amplitude to the state $| E_j^0 \rangle$, where $j \neq i$, is given to first order by

$$\lambda c_f^{(1)}(t) = -\frac{i\lambda}{\hbar}\langle E_\text{f}^0 | \hat{V} | E_\text{i}^0 \rangle \int_0^t \text{d}\tau\, e^{i\omega_{fi}\tau} = -\frac{2i\lambda}{\hbar\omega_{fi}}\langle E_j^0 | \hat{V} | E_\text{i}^0 \rangle e^{\frac{1}{2}i\omega_{fi}t} \sin\left(\frac{1}{2}\omega_{fi}t\right). \tag{8.118}$$

The transition probability $P_{i\to f}$ is given by

$$P_{i\to f}(t) = \lambda^2 |c_f^{(1)}(t)|^2 = \frac{4\lambda^2}{\hbar^2} |\langle E_\text{f}^0 | \hat{V} | E_\text{i}^0 \rangle|^2 \left(\frac{\sin^2\left(\frac{1}{2}\omega_{fi}t\right)}{\omega_{fi}^2} \right). \tag{8.119}$$

In order to be consistent with first order perturbation theory, this probability should be small. The probability (8.119) is oscillatory, and so it is useful to average the probability over a long time interval T. Using (2.17) this is given by

$$\begin{aligned} \bar{P}_{i\to f} &= \lim_{T\to\infty} \frac{1}{T}\int_0^T \text{d}t\, P_{i\to f}(t) = \frac{4\lambda^2}{\hbar^2\omega_{fi}^2} |\langle E_\text{f}^0 | \hat{V} | E_\text{i}^0 \rangle|^2 \lim_{T\to\infty}\int_0^T \text{d}t\, \sin^2\left(\frac{1}{2}\omega_{fi}t\right) \\ &= \frac{2\lambda^2}{\hbar^2\omega_{fi}^2} |\langle E_\text{f}^0 | \hat{V} | E_\text{i}^0 \rangle|^2. \end{aligned} \tag{8.120}$$

This is understood as the probability of finding the particle in the $| E_\text{f}^0 \rangle$ state after a long period of time has elapsed subsequent to the perturbation being switched on.

An application using the simple harmonic oscillator illustrates several aspects of this result. The quartic interaction (8.16) is switched on at $t = 0$ when the oscillator is in its ground state. It is straightforward to find the transition probability into the next two higher energy levels induced by the interaction. Using the normalized eigenfunctions of (5.92) the transition element to the first excited state is given by

$$\langle E_1^0 | \hat{V} | E_0^0 \rangle = \langle E_1^0 | \frac{1}{4}\hat{Q}^4 | E_0^0 \rangle = \frac{1}{4}\int_{-\infty}^{\infty} \text{d}x\, \psi_1^*(x)\, x^4\, \psi_0(x) = \frac{\alpha^2}{4\sqrt{2\pi}}\int_{-\infty}^{\infty} \text{d}x\, x^5 e^{-\alpha^2 x^2} = 0. \tag{8.121}$$

This transition is suppressed since the first excited eigenfunction has the property $\psi_1(-x) = -\psi_1(x)$, while the ground state eigenfunction has the property $\psi_0(-x) = \psi_0(x)$. Because the perturbation has the property that $\hat{V}(-\hat{Q}) = \hat{V}(\hat{Q})$, only transitions to states with the property $\psi_n(-x) = \psi_n(x)$ will be *allowed*. This is an example of a *selection rule* determined from the properties of the perturbation. The transition element to the second excited state is found identically, and is given by

$$\langle E_2^0 \mid \hat{V} \mid E_0^0 \rangle = \frac{1}{4}\int_{-\infty}^{\infty} \mathrm{d}x\, \psi_2^*(x)\, x^4\, \psi_0(x) = \frac{1}{4}\sqrt{\frac{\alpha^2}{2\pi}}\int_{-\infty}^{\infty} \mathrm{d}x\, x^4(2\alpha^2x^2 - 1)e^{-\alpha^2x^2} = \frac{3\sqrt{2}}{8\alpha^4}. \quad (8.122)$$

Combining (8.122) with (8.120) gives the first order time-averaged probability of a transition from the ground state to the second excited state caused by the perturbation (8.16),

$$\bar{P}_{0\to2} = \frac{9\lambda^2\hbar^2}{64\, m^4\omega^6}, \quad (8.123)$$

where $\alpha = \sqrt{m\omega/\hbar}$ and $\omega_{20} = 2\omega$ was used. Result (8.123) is small as long as

$$\lambda \ll \frac{8m^2\omega^3}{3\hbar}. \quad (8.124)$$

For a typical atomic oscillator $\omega \approx 10^{10}$ Hz and $m \approx 10^{-25}$ kg, so that $\lambda \approx 0.01$ eV nm^{-4} satisfies (8.124).

8.3.3 The density of states and Fermi's golden rule

It is important to note that the last factor in (8.119), when viewed as a function of ω_{fi}, is strongly peaked around $\omega_{fi} \approx 0$. The small argument form $\sin x \approx x$ shows that this factor has a central peak of $t^2/4$ at $\omega_{fi} \to 0$, and this peak lies between the two zeroes of the sine function that occur at

$$-\frac{2\pi}{t} \leqslant \omega_{fi} \leqslant \frac{2\pi}{t}. \quad (8.125)$$

This factor is displayed in figure 8.3, showing that the central peak grows taller and more narrow, sharing the behavior of the sinc squared version (3.91) of the Dirac delta. Using the definition of ω_{fi} (8.110) shows that the largest contribution to the probability (8.119) occurs for the range of final energies given by

$$E_\mathrm{i}^0 - \frac{2\pi\hbar}{t} \leqslant E_\mathrm{f}^0 \leqslant E_\mathrm{i}^0 + \frac{2\pi\hbar}{t}. \quad (8.126)$$

For the case that $\omega_{fi} = 0$, this shows that the transition probability (8.119) is dominated by the group of final states which have the energies $E_\mathrm{f} = E_\mathrm{i} \pm \Delta E$, where $\Delta E = 2\pi\hbar/t$. It is useful to note that the relevant area under the function $\sin^2(\frac{1}{2}\omega_{fi}t)/\omega_{fi}^2$ can be approximated as a triangle with height $t^4/4$ and base $4\pi/t$, which gives

$$\text{Area} = \frac{1}{2}\left(\frac{4\pi}{t}\right)\left(\frac{t^2}{4}\right) = \frac{1}{2}\pi t, \quad (8.127)$$

showing that it grows linearly with time.

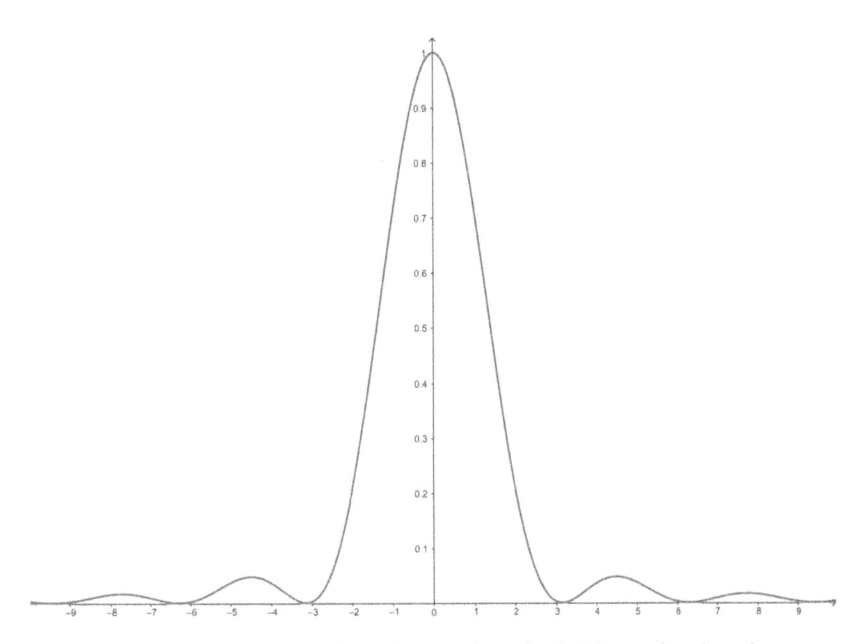

Figure 8.3. The sine squared factor in parentheses in (8.119) as a function of ω_{fi}.

However, if the energy eigenstates are discrete, then as time passes this interval of energies will cease to include any eigenstates except the original state $| E_{\mathrm{i}}^0 \rangle$. As a result, eventually there will be no transition to a different discrete state in the central peak range. There is still a probability of transition to a state outside this region of energies, as demonstrated by (8.123). However, in the case of a set of continous energy eigenstates, there can be transitions to other states that are energetically degenerate with the original state and therefore lie in the central peak range. For example, in the case of a free particle it is possible to transition to another momentum eigenstate with the same magnitude of momentum but directed differently. This is the mechanism of elastic scattering in quantum mechanics, and is discussed in the next chapter. For such a case, the transition probability *from the initial state to an energetically degenerate state* is dominated by the contribution of the central peak,

$$P = \sum_f P_{i\to f}(t) = \frac{4\lambda^2}{\hbar^2} \sum_f |\langle E_{\mathrm{f}}^0 | \hat{V} | E_{\mathrm{i}}^0 \rangle|^2 \frac{\sin^2\left(\frac{1}{2}\omega_{fi}t\right)}{\omega_{fi}^2}, \tag{8.128}$$

where the sum is over the set of energy states that satisfy (8.126), so that $E_{\mathrm{f}}^0 = E_{\mathrm{i}}^0 \pm \Delta E$, where $\Delta E = 2\pi\hbar/t$.

It is convenient to convert this sum into an integral by introducing *the density of states* $\rho(E_n^0)$, which gives the number of states within the range of energies $\mathrm{d}E_n^0$. Of course, the density of states depends on the details of the system under consideration, but it is typically straightforward to find. This process begins by finding a function of energy $f(E_n^0)$ that gives the *total* number of states with energy less than or equal to E_n^0. The density of states available in the range $\mathrm{d}E_k^0$ is then given by

$$\rho(E_n^0) = \frac{\partial f(E_n^0)}{\partial E_n^0}. \tag{8.129}$$

It follows from its definition that

$$\int_{E_j^0}^{E_k^0} \mathrm{d}E_n^0\, \rho(E_n^0) = \int_{E_j^0}^{E_k^0} \mathrm{d}E_n^0\, \frac{\partial f(E_n^0)}{\partial E_n^0} = f(E_k^0) - f(E_j^0), \tag{8.130}$$

which gives the number of final states in the energy interval E_j^0 to E_k^0.

Finding the function $f(E_n^0)$ is particularly straightforward if the energy states are characterized by a single quantum number n which corresponds to the number of states above the ground state energy. As an example, a particle in a one-dimensional infinite well has the energy levels given by (4.89),

$$E_n^0 = \frac{n^2\pi^2\hbar^2}{2\,mL^2} = n^2E_0 \implies n = \sqrt{\frac{E_n^0}{E_0}} = f(E_n^0). \tag{8.131}$$

The density of states for the one-dimensional infinite well is therefore given by

$$\rho(E_n) = \frac{\partial f(E_n^0)}{\partial E_n^0} = \frac{1}{2\sqrt{E_n^0E_0}}. \tag{8.132}$$

A similar treatment can be given to the one-dimensional simple harmonic oscillator, where (5.54) gives

$$E_n^0 = (n + \frac{1}{2})\hbar\omega \implies n = \frac{E_n^0}{\hbar\omega} + \frac{1}{2} = f(E_n^0), \tag{8.133}$$

which gives the density of states for the one-dimensional harmonic oscillator,

$$\rho(E_n^0) = \frac{\partial f(E_n^0)}{\partial E_n^0} = \frac{1}{\hbar\omega}, \tag{8.134}$$

showing that it is independent of the energy.

This process is somewhat more complicated for the hydrogen atom, where (5.205) shows that each energy level E_j has $2j^2$ energetically degenerate states when the spin is included. Using the energy eigenvalue $E_n^0 = E_0/n^2$, where $E_0 < 0$, shows that the total number of states N for the hydrogen atom from the ground state to E_n is given by

$$N = \sum_{j=1}^{n} 2j^2 = \frac{2}{3}\left(n^3 + \frac{3}{2}n^2 + \frac{1}{2}n\right) = \frac{2}{3}\left(\left(\frac{E_0}{E_n^0}\right)^{3/2} + \frac{3}{2}\left(\frac{E_0}{E_n^0}\right) + \frac{1}{2}\left(\frac{E_0}{E_n^0}\right)^{1/2}\right) = f(E_n^0), \tag{8.135}$$

which gives the density of states for hydrogen,

$$\rho(E_n^0) = \frac{\partial f(E_n^0)}{\partial E_n^0} = -\frac{1}{E_0}\left(\frac{E_0}{E_n^0}\right)^{5/2} - \frac{1}{E_0}\left(\frac{E_0}{E_n^0}\right)^2 - \frac{1}{6}\frac{1}{E_0}\left(\frac{E_0}{E_n^0}\right)^{3/2}. \tag{8.136}$$

The negative sign for E_0 renders (8.136) positive. This is useful for atomic transitions into high Rydberg states, where $E_n^0 \to 0$ and the density of states diverges.

The density of states derivations have so far been limited to cases where the energy levels are discrete. It is often the case, particularly in scattering processes, that the energy eigenvalues are continuous. This can be understood by placing a free particle in a very large cubic box of side L. Such a situation leads the particle to possess the three-dimensional momentum characterized by the set of integers $\boldsymbol{n} = (n_x, n_y, n_z)$, where the associated momentum of each integer is given by three copies of the formula (4.88),

$$p_i = \frac{n_i \pi \hbar}{L}. \tag{8.137}$$

For the case that L becomes large, these momenta become continuous. This gives the associated energy as

$$E_n = \sum_{i=1}^{3} \frac{n_i^2 \pi^2 \hbar^2}{2\,mL^2} = \frac{n^2 \pi^2 \hbar^2}{2\,mL^2} = n^2 E_0, \tag{8.138}$$

where $n^2 = n_x^2 + n_y^2 + n_z^2$. Similarly to result (2.28) the value of n functions as a radius in n-space. However, like the standing waves in blackbody radiation, the energies with $n_i \to -n_i$ are identical. The spherical volume of numbers between $n = 0$ and n must be divided by 8 in order to count only the distinct energy states. Using (8.138) to find n, the number of distinct energy states is given by

$$N = \frac{1}{8}\left(\frac{4}{3}\pi n^3\right) = \frac{\pi}{6}\left(\frac{E_n}{E_0}\right)^{3/2}. \tag{8.139}$$

This gives the density of states

$$\rho(E_n) = \frac{\partial N}{\partial E_n} = \frac{\pi}{4}\sqrt{\frac{E_n}{E_0^3}} = \frac{mL^3}{2\pi^2\hbar^3}\sqrt{2\,mE_n} = \frac{mp_n L^3}{2\pi^2\hbar^3}, \tag{8.140}$$

where p_n is the magnitude of the momentum associated with E_n. It is common to divide by the volume of space $V = L^3$ to obtain the density of states per unit volume. If the particles are electrons, it is necessary to multiply by a factor of two in order to account for the two spin degrees of freedom.

The density of states can be used in the transition probability (8.128) to convert the sum into an integral by the procedure just outlined. Earlier analysis showed that the final states dominating the transition probability lie in the energy interval $E_\mathrm{i}^0 - \Delta E$ to $E_\mathrm{i}^0 + \Delta E$, so that the total probability for transition from E_i^0 into that group of states is given by an integral,

$$P = \frac{4\lambda^2}{\hbar^2} \int_{E_i^0-\Delta E}^{E_i^0+\Delta E} \mathrm{d}E_f^0\, \rho(E_f^0)\, |\langle E_f^0 \mid \hat{V} \mid E_i^0 \rangle|^2 \left(\frac{\sin^2\left(\frac{1}{2}\omega_{fi} t\right)}{\omega_{fi}^2} \right). \tag{8.141}$$

The integral in (8.141) can be understood by comparing it to the integral (3.92) appearing in the analysis of the sinc squared Dirac delta (3.91). Rewriting the integral to conform with the sinc squared delta gives the large t result,

$$\begin{aligned} P &= \frac{2\pi t\lambda^2}{\hbar^2} \int_{E_i^0-\Delta E}^{E_i^0+\Delta E} \mathrm{d}E_f^0\, \rho(E_f^0)\, |\langle E_f^0 \mid \hat{V} \mid E_i^0 \rangle|^2 \left(\frac{\sin^2\left(\frac{1}{2}\omega_{fi} t\right)}{\frac{1}{2}\pi t \omega_{fi}^2} \right) \\ &\approx \frac{2\pi t\lambda^2}{\hbar} \int_{\omega_{fi}-\Delta E/\hbar}^{\omega_{fi}+\Delta E/\hbar} \mathrm{d}\omega_{fi}\, \rho(E_f^0)\, |\langle E_f^0 \mid \hat{V} \mid E_i^0 \rangle|^2\, \delta(\omega_{fi}) \\ &= \frac{2\pi t\lambda^2}{\hbar} \rho(E_f^0)\, |\langle E_f^0 \mid \hat{V} \mid E_i^0 \rangle|^2 \big|_{\omega_{fi}=0}. \end{aligned} \tag{8.142}$$

This is exactly consistent with the estimate (8.127). Using this result allows the *transition rate* into the energetically degenerate states to be defined as

$$\Gamma_{i\to j} \equiv \frac{P}{t} = \frac{2\pi\lambda^2}{\hbar} \rho(E_f^0)\, |\langle E_f^0 \mid \hat{V} \mid E_i^0 \rangle|^2 \big|_{E_f^0=E_i^0}, \tag{8.143}$$

which has the units of inverse time. It defines the transition rate as the probability per unit time for a transition from the initial state $| E_i^0 \rangle$ to the final state $| E_f^0 \rangle$ whose energy is degenerate with the initial state. This allows the estimate of a *half-life* for the initial state, which is defined as the time required for the probability to be 1/2. Result (8.143) is referred to as *Fermi's golden rule*, although it was first derived by Dirac. It is important to remember that result (8.143) applies to the case (8.117), which is that of a static perturbation switched on at time $t = 0$. Different forms of perturbation will give rise to different forms of Fermi's golden rule.

8.3.4 Harmonic perturbations

Rather than simply switching on a static perturbation, it is possible to apply a truly time-dependent perturbation the system. An example is a harmonic perturbation brought about by placing a charge in an external oscillating electric field, so that the energy of the particle is modified by

$$V(\boldsymbol{x}, t) = q\, \Phi(\boldsymbol{x}) \sin(\omega_o t), \tag{8.144}$$

where $\Phi(\boldsymbol{x})$ is a real valued function that gives the spatial shape of the electric field, $\mathbf{E} = -\nabla\Phi(\boldsymbol{x})$, and ω_o is the frequency of the field. From the discussion around (2.48), this applied electric field is associated with photons of energy $E = \hbar\omega_o$. As a result, the charged particle is expected to interact with the photons much in the same way as electrons in the metal were ejected by the applied electromagnetic field.

The first order solution (8.115) for this situation is given by

$$\begin{aligned} qc_f^{(1)}(t) &= -\frac{i}{\hbar}\langle\, E_{\rm f}^0 \mid q\Phi(\boldsymbol{x}) \mid E_{\rm i}^0\,\rangle \int_0^t \mathrm{d}\tau\, \sin(\omega_o\tau) e^{i\omega_{fi}\tau} \\ &= \frac{1}{2\hbar}\langle\, E_{\rm f}^0 \mid q\Phi(\boldsymbol{x}) \mid E_{\rm i}^0\,\rangle \int_0^t \mathrm{d}\tau\, \left(e^{i(\omega_{fi}-\omega_o)\tau} - e^{i(\omega_{fi}+\omega_o)\tau}\right), \end{aligned} \tag{8.145}$$

where the coupling constant λ has been replaced by q. Performing the integrations and using basic trigonometric identities gives

$$qc_f^{(1)}(t) = \frac{1}{\hbar}\langle\, E_{\rm f}^0 \mid q\Phi(\boldsymbol{x}) \mid E_{\rm i}^0\,\rangle \left(\frac{e^{i\frac{1}{2}(\omega_{fi}+\omega_o)t}}{(\omega_{fi}+\omega_o)} \sin(\frac{1}{2}(\omega_{fi}+\omega_o)t) - \frac{e^{i\frac{1}{2}(\omega_{fi}-\omega_o)t}}{(\omega_{fi}-\omega_o)} \sin(\frac{1}{2}(\omega_{fi}-\omega_o)t) \right). \tag{8.146}$$

It follows that the transition probability is given by

$$\begin{aligned} P_{i\to f}(t) = q^2|c_f^{(1)}(t)|^2 = \frac{q^2}{\hbar^2}|\langle\, E_{\rm f}^0 \mid \Phi(\boldsymbol{x}) \mid E_{\rm i}^0\,\rangle|^2 &\left(\frac{\sin^2(\frac{1}{2}(\omega_{fi}+\omega_o)t)}{(\omega_{fi}+\omega_o)^2} + \frac{\sin^2(\frac{1}{2}(\omega_{fi}-\omega_o)t)}{(\omega_{fi}-\omega_o)^2} \right. \\ &\left. - \frac{2\cos(\omega_o t)\sin(\frac{1}{2}(\omega_{fi}-\omega_o)t)\sin(\frac{1}{2}(\omega_{fi}+\omega_o)t)}{(\omega_{fi}^2-\omega_o^2)} \right). \end{aligned} \tag{8.147}$$

This result exhibits the presence of the sinc squared function familiar from the perturbation (8.117). Now, however, the transition probability is strongly peaked around the two values $\omega_{fi} = \pm\omega_o$. In particular, the first two terms in the parentheses both become $t^2/4$ for the respective cases of $\omega_{fi} = -\omega_o$ and $\omega_{fi} = \omega_o$. In contrast, the third term is given by $-t\,\sin(2\omega_o t)/2\omega_o$ for either case.

As t grows large the third term therefore becomes negligible compared to the first two terms, and the range of energies corresponding to maximum transition probability breaks into two pieces. The first corresponds to the case $\omega_{fi} = \omega_o$, which is the situation where the final energy satisfies

$$E_{\rm f}^0 = E_{\rm i}^0 + \hbar\omega_o. \tag{8.148}$$

This is the situation where the system absorbs a photon of energy $\hbar\omega_o$ from the oscillating electric field and transits to the state with the energy $E_{\rm i}^0 + \hbar\omega_o$, if such a state is available. The transition rate into the group of states with this energy is given by an expression identical to (8.143), so that Fermi's golden rule gives the transition rate for absorption,

$$\Gamma_{i\to f} \equiv \frac{P}{t} = \frac{\pi q^2}{2\hbar}\rho(E_{\rm f}^0)\,|\langle\, E_{\rm f}^0 \mid \Phi(\boldsymbol{x}) \mid E_{\rm i}^0\,\rangle|^2|_{E_{\rm f}^0=E_{\rm i}^0+\hbar\omega_o}, \tag{8.149}$$

The second case is given by $\omega_{fi} = -\omega_o$, so that the final energy satisfies

$$E_{\mathrm{f}}^{0} = E_{\mathrm{i}}^{0} - \hbar\omega_o. \tag{8.150}$$

This corresponds to a drop to a lower energy state whose energy difference matches $\hbar\omega_o$. If the system is in an elevated energy state E_{i}^{0} and the lower energy state E_{f}^{0} is available, then the system is *stimulated* to emit a photon whose energy matches that of the photons in the applied field and drop to the lower energy state. Fermi's golden rule then gives the transition rate for *stimulated emission*,

$$\Gamma_{i\to f} \equiv \frac{P}{t} = \frac{\pi q^2}{2\hbar}\rho(E_{\mathrm{f}}^{0})\,|\langle E_{\mathrm{f}}^{0} \mid \Phi(\boldsymbol{x}) \mid E_{\mathrm{i}}^{0} \rangle|^2|_{E_{\mathrm{f}}^{0}=E_{\mathrm{i}}^{0}-\hbar\omega_o}. \tag{8.151}$$

This is the mechanism employed in a laser, an acronym for light amplification by stimulated emission of radiation.

The presence of both absorption and stimulated emission rates is an outgrowth of *time reversability invariance* in quantum mechanics. The stimulated emission process is the time-reversed version of the absorption process, and is therefore quantum mechanically allowed. The two rates will differ if the density of states $\rho(E_{\mathrm{i}}^{0} + \hbar\omega_o)$ and $\rho(E_{\mathrm{i}}^{0} - \hbar\omega_o)$ are different or the matrix element is different for the two cases.

8.4 The sudden approximation

It is occasionally the case that the parameters governing a quantum system undergo a sudden change. For example, a particle in an infinite well of width L may have the size of the well suddenly changed to L', or the nucleus of an atom may undergo a sudden change in its charge due to radioactive decay. In both of these cases the time-independent basis Hamiltonian $\hat{H}_0$ is approximately solvable before and after the change occurs. However, the spectrum of energies and the states of the system will be different after the change occurs. It can be the case that the sudden change results in a transition probability to a new state.

The analysis begins by denoting the Hamiltonians before and after the sudden change to be $\hat{H}_0$ and $\hat{H}_0'$. Since both of these Hamiltonians are exactly solvable, they possess a complete set of eigenstates, which are denoted

$$\begin{aligned}\hat{H}_0| E_n \rangle &= E_n| E_n \rangle,\\ \hat{H}_0'| E_n' \rangle &= E_n'| E_n' \rangle.\end{aligned} \tag{8.152}$$

The energy eigenstates of both these Hamiltonians are complete, so that the state of the system can be expanded in terms of either one at $t = 0$. Arbitrarily choosing the change to occur at $t = 0$, Born's continuity conditions, discussed in chapter 4, require that the two versions of the state are continuous at $t = 0$, so that

$$| \Psi, 0 \rangle = \sum_n a_n| E_n \rangle = \sum_n b_n| E_n' \rangle. \tag{8.153}$$

Using the orthonormality of the eigenstates of the new basis allows the coefficients b_n to be determined in terms of the coefficients a_n. The inner product with $\langle E_{\mathrm{f}}' \mid$ gives

$$\sum_n a_n \langle E_f' \mid E_n \rangle = \sum_n b_n \langle E_f' \mid E_n' \rangle = \sum_n b_n \delta_{fn} = b_f. \tag{8.154}$$

This result simplifies if the initial state of the system corresponds to $a_n = \delta_{ni}$. The final coefficients are then given by

$$b_f = \langle E_f' \mid E_i \rangle. \tag{8.155}$$

The coefficient b_f is the probability amplitude for the system transiting from the initial state $\mid E_i \rangle$ into the final state $\mid E_f' \rangle$. This follows since the completeness of the states $\mid E_n' \rangle$ results in the sum of these probabilities satisfying

$$\sum_f |b_f|^2 = \sum_f \langle E_i \mid E_f' \rangle \langle E_f' \mid E_i \rangle = \langle E_i \mid E_i \rangle = 1, \tag{8.156}$$

which is simply an expression of Born's rule.

A simple example can be found for the case of a particle in the ground state of a one-dimensional infinite well of width L. For the case that the well width is suddenly doubled, the probability amplitude for the particle to be found in the nth state of the new well is given by

$$\begin{aligned} b_n = \langle E_n' \mid E_0 \rangle &= \int_0^L \mathrm{d}x \, \langle E_n' \mid x \rangle \langle x \mid E_0 \rangle = \frac{2}{\sqrt{2L^2}} \int_0^L \mathrm{d}x \sin\left(\frac{n\pi x}{2L}\right) \sin\left(\frac{\pi x}{L}\right) \\ &= -\frac{4\sqrt{2}}{\pi(n^2 - 4)} \sin\left(\frac{n\pi}{2}\right). \end{aligned} \tag{8.157}$$

It is important to note that the integration is from 0 to L since the wave function $\langle x \mid E_0 \rangle$ vanishes outside that interval. It appears that (8.157) is singular at $n = 2$, which corresponds to the new energy level given by

$$E_2' = \frac{2^2 \pi^2 \hbar^2}{2\,m(2L)^2} = \frac{\pi^2 \hbar^2}{2\,mL^2} = E_1, \tag{8.158}$$

exactly matching the previous ground state energy. However, it follows that

$$b_2 = \lim_{\epsilon \to 0} b_{2+\epsilon} = \lim_{\epsilon \to 0} \frac{4\sqrt{2}}{\pi(4 - (2 + \epsilon)^2)} \sin\left(\pi + \frac{1}{2}\epsilon\pi\right) = \lim_{\epsilon \to 0} \frac{\sqrt{2}}{\pi\epsilon} \sin\left(\frac{1}{2}\epsilon\pi\right) = \frac{\sqrt{2}}{2}. \tag{8.159}$$

This is the coefficient that gives the largest probability, $|b_2|^2 = 1/2$, with all other coefficients for even n vanishing.

8.5 The interaction picture and the evolution operator

Another approach to time-dependent perturbation theory is given by using the *interaction* or *Dirac picture*. This picture of quantum mechanics lies between the Heisenberg and Schrödinger picture and isolates the effects of a perturbation for analysis.

8.5.1 The interaction picture

To begin with, the Hamiltonian is broken into the usual two pieces,

$$\hat{H} = \hat{H}_0 + \lambda\hat{V}, \tag{8.160}$$

where the Hamiltonian $\hat{H}_0$ is time-independent with a complete set of energy eigenstates $| E_n^0 \rangle$ as before. The perturbation potential $\lambda\hat{V}$ may or may not be explicitly time-dependent. The interaction picture states $| \Psi, t \rangle_{\rm I}$ are defined in terms of the Schrödinger picture states $| \Psi, t \rangle_{\rm S}$ as

$$| \Psi, t \rangle_{\rm I} = e^{i\hat{H}_0 t/\hbar} | \Psi, t \rangle_{\rm S}. \tag{8.161}$$

If the perturbation vanishes, $\lambda\hat{V} = 0$, then the interaction picture states are time-independent and correspond to the Schrödinger picture states at $t = 0$ or, equivalently, the Heisenberg picture states. Definition (8.161) shows that this equality also holds at $t = 0$, so that the interaction picture states coincide with the Schrödinger and Heisenberg picture states at $t = 0$.

The interaction picture states satisfy a variant of the Schrödinger equation, given by

$$\begin{aligned} i\hbar\frac{\partial}{\partial t}| \Psi, t \rangle_{\rm I} &= -e^{i\hat{H}_0 t/\hbar}\hat{H}_0 | \Psi, t \rangle_{\rm S} + e^{i\hat{H}_0 t/\hbar}\, i\hbar\frac{\partial}{\partial t}| \Psi, t \rangle_{\rm S} = e^{i\hat{H}_0 t/\hbar}(\hat{H} - \hat{H}_0) | \Psi, t \rangle_{\rm S} \\ &= e^{i\hat{H}_0 t/\hbar}\lambda\hat{V} | \Psi, t \rangle_{\rm S} = e^{i\hat{H}_0 t/\hbar}\lambda\hat{V}e^{-i\hat{H}_0 t/\hbar} | \Psi, t \rangle_{\rm I} \equiv \lambda\hat{V}_{\rm I}(t) | \Psi, t \rangle_{\rm I}. \end{aligned} \tag{8.162}$$

The resulting interaction picture perturbation,

$$\lambda\hat{V}_{\rm I}(t) = e^{i\hat{H}_0 t/\hbar}\lambda\hat{V}e^{-i\hat{H}_0 t/\hbar}, \tag{8.163}$$

will typically be time-dependent even if the Schrödinger picture operator $\hat{V}$ is not. In general, operators in the interaction picture are given by

$$\hat{O}_{\rm I}(t) = e^{i\hat{H}_0 t/\hbar}\hat{O}_{\rm S}e^{-i\hat{H}_0 t/\hbar}, \tag{8.164}$$

where $\hat{O}_{\rm S}$ is the operator in the Schrödinger picture. However, it is often the case that the form of the time-dependent interaction picture operator is relatively easy to find.

8.5.2 The evolution operator

Solving (8.162) begins by assuming the interaction picture state at time t is found from the state at t_o by a unitary operator, denoted $\hat{U}(t, t_o)$, so that

$$| \Psi, t \rangle_{\rm I} = \hat{U}(t, t_o) | \Psi, t_o \rangle_{\rm I}. \tag{8.165}$$

Unitarity is required to ensure that the state $| \Psi, t \rangle_{\rm I}$ remains normalized. The unitary operator must also satisfy the boundary condition

$$\hat{U}(t_o, t_o) = 1. \tag{8.166}$$

This unitary operator is often referred to as the *evolution operator*.

Solving (8.162) reduces to solving

$$i\hbar\frac{\partial}{\partial t}\hat{U}(t,\,t_o) = \lambda\hat{V}_{\mathrm{I}}(t)\,\hat{U}(t,\,t_o), \tag{8.167}$$

subject to the boundary condition (8.166). This begins by integrating both sides of (8.167) and using (8.166), which gives

$$\int_{t_o}^{t} \mathrm{d}t_1\,\frac{\partial}{\partial t_1}\hat{U}(t_1,\,t_o) = \hat{U}(t,\,t_o) - 1 = \frac{\lambda}{i\hbar}\int_{t_o}^{t} \mathrm{d}t_1\,\hat{V}_{\mathrm{I}}(t_1)\,\hat{U}(t_1,\,t_o). \tag{8.168}$$

Result (8.168) is self-referential and can be solved by *iteration*. The first iteration is given by simply substituting the expression for $\hat{U}(t_1,\,t_o)$ obtained from (8.168),

$$\hat{U}(t_1,\,t_o) = 1 + \frac{\lambda}{i\hbar}\int_{t_o}^{t_1} \mathrm{d}t_2\,\hat{V}_{\mathrm{I}}(t_2)\,\hat{U}(t_2,\,t_o), \tag{8.169}$$

back into equation (8.168). This gives the first order result,

$$\hat{U}(t,\,t_o) = 1 + \frac{\lambda}{i\hbar}\int_{t_o}^{t} \mathrm{d}t_1\,\hat{V}_{\mathrm{I}}(t_1) + \frac{\lambda^2}{(i\hbar)^2}\int_{t_o}^{t} \mathrm{d}t_1\hat{V}_{\mathrm{I}}(t_1)\int_{t_o}^{t_1} \mathrm{d}t_2\,\hat{V}_{\mathrm{I}}(t_2)\,\hat{U}(t_2,\,t_o). \tag{8.170}$$

The process of iteration can be performed an infinite number of times, with the result

$$\hat{U}(t,\,t_o) = 1 + \frac{\lambda}{i\hbar}\int_{t_o}^{t} \mathrm{d}t_1\,\hat{V}_{\mathrm{I}}(t_1) + \cdots + \frac{\lambda^n}{(i\hbar)^n}\int_{t_o}^{t} \mathrm{d}t_1\hat{V}_{\mathrm{I}}(t_1)\int_{t_o}^{t_1} \mathrm{d}t_2\,\hat{V}_{\mathrm{I}}(t_2)\cdots\int_{t_o}^{t_{n-1}} \mathrm{d}t_n\hat{V}_{\mathrm{I}}(t_n) + \cdots, \tag{8.171}$$

where the first and nth order terms in the infinite series are explicitly displayed.

Expression (8.171) can be rewritten more compactly by introducing *time-ordering*. This consists of moving earlier operators to the right in a product. For the product of two operators, the time-ordering operator $\mathcal{T}$ gives

$$\mathcal{T}\left\{\hat{V}_{\mathrm{I}}(t_1)\hat{V}_{\mathrm{I}}(t_2)\right\} = \theta(t_1 - t_2)\hat{V}_{\mathrm{I}}(t_1)\hat{V}_{\mathrm{I}}(t_2) + \theta(t_2 - t_1)\hat{V}_{\mathrm{I}}(t_2)\hat{V}_{\mathrm{I}}(t_1), \tag{8.172}$$

where $\theta(t_1 - t_2)$ is the Heaviside step function defined in (3.70). Using the properties of the step function and relabeling dummy variables of integration it follows that

$$\frac{1}{2!}\int_{t_o}^{t} \mathrm{d}t_1\int_{t_o}^{t} \mathrm{d}t_2\,\mathcal{T}\left\{\hat{V}_{\mathrm{I}}(t_1)\hat{V}_{\mathrm{I}}(t_2)\right\} = \int_{t_o}^{t} \mathrm{d}t_1\hat{V}_{\mathrm{I}}(t_1)\int_{t_o}^{t_1} \mathrm{d}t_2\,\hat{V}_{\mathrm{I}}(t_2). \tag{8.173}$$

A similar expression follows for the third order. Since there are $n!$ possible time-orderings for n operators, the nth order term can be written

$$\begin{aligned}&\int_{t_o}^{t} \mathrm{d}t_1\hat{V}_{\mathrm{I}}(t_1)\int_{t_o}^{t_1} \mathrm{d}t_2\,\hat{V}_{\mathrm{I}}(t_2)\cdots\int_{t_o}^{t_{n-1}} \mathrm{d}t_n\hat{V}_{\mathrm{I}}(t_n)\\ &= \frac{1}{n!}\int_{t_o}^{t} \mathrm{d}t_1\int_{t_o}^{t} \mathrm{d}t_2\ldots\int_{t_o}^{t} \mathrm{d}t_n\mathcal{T}\left\{\hat{V}_{\mathrm{I}}(t_1)\hat{V}_{\mathrm{I}}(t_2)\cdots\hat{V}(t_n)\right\}.\end{aligned} \tag{8.174}$$

Because the time-ordering operator $\mathcal{T}$ acts only the operators in the product, it follows that the infinite sum of (8.171) can be written

$$\hat{U}(t, t_o) = \sum_{n=0}^{\infty} \frac{\lambda^n}{n!(i\hbar)^n} \int_{t_o}^{t} dt_1 \cdots \int_{t_o}^{t} dt_n \, \mathcal{T}\{\hat{V}_I(t_1) \cdots \hat{V}(t_n)\} = \mathcal{T}\left\{\exp\left(-\frac{i}{\hbar}\int_{t_o}^{t} d\tau \, \lambda \hat{V}_I(\tau)\right)\right\}. \quad (8.175)$$

This form for the evolution operator is manifestly unitary and satisfies the boundary condition (8.166). Because t is the latest time in the expression, it follows that

$$i\hbar\frac{\partial}{\partial t}\hat{U}(t, t_o) = \mathcal{T}\left\{\lambda \hat{V}_I(t) \exp\left(-\frac{i}{\hbar}\int_{t_o}^{t} d\tau \, \lambda \hat{V}_I(\tau)\right)\right\} = \lambda \hat{V}_I(t) \, \hat{U}(t, t_o), \quad (8.176)$$

demonstrating that it solves (8.162).

8.5.3 Time-dependent transition amplitudes

It is instructive to show that this result is equivalent to the time-dependent perturbation theory results obtained earlier. This begins by noting that the coefficients (8.102) in the previous perturbative expansion are given by

$$c_f(t) = {}_S\langle E_f^0, t \mid \Psi, t \rangle_S, \quad (8.177)$$

where both states are in the Schrödinger picture. In the case of the unperturbed states $\mid E_n^0, t \rangle_S$, its time evolution gives

$$\mid E_f^0, t \rangle_S = e^{-i\hat{H}_0 t/\hbar} \mid E_f^0, 0 \rangle_S. \quad (8.178)$$

Combining this with the definition (8.161) of the interaction picture states gives

$$c_f(t) = {}_S\langle E_f^0, t \mid \Psi, t \rangle_S = \langle E_f^0, 0 \mid e^{i\hat{H}_0 t/\hbar} \mid \Psi, t \rangle_S = \langle E_f^0, 0 \mid \Psi, t \rangle_I = \langle E_f, 0 \mid \hat{U}(t, 0) \mid \Psi, 0 \rangle, \quad (8.179)$$

where the equivalence of all picture states at $t = 0$ was used.

For the case that the initial state of the system is $\mid E_i^0, 0 \rangle$ the expansion of the evolution operator to first order given by (8.170) results in the approximate value for $c_f(t)$,

$$c_f(t) = \langle E_n^0, 0 \mid \hat{U}(t, 0) \mid E_i^0, 0 \rangle \approx \langle E_f^0, 0 \mid E_i^0, 0 \rangle - \frac{i\lambda}{\hbar}\int_0^t d\tau \, \langle E_f^0, 0 \mid \hat{V}_I(t) \mid E_i^0, 0 \rangle. \quad (8.180)$$

Using the form of the interaction picture perturbation (8.163) gives

$$\begin{aligned} c_f(t) &= \delta_{fi} - \frac{i\lambda}{\hbar}\int_0^t d\tau \, \langle E_f^0, 0 \mid e^{i\hat{H}_0 t/\hbar} \hat{V} e^{i\hat{H}_0 t/\hbar} \mid E_i^0, 0 \rangle \\ &= \delta_{fi} - \frac{i}{\hbar}\int_0^t d\tau \, e^{i(E_f^0 - E_i^0)t/\hbar} \langle E_f^0, 0 \mid \lambda \hat{V} \mid E_i^0, 0 \rangle, \end{aligned} \quad (8.181)$$

which is identical to the previous first order results (8.116) and (8.115).

8.5.4 Environmental evolution and decoherence

In (7.355) it was argued that a mixed state may become entangled with its environment over time. In that regard it is instructive to examine the evolution operator associated with the model interaction of (7.356) between a mixed state and its environment. The interaction can be used to model an environmental particle scattering from the system without altering the state of the system. In the case of a mixed state for the system the potential encountered by the scattered particle will differ depending on the system state to which it is coupled.

The initial state $|\,\Psi\,\rangle$ under consideration is the normalized tensor product of a mixed state, referred to as the *system*, and an environmental state $|\,\varphi\,\rangle$,

$$|\,\Psi\,\rangle = \frac{1}{\sqrt{2}}\big(|\,\psi_1\,\rangle + |\,\psi_2\,\rangle\big) \otimes |\,\varphi\,\rangle, \tag{8.182}$$

where the environmental state is initially the same for both possible system states. The three states, $|\,\psi_j\,\rangle$ and $|\,\varphi\,\rangle$, are normalized. For simplicity, it is assumed that the two system states, $|\,\psi_j\,\rangle$, are eigenstates of the Hermitian Hamiltonian $\hat{H}_{0\mathrm{s}}$ with the eigenvalues E_j, while the environmental state $|\,\varphi\,\rangle$ is an eigenstate of the Hermitian Hamiltonian $\hat{H}_{0\mathrm{e}}$ with the eigenvalue E_φ. The two system states are therefore orthonormal, so that $\langle\,\psi_1\,|\,\psi_2\,\rangle = 0$. The full Hamiltonian of the system is assumed to take the form

$$\hat{H} = \hat{H}_{0\mathrm{s}} \otimes \hat{\mathbf{1}} + \hat{\mathbf{1}} \otimes \hat{H}_{0\mathrm{e}} + \hat{V} \,\equiv\, \hat{H}_0 + \hat{V}, \tag{8.183}$$

where the interaction takes the simple form

$$\hat{V} = |\,\psi_1\,\rangle\langle\,\psi_1\,| \otimes \hat{V}_1 + |\,\psi_2\,\rangle\langle\,\psi_2\,| \otimes \hat{V}_2. \tag{8.184}$$

The two potentials $\hat{V}_1$ and $\hat{V}_2$ are assumed to be Hermitian operators that depend on the quantum numbers of the system states. An example of such a situation is the possible presence of a magnetic field that differs depending on the relative velocity or momentum of the two system states. The environmental particle will then encounter a different magnetic field for $|\,\psi_1\,\rangle$ as opposed to $|\,\psi_2\,\rangle$.

The evolution operator is found using the interaction picture technique introduced in (8.163) and (8.165). The basis Hamiltonian $\hat{H}_0$ defined in (8.183) is used to find the interaction picture version of the potential,

$$\hat{V}_{\mathrm{I}}(t) = e^{i\hat{H}_0 t/\hbar}\hat{V}e^{-i\hat{H}_0 t/\hbar} = e^{i\hat{H}_0 t/\hbar}\Big(|\,\psi_1\,\rangle\langle\,\psi_1\,| \otimes \hat{V}_1 + |\,\psi_2\,\rangle\langle\,\psi_2\,| \otimes \hat{V}_2\Big)e^{-i\hat{H}_0 t/\hbar}. \tag{8.185}$$

This is simplified by noting that the system and environmental parts of $\hat{H}_0$ commute,

$$[\hat{H}_{0\mathrm{s}} \otimes \hat{\mathbf{1}},\, \hat{\mathbf{1}} \otimes \hat{H}_{0\mathrm{e}}] = \hat{H}_{0\mathrm{s}} \otimes \hat{H}_{0\mathrm{e}} - \hat{H}_{0\mathrm{s}} \otimes \hat{H}_{0\mathrm{e}} = 0, \tag{8.186}$$

so that

$$e^{i\hat{H}_0 t/\hbar} = e^{i(\hat{H}_{0\mathrm{s}}\otimes\hat{\mathbf{1}})t/\hbar}e^{i(\hat{\mathbf{1}}\otimes\hat{H}_{0\mathrm{e}})t/\hbar}. \tag{8.187}$$

The interaction picture potential (8.185) is therefore given by

$$\begin{aligned}\hat{V}_{\mathrm{I}}(t) &= e^{i\hat{H}_{0\mathrm{s}}t/\hbar}|\,\psi_1\,\rangle\langle\,\psi_1\,|e^{-i\hat{H}_{0\mathrm{s}}t/\hbar} \otimes e^{i\hat{H}_{0\mathrm{e}}t/\hbar}\hat{V}_1e^{-i\hat{H}_{0\mathrm{e}}t/\hbar}\\ &\quad + e^{i\hat{H}_{0\mathrm{s}}t/\hbar}|\,\psi_2\,\rangle\langle\,\psi_2\,|e^{-i\hat{H}_{0\mathrm{s}}t/\hbar} \otimes e^{i\hat{H}_{0\mathrm{e}}t/\hbar}\hat{V}_2e^{-i\hat{H}_{0\mathrm{e}}t/\hbar}\\ &= |\,\psi_1\,\rangle\langle\,\psi_1\,| \otimes \hat{V}_{1\mathrm{I}}(t) + |\,\psi_2\,\rangle\langle\,\psi_2\,| \otimes \hat{V}_{2\mathrm{I}}(t),\end{aligned}$$

where the time-development of the system states cancels. As a result, only the potentials acting on the environment are time-dependent in the interaction picture.

Due to the orthonormality of the system states the interaction picture potential has the property that

$$\hat{V}_{\mathrm{I}}(t_1)\cdots\hat{V}_{\mathrm{I}}(t_n) = |\,\psi_1\,\rangle\langle\,\psi_1\,| \otimes \hat{V}_{1\mathrm{I}}(t_1)\cdots\hat{V}_{1\mathrm{I}}(t_n) + |\,\psi_2\,\rangle\langle\,\psi_2\,| \otimes \hat{V}_{2\mathrm{I}}(t_1)\cdots\hat{V}_{2\mathrm{I}}(t_n), \quad (8.188)$$

so that the evolution operator breaks into the sum of two distinct evolution operators,

$$\hat{U}(t_f,\, t_i) = |\,\psi_1\,\rangle\langle\,\psi_1\,| \otimes \hat{U}_1(t_f,\, t_i) + |\,\psi_2\,\rangle\langle\,\psi_2\,| \otimes \hat{U}_2(t_f,\, t_i), \quad (8.189)$$

where result (8.171) follows for both terms,

$$\hat{U}_j(t_f,\, t_i) = \sum_{n=0}^{\infty}\left(\int_{t_i}^{t_f} \mathrm{d}t_1 \hat{V}_{j\mathrm{I}}(t_1)\cdots\int_{t_i}^{t_{n-1}} \mathrm{d}t_n \hat{V}_{j\mathrm{I}}(t_n)\right). \quad (8.190)$$

The action of the potential (8.184) has caused the initial environmental state to evolve into two different environmental states,

$$\begin{aligned} |\,\Psi,\, t_f\,\rangle = \hat{U}(t_f,\, t_i)|\,\Psi,\, t_i\,\rangle &= \frac{1}{\sqrt{2}}|\,\psi_1\,\rangle \otimes \hat{U}_1(t_f,\, t_i)|\,\varphi,\, t_i\,\rangle + \frac{1}{\sqrt{2}}|\,\psi_2\,\rangle \otimes \hat{U}_2(t_f,\, t_i)|\,\varphi,\, t_i\,\rangle \\ &= \frac{1}{\sqrt{2}}|\,\psi_1\,\rangle \otimes |\,\varphi_1,\, t_f\,\rangle + \frac{1}{\sqrt{2}}|\,\psi_2\,\rangle \otimes |\,\varphi_2,\, t_f\,\rangle, \end{aligned} \quad (8.191)$$

where the environmental states at a later time are given by

$$|\,\varphi_j,\, t_f\,\rangle = \hat{U}_j(t_f,\, t_i)|\,\varphi,\, t_i\,\rangle. \quad (8.192)$$

By virtue of the unitary nature of the operators $\hat{U}_j(t_f,\, t_i)$ the two environmental states remain normalized,

$$\langle\,\varphi_j,\, t_f\,|\,\varphi_j,\, t_f\,\rangle = \langle\,\varphi,\, t_i\,|\,\hat{U}_j^{\dagger}(t_f,\, t_i)\hat{U}_j(t_f,\, t_i)\,|\,\varphi,\, t_i\,\rangle = \langle\,\varphi,\, t_i\,|\,\varphi,\, t_i\,\rangle = 1. \quad (8.193)$$

Since the potentials acting on the environmental state are different, it follows that the *pure state* has become entangled with its environment with the subsequent possibility that the quantum interference pattern associated with the measurement of the system may be suppressed depending on the details of the potentials. These potentials determine the inner product

$$\langle\,\varphi_1,\, t_f\,|\,\varphi_2,\, t_f\,\rangle = \langle\,\varphi,\, t_i\,|\,\hat{U}_1^{\dagger}(t_f,\, t_i)\hat{U}_2(t_f,\, t_i)\,|\,\varphi,\, t_i\,\rangle, \quad (8.194)$$

which in turn determines the suppression of the quantum interference between the two system states during measurement. The Cauchy–Schwarz inequality (3.44) assures that

$$|\langle\,\varphi_1,\, t_f\,|\,\varphi_2,\, t_f\,\rangle|^2 \leqslant 1, \quad (8.195)$$

so that the difference in the potentials can only induce decoherence for the two system states. Environmental decoherence and entanglement is a vigorous area of research and not without controversy. The interested reader is recommended to the references.

8.6 The path integral transition amplitude

An alternative approach to quantum mechanical transition amplitudes was developed extensively by Feynman based on an earlier observation by Dirac that the propagator, defined by (6.363), can viewed in terms of classical motion. This allows the results of classical analysis to be adapted to the quantum mechanical transition amplitude.

8.6.1 The Schrödinger picture evolution operator

Before developing the path integral formalism, it is necessary to find the evolution operator in the Schrödinger picture for an arbitrary time-dependent Hamiltonian, denoted $\hat{H}(t)$. The Schrödinger picture states are assumed to undergo unitary time evolution, so that

$$| \Psi, t \rangle_S = \hat{U}(t, t_o)| \Psi, t_o \rangle_S. \tag{8.196}$$

In order for this state to satisfy the Schrödinger equation, the unitary operator must obey

$$\imath\hbar\frac{\partial}{\partial t}\hat{U}(t, t_o) = \hat{H}(t)\hat{U}(t, t_o), \tag{8.197}$$

as well as the boundary condition $\hat{U}(t_o, t_o) = 1$.

The form of the unitary operator in the Schrödinger picture is found identically to the procedure followed in the case of the interaction picture to solve (8.162). The evolution operator is therefore obtained by simply replacing $\lambda\hat{V}_I(t)$ with $\hat{H}(t)$ in (8.175),

$$\hat{U}(t, t_o) = \mathcal{T}\left\{\exp\left(-\frac{i}{\hbar}\int_{t_o}^{t} d\tau\ \hat{H}(\tau)\right)\right\}. \tag{8.198}$$

The correctness of this result is verified by considering the case that $\hat{H}$ is time-independent. For such a case, the evolution operator becomes

$$\hat{U}(t, t_o) = \exp\left(-\frac{i}{\hbar}\hat{H}(t - t_o)\right), \tag{8.199}$$

which is the unitary time evolution operator (4.57) found earlier for the case of a time-independent Hamiltonian.

8.6.2 The infinitesimal transition amplitude

The starting point for deriving the path integral is an observation, first made by Dirac, regarding the nature of the quantum mechanical transition amplitude for the case of an infinitesimal time interval. For simplicity of notation initial attention will be restricted to the transition element in one dimension for a time-dependent Hamiltonian $\hat{H}(\hat{P}, \hat{Q}, t)$. The infinitesimal transition element $G(x', t_o + \epsilon, x, t_o)$ is defined as the propagator (6.363) for an infinitesimal time interval $\epsilon = t' - t \approx 0$, so that (8.198) shows that it is given by

$$G(x', t_o + \epsilon, x, t_o) = \langle x' \mid \hat{U}(t_o + \epsilon, t_o) \mid x \rangle = \langle x' \mid e^{-i\epsilon \hat{H}(\hat{P}, \hat{Q}, t_o)/\hbar} \mid x \rangle, \quad (8.200)$$

where the time-ordered exponential integral in (8.198) reduces to the single factor appearing in (8.200). In the Copenhagen interpretation of quantum mechanics (8.200) gives the probability amplitude for a particle initially at the position x to be observed at the position x' after the infinitesimal time interval ϵ has elapsed.

A common general form for the Hamiltonian is

$$\hat{H}(\hat{P}, \hat{Q}, t) = \frac{1}{2m}\hat{P}^2 + V(\hat{Q}, t), \quad (8.201)$$

where the potential $\hat{V}$ is a function solely of position but may be time-dependent. For the purposes of evaluating (8.200) the Baker-Campbell-Hausdorff (BCH) theorem (6.279) allows the separation of the time evolution operator into exponential factors, which gives

$$e^{-i\epsilon \hat{H}(t)/\hbar} = e^{-i\epsilon \hat{P}^2/2m\hbar} e^{-i\epsilon V(\hat{Q}, t)/\hbar} e^{-\epsilon^2 [\hat{P}^2, V(\hat{Q}, t)]/2m\hbar^2} e^{O(\epsilon^3)} \dots \quad (8.202)$$

The arguments of the higher order exponentials in the BCH theorem, including the third exponential factor in (8.202), are suppressed since the commutators are $O(\epsilon^2)$ and higher, assuming the potential $V(\hat{Q})$ is *not singular*. Singular potentials, including the Coulomb potential, require a more careful approach. Under this assumption, discarding these higher order exponentials as unity in the limit $\epsilon \to 0$ allows (8.200) to be evaluated by using the unit momentum projection operator of (6.58) and the inner product (6.67), giving

$$\begin{aligned} G(x', t_o + \epsilon, x, t_o) &= \int_{-\infty}^{\infty} dp \, \langle x' \mid e^{-i\epsilon \hat{P}^2/2m\hbar} \mid p \rangle \langle p \mid e^{-i\epsilon V(\hat{Q}, t_o)/\hbar} \mid x \rangle \\ &= \int_{-\infty}^{\infty} dp \, e^{-i\epsilon(p^2/2m + V(x,t_o))/\hbar} \langle x' \mid p \rangle \langle p \mid x \rangle \\ &= \int_{-\infty}^{\infty} \frac{dp}{2\pi\hbar} e^{-i\epsilon(p^2/2m + V(x,t_o))/\hbar} e^{ip(x'-x)/\hbar} \\ &= \int_{-\infty}^{\infty} \frac{dp}{2\pi\hbar} \exp\left\{ \frac{i}{\hbar}\epsilon\left(p\frac{(x'-x)}{\epsilon} - \frac{p^2}{2m} - V(x, t_o) \right) \right\}. \end{aligned} \quad (8.203)$$

If the particle, initially at x, is observed at x' after the time interval ϵ, then it can be inferred that the *average speed* over that time interval would have been $v = (x' - x)/\epsilon$. As a result, the term appearing in (8.203) is denoted

$$\frac{(x' - x)}{\epsilon} = \dot{x}. \quad (8.204)$$

Using this formal notation the argument of the exponential in the last line of (8.203) becomes

$$\frac{i\epsilon}{\hbar}\big(p\dot{x} - H(p, x)\big) = \frac{i\epsilon}{\hbar}\mathcal{L}(p, x), \tag{8.205}$$

where the *classical* Lagrangian (1.21), expressed in terms of p and x, has formally emerged in the context of the infinitesimal quantum mechanical transition element. This analogy can be further developed by using Fresnel's theorem to evaluate the Gaussian integral over p, which gives

$$G(x', t_o + \epsilon, x, t_o) = \sqrt{\frac{m}{2\pi i\epsilon\hbar}}\exp\left\{\frac{i\epsilon}{\hbar}\left(\frac{1}{2}m\dot{x}^2 - V(x, t_o)\right)\right\}, \tag{8.206}$$

where, once again, $\dot{x}$ stands for the *purely formal identification* $\dot{x} = (x' - x)/\epsilon$. The infinitesimal transition element is proportional to $\exp(i\epsilon\mathcal{L}(x, \dot{x})/\hbar)$, where the Lagrangian (1.11) in terms of x and $\dot{x}$ has formally appeared.

It is stressed that the identification of $\mathcal{L}(x, \dot{x})$ in (8.206) is purely *formal* in the sense that x' and x are *not* related through a time-dependent function $x(t)$. Instead, they are arbitrary initial and final possible positions for the particle. Quantum mechanics makes no mention of classical trajectories in its assumptions, instead presenting a method to calculate the probability that the particle is observed at the two positions over the time interval. Nevertheless, result (8.206) shows that the particle's *quantum motion* is related to *classical mechanics,* in particular the value of the classical action $\epsilon\ \mathcal{L}(x, \dot{x})$ associated with the time interval ϵ. This association is most apparent for infinitesimal time intervals, since for that case the average velocity of the particle is inferred to have been $v \approx (x' - x)/\epsilon$. The appearance of the classical action is arguably the feature of path integrals most often used in applications.

8.6.3 The finite time path integral

For the case of a *time-dependent* potential $V(\hat{Q}, t)$ and a *finite* time interval, the path integral is derived by partitioning $T = t_f - t_o$ into N intervals, so that $\epsilon = T/N$, where N is an arbitrarily large integer. The Schrödinger picture evolution operator $\hat{U}(t_f, t_o)$ given by (8.198) becomes

$$\hat{U}(t_f, t_o) = \lim_{N\to\infty}\mathcal{T}\left\{\exp\left(-\frac{i}{\hbar}\sum_{j=0}^{N-1}\epsilon\hat{H}(t_j)\right)\right\}. \tag{8.207}$$

The time argument of the Hamiltonian has been indexed by the integer j, so that $t_j = t_o + j\epsilon$. Expression (8.207) can be simplified using the following property of time-ordering,

$$\mathcal{T}\left\{e^{\hat{A}(t_1)+\hat{B}(t_2)}\right\} = \theta(t_1 - t_2)\, e^{\hat{A}(t_1)}e^{\hat{B}(t_2)} + \theta(t_2 - t_1)\, e^{\hat{B}(t_2)}e^{\hat{A}(t_1)} = \mathcal{T}\left\{e^{\hat{A}(t_1)}e^{\hat{B}(t_2)}\right\}. \tag{8.208}$$

Result (8.208) is proved by expanding the first term in a power series and then time-ordering the products of the operators. Since the later time, *whichever it is*, is moved to the left, the time-ordering reproduces the time-ordering of the product of the two exponentials. The time-ordering does not generate any of the commutators present in the BCH theorem.

Applying (8.208) allows the evolution operator to be written

$$\begin{aligned}\hat{U}(t_f, t_o) &= \lim_{N\to\infty} \mathcal{T}\left\{e^{-i\epsilon[\hat{H}(t_0)+\hat{H}(t_1)+\cdots+\hat{H}(t_{N-1})]/\hbar}\right\} \\ &= \lim_{N\to\infty} e^{-i\epsilon\hat{H}(t_{N-1})/\hbar}e^{-i\epsilon\hat{H}(t_{N-2})/\hbar}\cdots e^{-i\epsilon\hat{H}(t_0)/\hbar}.\end{aligned} \tag{8.209}$$

Result (8.209) shows that the evolution operator for finite time intervals can be viewed as the product of N infinitesimal evolution operators. This is consistent with the Schrödinger equation, which gives the evolution of the wave function as

$$e^{-i\epsilon\hat{H}(t_j)/\hbar}\psi(x, t_j) = e^{\epsilon\partial/\partial t_j}\psi(x, t_j) = \psi(x, t_j + \epsilon). \tag{8.210}$$

The process of dividing the finite time interval into a sequence of N infinitesimal time intervals is referred to as *time slicing* and provides the most common method to derive the path integral.

The path integral representation of the propagator for a finite time interval is obtained by using $N - 1$ configuration space projection operators indexed for convenience of notation to match the indices of the times appearing in (8.209). This gives

$$\begin{aligned}G(x', t_f, x, t_o) = \int \mathrm{d}x_1\cdots\mathrm{d}x_{N-1}\, &\langle\, x' \,|\, e^{-i\epsilon\hat{H}(t_{N-1})/\hbar} \,|\, x_{N-1}\,\rangle\langle\, x_{N-1} \,|\, e^{-i\epsilon\hat{H}(t_{N-2})/\hbar} \,|\, x_{N-2}\,\rangle\cdots \\ &\cdots\langle\, x_2 \,|\, e^{-i\epsilon\hat{H}(t_1)/\hbar} \,|\, x_1\,\rangle\langle\, x_1 \,|\, e^{-i\epsilon\hat{H}(t_o)/\hbar} \,|\, x\,\rangle.\end{aligned} \tag{8.211}$$

Each of the N infinitesimal transition elements can be replaced using result (8.206) indexed by the appropriate j, with the result that

$$G(x', t_f, x, t_o) = \int [\mathcal{D}x]\exp\left\{\frac{i}{\hbar}\sum_{j=0}^{N-1}\epsilon\left(\frac{1}{2}m\dot{x}_j^2 - V(x_j, t_j)\right)\right\} \tag{8.212}$$

where the *measure* is given by

$$[\mathcal{D}x] = \left(\frac{m}{2\pi i\epsilon\hbar}\right)^{N/2}\prod_{j=1}^{N-1}\mathrm{d}x_j, \tag{8.213}$$

and the identifications $x_0 = x$ and $x_N = x'$ are made. There are N square root factors since there are N infinitesimal transition elements. The formal time derivative notation stands for

$$\dot{x}_j = \frac{(x_{j+1} - x_j)}{\epsilon}, \tag{8.214}$$

and as before is not a real time derivative of a function. The sum appearing in (8.212) has the form of a Riemann integral with the classical Lagrangian formally appearing as the integrand. As a result, (8.212) is commonly written as the *path integral*,

$$G(x', t_f, x, t_o) = \int [\mathcal{D}x]\exp\left\{\frac{i}{\hbar}\int_{t_o}^{t_f}\mathrm{d}\tau\left(\frac{1}{2}m\dot{x}^2 - V(x, t)\right)\right\} = \int [\mathcal{D}x]\exp\left(\frac{i}{\hbar}S[x, \dot{x}, T]\right), \tag{8.215}$$

where $S[x, \dot{x}, T]$ is the classical action functional for the system defined in (1.11).

Because the path integral includes all possible intermediate positions for the particle, the path integral is often described as a *sum over paths or histories*, each of which are weighted by the exponential of the associated value of the classical action for the path. While this description of quantum mechanical processes is intuitively appealing, it has proved a formidable mathematical endeavor to define exactly how to implement a sum over all paths. In the next section, the path integral will be used to derive an approximate form for the propagator derived from the particle trajectory that extremizes the classical action. In addition, there are numerous subtleties that arise in formulating the path integral for discrete systems, higher dimensional spherical coordinate systems, and systems with a singular potential. The interested reader is recommended to the references.

8.6.4 The path integral and the WKB approximation

There is a circumstance which allows the formal time derivative (8.214) to be implemented. This is the case where the variables of integration in (8.212) are translated by a classical solution $x_c(t)$ satisfying the boundary conditions $x_c(t_f) = x'$ and $x_c(t_o) = x$. This assumes that the range of integrations are $[-\infty, \infty]$. The new variables are given by

$$x_j \rightarrow x_j + x_c(t_j). \tag{8.216}$$

For such a case it follows that

$$\frac{(x_{j+1} - x_j)}{\epsilon} \rightarrow \frac{(x_{j+1} - x_j)}{\epsilon} + \frac{x_c(t_j + \epsilon) - x_c(t_j)}{\epsilon} = \frac{(x_{j+1} - x_j)}{\epsilon} + \dot{x}_c(t_j). \tag{8.217}$$

The Lagrangian appearing in (8.215) is then expanded around the classical solution. Denoting $\mathcal{L}_c = \mathcal{L}(x_c, \dot{x}_c)$, and using the specific form (8.201) and the associated Lagrangian, the result is

$$\int_{t_o}^{t_f} \mathrm{d}t\, \mathcal{L}(x_c(t) + x, \dot{x}_c(t) + \dot{x}) = \int_{t_o}^{t_f} \mathrm{d}t \left(\mathcal{L}_c + x\frac{\partial \mathcal{L}_c}{\partial x_c} + \dot{x}\frac{\partial \mathcal{L}_c}{\partial \dot{x}_c} - \frac{1}{2}x^2\frac{\partial^2 V(x_c, t)}{\partial x_c^2} + \frac{1}{2}m^2\dot{x}^2 + \cdots \right). \tag{8.218}$$

The third term can be rewritten using the adjacent terms in the series and the formal time derivative,

$$\begin{aligned} &\frac{(x_{j+1} - x_j)}{\epsilon}\frac{\partial \mathcal{L}_c}{\partial \dot{x}_c(t_j)} + \frac{(x_j - x_{j-1})}{\epsilon}\frac{\partial \mathcal{L}_c}{\partial \dot{x}_c(t_{j-1})} \\ &= \frac{x_{j+1}}{\epsilon}\frac{\partial \mathcal{L}_c}{\partial \dot{x}_c(t_j)} - \frac{x_j}{\epsilon}\left(\frac{\partial \mathcal{L}_c}{\partial \dot{x}_c(t_j)} - \frac{\partial \mathcal{L}_c}{\partial \dot{x}_c(t_{j-1})} \right) - \frac{x_{j-1}}{\epsilon}\frac{\partial \mathcal{L}_c}{\partial \dot{x}_c(t_{j-1})} \\ &= \frac{x_{j+1}}{\epsilon}\frac{\partial \mathcal{L}_c}{\partial \dot{x}_c(t_j)} - x_j\frac{\mathrm{d}}{\mathrm{d}t}\frac{\partial \mathcal{L}_c}{\partial \dot{x}_c(t_{j-1})} - \frac{x_{j-1}}{\epsilon}\frac{\partial \mathcal{L}_c}{\partial \dot{x}_c(t_{j-1})}. \end{aligned} \tag{8.219}$$

The remaining two terms join with the adjacent terms in the series to generate further time derivatives. The upshot is the equivalent of an integration by parts, so that

$$\int_{t_o}^{t_f} \mathrm{d}t\, \dot{x}\frac{\partial \mathcal{L}_c}{\partial \dot{x}_c} = -\int_{t_o}^{t_f} \mathrm{d}t\, x\frac{\mathrm{d}}{\mathrm{d}t}\frac{\partial \mathcal{L}_c}{\partial \dot{x}_c}. \tag{8.220}$$

As a result of the classical solution satisfying the Euler–Lagrange equation (1.14), the terms linear in x and $\dot{x}$ vanish from the expansion just as they did in the classical derivation (1.12) of the Euler–Lagrange equation. The remaining action in the path integral is given by

$$\int_{t_o}^{t_f} \mathrm{d}t\, \mathcal{L}(x_c(t) + x, \dot{x}_c(t) + \dot{x}) = S[x_c, \dot{x}_c, T] + \int_{t_o}^{t_f} \mathrm{d}t \left(\frac{1}{2}m^2\dot{x}^2 - \frac{1}{2}\frac{\partial^2 V(x_c, t)}{\partial x_c^2}x^2 + \cdots\right). \tag{8.221}$$

The remaining quadratic and higher terms are referred to as *fluctuations*, since they represent quantum deviations from the classical path $x_c(t)$. The path integral factors into the classical contribution and the fluctuation contribution,

$$G(x', t_f, x, t_o) = \exp\left(\frac{i}{\hbar}S[x_c, \dot{x}_c, T]\right)\int [\mathcal{D}x]\exp\left\{\frac{i}{\hbar}\int_{t_o}^{t_f} \mathrm{d}t\left(\frac{1}{2}m\dot{x}^2 - \frac{1}{2}\frac{\partial^2 V(x_c, t)}{\partial x_c^2}x^2 + \cdots\right)\right\}. \tag{8.222}$$

Evaluating the quantum fluctuation contribution requires a specific form for the potential as well as a classical solution for the system. However, there is a way to obtain a general form for the result in the event that the higher order terms in (8.222) are ignored. This starts by noting that the classical action is an implicit function of the initial and final positions, x and x', so that

$$S[x_c, \dot{x}_c, T] = S_c(x', x). \tag{8.223}$$

This is true as well for the truncated version of the fluctuation contribution, so that it is written

$$P(x', x) = \int [\mathcal{D}x]\exp\left\{\frac{i}{\hbar}\int_{t_o}^{t_f} \mathrm{d}t\left(\frac{1}{2}m\dot{x}^2 - \frac{1}{2}\frac{\partial^2 V(x_c, t)}{\partial x_c^2}x^2\right)\right\}, \tag{8.224}$$

and is referred to as the *prefactor*. It is possible to use the eigenvalues of the differential operator appearing in (8.224),

$$D = \frac{\mathrm{d}^2}{\mathrm{d}t^2} + \frac{\partial^2 V(x_c, t)}{\partial x_c^2}, \tag{8.225}$$

to analyse the prefactor, and this approach forms the basis of the *Gelfand–Yaglom theorem*. Their theorem shows that the prefactor can be found from the eigenfunctions of the differential operator (8.225) which have eigenvalues of zero. The interested reader is recommended to the references.

However, there is a simpler method to infer the modulus of the prefactor. The key step in this derivation is to note that the propagator must satisfy the condition

$$\int_{-\infty}^{\infty} \mathrm{d}x\, G(x', t_f, x, t_0) G^*(x'', t_f, x, t_0) = \int_{-\infty}^{\infty} \mathrm{d}x\, \langle x' \mid \hat{U}(t_f, t_0) \mid x \rangle \langle x \mid \hat{U}^\dagger(t_f, t_0) \mid x'' \rangle$$
$$= \langle x' \mid \hat{U}(t_f, t_0) \hat{U}^\dagger(t_f, t_0) \mid x'' \rangle = \langle x' \mid x'' \rangle = \delta(x' - x''). \tag{8.226}$$

Using (8.223) and (8.224) in (8.226) gives the condition

$$\int_{-\infty}^{\infty} \mathrm{d}x \exp\left(\frac{i}{\hbar} S_c(x', x) - \frac{i}{\hbar} S_c(x'', x)\right) P(x', x) P^*(x'', x) = \delta(x' - x''). \tag{8.227}$$

The next step is to note that relation (8.227) must hold even if $x' - x''$ is arbitrarily small due to the singular nature of the Dirac delta. Expanding the action using $x'' = x' - (x' - x'')$ gives

$$S_c(x', x) - S_c(x'', x) \approx \frac{\partial S_c(x', x)}{\partial x'}(x' - x''), \tag{8.228}$$

which becomes exact as $x' - x'' \to 0$. Similarly, the prefactor product can be written

$$P(x', x) P^*(x'', x) = |P(x', x)|^2 - (x' - x'') P(x', x) \frac{\partial}{\partial x'} P^*(x', x). \tag{8.229}$$

For $x' - x''$ infinitesimal the second term can be suppressed. For $O(x' - x'')$ condition (8.227) therefore becomes

$$\int_{-\infty}^{\infty} \mathrm{d}x \exp\left(\frac{i}{\hbar} \frac{\partial S_c(x', x)}{\partial x'}(x' - x'')\right) |P(x', x)|^2 = \delta(x' - x''). \tag{8.230}$$

The next step is to change the variable of integration to

$$p = \pm \frac{\partial S_c(x', x)}{\partial x'} \implies \mathrm{d}p = \pm \frac{\partial^2 S_c(x', x)}{\partial x' \, \partial x} \mathrm{d}x, \tag{8.231}$$

where the choice of sign will be dictated by matching the required positive modulus of the prefactor. Result (8.230) becomes

$$\int_{-\infty}^{\infty} \mathrm{d}p \left(\pm \frac{\partial^2 S_c(x', x)}{\partial x' \, \partial x} \right)^{-1} |P(x', x)|^2 \, e^{\pm ip(x' - x'')/\hbar} = \delta(x' - x''). \tag{8.232}$$

The left side of result (8.232) gives the integral representation of the Dirac delta (3.95) if the following equality holds,

$$\pm \left(\frac{\partial^2 S_c(x', x)}{\partial x' \, \partial x} \right)^{-1} |P(x', x)|^2 = \frac{1}{2\pi\hbar}. \tag{8.233}$$

It follows that the *modulus* of the prefactor is inferred to be

$$|P(x', x)| = \sqrt{\frac{1}{2\pi\hbar} \left| \frac{\partial^2 S_c(x', x)}{\partial x' \, \partial x} \right|}, \tag{8.234}$$

where the absolute value in the square root occurs since the sign in (8.233) can be chosen to ensure that the second derivative factor is positive.

Unfortunately, this derivation cannot supply the overall phase factor of $P(x', x)$. This requires a more careful analysis using the Gelfand–Yaglom theorem mentioned earlier, and is available in the references. For a large variety of classical trajectories this simply adds an additional factor of $\sqrt{-i}$ to the prefactor, so that

$$P(x', x) = \sqrt{\frac{1}{2\pi i\hbar}\left|\frac{\partial^2 S_c(x', x)}{\partial x'\,\partial x}\right|}. \tag{8.235}$$

Returning this to (8.222) gives the final result for the propagator,

$$G(x', t_f, x, t_o) = \sqrt{\frac{1}{2\pi i\hbar}\left|\frac{\partial^2 S_c(x', x)}{\partial x'\,\partial x}\right|}\,\exp\left(\frac{i}{\hbar}S_c(x', x)\right), \tag{8.236}$$

where the classical action is evaluated using the classical trajectory from x at time t_o to x' at time t_f. Result (8.236) is referred to as the *Wentzel–Kramers–Brillouin (WKB) approximation.* The original derivation of the WKB approximation uses wave mechanics to find the approximate wave function in the region where the classical momentum is changing slowly. The original derivation is also available in the references.

It is instructive to apply result (8.236) to the specific case of a free particle in one dimension. The classical Lagrangian density is

$$\mathcal{L} = \frac{1}{2}m\dot{x}^2, \tag{8.237}$$

which gives the classical equation of motion

$$\ddot{x}_c = 0. \tag{8.238}$$

The classical trajectory from x to x' in the time interval $T = t_f - t_o$ that satisfies (8.238) is given by

$$x_c(t) = x + \left(\frac{x' - x}{t_f - t_o}\right)(t - t_o). \tag{8.239}$$

The classical action for this trajectory is given by

$$S_c(x', x) = \int_{t_o}^{t_f} \mathrm{d}t\,\frac{1}{2}m\dot{x}_c^2 = \frac{1}{2}m\left(\frac{x' - x}{t_f - t_o}\right)^2\Bigg|_{t_o}^{t_f} = \frac{1}{2}m\frac{(x' - x)^2}{(t_f - t_o)}, \tag{8.240}$$

so that

$$\left|\frac{\partial^2 S_c(x', x)}{\partial x'\,\partial x}\right| = \frac{m}{t_f - t_o}. \tag{8.241}$$

Using these results in (8.236) gives the WKB approximation for the free particle propagator,

$$G(x', t_f, x, t_o) = \sqrt{\frac{m}{2\pi i\hbar(t_f - t_o)}} \exp\left(\frac{im(x' - x)^2}{2\hbar(t_f - t_o)}\right). \tag{8.242}$$

This is precisely the previous result (6.364) for the free propagator in one dimension that was found using operator techniques and Fresnel's theorem. It follows from the derivation presented here that the WKB approximation is exact for quadratic Lagrangians since there are no higher order terms in (8.222) to be ignored.

Because of its connection to the classical analysis of a system, the path integral is a powerful tool for analysing a variety of systems in both quantum mechanics and quantum field theory. The interested reader is recommended to the numerous excellent texts that explore the myriad applications and subtleties of the path integral formalism.

References and recommended further reading

Texts that develop perturbation theory in quantum mechanics include

- Schiff L 1968 *Quantum Mechanics* 3rd edn (New York: McGraw-Hill)
- Greiner W 2001 *Quantum Mechanics: An Introduction* 4th edn (New York: Springer)
- Gasiorowicz S 2003 *Quantum Physics* 3rd edn (New York: Wiley)
- Schwabl F 2007 *Quantum Mechanics* 4th edn (New York: Springer)
- Mahan D 2009 *Quantum Mechanics in a Nutshell* (Princeton, NJ: Princeton University Press)
- Messiah A 2014 *Quantum Mechanics* (New York: Dover)
- Griffiths D and Schroeter D 2018 *Introduction to Quantum Mechanics* 3rd edn (New York: Cambridge University Press)
- Cohen–Tannoudji C, Diu B, and Laloë F 2019 *Quantum Mechanics* vol 1 2nd edn (New York: Wiley)
- Sakurai J and Napolitano J 2020 *Modern Quantum Mechanics* (New York: Cambridge University Press)

Rayleigh developed perturbation theory for sound waves in

- Rayleigh J 1894 *Theory of Sound* vol 1, 2nd edn (London: McMillan)

Schrödinger's adaptation of Rayleigh's perturbative methods for quantum mechanics is found in

- Schrödinger E 1926 Quantisierung als Eigenwertproblem *Ann. Phys.* **80** 437 (https://doi.org/10.1002/andp.19263851302)

The relativistically correct spin–orbit coupling is derived in

- Thomas L 1926 The motion of the spinning electron *Nature* **117** 2945 (https://www.nature.com/articles/117514a0)

- Jackson J 1999 *Classical Electrodynamics* 3rd edn (Hoboken, NJ: Wiley)

The Zeeman effect was reported in

- Zeeman P 1897 The effect of magnetisation on the nature of light emitted by a substance *Nature* **55** 347 (https://doi.org/10.1038/055347a0)

The distinction between the normal and anomalous Zeeman effects was first reported in

- Preston T 1898 Radiation Phenomena in a Strong Magnetic Field *Sci. Trans. R. Dubl. Soc.* **6** 385

The derivation of Kramer's relation from the Feynman–Hellmann theorem and the virial theorem as well as the problem with $\ell = 0$ in the spin–orbit coupling is discussed in the excellent text

- Schwabl F 2007 *Quantum Mechanics* 4th edn (New York: Springer)

Dirac developed time-dependent perturbation theory to study radioactive decay in

- Dirac P 1927 The quantum theory of the emission and absorption of radiation *Proc. R. Soc.* A **114** 767 (https://doi.org/10.1098/rspa.1927.0039)

Much of Fermi's golden rule was developed by Dirac in the previous reference. Fermi named it the golden rule in

- Fermi E 1950 *Nuclear Physics* (Chicago: University of Chicago Press)

Dirac presented the connection of the classical Lagrangian to quantum mechanics in

- Dirac P 1933 The Lagrangian in Quantum Mechanics *Phys. Z. Sowjetunion* **3** 64

Feynman developed the path integral approach to quantum mechanics in

- Feynman R 1948 Space-time approach to non-relativistic quantum mechanics *Rev. Mod. Phys.* **20** 367 (https://doi.org/10.1103/RevModPhys.20.367)

There are many excellent texts that develop path integral approaches to quantum mechanics. They include

- Feynman R, Hibbs A and Styer D 2010 *Quantum Mechanics and Path Integrals* (New York: Dover)
- Schulman L 2005 *Techniques and Applications of Path Integration* (New York: Dover)
- Kleinert H 2004 *Path Integrals in Quantum Mechanics, Statistics, Polymer Physics, and Financial Markets* (Singapore: World Scientific)

The WKB approximation was published in

- Brillouin L 1926 La mécanique ondulatoire de Schrödinger: une méthode générale de resolution par approximations successives *Compt. Rend. Acad. Sci.* **183** 24

- Kramers H 1926 Wellenmechanik und halbzahlige Quantisierung *Z. Phys.* **39** 828 (https://doi.org/10.1007/BF01451751)
- Wentzel G 1926 Eine Verallgemeinerung der Quantenbedingungen für die Zwecke der Wellenmechanik *Z. Phys.* **38** 518 (https://doi.org/10.1007/BF01397171)

IOP Publishing

A Concise Introduction to Quantum Mechanics (Second Edition)

Mark S Swanson

Chapter 9

Scattering

One of the most important experimental methods for understanding quantum level behavior is scattering. The use of scattering to determine the nature of atomic level phenomena began with Rutherford's experiments to determine the nature of the atom. His experiment used alpha particles, now known as helium nuclei, generated by radium decay. These were aimed at a thin foil of gold around one hundred atoms thick. Most of the alpha particles passed through the foil with only a small to imperceptible deviation, but occasionally one collided with an object and underwent a dramatic deviation in direction. From the rate of collisions and the angles of deflection, Rutherford was able to infer the size of the gold nucleus, revealing that it was extremely compact and dense, while the rest of the atom created little to no impediment to the alpha particles. This experiment supported the Keplerian model of the atom as a cloud of light electrons surrounding a dense positive nucleus, helping to launch modern quantum mechanics. This is discussed again later.

9.1 Basic concepts of scattering

In order to quantify scattering at the quantum mechanical level, it is necessary to define the basic concepts involved.

9.1.1 Cross-sections and scattering rates

In a simple example, a collection of n identical metal plate targets, each with cross-sectional area σ_T, are hung on a flat wall of area A so that they do not overlap and do not cover the entire wall. Small projectiles, such as bullets, are shot perpendicularly but otherwise randomly against the wall at the rate R_o, which is measured in *projectiles per second*. Each time a projectile strikes one of the metal plates it makes an audible ringing sound, but if it misses there is no noise. The fraction of the wall covered by the plates is $n\sigma_T/A$. The *average rate* R at which collisions between the projectiles and the metal plates are heard is given by

$$R = \frac{n\sigma_T}{A} R_o. \tag{9.1}$$

For the purposes of defining scattering in quantum mechanics it is very useful to note that R_o/A has the units of projectiles *or particles* per area per unit time, and so it is designated as the *flux* J of projectiles. Result (9.1) is easily reversed to find the area σ_T of one of the metal plates from the number of plates n and the rate R at which the flux results in collisions,

$$\sigma_T = \frac{R}{nJ}. \tag{9.2}$$

However, in a typical experiment it is impossible to measure all the collisions the incident particles make with the targets.

9.1.2 Differential cross-sections

In addition, detecting the scattered particles is limited by the detector size, which is typically characterized by the infinitesimal solid angle $\mathrm{d}\Omega = \sin\theta \, \mathrm{d}\theta \, \mathrm{d}\phi$ that the detector subtends. This is depicted in figure 9.1. The detected flux of scattered particles occurs at the spherical angles θ and ϕ measured with respect to the axis of the incident flux. The number of particles scattered per unit time into the solid angle $\mathrm{d}\Omega$ is denoted $\mathrm{d}\mathcal{N}$ and will be proportional to the number of targets n, the incident flux J, and the solid angle $\mathrm{d}\Omega$ of the detector. The constant of proportionality will depend on the angles θ and ϕ and is denoted $\sigma(\theta, \phi)$, so that

$$\mathrm{d}\mathcal{N} = \sigma(\theta, \phi) \, nJ \, \mathrm{d}\Omega, \tag{9.3}$$

where $\sigma(\theta, \phi)$ has the units of area and is known as the *differential cross-section* of the target. This relation is readily inverted to give

$$\sigma(\theta, \phi) = \frac{1}{nJ} \frac{\mathrm{d}\mathcal{N}}{\mathrm{d}\Omega}. \tag{9.4}$$

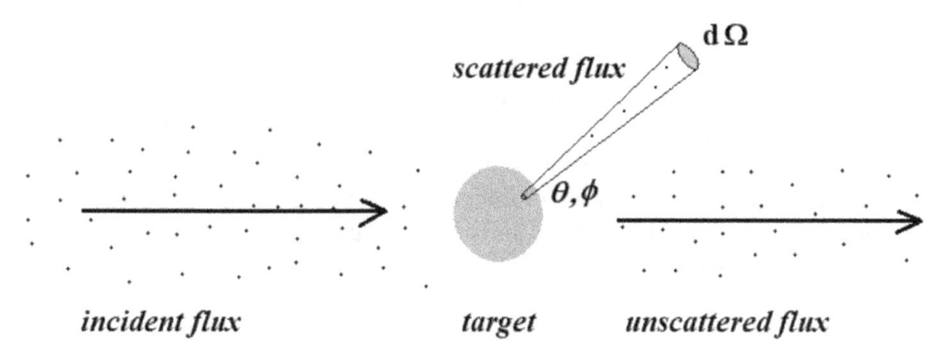

Figure 9.1. An idealized scattering process.

The definitions (9.2) and (9.4) can be related to each other by integrating (9.4) over the sphere surrounding the target and treating the flux as a constant,

$$\oint \sigma(\theta, \phi)\, d\Omega = \oint \frac{1}{nJ}\frac{d\mathcal{N}}{d\Omega}\, d\Omega = \frac{1}{nJ}\oint d\mathcal{N} = \frac{R}{nJ} = \sigma_T, \tag{9.5}$$

where the sum of the particles emerging from the sphere per unit time is the rate R of scattering defined earlier. As a result, the differential cross-section can be written

$$\sigma(\theta, \phi) = \frac{d\sigma_T}{d\Omega}, \tag{9.6}$$

showing that it is the contribution at the angles θ and ϕ to the total cross-section.

9.2 Basic aspects of quantum mechanical scattering

In chapter 5 several basic examples of quantum mechanical scattering were encountered. The first was reflection by a one-dimensional quantum well, which consisted of finding a wave mechanical solution that represented an incoming momentum wave function as well as two outgoing momentum wave functions. For that simple one-dimensional system result (5.32) corresponds to the reflection or *scattering* of the incoming wave while (5.33) corresponds to the transmitted or *unscattered* component of the incoming wave. The second more realistic example was the presence of the positive energy spherical wave (5.215) for the hydrogen atom, which represents the scattering of an electron from the Coulomb potential of the nucleus for the case of zero angular momentum. The hydrogen atom result demonstrates the basic aspects of quantum mechanical scattering. The incoming flux is characterized in terms of a momentum wave, while the scattered flux is characterized in terms of a spherical wave.

9.2.1 Simplifying assumptions

There are a number of assumptions made regarding the scattering process in order to simplify analysis. The incident particles are assumed not to interact among themselves. The incident particles are also assumed to undergo a single scattering interaction with one of the target particles. This can be stated as the requirement that the wave functions of the incident particles do not overlap two targets simultaneously. It is assumed that the detector is positioned to interact only with scattered particles, and that the scattered particles do not interact with the incident particles. It is assumed the potential is such that *elastic scattering* occurs, so that the total energy of the incident particle and its target remains constant. Finally, it is assumed that the potential V governing the incident particle and target interaction is spherically symmetric and satisfies

$$\lim_{r\to\infty} rV(r) = 0, \tag{9.7}$$

where r is the distance between the incident particle and its target. Requirement (9.7) excludes the Coulomb potential in what follows. In that regard, there are alternate methods which allow such a long-range potential to be analysed.

9.2.2 The Schrödinger equation and scattering

Scattering is an intrinsically two-body interaction and therefore requires that the Schrödinger equation deal with both the incident particle and the target. The analysis of scattering for such a system therefore begins by using the center of mass coordinate system (5.106) and (5.108) used earlier to analyse the two-body problem of the hydrogen atom. The mass of the incident particle and target are denoted m_1 and m_2, respectively, their positions are $\boldsymbol{x}_1$ and $\boldsymbol{x}_2$, and their momenta are $\boldsymbol{p}_1$ and $\boldsymbol{p}_2$. Their total mass is $M = m_1 + m_2$ while the relative position is $\boldsymbol{r} = \boldsymbol{x}_1 - \boldsymbol{x}_2$. The remaining center of mass coordinates are repeated below,

$$
\begin{aligned}
\boldsymbol{p} &= \frac{m_2\boldsymbol{p}_1 - m_1\boldsymbol{p}_2}{m_1 + m_2}, \\
\mu &= \frac{m_1 m_2}{m_1 + m_2}, \\
\boldsymbol{P} &= \boldsymbol{p}_1 + \boldsymbol{p}_2, \\
\boldsymbol{R} &= \frac{m_1\boldsymbol{x}_1 + m_2\boldsymbol{x}_2}{m_1 + m_2}.
\end{aligned}
\tag{9.8}
$$

For the case that the V is a central potential, the Hamiltonian of the two-particle system is

$$H = \frac{P^2}{2M} + \frac{p^2}{2\mu} + V(r). \tag{9.9}$$

For the case that $m_1 \ll m_2$, the coordinates $\boldsymbol{p}$ and $\boldsymbol{r}$ coincide with the momentum and position of the incident particle and the reduced mass is given by $\mu = m_1$, but such an assumption regarding the relative masses is not necessary. Like the analysis of the hydrogen atom, the center of mass term will be ignored since it corresponds to free motion, while the reduced mass corresponds to scattering.

Elevating the remaining part of the Hamiltonian to a differential operator and using the Laplacian in spherical coordinates (1.52) results in the version of the Schrödinger equation to be applied to scattering,

$$-\frac{\hbar^2}{2\mu}\left(\frac{\partial^2}{\partial r^2} + \frac{2}{r}\frac{\partial}{\partial r} + \frac{\cot\theta}{r^2}\frac{\partial}{\partial\theta} + \frac{1}{r^2}\frac{\partial^2}{\partial\theta^2} + \frac{1}{r^2\sin^2\theta}\frac{\partial^2}{\partial\phi^2}\right)\psi(\boldsymbol{r}) + V(r)\psi(\boldsymbol{r}) = E\psi(\boldsymbol{r}). \tag{9.10}$$

Since the potential satisfies condition (9.7), a solution to (9.10) is sought that corresponds to a positive energy eigenvalue given by

$$E = \frac{p^2}{2\mu}, \tag{9.11}$$

where p is the magnitude of the *asymptotic momentum* of the scattered particle. The form (9.11) corresponds to the assumption that the particle becomes a free particle *asymptotically*. Since the collision process is assumed to be elastic, the incident particle must also possess the same magnitude of momentum. For consistency with the scattering diagram figure 9.1, the vector momentum $\boldsymbol{p}$ of the incident particle will be chosen to lie along the z-axis.

Because the scattered particle will be characterized by a static spatially dependent wave, it is typically more convenient to work with the wave vector $\boldsymbol{k}$, which has the units of inverse length. This means that the magnitude of the asymptotic momentum is given by $p = \hbar k$, while the asymptotic energy is given by

$$E = \frac{\hbar^2 k^2}{2\mu}. \tag{9.12}$$

This has the useful property of obviating the need for factors of $\hbar$ in the wave states. It is also useful to introduce a *wave number state*, denoted by $|\, \boldsymbol{k} \,\rangle$. These states, like the momentum eigenstates, are orthonormal in the continuum sense,

$$\langle\, \boldsymbol{k} \mid \boldsymbol{k}' \,\rangle = \delta^3(\boldsymbol{k} - \boldsymbol{k}'). \tag{9.13}$$

As a result, their inner product with the position states is given by

$$\varphi_k(\boldsymbol{x}) = \langle\, \boldsymbol{x} \mid \boldsymbol{k} \,\rangle = \frac{1}{(2\pi)^{3/2}} e^{i\boldsymbol{k}\cdot\boldsymbol{x}}, \tag{9.14}$$

where $\varphi_k(\boldsymbol{x})$ can be thought of as a wave with momentum $\boldsymbol{p} = \hbar\boldsymbol{k}$.

There is often a need to work with *normalized wave states* over a volume V. These are written as

$$|\, \bar{\boldsymbol{k}} \,\rangle = N|\, \boldsymbol{k} \,\rangle, \tag{9.15}$$

which have the property that

$$\langle\, \boldsymbol{x} \mid \bar{\boldsymbol{k}} \,\rangle = N\langle\, \boldsymbol{x} \mid \boldsymbol{k} \,\rangle = N\frac{e^{i\boldsymbol{k}\cdot\boldsymbol{x}}}{(2\pi)^{3/2}}. \tag{9.16}$$

The constant N is determined by normalization for the choice that the volume V is a cubic box of side L,

$$\langle\, \bar{\boldsymbol{k}} \mid \bar{\boldsymbol{k}} \,\rangle = \int_V \mathrm{d}^3x \,\langle\, \bar{\boldsymbol{k}} \mid \boldsymbol{x} \,\rangle\langle\, \boldsymbol{x} \mid \bar{\boldsymbol{k}} \,\rangle = \int_V \mathrm{d}^3x \,\frac{|N|^2}{(2\pi)^3} = \frac{L^3|N|^2}{(2\pi)^3} = 1, \tag{9.17}$$

which gives what is known as *box normalization*,

$$N = \sqrt{\frac{(2\pi)^3}{L^3}}. \tag{9.18}$$

The resulting wave function is familiar from the momentum wave of (3.127),

$$\langle\, \boldsymbol{x} \mid \bar{\boldsymbol{k}} \,\rangle = \frac{1}{\sqrt{L^3}} e^{i\boldsymbol{k}\cdot\boldsymbol{x}}. \tag{9.19}$$

For the case of *continuum normalization* N is chosen to be one, so that

$$\langle\, \boldsymbol{x} \mid \boldsymbol{k} \,\rangle = \frac{1}{(2\pi)^{3/2}} e^{i\boldsymbol{k}\cdot\boldsymbol{x}}. \tag{9.20}$$

The choice of normalization will be indicated when relevant.

9.2.3 The general asymptotic scattering solution

The analysis begins by examining the large r behavior of the wave function. To that end, the wave function for large r that is consistent with the assumptions made so far is given by the general form

$$\lim_{r\to\infty} \psi(\boldsymbol{r}) = N\left(e^{i\boldsymbol{k}\cdot\boldsymbol{r}} + \frac{f(\theta,\, \phi)}{r} e^{ikr} \right), \tag{9.21}$$

where N is the normalization constant, either box normalization (9.19) or continuum normalization (9.20). It is important to note that, for simplicity of notation, the factor of $1/(2\pi)^{3/2}$ has been absorbed into the normalization factor N. As a result, continuum normalization corresponds to $|N|^2 = 1/(2\pi)^3$, while box normalization corresponds to $|N|^2 = 1/L^3$. The wave function (9.21) is the superposition of a momentum wave corresponding to *both* the incident and unscattered particles. It also has an outgoing spherical wave corresponding to the spatial probability amplitude of the scattered particle, which has the same magnitude of momentum as the incident particle.

The first term clearly satisfies

$$-\frac{\hbar^2}{2\mu} \nabla^2 (N e^{i\boldsymbol{k}\cdot\boldsymbol{r}}) = \frac{\hbar^2 k^2}{2\mu} (N e^{i\boldsymbol{k}\cdot\boldsymbol{r}}). \tag{9.22}$$

In what follows it will be assumed that $\boldsymbol{k}$ is directed along the z-axis, so that the first term becomes

$$N e^{i\boldsymbol{k}\cdot\boldsymbol{r}} = N e^{ikr\cos\theta}. \tag{9.23}$$

For $\theta = 0$, this expression becomes e^{ikr}, so that it corresponds to an outgoing wave, while for $\theta = \pi$ the expression becomes e^{-ikr}, so that it corresponds to an incoming wave. The presence of both is required to model *both* the incident and the *unscattered* particles.

The second term gives

$$-\frac{\hbar^2}{2\mu} \nabla^2 \left(N \frac{f(\theta,\, \phi)}{r} e^{ikr} \right) = \frac{\hbar^2 k^2}{2\mu} \left(N \frac{f(\theta,\, \phi)}{r} e^{ikr} \right) + \mathcal{O}\left(\frac{1}{r^3} \right). \tag{9.24}$$

The terms proportional to $1/r^3$ can be ignored in the large r limit. The last term in (9.10) is proportional to $V(r)$, which is assumed to satisfy condition (9.7). This means that the potential term in the Schrödinger eqution is assumed to drop off *faster* than the $1/r$ term in (9.21), and can also be ignored for large r. The combination of (9.22) and (9.24) therefore gives the asymptotic large r result

$$-\frac{\hbar^2}{2\mu}\nabla^2\psi(\boldsymbol{r}) \approx \frac{\hbar^2 k^2}{2\mu}\psi(\boldsymbol{r}) = E\psi(\boldsymbol{r}). \tag{9.25}$$

The form (9.21) therefore satisfies the boundary conditions appropriate to *elastic scattering* since it corresponds to the superposition of an incident and outgoing unscattered momentum wave and a scattered spherical wave, both of which have the same energy by virtue of having the same magnitude of momentum. This is identical to the scattering result (5.215) found earlier for the Coulomb potential.

9.2.4 The scattering probability current density

Now that the general form of the large r solution (9.21) consistent with scattering boundary conditions has been found, the next step is to determine how the spherical wave function $f(\theta, \phi)$ associated with scattering is related to the differential cross-section. The key to this is finding the probability current density or *flux* associated with the solution (9.21). Assuming that $\boldsymbol{k}$ is directed along the z-axis, the solution is broken into two pieces,

$$\psi_k(\boldsymbol{r}) = Ne^{i\boldsymbol{k}\cdot\boldsymbol{x}} = Ne^{ikr\cos\theta} = Ne^{ik_z z}, \tag{9.26}$$

$$\psi_s(\boldsymbol{r}) = N\frac{f(\theta, \phi)}{r}e^{ikr}, \tag{9.27}$$

where ψ_k correspond to the incident and unscattered particles and ψ_s corresponds to the scattered particles at a large distance r from the scattering center. The probability current density for each is calculated using the expression (4.27) derived earlier,

$$\boldsymbol{J} = -\frac{i\hbar}{2\mu}(\psi^*(\boldsymbol{r})\nabla\psi(\boldsymbol{r}) - \nabla\psi^*(\boldsymbol{r})\psi(\boldsymbol{r})). \tag{9.28}$$

Using the gradient in spherical coordinates given by (1.51) shows that (9.26) corresponds to the incident current density

$$\boldsymbol{J}_k = -i|N|^2\hbar(ik\cos\theta\,\hat{\boldsymbol{e}}_r - ik\sin\theta\,\hat{\boldsymbol{e}}_\theta) = |N|^2\frac{\hbar k}{\mu}\hat{\boldsymbol{e}}_z = |N|^2\frac{\boldsymbol{p}}{\mu} = |N|^2\boldsymbol{v}, \tag{9.29}$$

where $\boldsymbol{v} = \boldsymbol{p}/\mu$ is the velocity of the incoming and outgoing reduced mass particle obtained from its momentum $\boldsymbol{p} = \hbar\boldsymbol{k}$. For the case of box normalization $|N|^2 = 1/L^3$ and it follows that J_k has the units of probability per unit area per unit time

$$J_k = \frac{\hbar k}{\mu L^3}. \tag{9.30}$$

Similarly, the spherical wave component (9.27) has the asymptotic property

$$\begin{aligned}\nabla\psi_s(\boldsymbol{r}) &= \hat{\boldsymbol{e}}_r N\left(ik\frac{f(\theta,\phi)}{r} - \frac{f(\theta,\phi)}{r^2}\right)e^{ikr} + \hat{\boldsymbol{e}}_\theta N\frac{1}{r^2}\frac{\partial f(\theta,\phi)}{\partial\theta} + \hat{\boldsymbol{e}}_\phi N\frac{1}{r^2\sin\theta}\frac{\partial f(\theta,\phi)}{\partial\phi} \\ &\to \hat{\boldsymbol{e}}_r Nik\frac{f(\theta,\phi)}{r},\end{aligned} \tag{9.31}$$

where the terms proportional to $1/r^2$ were dropped in the large r limit. Using this in (9.28) gives the probability current density for the scattered particles,

$$\boldsymbol{J}_s = |N|^2\frac{\hbar k\hat{\boldsymbol{e}}_r}{\mu}\frac{|f(\theta,\phi)|^2}{r^2}. \tag{9.32}$$

The cross-terms between $\psi_k(\boldsymbol{r})$ and $\psi_s(\boldsymbol{r})$ in $\boldsymbol{J}$ are ignored using the assumption that the incident and the unscattered particles do not interact with the scattered particles.

9.2.5 The scattering amplitude and the differential cross-section

Because it is directed radially from the scattering center, the physical interpretation of the box normalized version of $\boldsymbol{J}_s$ is the probability per unit area per unit time for the scattering event. If a detector with the angular size of $\mathrm{d}\Omega$ is placed at a radial position a distance r from the target, the probability per unit time for the particle to strike the detector is given by the area $r^2\mathrm{d}\Omega$ of the detector and the scattered particle current density (9.32), which gives the rate at which particles will strike the detector,

$$\mathrm{d}\mathcal{N} = J_s r^2\mathrm{d}\Omega = |N|^2\frac{\hbar k}{\mu}|f(\theta,\phi)|^2\mathrm{d}\Omega. \tag{9.33}$$

This result can be combined with (9.3) by setting $n = 1$ to represent a single target and using the incident flux (9.29) to obtain

$$\mathrm{d}\mathcal{N} = \sigma(\theta,\phi)|N|^2\frac{\hbar k}{\mu}\mathrm{d}\Omega. \tag{9.34}$$

Equating (9.33) and (9.34) gives the differential cross-section,

$$\sigma(\theta,\phi) = |f(\theta,\phi)|^2. \tag{9.35}$$

As a result, finding the function $f(\theta,\phi)$, referred to as the *scattering amplitude*, will predict the differential cross-section and from it the total cross-section for scattering.

9.3 Partial wave analysis

Finding the scattering amplitude $f(\theta,\phi)$ can be achieved by solving, at least approximately, the Schrödinger equation (9.10) for the case that the energy eigenvalue E is given by (9.12) and the *effect* of the potential is included in an indirect manner. The latter statement is understood to mean that the wave function outside of the scattering region will exhibit aspects of the interaction. This will be clarified by constructing the solution for large r using two general methods.

9.3.1 Separation of variables

Like the solution to the hydrogen atom, this approach uses separation of variables in spherical coordinates, so that the stationary wave function is written

$$\psi(\boldsymbol{r}) = R(r)\,\Theta(\theta,\,\phi). \tag{9.36}$$

Substituting (9.36) into (9.10) and dividing by $\psi(\boldsymbol{r})$ and multiplying by r^2 gives

$$\begin{aligned}&-\frac{\hbar^2 r^2}{2\mu R(r)}\left(\frac{\partial^2}{\partial r^2}+\frac{2}{r}\frac{\partial}{\partial r}\right)R(r)+r^2(V(r)-E)\\&-\frac{\hbar^2}{2\mu\Theta(\theta,\,\phi)}\left(\cot\theta\frac{\partial}{\partial\theta}+\frac{\partial^2}{\partial\theta^2}+\frac{1}{\sin^2\theta}\frac{\partial^2}{\partial\phi^2}\right)\Theta(\theta,\,\phi)=0.\end{aligned} \tag{9.37}$$

Since the first line is a function solely of r and the second line is a function solely of θ and ϕ, the first line must be a constant, denoted $-C$, while the second line must be the constant C.

The second line of (9.37) therefore gives the eigenvalue equation

$$-\frac{\hbar^2}{2\mu}\left(\cot\theta\frac{\partial}{\partial\theta}+\frac{\partial^2}{\partial\theta^2}+\frac{1}{\sin^2\theta}\frac{\partial^2}{\partial\phi^2}\right)\Theta(\theta,\,\phi)=C\Theta(\theta,\,\phi). \tag{9.38}$$

The analysis of the hydrogen atom gave result (5.135), which shows that the differential operator in the parentheses of (9.38) is given by $\hat{L}^2/\hbar^2$, where $\hat{L}$ is the orbital angular momentum operator. The eigenvalue equation (9.38) therefore becomes

$$\frac{\hat{L}^2}{2\mu}\,\Theta(\theta,\,\phi)=C\,\Theta(\theta,\,\phi). \tag{9.39}$$

The solutions to (9.39) were found in (5.177) and are the spherical harmonics $Y_{\ell m}(\theta,\,\phi)$ with the eigenvalue

$$C=\frac{\ell(\ell+1)\hbar^2}{2\mu}, \tag{9.40}$$

where ℓ is a non-negative integer.

Returning this constant to the first line of (9.37) gives the radial equation,

$$\left(-\frac{\hbar^2}{2\mu}\left(\frac{\partial^2}{\partial r^2}+\frac{2}{r}\frac{\partial}{\partial r}\right)+V(r)+\frac{\ell(\ell+1)\hbar^2}{2\mu r^2}-\frac{\hbar^2k^2}{2\mu}\right)R_\ell(r)=0, \tag{9.41}$$

where $R_\ell(r)$ has been labeled for its dependence on ℓ. This is simplified by scaling the potential,

$$U(r)=\frac{2\mu}{\hbar^2}V(r), \tag{9.42}$$

so that (9.41) becomes

$$\left(\frac{d^2}{dr^2} + \frac{2}{r}\frac{d}{dr} + k^2 - U(r) - \frac{\ell(\ell+1)}{r^2}\right)R_\ell(r) = 0. \tag{9.43}$$

The last two terms combine into an *effective potential*,

$$U_{\text{eff}}(r) = U(r) + \frac{\ell(\ell+1)}{r^2}, \tag{9.44}$$

where the second term is the centrifugal barrier familiar from classical motion in a central field. The condition $rU_{\text{eff}}(r) = 0$ for r large still clearly holds by virtue of (9.7).

The wave function (9.36) can now be written

$$\psi(\mathbf{r}) = \sum_{m=-\ell}^{\ell}\sum_{\ell=0}^{\infty} C_{\ell m}R_\ell(r)Y_{\ell m}(\theta, \phi), \tag{9.45}$$

where $R_\ell(r)$ satisfies the radial equation (9.43). For the case that original potential $V(r)$ is spherically symmetric, referred to as a *central potential*, (9.47) simplifies since it must be independent of the azimuthal angle ϕ. As a result, only $Y_{\ell 0}$ may be present. Recalling that the spherical harmonics $Y_{\ell m}$ become the Legendre polynomials $P_\ell(\cos\theta)$ for $m = 0$,

$$Y_{\ell 0}(\theta, \phi) = \sqrt{\frac{2\ell+1}{4\pi}}\,P_\ell(\cos\theta), \tag{9.46}$$

the expansion consistent with spherical symmetry is written

$$\psi(\mathbf{r}) = N\sum_{\ell=0}^{\infty} i^{\ell+1}(2\ell+1)R_\ell(r)P_\ell(\cos\theta), \tag{9.47}$$

where N is the normalization factor used in (9.21). Introducing N requires that $R_\ell(kr)$ is a dimensionless function of the dimensionless variable kr. The form for (9.47) also reflects the absorption of $C_{\ell 0}$ into the function $R_\ell(r)$, with the other factors $i^{\ell+1}(2\ell+1)$ introduced for later convenience. Equation (9.47) has broken the solution into the sum of what are referred to as *partial waves*. Each partial wave is characterized by the quantum number ℓ, corresponding to the angular momentum of the incident and scattered particle. Because angular momentum is conserved, the two angular momenta must be the same. In that regard, ℓ subsumes the role of the *impact parameter* familiar from classical mechanics. In classical scattering, the incident and scattered particles are characterized by their angular momentum $L = pb$ in the scattering frame, where p is the momentum while b measures the minimum perpendicular distance from the z-axis that the incident particle will obtain, known as the impact parameter. Partial wave scattering therefore borrows the nomenclature of atomic angular momentum, referring to $\ell = 0$ as s-wave scattering, $\ell = 1$ as p-wave scattering, and so on.

9.3.2 Spherical functions and the scattering solution

As discussed earlier, it is useful to model the outcome of scattering in terms of the solutions to (9.43) for the case $U(r)$ can be suppressed for large values of r. This assumes that the asymptotic form of the radial wave function $R_\ell(r)$ reflects the vestiges of the interaction with the potential, but avoids requiring detailed knowledge of the wave function for small r. The radial equation (9.43) for such a case becomes

$$\left(\frac{\mathrm{d}^2}{\mathrm{d}r^2} + \frac{2}{r}\frac{\mathrm{d}}{\mathrm{d}r} + k^2 - \frac{\ell(\ell+1)}{r^2}\right)R_\ell(r) = 0. \tag{9.48}$$

The real valued oscillatory solutions to (9.48) are given by the *spherical Bessel functions*, denoted $j_\ell(kr)$, and the *spherical Neumann functions*, denoted $n_\ell(kr)$. Setting $kr = x$, these are given by

$$\begin{aligned} j_\ell(x) &= (-1)^\ell x^\ell\left(\frac{1}{x}\frac{\mathrm{d}}{\mathrm{d}x}\right)^\ell\left(\frac{\sin x}{x}\right) \Longrightarrow j_0(x) = \frac{\sin x}{x}, \quad j_1(x) = \frac{\sin x}{x^2} - \frac{\cos x}{x}, \\ n_\ell(x) &= -(-1)^\ell x^\ell\left(\frac{1}{x}\frac{\mathrm{d}}{\mathrm{d}x}\right)^\ell\left(\frac{\cos x}{x}\right) \Longrightarrow n_0(x) = -\frac{\cos x}{x}, \quad n_1(x) = \frac{\sin x}{x} - \frac{\cos x}{x^2}. \end{aligned} \tag{9.49}$$

It should be clear that the spherical Bessel function $j_\ell(kr)$ is well behaved as $r \to 0$, while the spherical Neumann function $n_\ell(kr)$ diverges as $r \to 0$. Because the latter diverges for small r, it can asymptotically manifest the effect of the particle's interaction with a central potential. However, usage of the spherical Neumann function must avoid the singularity at $r = 0$. Because these functions are oscillatory, both functions possess an infinite sequence of zeroes that is useful for matching boundary conditions.

Like the sine and cosine functions, these two functions form a complete set suitable for expanding the radial scattering wave function in the large r region, while retaining the effects of the small r interaction of the particle with the scattering potential. The ℓ th partial wave radial solution can then be written

$$R_\ell(kr) = A_\ell\, j_\ell(kr) + B_\ell\, n_\ell(kr), \tag{9.50}$$

where A_ℓ and B_ℓ are *dimensionless constants.* These constants are written more commonly by defining the following equivalent expressions,

$$C_\ell = \sqrt{A_\ell^2 + B_\ell^2}, \qquad \cos\delta_\ell = \frac{A_\ell}{C_\ell}, \quad \sin\delta_\ell = -\frac{B_\ell}{C_\ell}, \tag{9.51}$$

where the minus sign is for later convenience. In order for the sine and cosine to be real valued the two original constants must have the same the complex phase. Expression (9.50) then becomes

$$R_\ell(kr) = C_\ell\left(\cos\delta_\ell\, j_\ell(kr) - \sin\delta_\ell\, n_\ell(kr)\right). \tag{9.52}$$

The advantage of this parameterization occurs when the large r forms of the two functions are used. It can be shown that

$$\lim_{r\to\infty} j_\ell(kr) = \frac{1}{kr}\sin\left(kr - \frac{1}{2}\ell\pi\right) = \frac{1}{2ikr}\left(e^{ikr-\frac{1}{2}\ell\pi} - e^{-ikr+\frac{1}{2}\ell\pi}\right),$$
$$\lim_{r\to\infty} n_\ell(kr) = -\frac{1}{kr}\cos\left(kr - \frac{1}{2}\ell\pi\right) = -\frac{1}{2kr}\left(e^{ikr-\frac{1}{2}\ell\pi} + e^{-ikr+\frac{1}{2}\ell\pi}\right). \tag{9.53}$$

Using these in (9.52) with some basic trigonometry gives the large r form of the partial wave radial solution,

$$\lim_{r\to\infty} R_\ell(kr) = \frac{C_\ell}{kr}\sin\left(kr - \frac{1}{2}\ell\pi + \delta_\ell\right). \tag{9.54}$$

The quantity δ_ℓ is known as the *phase shift* of the ℓ th *partial wave* and represents the deviation of the scattered wave from the spherical Bessel function. It will be shown that the presence of a non-zero value for δ_ℓ represents the effects of scattering from the potential. The partial wave superposition (9.47) can now be written

$$\psi(\mathbf{r}) = N\sum_{\ell=0}^{\infty} i^{\ell+1}(2\ell + 1)\, C_\ell\big(\cos\delta_\ell\, j_\ell(kr) - \sin\delta_\ell\, n_\ell(kr)\big)P_\ell(\cos\theta). \tag{9.55}$$

9.3.3 Spherical Hankel functions, cross-sections, and the Rayleigh expansion

The next step is to find a *second solution* to (9.48) with an asymptotic form that corresponds to the earlier deduced form (9.21), a combination of momentum wave and outgoing spherical wave. Neither $j_\ell(kr)$ nor $n_\ell(kr)$ can model the *solely outgoing spherical wave* present in the desired form (9.21) since (9.53) shows that they represent both incoming and outgoing spherical waves asymptotically. The key step is to combine them into the *spherical Hankel function of the first kind, $h_\ell(kr)$*, which is defined as

$$h_\ell(kr) = j_\ell(kr) + in_\ell(kr). \tag{9.56}$$

Since both $j_\ell(kr)$ and $n_\ell(kr)$ solve (9.48), so does $h_\ell(kr)$. Combining the two formulas of (9.49) shows that these complex-valued functions are given by the formula

$$h_\ell(x) = -(-x)^\ell\left(\frac{1}{x}\frac{\mathrm{d}}{\mathrm{d}x}\right)^\ell\left(\frac{i}{x}e^{ix}\right), \tag{9.57}$$

with the first three given explicitly by

$$h_0(kr) = -\frac{i}{kr}e^{ikr},$$
$$h_1(kr) = -\frac{(kr + i)}{(kr)^2}e^{ikr}, \tag{9.58}$$
$$h_2(kr) = \frac{(i(kr)^2 - kr - 3i)}{(kr)^3}e^{ikr}.$$

Up to a physically irrelevant phase factor, all three functions of (9.58) have the correct large r behavior required for the outgoing spherical wave in (9.21). The asymptotic behavior of an arbitrary spherical Hankel function can be written

$$\lim_{r\to\infty} h_\ell(kr) = (-i)^{\ell+1}\frac{e^{ikr}}{kr}, \tag{9.59}$$

with the three functions of (9.58) clearly demonstrating this general result.

The asymptotic behavior of the spherical Hankel function given by (9.59) allows the scattered spherical wave of (9.21) to be modeled while simultaneously solving (9.48) in the asymptotic region where $U(r) \to 0$. This begins by writing the scattered spherical wave solution of (9.48) as

$$\psi_s(r, \theta) = N\sum_{\ell=0}^{\infty} i^{\ell+1}(2\ell + 1)D_\ell\, h_\ell(kr)P_\ell(\cos\theta), \tag{9.60}$$

where the D_ℓ are *dimensionless constants* and the choice of the remaining coefficent factors is for convenience. The asymptotic behavior (9.59) gives

$$\lim_{r\to\infty} \psi_s(r, \theta) = N\frac{e^{ikr}}{kr}\sum_{\ell=0}^{\infty}(2\ell + 1)D_\ell\, P_\ell(\cos\theta), \tag{9.61}$$

which gives the desired outgoing spherical wave postulated in (9.21). Comparing this to (9.21) immediately identifies the scattering amplitude,

$$f(\theta) = \frac{1}{k}\sum_{\ell=0}^{\infty}(2\ell + 1)D_\ell\, P_\ell(\cos\theta). \tag{9.62}$$

Using relation (9.35) gives the differential cross-section,

$$\sigma(\theta) = |f(\theta)|^2 = \frac{1}{k^2}\sum_{\ell,\ell'=0}^{\infty}(2\ell + 1)(2\ell' + 1)D_\ell^* D_{\ell'} P_\ell(\cos\theta)P_{\ell'}(\cos\theta). \tag{9.63}$$

The Legendre polynomials satisfy the orthogonality relation (5.160), which allows the total cross-section to be found using relation (9.5),

$$\begin{aligned}\sigma_T = \oint d\Omega\, |f(\theta)|^2 &= \frac{1}{k^2}\sum_{\ell,\ell'=0}^{\infty}(2\ell + 1)(2\ell' + 1)D_\ell^* D_{\ell'}\int_0^{2\pi} d\phi \int_0^{\pi} d\theta \sin\theta\, P_\ell(\cos\theta)P_{\ell'}(\cos\theta)\\ &= \frac{1}{k^2}\sum_{\ell,\ell'=0}^{\infty}(2\ell + 1)(2\ell' + 1)D_\ell^* D_{\ell'}\left(\frac{4\pi\delta_{\ell\ell'}}{2\ell + 1}\right) = \frac{4\pi}{k^2}\sum_{\ell=0}^{\infty}(2\ell + 1)|D_\ell|^2.\end{aligned} \tag{9.64}$$

Determining the coefficients D_ℓ is therefore the key to evaluating both (9.63) and (9.64).

The next step is to relate the earlier solution for $\psi(\mathbf{r})$, given by (9.55), to the scattering solution (9.60). This begins by noting that the momentum wave component of (9.21) is also a solution of (9.48). As a result, combining it with

(9.60) gives the solution to (9.48) consistent with the scattering boundary conditions of (9.21). Doing this in a useful way requires the *Rayleigh lemma*,

$$\int_{-1}^{1} \mathrm{d}x\, e^{ikrx} P_\ell(x) = 2i^\ell j_\ell(kr). \tag{9.65}$$

This result is easy to see for $P_0(\cos\theta) = 1$ and $P_1(\cos\theta) = \cos\theta$ by comparing the result of the integration to the spherical Bessel function defined by (9.49). For example,

$$\int_{-1}^{1} \mathrm{d}x\, e^{ikrx} P_0(x) = \frac{1}{ikr}(e^{ikr} - e^{-ikr}) = \frac{2}{kr}\sin(kr) = 2i^0 j_0(kr). \tag{9.66}$$

The general proof of (9.65) is available in the references. The next step is to use the completeness (5.165) of the Legendre polynomials to invert (9.65). This gives

$$\begin{aligned}\sum_{\ell=0}^{\infty} i^\ell(2\ell+1) j_\ell(kr) P_\ell(\cos\theta) &= \int_{-1}^{1} \mathrm{d}x\, e^{ikrx} \sum_{\ell=0}^{\infty}\left(\frac{2\ell+1}{2}\right) P_\ell(x) P_\ell(\cos\theta) \\ &= \int_{-1}^{1} \mathrm{d}x\, e^{ikrx}\delta(x - \cos\theta) = e^{ikr\cos\theta}.\end{aligned} \tag{9.67}$$

The left side of (9.67) is known as the *Rayleigh expansion* for the incident and unscattered momentum wave $e^{ikr\cos\theta}$. The Rayleigh expansion can be combined with (9.60) to give a form for the wave function which becomes the form of the scattered wave given by (9.21) for large r,

$$\psi(\mathbf{r}) = N\sum_{\ell=0}^{\infty} i^\ell(2\ell+1)\big(j_\ell(kr) + iD_\ell h_\ell(kr)\big)P_\ell(\cos\theta). \tag{9.68}$$

The final step is to find the conditions for which the Rayleigh expansion (9.68) for the wave function is identical to the earlier expansion (9.55). Comparing the two expansions and using the form of the spherical Hankel function (9.56) shows that this equality requires

$$j_\ell(kr) + iD_\ell h_\ell(kr) = (1 + iD_\ell)j_\ell - D_\ell n_\ell(kr) = C_\ell\cos\delta_\ell\, j_\ell(kr) - C_\ell\sin\delta_\ell\, n_\ell(kr). \tag{9.69}$$

Because the two eigenfunctions are independent, this gives two equations,

$$\begin{aligned} C_\ell\cos\delta_\ell &= 1 + iD_\ell, \\ C_\ell\sin\delta_\ell &= D_\ell. \end{aligned} \tag{9.70}$$

These two equations have the simultaneous solutions

$$C_\ell = e^{i\delta_\ell}, \quad D_\ell = e^{i\delta_\ell}\sin\delta_\ell. \tag{9.71}$$

It follows from (9.62) that the scattering amplitude $f(\theta)$ is given by

$$f(\theta) = \frac{1}{k}\sum_{\ell=0}^{\infty}(2\ell+1)e^{i\delta_\ell}\sin\delta_\ell\, P_\ell(\cos\theta). \tag{9.72}$$

The total cross-section (9.64) can then be written

$$\sigma_T = \frac{4\pi}{k^2}\sum_{\ell=0}^{\infty}(2\ell + 1)\sin^2\delta_\ell, \tag{9.73}$$

which shows that $\delta_\ell = 0$ for all ℓ corresponds to the *absence* of scattering. As a result, the phase shifts carry the information regarding scattering.

A useful theorem can be proved by examining *forward scattering*, which corresponds to $\theta = 0$ or $\cos\theta = 1$. For such a case, $P_\ell(1) = 1$ for all ℓ. Using the Euler relation (3.19) it follows from (9.72) that

$$f(0) = \frac{1}{k}\sum_{\ell=0}^{\infty}(2\ell + 1)\cos\delta_\ell\sin\delta_\ell + \frac{i}{k}\sum_{\ell=0}^{\infty}(2\ell + 1)\sin^2\delta_\ell. \tag{9.74}$$

This shows that the imaginary part of the forward scattering amplitude $f(0)$ is given by

$$\operatorname{Im} f(0) = \frac{1}{k}\sum_{\ell=0}^{\infty}(2\ell + 1)\sin^2\delta_\ell = \frac{k}{4\pi}\left(\frac{4\pi}{k^2}\sum_{\ell=0}^{\infty}(2\ell + 1)\sin^2\delta_\ell\right) = \frac{k}{4\pi}\sigma_T, \tag{9.75}$$

where (9.73) was used. Result (9.75) is known as the *optical theorem*.

9.3.4 Finding the phase shifts

The effects of scattering manifest themselves in the radial component $R_\ell(r)$ of the wave function in the form of phase shifts that represent a mixture of spherical Bessel and Neumann functions. These phase shifts are directly related to the potential responsible for the scattering. Understanding the relationship between the phase shifts and the potential requires consideration of radial equation (9.43) where the potential is present.

Analysing (9.43) is expedited by redefining its solution as

$$\chi_\ell(r) = rR_\ell(r), \tag{9.76}$$

so that (9.43) becomes

$$\left(\frac{d^2}{dr^2} + k^2 - U(r) - \frac{\ell(\ell + 1)}{r^2}\right)\chi_\ell(r) = 0. \tag{9.77}$$

For the case that $U(r) = 0$ everywhere, it follows from previous analysis that the solution is given by $u_\ell(r) = rj_\ell(kr)$, which vanishes at $r = 0$ and solves

$$\left(\frac{d^2}{dr^2} + k^2 - \frac{\ell(\ell + 1)}{r^2}\right)u_\ell(r) = 0. \tag{9.78}$$

Multiplying (9.77) by $u_\ell(r)$ and (9.78) by $\chi_\ell(r)$ and subtracting the two results gives

$$u_\ell(r)\frac{d^2\chi_\ell(r)}{dr^2} - \chi_\ell(r)\frac{d^2u_\ell(r)}{dr^2} = \frac{d}{dr}\left(u_\ell(r)\frac{d\chi_\ell(r)}{dr} - \chi_\ell(r)\frac{du_\ell(r)}{dr}\right) = u_\ell(r)U(r)\chi_\ell(r). \quad (9.79)$$

Integrating the left side gives

$$\begin{aligned}\int_0^\infty dr\,\frac{d}{dr}\left(u_\ell(r)\frac{d\chi_\ell(r)}{dr} - \chi_\ell(r)\frac{du_\ell(r)}{dr}\right) &= \left(u_\ell(r)\frac{d\chi_\ell(r)}{dr} - \chi_\ell(r)\frac{du_\ell(r)}{dr}\right)\Bigg|_0^\infty \\ &= \lim_{r\to\infty}\left(u_\ell(r)\frac{d\chi_\ell(r)}{dr} - \chi_\ell(r)\frac{du_\ell(r)}{dr}\right),\end{aligned} \quad (9.80)$$

where $u_\ell(0) = 0$ and $du_\ell(0)/dr = 0$ were used.

This limit can be rewritten using the earlier result (9.53), which shows that

$$\begin{aligned}\lim_{r\to\infty} u_\ell(r) &= \lim_{r\to\infty} rj_\ell(kr) = \frac{1}{k}\sin(kr - \frac{1}{2}\ell\pi), \\ \lim_{r\to\infty}\frac{d}{dr}u_\ell(r) &= \frac{d}{dr}\left(\lim_{r\to\infty}(rj_\ell(kr))\right) = \cos(kr - \frac{1}{2}\ell\pi).\end{aligned} \quad (9.81)$$

Similarly, combining (9.54) with (9.71) gives

$$\begin{aligned}\lim_{r\to\infty}\chi_\ell(r) &= \lim_{r\to\infty} rR_\ell(kr) = \frac{e^{i\delta_\ell}}{k}\sin(kr - \frac{1}{2}\ell\pi + \delta_\ell), \\ \lim_{r\to\infty}\frac{d}{dr}\chi_\ell(r) &= \frac{d}{dr}\left(\lim_{r\to\infty}(rR_\ell(kr))\right) = e^{i\delta_\ell}\cos(kr - \frac{1}{2}\ell\pi + \delta_\ell).\end{aligned} \quad (9.82)$$

Using elementary trigonometry shows that (9.80) is given by

$$\begin{aligned}&\lim_{r\to\infty}\left(u_\ell(r)\frac{d\chi_\ell(r)}{dr} - \chi_\ell(r)\frac{du_\ell(r)}{dr}\right) \\ &= \frac{e^{i\delta_\ell}}{k}\left(\sin(kr - \frac{1}{2}\ell\pi)\cos(kr - \frac{1}{2}\ell\pi + \delta_\ell) - \cos(kr - \frac{1}{2}\ell\pi)\sin(kr - \frac{1}{2}\ell\pi + \delta_\ell)\right) = -\frac{e^{i\delta_\ell}}{k}\sin\delta_\ell.\end{aligned} \quad (9.83)$$

Equating this result to the integral of the right-hand side of (9.79) gives

$$e^{i\delta_\ell}\sin\delta_\ell = -k\int_0^\infty dr\,\chi_\ell(r)U(r)u_\ell(r) = -\frac{2mk}{\hbar^2}\int_0^\infty dr\,\chi_\ell(r)V(r)u_\ell(r). \quad (9.84)$$

Result (9.84) is exact, but has the drawback of requiring the exact solution $\chi_\ell(r)$.

The right-hand side of (9.84) can be approximated for the case that $V(r) \approx 0$ for $r > r_o$ and k is such that $kr \ll 1$ for $r < r_o$. The latter condition is referred to as *quantum mechanically slow*. The next assumption is that $\chi_\ell(r)$ can be approximated by its form outside of $r > r_o$, which is given by (9.52), so that

$$\chi_\ell(r) \approx e^{i\delta_\ell}r\left(\cos\delta_\ell\, j_\ell(kr) - \sin\delta_\ell\, n_\ell(kr)\right). \quad (9.85)$$

This shows that the factor $e^{i\delta_\ell}$ drops out of (9.84). If the potential is weak, then it is plausible to assume that δ_ℓ is small, so that $\cos\delta_\ell \approx 1$ and $\sin\delta_\ell \approx \delta_\ell$. For such a case, the spherical Neumann function can be suppressed in the integral and expression (9.84) can be approximated by

$$\delta_\ell \approx -\frac{2\,mk}{\hbar^2}\int_0^{r_o} \mathrm{d}r\; r^2 j_\ell^2(kr)V(r). \tag{9.86}$$

For large ℓ the integrand tends to zero since for $kr \ll 1$ the spherical Bessel function can be shown to behave as

$$\lim_{kr\to 0} j_\ell(kr) \approx \frac{(kr)^\ell}{(2\ell+1)!!}, \\ (2\ell+1)!! \equiv (2\ell+1)(2\ell-1)(2\ell-3)\cdots 1. \tag{9.87}$$

This is consistent with the earlier observation that the parameter ℓ functions as an effective impact parameter. For a short-ranged potential, which is assumed in (9.86), it follows that the incident particle will have less interaction with the scattering center as ℓ increases.

9.3.5 A simple example

A relatively simple system consists of a spherical barrier potential of the form

$$V(r) = \begin{cases} V_o & \text{if } r < r_o \\ 0 & \text{if } r > r_o \end{cases}, \tag{9.88}$$

where V_0 is a low energy potential and r_o is small. Using (9.86) gives the s-wave phase shift,

$$\delta_0 \approx -\frac{2m}{\hbar^2 r_o}\int_0^{r_o} \mathrm{d}r\; r^2 V_o = -\frac{2\,mV_o r_o^2}{3\hbar^2}. \tag{9.89}$$

Using this result in (9.73) gives the s-wave version of the total scattering cross-section,

$$\sigma_T = \frac{4\pi}{k^2}\sum_{\ell=0}^{\infty}(2\ell+1)\sin^2\delta_\ell \approx \frac{4\pi}{k^2}\delta_0^2 = \frac{4\pi m^2 V_o^2 r_o^4}{\hbar^4 k^2} = \frac{2\pi m V_o^2 r_o^4}{\hbar^2 E_k}, \tag{9.90}$$

where $E_k = \hbar^2 k^2/2\,m$ is the energy of the elastically scattered particle. This result shows that scattering vanishes as $r_o \to 0$.

However, for the case that $V_0 \to \infty$ in (9.88), referred to as a *hard sphere potential*, the approximation (9.86) breaks down. Result (5.19) from the analysis of barrier penetration shows that there is no penetration of the wave function into the interior of the sphere for the case of an infinite barrier. As a result, the radial wave function outside the sphere is given by the free particle solution and must be such that it vanishes at the surface of the sphere. Using the free particle solution outside the hard sphere, given by (9.52), shows that the phase shifts satisfy

$$R_\ell(kr_o) = C_\ell\big(\cos\delta_\ell\, j_\ell(kr_o) - \sin\delta_\ell\, n_\ell(kr_o)\big) = 0, \tag{9.91}$$

which has the solution

$$\tan\delta_\ell = \frac{j_\ell(kr_o)}{n_\ell(kr_o)}. \tag{9.92}$$

For $\ell = 0$, the forms for the spherical Bessel and Neumann functions provided by (9.49) give

$$\tan\delta_0 = -\frac{\sin(kr_o)}{\cos(kr_o)} = -\tan(kr_o) \implies \delta_0 = -kr_o. \tag{9.93}$$

The s-wave component of the radial wave is therefore given by

$$R_0(kr) = C_0(\cos\delta_0 j_\ell(kr) - \sin\delta_0 n_0(kr)) = \frac{e^{i\delta_0}}{kr}\sin(kr + \delta_0) = \frac{e^{-ikr_o}}{kr}\sin(k(r - r_o)), \tag{9.94}$$

which clearly vanishes at the surface of the sphere. The s-wave contribution to the total scattering cross-section is then given by

$$\sigma_T = \frac{4\pi}{k^2}\sin^2\delta_0 = \frac{4\pi}{k^2}\sin^2(kr_o). \tag{9.95}$$

If the system is quantum mechanically slow, then $\sin(kr_o) \approx kr_o$ and the total s-wave cross-section becomes

$$\sigma_T \approx 4\pi r_o^2. \tag{9.96}$$

This is equal to the total surface of the sphere rather than the classical cross-section πr_o^2.

Similarly, using $j_1(kr)$ and $n_1(kr)$ from (9.49) in (9.92) gives the p-wave phase shift

$$\tan\delta_1 = \frac{\tan(kr_o) - kr_o}{kr_o\tan(kr_o) - 1}. \tag{9.97}$$

In the quantum mechanically slow limit the Taylor series representation of the tangent function gives

$$\delta_1 \approx -\frac{1}{3}(kr_o)^3. \tag{9.98}$$

This shows the p-wave phase shift is small compared to the s-wave phase shift in the quantum mechanically slow limit. This is once again the result of the effective impact parameter increasing with ℓ. For the quantum mechanically slow case, the differential cross-section can be found from the expansion (9.72), so that the scattering amplitude is given by

$$f(\theta) = \frac{e^{i\delta_0}}{k}\sin\delta_0 + 3\frac{e^{i\delta_1}}{k}\sin\delta_1\cos\theta \approx -e^{-ikr_o}r_o - \frac{1}{3}e^{-i(kr_o)^3/3}k^2r_o^3\cos\theta. \tag{9.99}$$

Using (9.35) gives the differential cross-section for quantum mechanically slow conditions,

$$\sigma(\theta) = |f(\theta)|^2 \approx r_o^2\left(1 + \frac{2}{3}k^2r_o^2\cos\theta\right), \tag{9.100}$$

where only lowest order terms in kr_o were kept and $\cos(kr_o) \approx 1$ was used.

9.3.6 Resonance scattering

For the case that kr_o is arbitrary there are unusual cases for the scattering that can occur. For example, if $kr_o = \pi/2$ then (9.93) shows that the s-wave phase shift is maximized. However, if $kr_o = n\pi$, where n is a positive integer, then the s-wave contribution to the total cross-section is zero. This occurs for the de Broglie wavelength $\lambda = 2\pi/k$ given by

$$\lambda = \frac{2r_o}{n}, \tag{9.101}$$

which corresponds to a wavelength such that $n\lambda$ matches the diameter of the sphere. While the scattering center becomes *invisible* to the s-wave component of the incident particles, result (9.97) shows that the values of $kr_o = n\pi$ give the p-wave phase shift

$$\tan\delta_1 = n\pi. \tag{9.102}$$

As n becomes large this gives $\delta_1 \approx \pi/2$, the maximum value of the phase shift.

The situation where the phase shift δ_ℓ becomes $\pi/2$ is referred to as a *scattering resonance*. Result (9.73) shows that such a phase shift maximizes the ℓth contribution to the total cross-section,

$$\sigma_T \approx \frac{4\pi}{k^2}(2\ell + 1)\sin^2\delta_\ell = \frac{4\pi}{k^2}(2\ell + 1). \tag{9.103}$$

It is common practice to parameterize the phase shift in the vicinity of the energy E_0 where the resonance occurs in the following way,

$$\delta_\ell = \tan^{-1}\left(\frac{\Gamma}{2(E - E_0)}\right), \tag{9.104}$$

where Γ has the units of energy. Simple trigonometry gives

$$\sigma_T \approx \frac{4\pi}{k^2}(2\ell + 1)\sin^2\delta_\ell = \frac{4\pi}{k^2}(2\ell + 1)\left(\frac{\tan^2\delta_\ell}{1 + \tan^2\delta_\ell}\right) = \frac{4\pi}{k^2}(2\ell + 1)\left(\frac{\Gamma^2}{4(E - E_0)^2 + \Gamma^2}\right). \tag{9.105}$$

A cross-section of the form (9.105) is referred to as a *Breit–Wigner resonance*. The quantity Γ corresponds to the energy width or, equivalently, to the range of energies where the resonance manifests itself. This follows by noting that an energy difference of $E - E_0 = \Gamma/2$ corresponds to a halving of the contribution of the ℓ th partial wave to the cross-section.

9.4 Green's function techniques

While partial wave analysis is a differential equation approach to scattering, Green's function techniques use integral representations to solve Schrödinger's equation. The method is quite general, but this section will specifically tailor this approach to describing quantum mechanical scattering.

9.4.1 Green's functions and the Schrödinger equation

As before, the goal is to extract the scattering cross-section from the nature of the potential energy V that represents the scattering center. The Hamiltonian is assumed to take the form

$$\hat{H} = \hat{H}_o + \hat{V}(\hat{Q}), \tag{9.106}$$

where $\hat{H}_o$ is an exactly solvable Hamiltonian, typically chosen to be the free particle Hamiltonian,

$$\hat{H}_o = \frac{\hat{P}^2}{2m}. \tag{9.107}$$

The goal is to identify the wave function consistent with incident and scattered particles with energy $E_k = \hbar^2 k^2/2m$ interacting elastically with the fixed potential $\hat{V}(\hat{Q})$. This must be done in a manner that allows the scattered part of the wave function to be identified, from which the relevant scattering information is obtained.

The stationary version of Schrödinger's equation for this situation can be written in configuration representation as

$$\left(-\frac{\hbar^2}{2m}\nabla_x^2 - E_k\right)\psi(\boldsymbol{x}) = -V(\boldsymbol{x})\psi(\boldsymbol{x}). \tag{9.108}$$

This equation is solved using the Green's function, $G_k^+(\boldsymbol{x} - \boldsymbol{y})$, which satisfies

$$\left(-\frac{\hbar^2}{2m}\nabla_x^2 - E_k\right)G_k^+(\boldsymbol{x} - \boldsymbol{y}) = \delta^3(\boldsymbol{x} - \boldsymbol{y}). \tag{9.109}$$

The designation G_k^+ is used to indicate that the Green's function must be consistent with the energy E_k and the *outgoing spherical wave* of (9.21) in order to satisfy scattering boundary conditions. The incoming and outgoing momentum wave of (9.21) is represented by the usual solution of the free Hamiltonian,

$$\varphi_k(\boldsymbol{x}) = Ne^{i\boldsymbol{k}\cdot\boldsymbol{x}}, \tag{9.110}$$

which satisfies

$$\left(-\frac{\hbar^2}{2m}\nabla_x^2 - E_k\right)\varphi_k(\boldsymbol{x}) = 0. \tag{9.111}$$

The solution to (9.108) is formally written as

$$\psi(\boldsymbol{x}) = \varphi_k(\boldsymbol{x}) - \int \mathrm{d}^3y\ G_k^+(\boldsymbol{x} - \boldsymbol{y})V(\boldsymbol{y})\psi(\boldsymbol{y}). \tag{9.112}$$

That this satisfies (9.108) follows from (9.109) and (9.111), which give

$$\begin{aligned}\left(-\frac{\hbar^2}{2m}\nabla_x^2 - E_k\right)\psi(\boldsymbol{x}) &= \left(-\frac{\hbar^2}{2m}\nabla_x^2 - E_k\right)\varphi_k(\boldsymbol{x}) - \left(-\frac{\hbar^2}{2m}\nabla_x^2 - E_k\right)\int \mathrm{d}^3y\ G_k^+(\boldsymbol{x} - \boldsymbol{y})V(\boldsymbol{y})\psi(\boldsymbol{y}) \\ &= -\int \mathrm{d}^3y\ \delta^3(\boldsymbol{x} - \boldsymbol{y})V(\boldsymbol{y})\psi(\boldsymbol{y}) = -V(\boldsymbol{x})\psi(\boldsymbol{x}).\end{aligned} \tag{9.113}$$

Because (9.110) is an arbitrary solution of the free particle equation it is referred to as a *complementary function*, while the integral is referred to as the *particular function* since it solves the full differential equation. Expression (9.112) is known as the *Lippmann–Schwinger equation* and reformulates the Schrödinger equation as an *integral equation*.

9.4.2 The scattering Green's function

At first glance, finding the Green's function appears to be straightforward. The Dirac delta appearing on the right-hand side of (9.109) has a representation given by the three-dimensional version of (3.95),

$$\delta^3(\boldsymbol{x} - \boldsymbol{y}) = \int \frac{\mathrm{d}^3p}{(2\pi)^3}\, e^{i\boldsymbol{p}\cdot(\boldsymbol{x}-\boldsymbol{y})}. \tag{9.114}$$

It follows that the function $G_k(\boldsymbol{x} - \boldsymbol{y})$, defined by

$$G_k(\boldsymbol{x} - \boldsymbol{y}) = \int \frac{\mathrm{d}^3p}{(2\pi)^3} \frac{e^{i\boldsymbol{p}\cdot(\boldsymbol{x}-\boldsymbol{y})}}{E_\mathrm{p} - E_k}, \tag{9.115}$$

has the property that it satisfies (9.109).

However, (9.115) does not satisfy the scattering boundary condition, which requires that it becomes an outgoing spherical wave as $r = |\boldsymbol{x}|$ becomes large. In order to expose this shortcoming and find a remedy, it is necessary to *attempt* to evaluate (9.115). The first step is to choose a coordinate system for $\boldsymbol{p}$ such that $\boldsymbol{x} - \boldsymbol{y}$ is directed along the z-axis in the $\boldsymbol{p}$ coordinate system. This is always possible since the integrand of (9.115) is rotationally invariant. In spherical coordinates for $\boldsymbol{p}$ the integral then becomes

$$\begin{aligned} G_k(\boldsymbol{x} - \boldsymbol{y}) &= \int \frac{\mathrm{d}^3p}{(2\pi)^3} \frac{e^{i\boldsymbol{p}\cdot(\boldsymbol{x}-\boldsymbol{y})}}{E_\mathrm{p} - E_k} = \frac{m}{4\pi^3\hbar^2}\int_0^\infty \mathrm{d}p\, \frac{p^2}{p^2 - k^2}\int_0^\pi \mathrm{d}\theta_p \sin\theta_p\, e^{ip|\boldsymbol{x}-\boldsymbol{y}|\cos\theta_p}\int_0^{2\pi}\mathrm{d}\phi_p \\ &= \frac{m}{\pi^2\hbar^2|\boldsymbol{x} - \boldsymbol{y}|}\int_0^\infty \mathrm{d}p\left(\frac{p}{p^2 - k^2}\right)\sin(p|\boldsymbol{x} - \boldsymbol{y}|). \end{aligned} \tag{9.116}$$

Because the integrand does not change sign if $p \to -p$, the range of integration can be extended to the whole real p-axis, so that

$$G_k(\boldsymbol{x} - \boldsymbol{y}) = \frac{m}{2\pi^2\hbar^2|\boldsymbol{x} - \boldsymbol{y}|}\int_{-\infty}^{\infty} \mathrm{d}p\left(\frac{p}{p^2 - k^2}\right)\sin(p\,|\boldsymbol{x} - \boldsymbol{y}|). \tag{9.117}$$

The quantity k is still positive-definite. However, the integral appearing in (9.117) is not defined since the integrand is singular at $p = \pm k$, possessing poles directly on the real p-axis. This prevents using a contour that lies directly along the real p-axis to evaluate the integral using Cauchy's theorem.

The solution is to adopt a prescription for moving the poles of the integrand off the real axis, thereby allowing the use of Cauchy's theorem (3.32) to evaluate the Green's function. It will be shown that the correct Green's function to describe scattering is defined by the following epsilon procedure,

$$
\begin{aligned}
G_k^+(\mathbf{x}-\mathbf{y}) &= \lim_{\epsilon\to 0^+} \frac{m}{2\pi^2\hbar^2|\mathbf{x}-\mathbf{y}|}\int_{-\infty}^{\infty} \mathrm{d}p\, p\, \frac{\sin(p\,|\mathbf{x}-\mathbf{y}|)}{(p^2-k^2-i\epsilon)} \\
&= \lim_{\epsilon\to 0^+} \frac{m}{4\pi^2 i\hbar^2|\mathbf{x}-\mathbf{y}|}\int_{-\infty}^{\infty} \mathrm{d}p\, p\, \frac{(e^{ip|x-y|}-e^{-ip|x-y|})}{(p-\sqrt{k^2+i\epsilon})(p+\sqrt{k^2+i\epsilon})},
\end{aligned} \tag{9.118}
$$

where ϵ is an infinitesimal *positive quantity*. Because k^2 and ϵ are positive, the quantity $\sqrt{k^2+i\epsilon}$ is a complex number given by

$$
\sqrt{k^2+i\epsilon} = k\sqrt{1+i\frac{\epsilon}{k^2}} = k + i\frac{\epsilon}{2k} \equiv k + i\alpha, \tag{9.119}
$$

where both k and α are both positive real numbers, with α infinitesimal since it is proportional to ϵ. As a result, the Green's function of (9.118) can be written

$$
G_k^+(\mathbf{x}-\mathbf{y}) = \lim_{\alpha\to 0^+} \frac{m}{4\pi^2 i\hbar^2|\mathbf{x}-\mathbf{y}|}\int_{-\infty}^{\infty} \mathrm{d}p\, p\, \frac{(e^{ip|x-y|}-e^{-ip|x-y|})}{(p-k-i\alpha)(p+k+i\alpha)}. \tag{9.120}
$$

The integrand now possesses simple poles at $p = k + i\alpha$ and $p = -k - i\alpha$.

The integral of (9.120) is readily evaluated using Cauchy's theorem (3.32) when coupled with Jordan's lemma as applied to (3.38) and (3.39). The contours for evaluating the two components of (9.118) are depicted in figure 9.2. The upper right-handed contour enclosing the simple pole $p = k + i\alpha$ gives

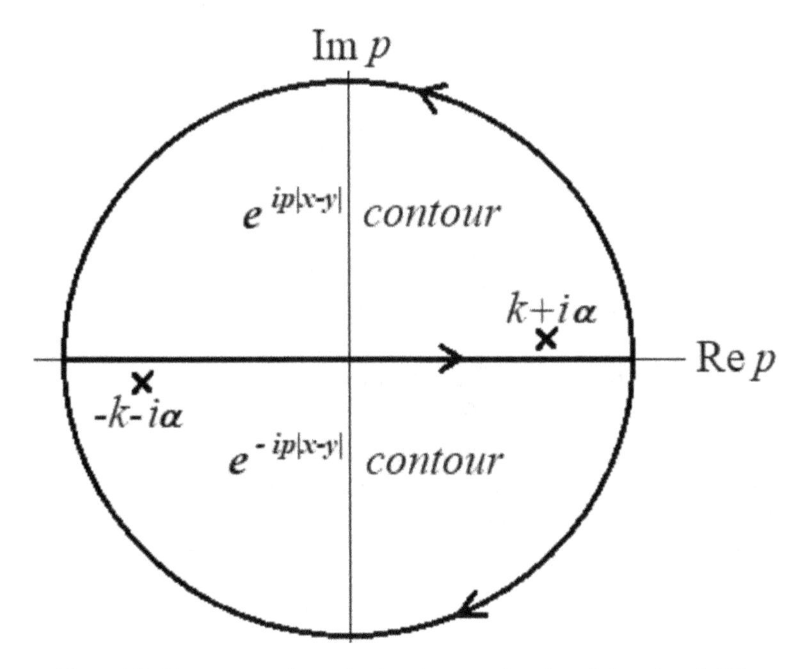

Figure 9.2. The contours for evaluating the scattering Green's function (9.118).

$$\lim_{\alpha\to 0^+}\int_{-\infty}^{\infty} \mathrm{d}p\, p\, \frac{e^{ip|x-y|}}{(p-k-i\alpha)(p+k+i\alpha)} = \pi i e^{ik|x-y|}, \tag{9.121}$$

while the lower left-handed contour enclosing the simple pole $p = -k - i\alpha$ gives

$$-\lim_{\alpha\to 0^+}\int_{-\infty}^{\infty} \mathrm{d}p\, p\, \frac{e^{-ip|x-y|}}{(p-k-i\alpha)(p+k+i\alpha)} = \pi i e^{ik|x-y|}. \tag{9.122}$$

Since the lower contour is left-handed the residue is multiplied by $-2\pi i$, which accounts for the positive sign. The result for the scattering Green's function is

$$G_k^+(\boldsymbol{x}-\boldsymbol{y}) = \frac{m}{2\pi\hbar^2}\frac{e^{ik|x-y|}}{|\boldsymbol{x}-\boldsymbol{y}|}. \tag{9.123}$$

The final step is to demonstrate that (9.123) gives the desired outgoing spherical wave for large $r = |\boldsymbol{x}|$. This begins by noting that $\boldsymbol{x}$ can be thought of as the location of the particle detector, so that $\boldsymbol{x}$ corresponds to the location of interest for observing the scattering. It follows that, to $\mathcal{O}(1/r)$,

$$\lim_{r\to\infty}|\boldsymbol{x}-\boldsymbol{y}| = \lim_{r\to\infty}\sqrt{|\boldsymbol{x}|^2 - 2\boldsymbol{x}\cdot\boldsymbol{y} + |\boldsymbol{y}|^2} = \lim_{r\to\infty} r\sqrt{1 - \frac{2|\boldsymbol{y}|\cos\alpha}{r} + \frac{|\boldsymbol{y}|^2}{r^2}} \approx r - |\boldsymbol{y}|\cos\alpha, \tag{9.124}$$

where α is the angle that the variable of integration $\boldsymbol{y}$ forms with $\boldsymbol{x}$. For the case that $\boldsymbol{y}$ is being integrated using spherical coordinates, the angle θ_y can be chosen to be the polar angle of $\boldsymbol{y}$. Similarly, it follows that, to $\mathcal{O}(1/r)$,

$$\lim_{r\to\infty}\frac{1}{|\boldsymbol{x}-\boldsymbol{y}|} = \lim_{r\to\infty}\frac{1}{r-|\boldsymbol{y}|\cos\alpha} \approx \lim_{r\to\infty}\frac{1}{r}\left(1+\frac{|\boldsymbol{y}|}{r}\cos\alpha\right) \approx \frac{1}{r}. \tag{9.125}$$

The large r limit of (9.123) is therefore given by

$$\lim_{r\to\infty} G_k^+(\boldsymbol{x}-\boldsymbol{y}) \approx \frac{m}{2\pi\hbar^2}\frac{e^{ikr}}{r}e^{-ik|\boldsymbol{y}|\cos\alpha}, \tag{9.126}$$

which demonstrates the emergence of the desired outgoing spherical wave. Returning this result to (9.112) gives

$$\begin{aligned}\lim_{r\to\infty}\psi_+(\boldsymbol{x}) &= \varphi_k(\boldsymbol{x}) - \lim_{r\to\infty}\int \mathrm{d}^3y\, G_k^+(\boldsymbol{x}-\boldsymbol{y})V(\boldsymbol{y})\psi_+(\boldsymbol{y}) \\ &= Ne^{i\boldsymbol{k}\cdot\boldsymbol{x}} - N\left(\frac{m}{2N\pi\hbar^2}\int \mathrm{d}^3y\, e^{-ik|\boldsymbol{y}|\cos\alpha}V(\boldsymbol{y})\psi_+(\boldsymbol{y})\right)\frac{e^{ikr}}{r},\end{aligned} \tag{9.127}$$

where $\psi_+(\boldsymbol{x})$ is the *scattered wave function*. It is important to note that this form assumes that the integrand becomes zero for large $|\boldsymbol{y}|$ since otherwise the approximation (9.124) breaks down. This condition therefore requires that $V(\boldsymbol{y})$ is a short-ranged potential. Comparing this result to (9.21) identifies the scattering amplitude,

$$f(\theta,\phi) = -\frac{m}{2N\pi\hbar^2}\int \mathrm{d}^3y\, e^{-ik|\boldsymbol{y}|\cos\alpha}\, V(\boldsymbol{y})\,\psi_+(\boldsymbol{y}). \tag{9.128}$$

The remaining step is to solve (9.112) for $\psi_+(\boldsymbol{x})$ in order to evaluate (9.128).

9.4.3 Incoming spherical waves and bound states

Before proceeding to the solution of (9.128) it is useful to note that the choice of epsilon procedure employed in (9.118) can be reversed to define an alternative Green's function,

$$G_k^-(\boldsymbol{x} - \boldsymbol{y}) = \lim_{\epsilon \to 0^+} \frac{m}{2\pi^2\hbar^2|\boldsymbol{x} - \boldsymbol{y}|} \int_{-\infty}^{\infty} \mathrm{d}p\, p \frac{\sin(p\,|\boldsymbol{x} - \boldsymbol{y}|)}{(p^2 - k^2 + i\epsilon)} = (G_k^+(\boldsymbol{x} - \boldsymbol{y}))^*. \quad (9.129)$$

Performing the integrations using Cauchy's theorem gives

$$G_k^-(\boldsymbol{x} - \boldsymbol{y}) = \frac{m}{2\pi\hbar^2} \frac{e^{-ik|\boldsymbol{x}-\boldsymbol{y}|}}{|\boldsymbol{x} - \boldsymbol{y}|}, \quad (9.130)$$

which results in the large r limit

$$\lim_{r\to\infty} G_k^-(\boldsymbol{x} - \boldsymbol{y}) \approx \frac{m}{2\pi\hbar^2} \frac{e^{-ikr}}{r} e^{ik|\boldsymbol{y}|\cos\alpha}. \quad (9.131)$$

The Green's function (9.132) results in an *incoming spherical wave.* While this does not match the scattering boundary conditions, it will be important in formal scattering approaches since it is the combination of incoming and outgoing spherical waves that forms a *complete set* in the *absence of bound states.*

The presence of possible negative energy bound states in the spectrum of the Hamiltonian can be accommodated in the formalism by altering the Lippmann–Schwinger equation to include only the particular function. This corresponds to the absence of a free particle *asymptotically* for negative energies or bound states. For the case that there are bound states corresponding to $E_n < 0$, the Green's function appearing in the Lippmann–Schwinger equation is altered to

$$G_n^B(\boldsymbol{x} - \boldsymbol{y}) = \frac{m}{2\pi^2\hbar^2|\boldsymbol{x} - \boldsymbol{y}|} \int_{-\infty}^{\infty} \mathrm{d}p\, p \frac{\sin(p\,|\boldsymbol{x} - \boldsymbol{y}|)}{(p^2 + \kappa_n^2)}, \quad (9.132)$$

which satisfies

$$\left(\frac{\hbar^2}{2m}\nabla^2 + E_n\right) G_n^B(\boldsymbol{x} - \boldsymbol{y}) = \delta^3(\boldsymbol{x} - \boldsymbol{y}), \quad (9.133)$$

where E_n is the positive quantity given by

$$E_n = \frac{\hbar^2\kappa_n^2}{2m}. \quad (9.134)$$

This corresponds to the negative energy eigenvalue $-E_n$, thereby incorporating the bound state into the Lippmann–Schwinger equation, which is given by

$$\psi_n(\boldsymbol{x}) = -\int \mathrm{d}^3y\; G_n^B(\boldsymbol{x} - \boldsymbol{y}) V(\boldsymbol{y}) \psi_n(\boldsymbol{y}). \quad (9.135)$$

Evaluating the bound state Green's function is similar to that of the positive energy scattering Green's function. It possesses two imaginary poles at $p = \pm i\kappa_n$, with the $i\kappa$ in the upper half of the complex plane and $-i\kappa_n$ in the lower half of the complex plane. Using Cauchy's theorem gives

$$G_n^B(x - y) = \frac{m}{2\pi^2\hbar^2}\frac{e^{-k_n|x-y|}}{|x - y|}, \tag{9.136}$$

which shows that the wave function (9.135) is asymptotically damped by the presence of the bound state.

9.4.4 The Born approximation

A perturbative scattering solution to (9.112) is obtained by denoting

$$V(\boldsymbol{x}) = \lambda H_{\rm I}(\boldsymbol{x}). \tag{9.137}$$

The wave function $\psi(\boldsymbol{x})$ is then expanded in a power series in λ by writing

$$\psi_+(\boldsymbol{x}) = \sum_{n=0}^{\infty}\lambda^n\psi_+^{(n)}(\boldsymbol{x}). \tag{9.138}$$

Substituting (9.137) and (9.138) into (9.112) gives

$$\sum_{n=0}^{\infty}\lambda^n\psi_+^{(n)}(\boldsymbol{x}) = \varphi_k(\boldsymbol{x}) - \sum_{n=0}^{\infty}\lambda^{n+1}\int \mathrm{d}^3y\; G_k^+(\boldsymbol{x} - \boldsymbol{y})H_{\rm I}(\boldsymbol{y})\psi_+^{(n)}(\boldsymbol{y}). \tag{9.139}$$

Equating the powers of λ gives an infinite sequence of equations,

$$\begin{aligned}
\psi_+^{(0)}(\boldsymbol{x}) &= \varphi_k(\boldsymbol{x}) = Ne^{i\boldsymbol{k}\cdot\boldsymbol{x}},\\
\psi_+^{(1)}(\boldsymbol{x}) &= -\int \mathrm{d}^3y\; G_k^+(\boldsymbol{x} - \boldsymbol{y})H_{\rm I}(\boldsymbol{y})\psi_+^{(0)}(\boldsymbol{y}),\\
\psi_+^{(2)}(\boldsymbol{x}) &= -\int \mathrm{d}^3y\; G_k^+(\boldsymbol{x} - \boldsymbol{y})H_{\rm I}(\boldsymbol{y})\psi_+^{(1)}(\boldsymbol{y}),\\
&\ldots.
\end{aligned} \tag{9.140}$$

Each equation in (9.140) is determined uniquely by the result of the previous equation in the sequence.

To first order in λ the wave function $\psi_+(\boldsymbol{x})$ is given by

$$\psi_+(\boldsymbol{x}) = \psi_+^{(0)}(\boldsymbol{x}) + \lambda\psi_+^{(1)}(\boldsymbol{x}) = \varphi_k(\boldsymbol{x}) - \int \mathrm{d}^3y\; G_k^+(\boldsymbol{x} - \boldsymbol{y})\; V(\boldsymbol{y})\; \varphi_k(\boldsymbol{y}), \tag{9.141}$$

and is referred to as the *Born approximation.* To lowest order $\psi_+(\boldsymbol{x}) = \varphi_k(\boldsymbol{x})$, so that to first order in λ the scattering amplitude of (9.128) becomes

$$f(\theta, \phi) = -\frac{m}{2\pi\hbar^2}\int \mathrm{d}^3y\; e^{-ik|y|\cos\alpha}\; V(\boldsymbol{y})\; e^{i\boldsymbol{k}\cdot\boldsymbol{y}}, \tag{9.142}$$

where the normalization factor N present in both terms has canceled.

The angle α appearing in (9.142) was defined earlier in (9.124) as the angle between $\boldsymbol{x}$ and $\boldsymbol{y}$, while $\boldsymbol{k}$ was chosen to point along the z-axis of the $\boldsymbol{x}$-coordinate system. In order to make further progress in performing the integral over d^3y it is useful to introduce a second vector $\boldsymbol{k}'$ that is *parallel* to the position vector $\boldsymbol{x}$ and has the same magnitude as $\boldsymbol{k}$. The vector $\boldsymbol{k}'$ is therefore the wave vector for the case that the particle is scattered along the position vector $\boldsymbol{x}$. The restriction that its magnitude satisfies $k' = k$ insures that the scattering is *elastic*. For such a choice, it follows that the angle between $\boldsymbol{k}'$ and $\boldsymbol{y}$ is also α, which gives

$$k|\boldsymbol{y}|\cos\alpha = k'|\boldsymbol{y}|\cos\alpha = \boldsymbol{k}' \cdot \boldsymbol{y}. \tag{9.143}$$

Returning (9.143) to (9.142) gives

$$f(\theta, \phi) = -\frac{m}{2\pi\hbar^2}\int d^3y\, e^{i(\boldsymbol{k}-\boldsymbol{k}')\cdot\boldsymbol{y}}\, V(\boldsymbol{y}). \tag{9.144}$$

For the case that V is a central potential, so that $V(\boldsymbol{y}) = V(|\boldsymbol{y}|)$, the integral over $\boldsymbol{y}$ is spherically symmetric. This allows performing the integral in a $\boldsymbol{y}$ coordinate system such that $\boldsymbol{k} - \boldsymbol{k}'$ defines the polar angle of $\boldsymbol{y}$. This gives

$$(\boldsymbol{k} - \boldsymbol{k}') \cdot \boldsymbol{y} = |\boldsymbol{k} - \boldsymbol{k}'||\boldsymbol{y}|\cos\theta_y, \tag{9.145}$$

where θ_y is now the polar variable of integration in spherical coordinates for $\boldsymbol{y}$. It is important to note that the scattering amplitude (9.144) will depend only upon the polar angle of $\boldsymbol{k}'$, which is the same as the angle θ appearing in the scattering amplitude. Defining the vector

$$\boldsymbol{p} = \boldsymbol{k} - \boldsymbol{k}', \tag{9.146}$$

and using $k = k'$ gives

$$p^2 = k^2 - 2kk'\cos\theta + k'^2 = 2k^2(1 - \cos\theta) = 4k^2\sin^2\left(\frac{1}{2}\theta\right). \tag{9.147}$$

The vector $\boldsymbol{p}$ points in the direction of the change in the momentum of the scattered particle. It follows that

$$e^{i(\boldsymbol{k}-\boldsymbol{k}')\cdot\boldsymbol{y}} = e^{i|\boldsymbol{k}-\boldsymbol{k}'|y\cos\theta_y} = e^{ipy\cos\theta_y} = e^{i2k\sin(\frac{1}{2}\theta)y\cos\theta_y}. \tag{9.148}$$

Using spherical coordinates for $\boldsymbol{y}$ and denoting the dummy variable of integration as $y = r$, the integrations over θ_y and ϕ_y in (9.142) give

$$\begin{aligned} f(\theta) &= -\frac{m}{2\pi\hbar^2}\int_0^\infty r^2 dr\, V(r)\int_0^\pi d\theta_y \sin\theta_y\, e^{ipr\cos\theta_y}\int_0^{2\pi} d\phi_y \\ &= -\frac{2m}{\hbar^2 p}\int_0^\infty dr\, r\sin(pr)V(r), \end{aligned} \tag{9.149}$$

where the factor p in (9.149) is given by (9.147),

$$p = 2k \sin\left(\frac{1}{2}\theta\right). \tag{9.150}$$

It should be noted that the angle between $\boldsymbol{p}$ and $\boldsymbol{y}$ can be used as the polar angle of $\boldsymbol{y}$ for the purposes of integration as long as the potential is spherically symmetric, a condition met by a central potential. Result (9.149) is also often referred to as the Born approximation.

9.4.5 Scattering from the Yukawa potential

A simple example of applying (9.149) is the case of the attractive potential

$$V(r) = -\frac{g^2}{r}e^{-\mu r}, \tag{9.151}$$

where μ is a constant with dimensions of inverse length and g plays the role of a *charge* with the units of square root of energy times length. The form (9.151) is known as the *Yukawa potential*, and is used to model the attractive force between nuclei that is transmitted by the short-ranged exchange of *mesons* with mass $m = \hbar\mu/c$.

Before finding the differential cross-section for scattering, it is informative to apply the Heisenberg uncertainty principle to the phenomenon of force transmission by particle exchange. Borrowing Einstein's famous mass–energy relation, if a nucleon detaches a *virtual meson*, then the nucleon has an uncertainty in its mass–energy given by

$$\Delta E = mc^2 = \hbar\mu c. \tag{9.152}$$

This meson has a lifetime before the uncertainty in energy is measurable given by the uncertainty principle,

$$\Delta t = \frac{\hbar}{2\Delta E} = \frac{1}{2\mu c}. \tag{9.153}$$

In one lifetime, the meson can travel the maximum distance

$$R = c\,\Delta t = \frac{1}{2\mu}. \tag{9.154}$$

This explains the origin of the exponential factor in (9.151) and why it is governed by the constant μ. The mechanism of force transmission through particle exchange is the central feature of quantum field theory.

The Born approximation for scattering from the Yukawa potential gives the scattering amplitude

$$f(\theta) = \frac{2\,mg^2}{\hbar^2 p}\int_0^\infty \mathrm{d}r\,\sin(pr)\,e^{-\mu r} = \frac{2\,mg^2}{\hbar^2(p^2 + \mu^2)} = \frac{2\,mg^2}{\hbar^2\left(\mu^2 + 4k^2\sin^2(\frac{1}{2}\theta)\right)}, \tag{9.155}$$

so that the differential cross-section is given by

$$\sigma(\theta) = |f(\theta)|^2 = \frac{4m^2g^4}{\hbar^4(\mu^2 + 4k^2\sin^2(\frac{1}{2}\theta))^2}. \tag{9.156}$$

Using

$$E = \frac{\hbar^2k^2}{2m}, \tag{9.157}$$

the differential cross-section can be written

$$\sigma(\theta) = \frac{4m^2g^4}{(\hbar^2\mu^2 + 8\,mE\sin^2(\frac{1}{2}\theta))^2}. \tag{9.158}$$

The differential cross-section of (9.158) can be integrated to obtain the total cross-section,

$$\sigma_T = \int d\Omega\,\sigma(\theta) = \int_0^{2\pi} d\phi \int_0^{\pi} d\theta\,\frac{4m^2g^4\sin\theta}{(\hbar^2\mu^2 + 8\,mE\sin^2(\frac{1}{2}\theta))^2} = \frac{16\pi m^2g^4}{\hbar^2\mu^2(\hbar^2\mu^2 + 8\,mE)}. \tag{9.159}$$

The key to performing the integral of (9.159) is to change variables from θ to $x = \sin(\frac{1}{2}\theta)$, which renders the integral elementary.

Setting μ to zero and g to e in (9.151) results in the Yukawa potential becoming the Coulomb potential (1.35). The differential cross-section for Coulomb scattering can therefore be found by setting $\mu = 0$ and replacing g^2 by $Z_1Z_2e^2$ in (9.158), where Z_1 is the charge number of the incident particle and Z_2 is the charge number of the point-like target creating the Coulomb potential. This gives

$$\sigma(\theta) = \frac{Z_1^2Z_2^2e^4}{16E^2\sin^4(\frac{1}{2}\theta)}. \tag{9.160}$$

Result (9.160) is known as *Rutherford scattering* and was instrumental in understanding the nature of the atom. It is strongly peaked around angles near zero, which was the behavior of the alpha particles Rutherford passed through the gold foil to probe atomic structure. The differential cross-section observed for the alpha particles was therefore consistent with scattering from a Coulomb potential in the atom rather than from the diffuse positive cloud of the Thomson model discussed in chapter 1. However, setting μ to zero in (9.159) results in a divergent total cross-section for pure Coulomb scattering. This is a result of the long-ranged nature of the Coulomb potential since the potential does not obey $\lim_{r\to\infty} rV(r) = 0$. Obtaining a finite total cross-section requires retaining a non-zero value for μ in the Yukawa differential cross-section of (9.158), with the understanding that doing so represents a Coulomb potential that is *screened* by the surrounding electronic charges as well as by the finite shape of the target.

9.5 Formal scattering theory

In the first two approaches to finding the outcome of scattering it has been the outgoing spherical wave function that was calculated, either by solving the Schrödinger equation through differential methods or by integral methods. It is possible to formulate scattering in terms of operator expressions, and this leads to alternative but equivalent ways to examine scattering.

9.5.1 The Lippmann–Schwinger equation in Dirac notation

The first step is to state the Lippmann–Schwinger equation (9.112) for the scattered wave function in Dirac notation,

$$| \psi_{\boldsymbol{k}}^{+} \rangle = | \varphi_{\boldsymbol{k}} \rangle + \left(\frac{1}{E_k - \hat{H}_o + i\epsilon} \right) \hat{V}(\hat{Q}) | \psi_{\boldsymbol{k}}^{+} \rangle \equiv | \varphi_{\boldsymbol{k}} \rangle + \hat{G}_0^{+}(k)\, \hat{V}(\hat{Q}) | \psi_{\boldsymbol{k}}^{+} \rangle , \tag{9.161}$$

where the presence of the infinitesimal ϵ term insures the proper outgoing spherical wave boundary conditions for the operator version of the Green's function in the limit $\epsilon \to 0$. The operator $\hat{G}_0^{+}(k)$ is the *operator inverse* of $E_k - \hat{H}_o + i\epsilon$, so that

$$\left(E_k - \hat{H}_o + i\epsilon \right) \hat{G}_0^{+}(k) = \left(E_k - \hat{H}_o + i\epsilon \right) \left(\frac{1}{E_k - \hat{H}_o + i\epsilon} \right) = 1. \tag{9.162}$$

Because $\hat{G}_0^{+}(k)$ is also an operator it will not commute with the potential $\hat{V}(\hat{Q})$. The state $| \varphi_{\boldsymbol{k}} \rangle$ satisfies the operator equation

$$\lim_{\epsilon \to 0} (\hat{H}_o - E_k - i\epsilon) | \varphi_{\boldsymbol{k}} \rangle = 0, \tag{9.163}$$

so that it corresponds to a free particle momentum eigenstate $| \boldsymbol{k} \rangle$ if $\hat{H}_o = \hat{P}^2/2m$. Equation (9.161) is the Lippmann–Schwinger equation (9.112) in Dirac notation. This is demonstrated by using the position and momentum state projection operators (6.56) and (6.58), which give

$$\begin{aligned}
\psi_{\boldsymbol{k}}^{+}(\boldsymbol{x}) &= \langle \boldsymbol{x} | \psi_{\boldsymbol{k}}^{+} \rangle = \langle \boldsymbol{x} | \varphi_{\boldsymbol{k}} \rangle + \langle \boldsymbol{x} | \hat{G}_0^{+}(k)\, \hat{V}(\hat{Q}) | \psi_{\boldsymbol{k}}^{+} \rangle \\
&= \varphi_{\boldsymbol{k}}(\boldsymbol{x}) - \int \mathrm{d}^3 y \int \mathrm{d}^3 p\, \langle \boldsymbol{x} | \boldsymbol{p} \rangle \langle \boldsymbol{p} | \left(\frac{1}{\hat{H}_o - E_k - i\epsilon} \right) \hat{V}(\hat{Q}) | \boldsymbol{y} \rangle \langle \boldsymbol{y} | \psi_{\boldsymbol{k}}^{+} \rangle \\
&= \varphi_{\boldsymbol{k}}(\boldsymbol{x}) - \int \mathrm{d}^3 y \int \mathrm{d}^3 p\, \frac{e^{i\boldsymbol{p}\cdot\boldsymbol{x}/\hbar}}{(2\pi\hbar)^{3/2}} \left(\frac{2m}{p^2 - \hbar^2 k - i2\, m\epsilon} \right) \frac{e^{-i\boldsymbol{p}\cdot\boldsymbol{y}/\hbar}}{(2\pi\hbar)^{3/2}} V(\boldsymbol{y})\, \psi_{\boldsymbol{k}}^{+}(\boldsymbol{y}) \\
&= \varphi_{\boldsymbol{k}}(\boldsymbol{x}) - \int \mathrm{d}^3 y\, G_k^{+}(\boldsymbol{x} - \boldsymbol{y})\, V(\boldsymbol{y})\, \psi_{\boldsymbol{k}}^{+}(\boldsymbol{y}),
\end{aligned} \tag{9.164}$$

where continuum normalization is in use. The Green's function (9.118) is the configuration representation of $\hat{G}_0^+(k)$, which follows by changing the momentum integration variable $\boldsymbol{p}$ according to $\boldsymbol{p} \to \hbar\boldsymbol{q}$, where $\boldsymbol{q}$ is the wave vector. This gives

$$\begin{aligned}\langle\, \boldsymbol{x} \mid \hat{G}_0^+(k) \mid \boldsymbol{y}\,\rangle &= \int \mathrm{d}^3p\, \langle\, \boldsymbol{x} \mid \hat{G}_0^+(k) \mid \boldsymbol{p}\,\rangle\langle\, \boldsymbol{p} \mid \boldsymbol{y}\,\rangle \\ &= \int \mathrm{d}^3p\, \frac{e^{i\boldsymbol{p}\cdot\boldsymbol{x}/\hbar}}{(2\pi\hbar)^{3/2}}\left(\frac{2m}{p^2 - \hbar^2k^2 - i2\,m\epsilon}\right)\frac{e^{-i\boldsymbol{p}\cdot\boldsymbol{y}/\hbar}}{(2\pi\hbar)^{3/2}} \\ &= \frac{2m}{\hbar^2}\int \frac{\mathrm{d}^3q}{(2\pi)^3}\left(\frac{e^{i\boldsymbol{q}\cdot(\boldsymbol{x}-\boldsymbol{y})}}{q^2 - k^2 - i\alpha}\right) = G_k^+(\boldsymbol{x} - \boldsymbol{y}),\end{aligned} \tag{9.165}$$

where $\alpha = 2\,m\epsilon/\hbar^2$ is infinitesimal. Denoting the Fourier transform of $G_k^+(\boldsymbol{x} - \boldsymbol{y})$ as $\tilde{G}_k^+(\boldsymbol{p})$, it follows that

$$\tilde{G}_k^+(\boldsymbol{p}) = \frac{2m}{\hbar^2(p^2 - k^2 - i\alpha)}. \tag{9.166}$$

These results can be used to formulate scattering in momentum space. This starts by noting that the scattering amplitude $f(\theta, \phi)$, identified previously in (9.128), can be written

$$\begin{aligned}f(\theta, \phi) &= -\frac{m}{2N\pi\hbar^2}\int \mathrm{d}^3y\, e^{-ik|\boldsymbol{y}|\cos\theta}\, V(\boldsymbol{y})\, \psi_+(\boldsymbol{y}) = -\frac{m}{2N\pi\hbar^2}\int \mathrm{d}^3y\, e^{-i\boldsymbol{k}'\cdot\boldsymbol{y}}\, V(\boldsymbol{y})\, \psi_+(\boldsymbol{y})\Big|_{k'=k} \\ &= -\frac{m(2\pi)^{3/2}}{2N\pi\hbar^2}\int \mathrm{d}^3y\, \langle\, \boldsymbol{k}' \mid \boldsymbol{y}\,\rangle\langle\, \boldsymbol{y} \mid \hat{V} \mid \psi_{\boldsymbol{k}}^+\,\rangle\Big|_{k'=k} = -\frac{m\sqrt{2\pi}}{N\hbar^2}\langle\, \boldsymbol{k}' \mid \hat{V} \mid \psi_{\boldsymbol{k}}^+\,\rangle\Big|_{k'=k}.\end{aligned} \tag{9.167}$$

The vector $\boldsymbol{k}'$ was defined in (9.143) to have the same magnitude as $\boldsymbol{k}$ and to point along the radial direction of scattering, which is quantified by the two angles θ and ϕ. The continuum normalization factor was found earlier in (9.20), so that $N = 1/(2\pi)^{3/2}$, with the final result that the scattering amplitude can be written

$$f(\theta, \phi) = -\frac{4\pi^2\, m}{\hbar^2}\langle\, \boldsymbol{k}' \mid \hat{V} \mid \psi_{\boldsymbol{k}}^+\,\rangle\Big|_{k'=k}. \tag{9.168}$$

Of course, $f(\theta, \phi)$ also depends upon the magnitude of the momentum k.

9.5.2 The transition operator $\hat{T}$

A useful formal tool for analysing scattering is the *transition* or $\hat{T}$ operator. It is defined by its action on the free particle states $|\, \varphi_{\boldsymbol{k}}\,\rangle$,

$$\hat{T} \mid \varphi_{\boldsymbol{k}}\,\rangle = \hat{V} \mid \psi_{\boldsymbol{k}}^+\,\rangle. \tag{9.169}$$

Using this operator the scattering state appearing in (9.161) can be written

$$\mid \psi_{\boldsymbol{k}}^+\,\rangle = \mid \varphi_{\boldsymbol{k}}\,\rangle + \hat{G}_0^+(k)\, \hat{V} \mid \psi_{\boldsymbol{k}}^+\,\rangle = \mid \varphi_{\boldsymbol{k}}\,\rangle + \hat{G}_0^+(k)\, \hat{T} \mid \varphi_{\boldsymbol{k}}\,\rangle. \tag{9.170}$$

Multiplying both sides of (9.170) by $\hat{V}$ and using (9.169) gives

$$\hat{V}|\psi_k^+\rangle = \hat{T}|\varphi_k\rangle = \hat{V}|\varphi_k\rangle + \hat{V}\,\hat{G}_0^+(k)\,\hat{T}\,|\varphi_k\rangle = \left(\hat{V} + \hat{V}\,\hat{G}_0^+(k)\,\hat{T}\right)|\varphi_k\rangle, \quad (9.171)$$

from which follows the first operator equation involving $\hat{T}$,

$$\hat{T} = \hat{V} + \hat{V}\,\hat{G}_0^+(k)\,\hat{T}. \quad (9.172)$$

This shows that $\hat{T}$ has an implicit dependence on the choice of k. Equation (9.172) can be solved by iteration, so that after one iteration $\hat{T}$ is given by

$$\hat{T} = \hat{V} + \hat{V}\hat{G}_0^+(k)\hat{V} + \hat{V}\hat{G}_0^+(k)\hat{V}\hat{G}_0^+(k)\hat{T}. \quad (9.173)$$

After an infinite number of iterations the full result can be written

$$\hat{T} = \sum_{n=0}^{\infty}\left(\hat{V}\hat{G}_0^+(k)\right)^n\hat{V}. \quad (9.174)$$

For the case that the incident scattering particle can be treated as free, the $\hat{T}$ operator can be used in (9.168), which gives

$$f(\theta, \phi) = -\frac{4\pi^2 m}{\hbar^2}\langle \boldsymbol{k}' \,|\, \hat{T} \,|\, \varphi_{\boldsymbol{k}} \rangle\Big|_{k'=k} = -\frac{4\pi^2 m}{\hbar^2}\langle \boldsymbol{k}' \,|\, \hat{T} \,|\, \boldsymbol{k} \rangle\Big|_{k'=k}. \quad (9.175)$$

The scattering amplitude is therefore often written

$$f(\theta, \phi) = f(\boldsymbol{k}', \boldsymbol{k})|_{k'=k} = -\frac{4\pi^2 m}{\hbar^2}\langle \boldsymbol{k}' \,|\, \hat{T} \,|\, \boldsymbol{k} \rangle\Big|_{k'=k}, \quad (9.176)$$

so that the scattering amplitude is determined from the matrix elements of the $\hat{T}$ operator in a wave vector representation. Forward scattering occurs for $\boldsymbol{k}' = \boldsymbol{k}$, since it corresponds to the case that the polar angle θ of $\boldsymbol{k}'$ is zero. It is to be recalled that $k' = k$, which ensures that the collision process is elastic. It is also important to note that $|\,\boldsymbol{k}\,\rangle$ is a wave number state, so that $\langle\, \boldsymbol{k}' \,|\, \boldsymbol{k}\,\rangle = \delta^3(\boldsymbol{k}' - \boldsymbol{k})$ has the units of volume. Combining this with the units of energy for $\hat{T}$ insures that (9.176) is correctly measured in length.

It is possible to find a more compact expression for $\hat{T}$ by writing (9.172) as

$$\left(1 - \hat{V}\,\hat{G}_0^+(k)\right)\hat{T} = \hat{V}. \quad (9.177)$$

Taking care with the order of operators and using (9.162), it follows that

$$\begin{aligned} 1 - \hat{V}\,\hat{G}_0^+(k) &= 1 - \hat{V}\left(\frac{1}{E_k - \hat{H}_o + i\epsilon}\right) = (E_k - \hat{H}_o + i\epsilon)\left(\frac{1}{E_k - \hat{H}_o + i\epsilon}\right) - \hat{V}\left(\frac{1}{E_k - \hat{H}_o + i\epsilon}\right) \\ &= (E_k - \hat{H}_o - \hat{V} + i\epsilon)\left(\frac{1}{E_k - \hat{H}_o + i\epsilon}\right) = (E_k - \hat{H} + i\epsilon)\hat{G}_0^+(k), \end{aligned} \quad (9.178)$$

where $\hat{H} = \hat{H}_o + V$ is the exact Hamiltonian. Returning this result to (9.177) gives

$$\left(E_k - \hat{H} + i\epsilon\right)\hat{G}_0^+(k)\,\hat{T} = \hat{V}. \tag{9.179}$$

Multiplying both sides of (9.179) by the inverse of $E_k - \hat{H} + i\epsilon$ gives

$$\hat{G}_0^+(k)\,\hat{T} = \left(\frac{1}{E_k - \hat{H} + i\epsilon}\right)\hat{V}. \tag{9.180}$$

Using result (9.180) in (9.172) immediately gives

$$\hat{T} = \hat{V} + \hat{V}\left(\frac{1}{E_k - \hat{H} + i\epsilon}\right)\hat{V}, \tag{9.181}$$

which is known as the *Chew–Goldberger solution* to the Lippmann–Schwinger equation.

The inverted operator appearing in (9.181) can be expressed in terms of the complete set of eigenstates $|f_k\rangle$ which $\hat{H}$ is assumed to possess. For simplicity, it will also be assumed that the complete set of stationary states consists of a continuum indexed by a vector $\boldsymbol{k}$ with no negative energy states. The eigenvalues λ_k satisfy

$$\hat{H}|f_k\rangle = \lambda_k|f_k\rangle. \tag{9.182}$$

The eigenfunctions of $\hat{H}$ are given by

$$\langle \boldsymbol{x}|f_k\rangle = \frac{1}{(2\pi\hbar)^{3/2}}f_k(\boldsymbol{x}), \tag{9.183}$$

and these satisfy the configuration space representation of the eigenvalue equation,

$$\hat{H}f_k(\boldsymbol{x}) = \left(-\frac{\hbar^2}{2m}\nabla^2 + V(\boldsymbol{x})\right)f_k(\boldsymbol{x}) = \lambda_k f_k(\boldsymbol{x}). \tag{9.184}$$

The completeness of the eigenstates is expressed by

$$\int \mathrm{d}^3p\,|f_k\rangle\langle f_k| = \hat{1}, \tag{9.185}$$

or in configuration representation

$$\int \frac{\mathrm{d}^3p}{(2\pi\hbar)^3} f_k^*(\boldsymbol{x})f_k(\boldsymbol{y}) = \delta^3(\boldsymbol{x} - \boldsymbol{y}). \tag{9.186}$$

It follows that the configuration space representation of the inverted operator in (9.181) can be written

$$G^+(\boldsymbol{x} - \boldsymbol{y}) = \langle \boldsymbol{x}|\left(\frac{1}{E_k - \hat{H} + i\epsilon}\right)|\boldsymbol{y}\rangle = \int \frac{\mathrm{d}^3p}{(2\pi\hbar)^3}\frac{f_k^*(\boldsymbol{x})f_k(\boldsymbol{y})}{E_k - \lambda_k + i\epsilon}, \tag{9.187}$$

or in Dirac notation

$$\left(\frac{1}{E_k - \hat{H} + i\epsilon}\right) = \int \mathrm{d}^3p \, \frac{| f_k \rangle\langle f_k |}{E_k - \lambda_k + i\epsilon}. \tag{9.188}$$

In either expression (9.172) or (9.181) for the $\hat{T}$ operator, the lowest order approximation is given by

$$\hat{T} \approx \hat{V}. \tag{9.189}$$

For such a case the scattering function (9.176) is given by

$$\begin{aligned} f(\boldsymbol{k}', \boldsymbol{k}) &= -\frac{4\pi^2 m}{\hbar^2} \langle \boldsymbol{k}' | \hat{T} | \boldsymbol{k} \rangle \approx -\frac{4\pi^2 m}{\hbar^2} \langle \boldsymbol{k}' | \hat{V} | \boldsymbol{k} \rangle = -\frac{4\pi^2 m}{\hbar^2} \int \mathrm{d}^3y \, \langle \boldsymbol{k}' | \boldsymbol{y} \rangle\langle \boldsymbol{y} | \hat{V}(\hat{Q}) | \boldsymbol{k} \rangle \\ &= -\frac{m}{2\pi\hbar^2} \int \mathrm{d}^3y \, e^{i(\boldsymbol{k}-\boldsymbol{k}')\cdot \boldsymbol{y}} \, V(\boldsymbol{y}), \end{aligned} \tag{9.190}$$

which is once again the Born approximation (9.142).

9.5.3 The scattering operator $\hat{S}$

The definition and derivation of the $\hat{T}$ operator in the previous section began by assuming the Lippmann–Schwinger equation and expressing it in Dirac notation. It is possible to formulate scattering by using the interaction picture evolution operator defined in (8.165) and developed in (8.175) without first postulating the Lippmann–Schwinger equation.

The derivation uses the interaction picture evolution operator to define the scattering process. For the case that the initial time $t_i \to -\infty$, it is assumed that the eigenstates of the Hamiltonian correspond to an asymptotic or *incoming* wave number state, designated $| \boldsymbol{k} \rangle_{\text{in}}$, which is an eigenstate of $\hat{H}_o$ with eigenvalue

$$\hat{H}_o| \boldsymbol{k} \rangle_{\text{in}} = \frac{\hbar^2 k^2}{2m} | \boldsymbol{k} \rangle_{\text{in}} \equiv E_k | \boldsymbol{k} \rangle_{\text{in}}. \tag{9.191}$$

As time passes the state evolves according to

$$| \boldsymbol{k}, t \rangle = \hat{U}(t, -\infty)| \boldsymbol{k} \rangle_{\text{in}}. \tag{9.192}$$

As $t \to \infty$, it is assumed that this state evolves into an asymptotic or *outgoing* wave number state. The evolution operator in this limit is referred to as the *scattering operator* $\hat{S}$,

$$\hat{S} = \lim_{t \to \infty} \hat{U}(t, -t). \tag{9.193}$$

Because the evolution operator is unitary the $\hat{S}$ operator satisfies

$$\hat{S}\hat{S}^\dagger = \lim_{t \to \infty} \hat{U}(t, -t)\hat{U}^\dagger(t, -t) = \lim_{t \to \infty} \hat{U}(t, -t)\hat{U}(-t, t) = 1, \tag{9.194}$$

so that the $\hat{S}$ operator is also unitary. The probability amplitude $\mathcal{M}$ that the incoming state has evolved into a specific eigenstate of $\hat{H}_o$, designated $| \boldsymbol{k}' \rangle_{\text{out}}$, is given by

$$\mathcal{M}(\boldsymbol{k}', \boldsymbol{k}) = {}_{\text{out}}\langle\, \boldsymbol{k}' \mid \hat{S} \mid \boldsymbol{k}\,\rangle_{\text{in}}. \tag{9.195}$$

In some approaches to calculating the $\hat{S}$-matrix element (9.195) the potential undergoes *adiabatic switching*, so that $\hat{V}$ is replaced with

$$\hat{V} \to \hat{V}e^{-\epsilon|t|}. \tag{9.196}$$

This has the effect of turning off the potential for asymptotic times, with the understanding that the limit $\epsilon \to 0$ is taken at the end of the calculation. This is similar to the epsilon procedure used in defining the Green's function (9.118), and will be discussed in more detail later.

In order to develop the relationship between the $\hat{S}$ and $\hat{T}$ operators, the matrix elements of the expansion of the evolution operator (8.171) are examined for $\lambda = 1$. The first two terms in the expansion of $\mathcal{M}$ are

$$\begin{aligned} \mathcal{M}_0(\boldsymbol{k}', \boldsymbol{k}) &= \langle\, \boldsymbol{k}' \mid \boldsymbol{k}\,\rangle = \delta^3(\boldsymbol{k}' - \boldsymbol{k}), \\ \mathcal{M}_1(\boldsymbol{k}', \boldsymbol{k}) &= \frac{1}{i\hbar}\int_{-\infty}^{\infty} \mathrm{d}t_1 \,\langle\, \boldsymbol{k}' \mid \hat{V}_{\mathrm{I}}(t_1) \mid \boldsymbol{k}\,\rangle, \end{aligned} \tag{9.197}$$

where the interaction picture potential is given by

$$\hat{V}_{\mathrm{I}}(t) = e^{i\hat{H}_o t/\hbar}\, \hat{V}\, e^{-i\hat{H}_o t/\hbar}. \tag{9.198}$$

The second integral expression gives

$$\begin{aligned} \mathcal{M}_1(\boldsymbol{k}', \boldsymbol{k}) &= \frac{1}{i\hbar}\int_{-\infty}^{\infty} \mathrm{d}t_1 \,\langle\, \boldsymbol{k}' \mid e^{i\hat{H}_o t_1/\hbar}\, \hat{V}\, e^{-i\hat{H}_o t_1/\hbar} \mid \boldsymbol{k}\,\rangle = \frac{1}{i\hbar}\langle\, \boldsymbol{k}' \mid \hat{V} \mid \boldsymbol{k}\,\rangle \int_{-\infty}^{\infty} \mathrm{d}t_1\, e^{i(E_{k'} - E_k)t_1/\hbar} \\ &= -\,2\pi i\,\langle\, \boldsymbol{k}' \mid \hat{V} \mid \boldsymbol{k}\,\rangle\, \delta(E_{k'} - E_k), \end{aligned} \tag{9.199}$$

so that, to lowest order, energy is conserved in the scattering process.

For the case that $n > 1$ a general term in (8.171) consists of n integrals and takes the form

$$\mathcal{M}_n(\boldsymbol{k}', \boldsymbol{k}) = \frac{1}{(i\hbar)^n}\langle\, \boldsymbol{k}' \,|\int_{-\infty}^{\infty} \mathrm{d}t_1\, \hat{V}_{\mathrm{I}}(t_1)\int_{-\infty}^{t_1} \mathrm{d}t_2\, \hat{V}_{\mathrm{I}}(t_2)\int_{-\infty}^{t_2} \mathrm{d}t_3\, \hat{V}_{\mathrm{I}}(t_3) \cdots \int_{-\infty}^{t_{n-1}} \mathrm{d}t_n\, \hat{V}_{\mathrm{I}}(t_n)|\, \boldsymbol{k}\,\rangle. \tag{9.200}$$

Using the definition of the interaction picture potential (9.198) allows (9.200) to be written

$$\begin{aligned} \mathcal{M}_n(\boldsymbol{k}', \boldsymbol{k}) = &\frac{1}{(i\hbar)^n}\langle\, \boldsymbol{k}' \,|\int_{-\infty}^{\infty} \mathrm{d}t_1\, e^{iE_{k'}t_1/\hbar}\, \hat{V} \int_{-\infty}^{t_1} \mathrm{d}t_2\, e^{iH_o(t_2 - t_1)/\hbar}\, \hat{V} \int_{-\infty}^{t_2} \mathrm{d}t_3\, e^{iH_o(t_3 - t_2)/\hbar}\, \hat{V} \ldots \\ &\ldots \int_{-\infty}^{t_{n-1}} \mathrm{d}t_n\, e^{iH_o(t_n - t_{n-1})/\hbar}\, \hat{V}\, e^{-iE_k t_n/\hbar}|\, \boldsymbol{k}\,\rangle. \end{aligned} \tag{9.201}$$

Since E_k commutes with $\hat{V}$, it follows that the last term can be written

$$e^{i\hat{H}_o(t_n - t_{n-1})/\hbar}\, \hat{V}\, e^{-iE_k t_n/\hbar} = e^{-iE_k t_{n-1}/\hbar}\, e^{i(\hat{H}_o - E_k)(t_n - t_{n-1})/\hbar}\, \hat{V}. \tag{9.202}$$

This process can be therefore be repeated sequentially for each of the n integrals, so that (9.201) becomes

$$\mathcal{M}_n(\boldsymbol{k}', \boldsymbol{k}) = \frac{1}{(i\hbar)^n} \langle \boldsymbol{k}' | \int_{-\infty}^{\infty} \mathrm{d}t_1\, e^{i(E_{k'}-E_k)t_1/\hbar}\, \hat{V} \int_{-\infty}^{t_1} \mathrm{d}t_2\, e^{i(\hat{H}_o-E_k)(t_2-t_1)/\hbar}\, \hat{V} \cdots$$
$$\cdots \int_{-\infty}^{t_{n-2}} \mathrm{d}t_{n-1}\, e^{i(\hat{H}_o-E_k)(t_{n-1}-t_{n-2})/\hbar}\, \hat{V} \int_{-\infty}^{t_{n-1}} \mathrm{d}t_n\, e^{i(\hat{H}_o-E_k)(t_n-t_{n-1})/\hbar}\, \hat{V} | \boldsymbol{k} \rangle. \tag{9.203}$$

The next step is to change variables of integration to

$$\tau_1 = t_1, \quad \tau_2 = t_2 - t_1, \quad \ldots, \tau_{n-1} = t_{n-1} - t_{n-2}, \quad \tau_n = t_n - t_{n-1}, \tag{9.204}$$

which yields a Jacobian of one. Since the upper limit on the t_j integral is t_{j-1}, it follows that the upper limit on the τ_j integral is $\tau_{j\,\max} = t_{j\,\max} - t_{j-1} = t_{j-1} - t_{j-1} = 0$. This has the effect of changing the limits on $n - 1$ of the integrals to the same value,

$$\mathcal{M}_n(\boldsymbol{k}', \boldsymbol{k}) = \frac{1}{(i\hbar)^n} \langle \boldsymbol{k}' | \int_{-\infty}^{\infty} \mathrm{d}\tau_1\, e^{i(E_{k'}-E_k)\tau_1/\hbar}\, \hat{V} \int_{-\infty}^{0} \mathrm{d}\tau_2\, e^{i(\hat{H}_o-E_k)\tau_2/\hbar}\, \hat{V} \cdots$$
$$\cdots \int_{-\infty}^{0} \mathrm{d}\tau_{n-1}\, e^{i(\hat{H}_o-E_k)\tau_{n-1}/\hbar}\, \hat{V} \int_{-\infty}^{0} \mathrm{d}\tau_n\, e^{i(\hat{H}_o-E_k)\tau_n/\hbar}\, \hat{V} | \boldsymbol{k} \rangle. \tag{9.205}$$

The right-most $n - 1$ integrals in (9.205) are identical. However, they are singular in the event that $\hat{H}_o - E_k$ vanishes, and this singularity will manifest itself if a complete set of wave number states $| \boldsymbol{k} \rangle$ is inserted. This is remedied by an epsilon procedure that adiabatically switches off the potential as $\tau_j \to -\infty$. This is equivalent to the pole redefinition in the Green's function that was done in (9.118). Each of the $n - 1$ integrals is defined by an identical epsilon procedure,

$$\int_{-\infty}^{0} \mathrm{d}\tau_j\, e^{i(\hat{H}_o-E_k)\tau_j/\hbar}\, \hat{V} \to \lim_{\epsilon\to 0^+} \int_{-\infty}^{0} \mathrm{d}\tau_j\, e^{i(\hat{H}_o-E_k)\tau_j/\hbar}\, e^{\epsilon\tau_j/\hbar}\, \hat{V} = \lim_{\epsilon\to 0^+} \left(\frac{i\hbar}{E_k - \hat{H}_o + i\epsilon} \right) \hat{V}, \tag{9.206}$$

where the limit at $\tau_j \to -\infty$ vanishes due to the positive epsilon factor. It is worth noting that inserting the $n - 1$ regularizing factors can be written in terms of the original variables as

$$e^{\epsilon(\tau_2+\tau_3+\cdots+\tau_n)} = e^{\epsilon(t_n - t_1)}, \tag{9.207}$$

so that the limit $\epsilon \to 0$ remains well-defined as $n \to \infty$. Using this epsilon procedure results in the operator form of the Green's function (9.162) used in the Lippmann–Schwinger equation to appear in this representation of the $\hat{S}$-matrix. Returning result (9.206) to (9.205) gives

$$\mathcal{M}_n(\boldsymbol{k}', \boldsymbol{k}) = \frac{1}{(i\hbar)} \int_{-\infty}^{\infty} \mathrm{d}\tau_1\, e^{i(E_{k'}-E_k)\tau_1/\hbar} \langle \boldsymbol{k}' | \hat{V} \left(\frac{1}{E_k - \hat{H}_o + i\epsilon} \right) \hat{V} \cdots \left(\frac{1}{E_k - \hat{H}_o + i\epsilon} \right) \hat{V} | \boldsymbol{k} \rangle, \tag{9.208}$$

where there are n total factors of the potential $\hat{V}$. In what follows, the limit $\epsilon \to 0^+$ is understood without being explicitly stated.

The remaining integral over τ_1 once again yields a delta function in the energy, so that

$$\mathcal{M}_n(\boldsymbol{k}', \boldsymbol{k}) = -2\pi i\langle \boldsymbol{k}' | \hat{V} \left(\frac{1}{E_k - \hat{H}_o + i\epsilon} \right) \hat{V} \cdots \left(\frac{1}{E_k - \hat{H}_o + i\epsilon} \right) \hat{V} | \boldsymbol{k} \rangle \, \delta(E_{k'} - E_k). \quad (9.209)$$

Comparing result (9.209) to (9.174) shows that the matrix elements of the $\hat{T}$ operator have emerged from the $\hat{S}$ operator. The matrix elements of the two operators are therefore related by

$$\langle \boldsymbol{k}' | \hat{S} | \boldsymbol{k} \rangle = \sum_{n=0}^{\infty} \mathcal{M}_n(\boldsymbol{k}', \boldsymbol{k}) = \langle \boldsymbol{k}' | \boldsymbol{k} \rangle - 2\pi i \langle \boldsymbol{k}' | \hat{T} | \boldsymbol{k} \rangle \, \delta(E_{k'} - E_k). \quad (9.210)$$

This shows indirectly that the epsilon procedure of (9.206) for the evolution operator yields the Lippmann–Schwinger equation for scattering. Result (9.210) also has the important feature that the transition operator matrix element is evaluated at $k' = k$. This follows from the delta function, which can be written

$$\delta(E_{k'} - E_k) = \delta\left(\frac{\hbar^2}{2m}(k'^2 - k^2) \right) = \frac{2m}{\hbar^2}\delta((k' - k)(k' + k)) = \frac{m}{\hbar^2 k}\, \delta(k' - k). \quad (9.211)$$

It is possible to derive the Lippmann–Schwinger equation directly from the evolution operator by defining the outgoing scattering state $| \psi_{\boldsymbol{k}}^+ \rangle$ as

$$| \psi_{\boldsymbol{k}}^+ \rangle = \hat{U}(0, -\infty) | \boldsymbol{k} \rangle_{\text{in}}. \quad (9.212)$$

Using the expansion of the evolution operator (8.171) gives

$$| \psi_{\boldsymbol{k}}^+ \rangle = \left(1 + \sum_{n=1}^{\infty} \hat{\mathcal{E}}_n(0, -\infty) \right) | \boldsymbol{k} \rangle_{\text{in}}, \quad (9.213)$$

where

$$\hat{\mathcal{E}}_n(0, -\infty) | \boldsymbol{k} \rangle_{\text{in}} = \frac{1}{(i\hbar)^n} \int_{-\infty}^{0} \mathrm{d}t_1 \hat{V}_{\mathrm{I}}(t_1) \int_{-\infty}^{t_1} \mathrm{d}t_2\, \hat{V}_{\mathrm{I}}(t_2) \cdots \int_{-\infty}^{t_{n-1}} \mathrm{d}t_n\, \hat{V}_{\mathrm{I}}(t_n) | \boldsymbol{k} \rangle_{\text{in}}. \quad (9.214)$$

Following the same steps, (9.202) and (9.204), that led to (9.205) gives

$$\hat{\mathcal{E}}_n(0, -\infty) | \boldsymbol{k} \rangle_{\text{in}} = \int_{-\infty}^{0} \mathrm{d}\tau_1\, e^{i(\hat{H}_o - E_k)\tau_1/\hbar}\, \hat{V} \cdots \int_{-\infty}^{0} \mathrm{d}\tau_n\, e^{i(\hat{H}_o - E_k)\tau_n/\hbar}\, \hat{V} \, | \boldsymbol{k} \rangle_{\text{in}}, \quad (9.215)$$

where there is now a product of n identical integrals. Regulating each of the integrals in a manner identical to (9.206) gives

$$\hat{\mathcal{E}}_n(0, -\infty) | \boldsymbol{k} \rangle_{\text{in}} = \left(\left(\frac{1}{E_k - \hat{H}_o + i\epsilon} \right) \hat{V} \right)^n | \boldsymbol{k} \rangle_{\text{in}}. \quad (9.216)$$

It follows that the state defined by (9.212) can be written

$$| \psi_{\boldsymbol{k}}^+ \rangle = \sum_{n=0}^{\infty} \left(\left(\frac{1}{E_k - \hat{H}_o + i\epsilon} \right) \hat{V} \right)^n | \boldsymbol{k} \rangle_{\text{in}}. \quad (9.217)$$

This state can be rewritten to obtain

$$\begin{aligned}|\psi_k^+\rangle &= \sum_{n=0}^{\infty}\left(\left(\frac{1}{E_k - \hat{H}_o + i\epsilon}\right)\hat{V}\right)^n |\,\boldsymbol{k}\,\rangle_{\text{in}} = |\,\boldsymbol{k}\,\rangle_{\text{in}} + \sum_{n=1}^{\infty}\left(\left(\frac{1}{E_k - \hat{H}_o + i\epsilon}\right)\hat{V}\right)^n |\,\boldsymbol{k}\,\rangle_{\text{in}} \\ &= |\,\boldsymbol{k}\,\rangle_{\text{in}} + \left(\frac{1}{E_k - \hat{H}_o + i\epsilon}\right)\hat{V}\left(\sum_{n=0}^{\infty}\left(\left(\frac{1}{E_k - \hat{H}_o + i\epsilon}\right)\hat{V}\right)^n |\,\boldsymbol{k}\,\rangle_{\text{in}}\right) \\ &= |\,\boldsymbol{k}\,\rangle_{\text{in}} + \left(\frac{1}{E_k - \hat{H}_o + i\epsilon}\right)\hat{V}|\,\psi_k^+\,\rangle,\end{aligned} \tag{9.218}$$

which is the Dirac notation version of the Lippmann–Schwinger equation first introduced in (9.161).

9.5.4 Identities for $\hat{S}$ and $\hat{T}$

Based on the unitarity of the evolution operator it follows that the $\hat{S}$-matrix elements are unitary. It will now be shown that the $\hat{T}$-matrix elements obey an identity which is consistent with the unitarity of the $\hat{S}$-matrix.

The derivation begins by using result (9.210) to write

$$\begin{aligned}\langle\,\boldsymbol{k}\,|\,\hat{S}\,|\,\boldsymbol{k}'\,\rangle &= \delta^3(\boldsymbol{k} - \boldsymbol{k}') - 2\pi i\,\delta(E_k - E_{k'})\,\langle\,\boldsymbol{k}\,|\,\hat{T}\,|\,\boldsymbol{k}'\,\rangle, \\ \langle\,\boldsymbol{k}''\,|\,\hat{S}^\dagger\,|\,\boldsymbol{k}\,\rangle &= (\langle\,\boldsymbol{k}\,|\,\hat{S}\,|\,\boldsymbol{k}''\,\rangle)^* = \delta^3(\boldsymbol{k} - \boldsymbol{k}'') + 2\pi i\,\delta(E_k - E_{k''})\,\langle\,\boldsymbol{k}''\,|\,\hat{T}^\dagger\,|\,\boldsymbol{k}\,\rangle.\end{aligned} \tag{9.219}$$

Using the completeness of the $|\,\boldsymbol{k}\,\rangle$ states shows that

$$\int \mathrm{d}^3k\,\langle\,\boldsymbol{k}''\,|\,\hat{S}^\dagger\,|\,\boldsymbol{k}\,\rangle\langle\,\boldsymbol{k}\,|\,\hat{S}\,|\,\boldsymbol{k}'\,\rangle = \langle\,\boldsymbol{k}''\,|\,\hat{S}^\dagger\hat{S}\,|\,\boldsymbol{k}'\,\rangle. \tag{9.220}$$

Multiplying the right-hand expressions appearing in (9.219) and performing the integration over $\boldsymbol{k}$ gives

$$\begin{aligned}\langle\,\boldsymbol{k}''\,|\,\hat{S}^\dagger\hat{S}\,|\,\boldsymbol{k}'\,\rangle &= \delta^3(\boldsymbol{k}' - \boldsymbol{k}'') + 2\pi i\,\delta(E_{k'} - E_{k''})\langle\,\boldsymbol{k}''\,|\left(\hat{T}^\dagger - \hat{T}\right)|\,\boldsymbol{k}'\,\rangle \\ &\quad + 4\pi^2\,\delta(E_{k'} - E_{k''})\int \mathrm{d}^3k\,\delta(E_k - E_{k'})\langle\,\boldsymbol{k}''\,|\,\hat{T}^\dagger\,|\,\boldsymbol{k}\,\rangle\langle\,\boldsymbol{k}\,|\,\hat{T}\,|\,\boldsymbol{k}'\,\rangle,\end{aligned} \tag{9.221}$$

which used the delta function identity

$$\delta(E_k - E_{k''})\,\delta(E_k - E_{k'}) = \delta(E_{k'} - E_{k''})\,\delta(E_k - E_{k'}). \tag{9.222}$$

It should be noted that if $E_{k'} \neq E_{k''}$ the second and third terms on the right-hand side of (9.221) will vanish due to the delta function.

It will now be shown that the second and third terms on the right-hand side of (9.221) cancel even if $E_{k'} = E_{k''}$. This begins by using the definition of the $\hat{T}$ operator (9.169) to write the second term as

$$2\pi i\,\delta(E_{k'} - E_{k''})\langle\,\boldsymbol{k}''\,|\left(\hat{T}^\dagger - \hat{T}\right)|\,\boldsymbol{k}'\,\rangle = 2\pi i\,\delta(E_{k'} - E_{k''})\Big(\langle\,\psi_{k''}^+\,|\,\hat{V}\,|\,\boldsymbol{k}'\,\rangle - \langle\,\boldsymbol{k}''\,|\,\hat{V}\,|\,\psi_{k'}^+\,\rangle\Big), \tag{9.223}$$

where $\hat{V}^\dagger = \hat{V}$ was used. Using the rules for finding the adjoint of a state the Lippmann–Schwinger equation gives

$$
\begin{aligned}
|\, \boldsymbol{k}' \,\rangle &= |\, \psi_{k'}^{+} \,\rangle - \left(\frac{1}{E_{k'} - \hat{H}_o + i\epsilon} \right) \hat{V} |\, \psi_{k'}^{+} \,\rangle, \\
\langle\, \boldsymbol{k}'' \,| &= \langle\, \psi_{k''}^{+} \,| - \langle\, \psi_{k''}^{+} \,| \hat{V} \left(\frac{1}{E_{k''} - \hat{H}_o - i\epsilon} \right).
\end{aligned}
\tag{9.224}
$$

Using the states of (9.224) in (9.223) gives

$$
\begin{aligned}
&2\pi i\, \delta(E_{k'} - E_{k''}) \langle\, \boldsymbol{k}'' \,|\, (\hat{T}^\dagger - \hat{T}) \,|\, \boldsymbol{k}' \,\rangle \\
&= 2\pi i\, \delta(E_{k'} - E_{k''}) \langle\, \psi_{k''}^{+} \,|\, \hat{V} \left(\frac{1}{E_{k''} - \hat{H}_o - i\epsilon} - \frac{1}{E_{k'} - \hat{H}_o + i\epsilon} \right) \hat{V} \,|\, \psi_{k'}^{+} \,\rangle \\
&= 2\pi i\, \delta(E_{k'} - E_{k''}) \langle\, \psi_{k''}^{+} \,|\, \hat{V} \left(\frac{2i\epsilon}{(E_{k'} - \hat{H}_o)^2 + \epsilon^2} \right) \hat{V} \,|\, \psi_{k'}^{+} \,\rangle,
\end{aligned}
\tag{9.225}
$$

where the presence of the delta function in energy allows setting $E_{k''} = E_{k'}$ in the last step. Inserting a complete set of $|\, \boldsymbol{k} \,\rangle$ states gives

$$
\begin{aligned}
&2\pi i\, \delta(E_{k'} - E_{k''}) \langle\, \boldsymbol{k}'' \,|\, (\hat{T}^\dagger - \hat{T}) \,|\, \boldsymbol{k}' \,\rangle \\
&= 2\pi i\, \delta(E_{k'} - E_{k''}) \int \mathrm{d}^3k\, \langle\, \psi_{k''}^{+} \,|\, \hat{V} \,|\, \boldsymbol{k} \,\rangle \langle\, \boldsymbol{k} \,| \left(\frac{2i\epsilon}{(E_{k'} - \hat{H}_o)^2 + \epsilon^2} \right) \hat{V} \,|\, \psi_{k'}^{+} \,\rangle \\
&= 2\pi i\, \delta(E_{k'} - E_{k''}) \int \mathrm{d}^3k \left(\frac{2i\epsilon}{(E_{k'} - E_k)^2 + \epsilon^2} \right) \langle\, \psi_{k''}^{+} \,|\, \hat{V} \,|\, \boldsymbol{k} \,\rangle \langle\, \boldsymbol{k} \,|\, \hat{V} \,|\, \psi_{k'}^{+} \,\rangle.
\end{aligned}
\tag{9.226}
$$

Result (3.90) shows that

$$
\lim_{\epsilon \to 0^+} \frac{2i\epsilon}{(E_{k'} - E_k)^2 + \epsilon^2} = 2\pi i\, \delta(E_k - E_{k'}),
\tag{9.227}
$$

while using (9.169) once again gives

$$
\begin{aligned}
\langle\, \boldsymbol{k} \,|\, \hat{V} \,|\, \psi_{k'}^{+} \,\rangle &= \langle\, \boldsymbol{k} \,|\, \hat{T} \,|\, \boldsymbol{k}' \,\rangle, \\
\langle\, \psi_{k''}^{+} \,|\, \hat{V} \,|\, \boldsymbol{k} \,\rangle &= \langle\, \boldsymbol{k}'' \,|\, \hat{T}^\dagger \,|\, \boldsymbol{k} \,\rangle.
\end{aligned}
\tag{9.228}
$$

Returning (9.227) and (9.228) to (9.226) shows that

$$
\begin{aligned}
&2\pi i\, \delta(E_{k'} - E_{k''}) \langle\, \boldsymbol{k}'' \,|\, (\hat{T}^\dagger - \hat{T}) \,|\, \boldsymbol{k}' \,\rangle \\
&= -4\pi^2\, \delta(E_{k'} - E_{k''}) \int \mathrm{d}^3k\, \delta(E_k - E_{k'}) \langle\, \boldsymbol{k}'' \,|\, \hat{T}^\dagger \,|\, \boldsymbol{k} \,\rangle \langle\, \boldsymbol{k} \,|\, \hat{T} \,|\, \boldsymbol{k}' \,\rangle,
\end{aligned}
\tag{9.229}
$$

which exactly cancels the third term on the right-hand side of (9.221). The final result is that (9.221) becomes

$$\langle \boldsymbol{k}'' \mid \hat{S}^{\dagger}\hat{S} \mid \boldsymbol{k}' \rangle = \delta^3(\boldsymbol{k}'' - \boldsymbol{k}'), \tag{9.230}$$

which yields the unitarity of the $\hat{S}$ operator,

$$\hat{S}^{\dagger}\hat{S} = \hat{1}. \tag{9.231}$$

This result shows that the Lippmann–Schwinger equation is consistent with the unitarity of the evolution operator derived in (9.194).

The unitarity of the $\hat{S}$-matrix enables the probabilistic interpretation of the $\hat{S}$-matrix elements. In order to make contact with probability, it is necessary to use the box normalization (9.19) of the wave number state $|\, \boldsymbol{k} \,\rangle$, which is written

$$|\, \bar{\boldsymbol{k}} \,\rangle = N|\, \boldsymbol{k} \,\rangle = \sqrt{\left(\frac{2\pi}{L}\right)^3}\,|\, \boldsymbol{k} \,\rangle, \tag{9.232}$$

where L^3 is the spatial volume of the system. The transition probability from the normalized state $|\, \psi \,\rangle$ to the normalized wave number state $|\, \bar{\boldsymbol{k}} \,\rangle$ is defined in the usual manner as the amplitude squared of the S-matrix element,

$$P_{\psi k} = |N|^2 |\langle\, \boldsymbol{k} \mid \hat{S} \mid \psi \,\rangle|^2. \tag{9.233}$$

The units of the normalization factor render (9.233) dimensionless, which is required for a probability. The sum of all possible transition probabilities out of the state $|\, \psi \,\rangle$ is given by

$$\sum_k P_{\psi k} = \int \left(\frac{\mathrm{d}^3 k}{|N|^2}\right) |N|^2 |\langle\, \boldsymbol{k} \mid \hat{S} \mid \psi \,\rangle|^2 = \int \mathrm{d}^3 k \,\langle\, \psi \mid \hat{S}^{\dagger} \mid \boldsymbol{k} \,\rangle\langle\, \boldsymbol{k} \mid \hat{S} \mid \psi \,\rangle = \langle\, \psi \mid \psi \,\rangle = 1, \tag{9.234}$$

where the sum was replaced by the three-dimensional box normalized integration introduced in (3.107). The necessary condition for a probabilistic interpretation is therefore satisfied using box normalization.

9.5.5 Scattering rates and the $\hat{T}$ and $\hat{S}$ operators

The scattering process can be viewed as a problem in time-dependent perturbation theory. This begins by using the relation (9.219) for the scattering wave number $\boldsymbol{k}'$ differently directed from the incident wave number $\boldsymbol{k}$, so that $\delta^3(\boldsymbol{k}' - \boldsymbol{k}) = 0$,

$$\langle\, \boldsymbol{k}' \mid \hat{S} \mid \boldsymbol{k} \,\rangle = -2\pi i\, \delta(E_{k'} - E_k)\, \langle\, \boldsymbol{k}' \mid \hat{T} \mid \boldsymbol{k} \,\rangle. \tag{9.235}$$

The normalized probability for the transition from $\boldsymbol{k}$ to $\boldsymbol{k}'$ follows by combining this result with (9.233), which gives

$$P_{kk'} = |N|^4 |\langle\, \boldsymbol{k}' \mid \hat{S} \mid \boldsymbol{k} \,\rangle|^2 = 4\pi^2 |N|^4 (\delta(E_{k'} - E_k))^2 |\langle\, \boldsymbol{k}' \mid \hat{T} \mid \boldsymbol{k} \,\rangle|^2. \tag{9.236}$$

The square of the delta function can be written

$$(\delta(E_{k'} - E_k))^2 = \delta(E_{k'} - E_k) \lim_{\tau\to\infty} \int_{-\tau}^{\tau} \frac{\mathrm{d}t}{2\pi\hbar}\, e^{i(E_{k'}-E_k)t/\hbar}. \tag{9.237}$$

The presence of the first delta function gives a non-zero result only if $E_{k'} = E_k$, so that

$$(\delta(E_{k'} - E_k))^2 = \delta(E_{k'} - E_k)\lim_{\tau\to\infty}\int_{-\tau}^{\tau}\frac{\mathrm{d}t}{2\pi\hbar} = \lim_{\tau\to\infty}\frac{\tau}{\pi\hbar}\,\delta(E_{k'} - E_k). \tag{9.238}$$

Returning (9.238) to (9.236) gives

$$P_{\boldsymbol{kk}'} = \frac{4\pi\tau}{\hbar}|N|^4\delta(E_{k'} - E_k)\,|\langle\, \boldsymbol{k}' \mid \hat{T} \mid \boldsymbol{k}\,\rangle|^2, \tag{9.239}$$

where $\tau \to \infty$ is implicit. Recalling definition (9.193) shows that the probability of the transition in terms of the evolution operator is given by

$$P_{\boldsymbol{kk}'} = \lim_{\tau\to\infty}|N|^4\,|\langle\, \boldsymbol{k}' \mid \hat{U}(\tau, -\tau) \mid \boldsymbol{k}\,\rangle|^2, \tag{9.240}$$

which is the probability of the transition after the interval of time 2τ has passed. The transition rate $R_{\boldsymbol{kk}'}$ or probability per unit time for the scattering process to occur is therefore given by

$$R_{\boldsymbol{kk}'} = \frac{1}{2\tau}P_{\boldsymbol{kk}'}. \tag{9.241}$$

Combining this result with (9.239) gives the transition rate

$$R_{\boldsymbol{kk}'} = \frac{2\pi}{\hbar}|N|^4\delta(E_{k'} - E_k)|\langle\, \boldsymbol{k}' \mid \hat{T} \mid \boldsymbol{k}\,\rangle|^2. \tag{9.242}$$

This gives the total rate of scattering R by summing over all possible final momenta,

$$R = \sum_{\boldsymbol{k}'}R_{\boldsymbol{kk}'} \to \int\frac{\mathrm{d}^3k'}{|N|^2}R_{\boldsymbol{kk}'} = \frac{2\pi}{\hbar}|N|^2\int\mathrm{d}^3k'\,\delta(E_{k'} - E_k)\,|\langle\, \boldsymbol{k}' \mid \hat{T} \mid \boldsymbol{k}\,\rangle|^2, \tag{9.243}$$

where the recipe (3.107) was used to convert the sum into a dimensionless integral. For the case of a single scattering potential this result can be combined with the basic definition of the total cross-section (9.2) using the box normalized incident flux (9.30), $J_k = \hbar k/mL^3$, and $|N|^2 = (2\pi)^3/L^3$ to find the total cross-section in terms of the $\hat{T}$ operator matrix elements,

$$\sigma_T = \frac{R}{J_k} = \frac{16\pi^4 m}{\hbar^2 k}\int\mathrm{d}^3k'\,\delta(E_{k'} - E_k)|\langle\, \boldsymbol{k}' \mid \hat{T} \mid \boldsymbol{k}\,\rangle|^2. \tag{9.244}$$

It is useful to rewrite (9.244) by using the Dirac delta identity (9.211),

$$\delta(E_{k'} - E_k) = \frac{m}{\hbar^2 k}\delta(k' - k). \tag{9.245}$$

Noting that $\mathrm{d}^3k' = k'^2\,\mathrm{d}k'\,\mathrm{d}\Omega_{k'}$, the integration over k' in (9.244) gives

$$\sigma_T = \frac{R}{J_k} = \frac{16\pi^4 m^2}{\hbar^4}\oint\mathrm{d}\Omega_{k'}\;|\langle\, \boldsymbol{k}' \mid \hat{T} \mid \boldsymbol{k}\,\rangle|^2|_{k'=k}. \tag{9.246}$$

Result (9.246) can be compared to the one obtained from the scattering amplitude expression (9.176), which gives the differential cross-section

$$\sigma(\theta, \phi) = |f(\boldsymbol{k}', \boldsymbol{k})|^2|_{k'=k} = \frac{16\pi^4 m^2}{\hbar^4} |T_{k'k}|^2 \Big|_{k'=k}. \tag{9.247}$$

From definition (9.5) the total cross-section is then given by

$$\sigma_T = \oint \mathrm{d}\Omega_{k'}\, \sigma(\theta, \phi) = \frac{16\pi^4 m^2}{\hbar^4} \oint \mathrm{d}\Omega_{k'}\, |\langle\, \boldsymbol{k}' \mid \hat{T} \mid \boldsymbol{k}\,\rangle|^2|_{k'=k}, \tag{9.248}$$

where $\mathrm{d}\Omega_{k'} = \sin\theta\, \mathrm{d}\theta\, \mathrm{d}\phi$ is the solid angle of the scattered particle momentum $\boldsymbol{k}'$. This result is identical to (9.246), verifying the consistency of the $\hat{S}$-matrix method for calculating transition rates.

9.5.6 The optical theorem and formal scattering theory

It is possible to give a formal derivation of the optical theorem (9.75). This begins by noting that (9.176) gives the forward scattering amplitude

$$f(0) = f(\boldsymbol{k}, \boldsymbol{k}) = -\frac{4\pi^2 m}{\hbar^2} T_{\boldsymbol{k}\boldsymbol{k}} = -\frac{4\pi^2 m}{\hbar^2} \langle\, \boldsymbol{k} \mid \hat{T} \mid \boldsymbol{k}\,\rangle = -\frac{4\pi^2 m}{\hbar^2} \langle\, \boldsymbol{k} \mid \hat{V} \mid \psi_{\boldsymbol{k}}^+\,\rangle, \tag{9.249}$$

where definition (9.169) was used. Using the Lippmann–Schwinger result (9.224) to replace $\langle\, \boldsymbol{k} \mid$ gives

$$f(0) = -\frac{4\pi^2 m}{\hbar^2}\left(\langle\, \psi_{\boldsymbol{k}}^+ \mid \hat{V} \mid \psi_{\boldsymbol{k}}^+\,\rangle - \langle\, \psi_{\boldsymbol{k}}^+ \mid \hat{V}\left(\frac{1}{E_k - \hat{H}_o - i\epsilon}\right)\hat{V} \mid \psi_{\boldsymbol{k}}^+\,\rangle\right). \tag{9.250}$$

The imaginary part of the forward scattering amplitude (9.250) is given by

$$\begin{aligned} \mathrm{Im} f(0) &= -\frac{i}{2}(f(0) - f^*(0)) = -\frac{2\pi^2 mi}{\hbar^2}\langle\, \psi_{\boldsymbol{k}}^+ \mid \hat{V}\left(\frac{1}{E_k - \hat{H}_o - i\epsilon} - \frac{1}{E_k - \hat{H}_o + i\epsilon}\right)\hat{V} \mid \psi_{\boldsymbol{k}}^+\,\rangle \\ &= \frac{4\pi^2 m}{\hbar^2}\langle\, \psi_{\boldsymbol{k}}^+ \mid \hat{V}\left(\frac{\epsilon}{(E_k - \hat{H}_o)^2 + \epsilon^2}\right)\hat{V} \mid \psi_{\boldsymbol{k}}^+\,\rangle. \end{aligned} \tag{9.251}$$

Inserting a complete set of wave number states gives

$$\begin{aligned} \mathrm{Im} f(0) &= \frac{4\pi^2 m}{\hbar^2} \int \mathrm{d}^3k'\, \langle\, \psi_{\boldsymbol{k}}^+ \mid \hat{V}\left(\frac{\epsilon}{(E_k - \hat{H}_o)^2 + \epsilon^2}\right) \mid \boldsymbol{k}'\,\rangle\langle\, \boldsymbol{k}' \mid \hat{V} \mid \psi_{\boldsymbol{k}}^+\,\rangle \\ &= \frac{4\pi^2 m}{\hbar^2} \int \mathrm{d}^3k' \left(\frac{\epsilon}{(E_k - E_{k'})^2 + \epsilon^2}\right)\langle\, \psi_{\boldsymbol{k}}^+ \mid \hat{V} \mid \boldsymbol{k}'\,\rangle\langle\, \boldsymbol{k}' \mid \hat{V} \mid \psi_{\boldsymbol{k}}^+\,\rangle \\ &= \frac{4\pi^3 m}{\hbar^2} \int \mathrm{d}^3k'\, \delta(E_k - E_{k'})\, |\langle\, \boldsymbol{k}' \mid \hat{T} \mid \boldsymbol{k}\,\rangle|^2, \end{aligned} \tag{9.252}$$

where the Dirac delta (3.90) and the definition (9.169) were used in the final step. Using the Dirac delta identity (9.211) gives

$$\begin{aligned}\mathrm{Im} f(0) &= \frac{4\pi^3 m^2 k}{\hbar^4} \int \mathrm{d}k' \, \mathrm{d}\Omega_{k'} \, \delta(k - k') \, |\langle \boldsymbol{k}' \mid \hat{T} \mid \boldsymbol{k} \rangle|^2 \\ &= \frac{4\pi^3 m^2 k}{\hbar^4} \oint \mathrm{d}\Omega_{k'} \, |\langle \boldsymbol{k}' \mid \hat{T} \mid \boldsymbol{k} \rangle|^2 |_{k'=k}.\end{aligned} \tag{9.253}$$

Comparing this result to the expression (9.248) for the total cross-section gives

$$\mathrm{Im} f(0) = \frac{k}{4\pi} \sigma_T, \tag{9.254}$$

which is identical to the previous result (9.75) found from partial wave analysis.

References and recommended further reading

Texts that develop additional aspects of scattering in quantum mechanics include

- Schiff L 1968 *Quantum Mechanics* 3rd edn (New York: McGraw-Hill)
- Greiner W 2001 *Quantum Mechanics: An Introduction* 4th edn (New York: Springer)
- Gasiorowicz S 2003 *Quantum Physics* 3rd edn (New York: Wiley)
- Schwabl F 2007 *Quantum Mechanics* 4th edn (New York: Springer)
- Mahan D 2009 *Quantum Mechanics in a Nutshell* (Princeton, NJ: Princeton University Press)
- Messiah A 2014 *Quantum Mechanics* (New York: Dover)
- Griffiths D and Schroeter D 2018 *Introduction to Quantum Mechanics* 3rd edn (New York: Cambridge University Press)
- Cohen–Tannoudji C, Diu B, and Laloë F 2019 *Quantum Mechanics* vol 1, 2nd edn (New York: Wiley)
- Sakurai J and Napolitano J 2020 *Modern Quantum Mechanics* (New York: Cambridge University Press)

Rayleigh's partial wave analysis for sound scattering was first adapted to quantum scattering in

- Faxén H and Holtsmark J 1927 Beitrag zur Theorie des Durchganges langsamer Elektronen durch Gase *Z. Phys.* **45** 307 (https://doi.org/10.1007/BF01343053)

Rayleigh's lemma is proved in

- Arfken G, Weber H, and Harris F 2013 *Mathematical Methods for Physicists* 7th edn (Amsterdam: Elsevier)

Green's function techniques for solving the differential equations of quantum mechanics are detailed in

- Economou E 2006 *Green's Functions in Quantum Physics* 3rd edn (New York: Springer)

The Lippmann–Schwinger equation was first presented in

- Lippmann B and Schwinger J 1950 Variational principles for scattering processes. I *Phys. Rev. Lett.* **79** 469 (https://doi.org/10.1103/PhysRev.79.469)

The Born approximation was first presented in

- Born M 1926 Quantenmechanik der Stoßvorgänge *Z. Phys.* **38** 803 (https://doi.org/10.1007/BF01397184)

The electrodynamics version of the optical theorem first appeared in

- Strutt J 1871 XV. On the light from the sky, its polarization and colour *Phil. Mag.* **41** 107 (https://doi.org/10.1080/14786447108640452)

Printed in the USA
CPSIA information can be obtained
at www.ICGtesting.com
JSHW061341241223
54197JS00004B/72